HIGHLIGHTS OF ASTRONOMY

INTERNATIONAL ASTRONOMICAL UNION
UNION ASTRONOMIQUE INTERNATIONALE

HIGHLIGHTS OF ASTRONOMY

VOLUME 3

AS PRESENTED AT THE XVth GENERAL ASSEMBLY
AND THE EXTRA ORDINARY GENERAL ASSEMBLY OF THE I.A.U.

1973

EDITED BY

G. CONTOPOULOS

(General Secretary of the Union)

D. REIDEL PUBLISHING COMPANY

DORDRECHT-HOLLAND / BOSTON-U.S.A.

1974

Published on behalf of
the International Astronomical Union
by
D. Reidel Publishing Company, P.O. Box 17, Dordrecht, Holland

All Rights Reserved
Copyright © 1974 by the International Astronomical Union

Sold and distributed in the U.S.A., Canada, and Mexico
by D. Reidel Publishing Company, Inc.
306 Dartmouth Street, Boston,
Mass. 02116, U.S.A.

Library of Congress Catalog Card Number 71-159657
ISBN 90 277 0452 X

No part of this book may be reproduced in any form, by print, photoprint, microfilm, or any other means, without written permission from the publisher

Printed in The Netherlands by D. Reidel, Dordrecht

TABLE OF CONTENTS

PREFACE   IX

INVITED DISCOURSES

J. P. WILD / A New Look at the Sun	3
D. W. SCIAMA / Early Stages of the Universe	21
G. B. FIELD / The Physics of the Interstellar Matter	37
V. A. AMBARTSUMIAN / Galaxies and Their Nuclei	51
O. GINGERICH / The Astronomy and Cosmology of Copernicus	67

SELECTED PAPERS

M. J. REES / X-Ray Sources in Close Binary Systems	89
P. DELACHE / The Next Decade in Stellar Atmospheres Theory	109
L. V. KUHI / The Future of Observational Stellar Atmospheres	121
M. KUPERUS / The Next Decade of Theoretical Solar Physics	133
J. M. BECKERS / The Next Decade in Observational Solar Research	149
C. MAGNAN and J. C. PECKER / Asymmetry in Solar Spectral Lines	171

JOINT DISCUSSIONS

I. PRECESSION, PLANETARY EPHEMERIDES AND TIME SCALES
(Edited by J. Kovalevsky)

J. KOVALEVSKY / Introductory Remarks	209
W. FRICKE / Pro and Contra Changes in the Conventional Values of Precession	211
R. O. VICENTE / The Calculation of the Nutations	221
R. L. DUNCOMBE, P. K. SEIDELMANN, and P. M. JANICZEK / Planetary Ephemerides	223
G. A. WILKINS / Astronomical Units, Constants and Time-Scales	229

II. STELLAR INFRARED SPECTROSCOPY
(Edited by Y. Fujita)

Y. FUJITA / Introduction	235
D. L. LAMBERT / High Resolution Interferometry of Cool Stars	237
R. I. THOMPSON / Presentation and Interpretation of High Resolution Infrared Spectra of Late-Type Stars	255

J. P. MAILLARD / High Resolution Spectra of M and C Stars by Fourier Transform Spectroscopy — 269

R. F. WING / Scans and Narrow-Band Photometry of Late-Type Stars in the One-Micron Region — 285

A. R. HYLAND / Medium Resolution Stellar Spectra in the Two-Micron Region — 307

S. T. RIDGWAY / Fourier Transform Spectrophotometry and its Application to the Study of K-Giants — 327

F. QUERCI and M. QUERCI / Interpretation of Carbon Stars Spectra from Model Atmospheres Computations — 341

H. ZIRIN / Stellar Spectroscopy at $1.1\,\mu$ — 357

OPEN DISCUSSION — 361

III. KINEMATICS AND AGES OF STARS NEAR THE SUN
(Edited by L. Perek)

L. PEREK / Preface — 367

G. CAYREL DE STROBEL / The Ages of Stars in the Neighbourhood of the Sun — 369

P. O. LINDBLAD / Gould's Belt — 381

O. J. EGGEN / Red Variables and Their Main Sequence Progenitors — 387

W. J. LUYTEN / Low Luminosity Stars and White Dwarfs — 389

R. WIELEN / The Kinematics and Ages of Stars in Gliese's Catalogue — 395

R. H. MILLER / The Third and Fourth Moments of the Local Stellar Velocity Distribution — 409

R. E. S. CLEGG and R. A. BELL / The Abundance and Age Distribution of 500 F Stars in the Solar Neighbourhood — 415

J. H. OORT / The Space Density of Faint M-Dwarfs — 417

J. EINASTO / The Correlation Between Kinematical Properties and Ages of Stellar Populations — 419

H. WEAVER / Space Distribution and Motion of the Local H I Gas — 423

C. C. LIN, C. YUAN, and W. W. ROBERTS / Influence of a Spiral Gravitational Field on the Observational Determination of Galactic Structure — 441

T. A. AGEKJAN and K. F. OGORODNIKOV / Solar Neighbourhood as the Local Macroscopic Volume Element Within the Galaxy — 451

A. TOOMRE / How Can It All Be Stable? — 457

F. K. EDMONDSON / Closing Remarks — 465

IV. ORIGINS OF THE MOON AND SATELLITES
(Edited by G. Contopoulos)

F. L. WHIPPLE / On the Growth of the Earth-Moon System — 469

H. C. UREY / Evidence for Lunar-Type Objects in the Early Solar System — 475

T. GOLD / The Movement of Small Particulate Matter in the Early Solar System and the Formation of Satellites 483
R. B. LARSON / Gravitational Collapse and the Formation of the Solar Nebula 487
M. W. OVENDEN / The Principle of Least Interaction Action 489

V. JOVIAN RADIO BURSTS AND PULSARS
(Edited by F. G. Smith)

SUMMARY 493

VI. THE OUTER LAYERS OF NOVAE AND SUPERNOVAE
(Edited by C. de Jager)

L. I. ANTIPOVA / The Chemical Composition of the Envelopes of Novae 501
H. WEAVER / The Shell of V603 Aql and the Early Stages of the Nova Event 509
R. P. KIRSHNER / Spectrophotometry of Supernovae 533
E. R. MUSTEL / On the Physical Model of Supernovae Close to Light Maximum 545
S. VAN DEN BERGH / Supernova Remnants 559
J. C. ZARNECKI, J. L. CULHANE, A. C. FABIAN, C. G. RAPLEY, R. L. F. BOYD, J. H. PARKINSON, and R. SILK / Soft X-Ray Observations of Supernova Remnants 565
F. BERTOLA / Report on the Lecce Conference on Supernovae 573

PREFACE

The year 1973 marked the highest peak of IAU activity up to now. Besides the General Assembly in Sydney, and the Extraordinary General Assembly in Poland, there were held eleven IAU Symposia and one Colloquium.

Several IAU Publications cover this activity. The Proceedings of the Symposia are published in separate Volumes, while the Transactions of the General Assembly and of the Extraordinary General Assembly contain short reports of the Commission meetings, the administrative sessions, and the opening ceremonies.

The present Volume covers some of the scientific Highlights of the General Assembly and of the Extraordinary General Assembly. It contains five Invited Discourses given in Sydney and Poland, some selected papers, and the Joint Discussions at the General Assembly of Sydney.

Of course, there were many more papers of special interest presented in Sydney that could not be included in this Volume. Their titles can be seen in the reports of the various Commissions. It is regrettable that the Invited Discourses of C. H. Townes (Interstellar Molecules) and F. J. Low (Infrared Astronomy) were not submitted for publication. Also only five papers or abstracts of the Joint Discussion on the 'Origins of the Moon and Satellites' have been available.

Despite these minor shortcomings, I believe that the present Volume is faithful to its title: it gives a substantial part of the Highlights of Astronomy in 1973.

G. CONTOPOULOS
General Secretary

INVITED DISCOURSES

A NEW LOOK AT THE SUN

J. P. WILD

CSIRO, Division of Radiophysics, Epping, N.S.W., Australia

I have the feeling that to most astronomers the Sun is rather a nuisance. The reasons are quite complex. In the first place the Sun at once halves the astronomer's observing time from 24 to 12 hours, and then during most of the rest of the time it continues its perversity by illuminating the Moon. Furthermore I have met numerous astronomers who regard solar astronomy to be now, as always before, in a permanent state of decline – rather like Viennese music or English cricket. Nevertheless those who study the Sun and its planetary system occasionally make significant contributions. There were, for instance, Galileo and Newton who gave us mechanics and gravitation; Fraunhofer who gave us atomic spectra; Eddington and Bethe who pointed the way to nuclear energy; and Alfvén who gave us magneto-hydrodynamics. Perhaps the point to be recognized is that the Sun has more immediately to offer to physics rather than to astronomy. That is why it is quite rare that a solar man finds himself with a large captive audience of mainline astronomers: and so the responsibility weighs heavily on my shoulders tonight.

My theme will in fact be a very simple and clearly defined one. It is to describe the attempts of my colleagues and me to produce pictures – moving pictures – of the Sun in the metre wavelength part of the radio spectrum and to tell you about the results which have emerged from these observations. The instrument in question, known as a radioheliograph, is located 300 miles northwest of Sydney at Culgoora observatory which many of you will be visiting around this time. I would like you to consider my personal role in all this as that of a spokesman speaking on behalf of a research group. I should acknowledge in particular the instrumental work of Kevin Sheridan (without whom the thing would never have worked), Maston Beard, Keith McAlister and Warren Payten; and the research work of Steve Smerd, Norman Labrum, Donald McLean, Anthony Riddle, Ron Stewart; and also invaluable contributions by people from other countries – from Dr Morimoto, and Dr Kai of Tokyo Astronomical Observatory and Dr George Dulk from the University of Colorado. Those are a few names, but the list could be made actually much longer, of people who have played quite essential parts in this work.

1. The Culgoora Radioheliograph

The first germs of the idea of producing a radioheliograph came, as far as I was concerned, in the late 1950's when, after spending a decade or so studying the radio waves from the Sun by indirect and devious ways, the thought kept occurring: – wouldn't it be fun if we could actually look at or take pictures of the Sun, with a kind of radio movie camera. And to understand how to specify this radio camera we have to look first at what had already been established, what information was available at that time about the general properties of the radio emission.

As you all know it had been established that from time to time – significantly enough at the time of solar flares – the 'quiet' level of solar radio emission would suddenly increase by a factor of hundreds or even thousands; the intensity of such an outburst was found to be equivalent to an object of the size of the Sun radiating as a black body at hundreds of millions of degrees. One was clearly dealing with non-thermal processes and a lot of work and soul-searching went on during the forties and fifties to explore the mechanisms and processes responsible. The outcome is illustrated in Figure 1 which summarizes ten or more years of rather tough horse-trading between the observationalists and the theorists. It was concluded that there are three types of emission, all produced by electrons of different energy. First there are the thermal electrons in the solar atmosphere which produce bremsstrahlung of a comparatively very feeble intensity, merely that of a black body of the order of a million degrees. Then we have the supra-thermal electrons in the energy range

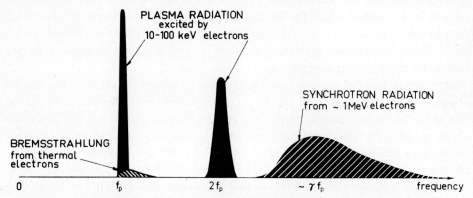

Fig. 1. Idealized spectrum of the three types of radiation believed to be responsible for solar radio emission.

10 to 100 keV which, by a series of processes which the theorists argue about at very great length, set the solar atmosphere into plasma oscillations which produce radiation at two sharp frequency resonances, one at the plasma frequency f_p, and another at the second harmonic, $2f_p$.

Such plasma radiation provides us with an inherently powerful tool for exploring the solar atmosphere because different frequencies are emitted from regions of different electron-density, N, ie. from different levels – according to the well – known formula

$$f_p = e\sqrt{\frac{N}{\pi m}}, \qquad (1)$$

where e and m denote electronic charge and mass. Typical isodensity contours of the solar corona are shown in Figure 2, plotted in terms of f_p rather than N. Finally we have the relativistic electrons with energies in the MeV range, and these generate synchrotron radiation. To those of you who study synchrotron radiation emitted by cosmic radio sources the spectrum shown in Figure 1 may look unfamiliar. The reason is that in the solar environment the low frequency cutoff is determined by the refractive index of the medium (corona) which suppresses the low frequencies; all that remains is a frequency tail situated around a frequency a few times the plasma frequency.

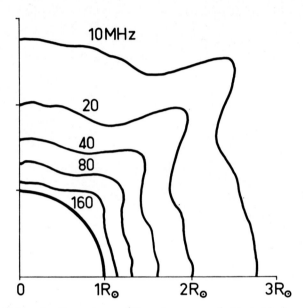

Fig. 2. Typical electron-density contours of the solar corona designated in terms of their plasma frequency in accordance with Equation (1).

This is the background knowledge on which the radioheliograph was designed. Our objective was to study the outer corona and so we chose a frequency as low as possible subject to limitations imposed by ionospheric refraction. The frequency was thus set at 80 MHz and other parameters as below:

Frequency: 80 MHz.

Diameter of telescope aperture: 3 km (=800 wavelengths – sufficient to obtain a reasonable scanning beam size of $3'.9$).

Field of view: 2° diameter (=4 solar radii).

Polarization: both left and right circular polarization (required to study the magnetic properties).

Picture rate: 1 picture per second, (actually 1 pair of pictures per second, one in each polarization).

Daily coverage: Noon $\pm 2\frac{1}{2}$ hours, a five hour observational day.

These requirements led to a rather novel design which, through its circular symmetry, seemed to be a very compact and quite elegant solution of the problem of

Fig. 3. Aerial photograph of the Culgoora radioheliograph.

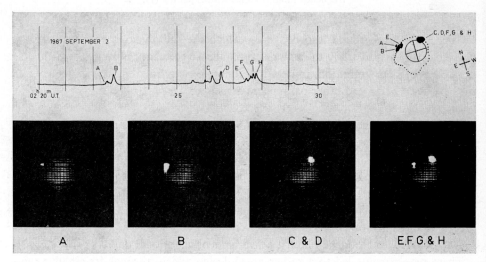

Fig. 4. 80 MHz radioheliograms of a group of weak bursts. Note that the bursts originated in two widely separated, though apparently connected, sources. (Morimoto *et al.*, 1967.)

obtaining so large an aperture. It consists of ninety-six dishes, each of them 13 m in diameter, arranged around a circle of diameter 3 km (Figure 3); the whole thing acts as though it is one huge dish 3 km in diameter. This instrument taught us all about Bessel functions, and we offer it in years to come as a humble monument to nineteenth century mathematics. The instrument took five years to build, between the years 1962 and 1967. Since then we have continually upgraded it and nowadays it operates on three frequencies; 43, 80, and 160 MHz.

First I would like to show you results obtained during the very first hours of operation in September 1967. There was actually no appreciable solar activity in progress but on the flux record at one period some tiny little bursts (A, B, C,..., H) appeared in isolation as shown on the record reproduced in the top left of Figure 4; it was the sort of insignificant record that was in the normal way destined for the waste-paper basket. What happened was quite extraordinary. As seen in the figure, bursts A, B and E appeared from one centre outside the optical limb, while bursts C, D, F, G and H appeared from a second, well separated centre. We were quite staggered at this extraordinary phenomenon of two distant, but apparently connected centres of activity. Surely it could not have been a coincidence. And, as it turned out, it was an indication of the shape of things to come.

2. Observations of the Coronal Magnetic Field

Now I would like to describe in a more or less systematic way some of the more interesting results which the instrument has revealed and, where possible, I will relate these coronal observations to what is seen optically in the photosphere and the chromosphere. It turns out not surprisingly, that one of the most important links between the photosphere and corona is the magnetic field which joins them.

So let us take a look at a sunspot group and its overlying region. In Figure 5 we see, at the top, an optical sunspot group, then the 160 MHz radiation of the associated bi-polar radio source and at the bottom the 80 MHz source. In this figure the thin dotted contours refer to photospheric fields and the heavy contours to radio emission. You see how separated are the bi-polar radio components, and how at 80 MHz they are a lot further apart than at 160 MHz: at 80 MHz there is a looser connection between the radio emission and the optical emission. The upper diagram of Figure 6 gives a three-dimensional model which directly interprets Figure 5 on the assumption that the emission is plasma radiation. Here we see the sunspot group and the loop fields drawn in such a way as to be consistent with the two polarized sources of radio emission. In fact it often happens that storms like this occur in very widespread parts of the Sun and, as it were, communicate with one another. It is quite clear that different centres of activity are connected by magnetic field lines and that storm centres become activated when electrons flow from one active centre to another. Occasionally one can see this in a very striking form with a transient burst of electrons which shows up first in one centre of activity and then in another. Indeed the observation shown in Figure 4 is the prototype example of this phenomenon.

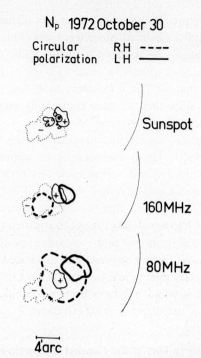

Fig. 5. A sunspot group observed in white light, 160 MHz and 80 MHz radiation. The outer (dotted) contour of the photospheric field is repeated in the other diagrams to provide positional reference. (Kai, 1974.)

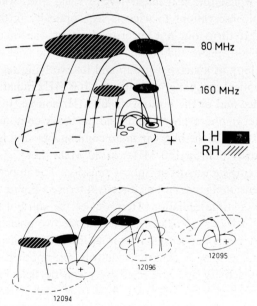

Fig. 6. *Upper*: Magnetic field configuration required to interpret records of the kind shown in Figure 5. The dark disks show the location of radio storm centres. *Lower*: Storm centres arising from magnetic fields that link distant centres of activity.

Now occasionally we get a more dramatic situation in which we can actually 'see' a magnetic arch through its radio emission, and see it erupt and expand. Figure 7 is an example of such an expanding arch, following a flare, recorded at 80 MHz. The four pictures of the sequence are taken at intervals of about 4 or 5 minutes. We believe the arch is made radio emitting by virtue of supra-thermal and relativistic electrons trapped within it. As the expansion proceeds the arch is seen as three discrete sources: one unpolarized source at the top due to synchrotron radiation from relativistic electrons, and a pair of oppositely polarized sources near the feet of the arch attributed to plasma radiation excited by the supra-thermal electrons (cf. Figure 1).

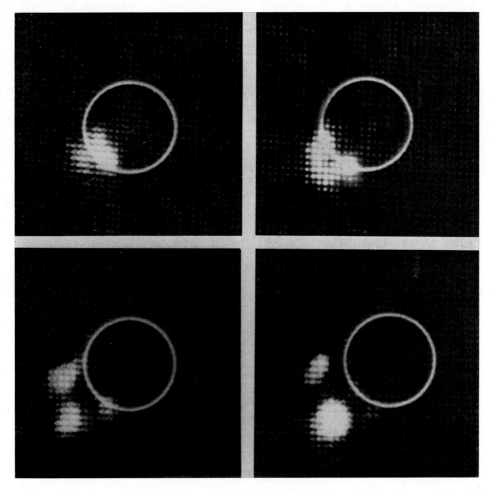

Fig. 7. A sequence $\begin{pmatrix} 1 & 2 \\ 3 & 4 \end{pmatrix}$ of 80 MHz heliograms showing an activated magnetic arch which erupts and expands. The circles indicate the optical disk. The pictures are taken at times separated typically by 5 min. (Wild, 1969.)

3. The Flash Phase of Solar Flares

What I have just been talking about sets the scene, the magnetic environment, in which our other phenomena perform, and perhaps now we should turn to the central problem of the solar flare. I think one of the main contributions of radio observations has been to recognize two distinct phases of the flare and to give a physical description of each of these phases. The first phase of a solar flare, generally called the flash phase, is epitomized by the occurrence of a so-called type III burst (Figure 8), in which high frequencies are emitted slightly before low frequencies. It is characteristic that the spectrum may show two frequency bands corresponding to a fundamental frequency and its second harmonic; this indicates that we are observing plasma radiation (see Figure 1). Then, because there is a frequency drift from high to low frequencies, we must be dealing with electrons which are moving out from the Sun,

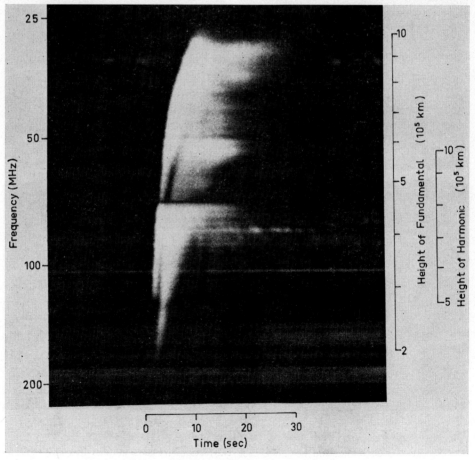

Fig. 8. Dynamic spectrum of a type III burst.

from low down in the solar atmosphere to high up (see Figure 2). And when we use this frequency drift to calculate the velocity of the electrons, we find a value of about one-third the velocity of light. In other words we are dealing with a burst of electrons in the range 10–100 keV. The ejection of such a burst of electron lies at the heart of the flash phase which occurs at the very start of a solar flare.

To explain one way in which a type III burst can be generated, let me now develop the previous picture of the magnetic arch, by observing that when the arch extends high enough into the corona the solar wind comes along and grabs hold of the field lines and drags them out into a helmet structure, as shown in Figure 9. We believe a type III electron burst originates in a region where fields of opposite polarity come very close together; here some kind of magneto-hydrodynamics instability is responsible for the ejection of the burst of electrons which escape out along the neutral plane of the helmet. One contribution by the radioheliograph to our understanding of these bursts is direct evidence that the sources are often located in the neutral plane between opposing magnetic fields.

More recently the type III electron bursts have been picked up directly by satellites in interplanetary space or close to the Earth. And appropriately enough, as one would predict, there is a time delay of about half an hour between their detection on the Sun and at the Earth. The satellite observations have been studied especially by Bob Linn and his colleagues at the Space Science Laboratories at Berkeley. Now it is very

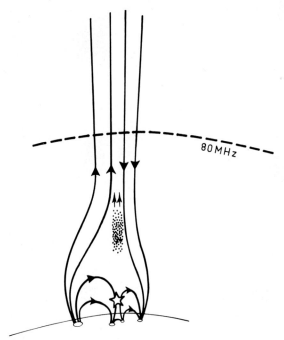

Fig. 9. Helmet structured magnetic field suited to the generation of electron pulses (cluster of dots) which generate type III bursts. The electrons may be generated by instabilities located where opposing lines of force are in close proximity (e.g. at star).

interesting to see the radioheliograph pictures of the type III bursts that correspond to detectable electron bursts. Some are shown in Figure 10 where one is actually seeing the pulses of electrons as they move out through the solar corona. You notice the remarkable fact that every one of them is on the western limb of the Sun. This makes very good sense when one takes into consideration the Archimedes spiral form of the interplanetary field (Figure 11): for particles to arrive near the Earth they have to be ejected from the west limb. Nowadays, type III bursts are studied very effectively by means of very low frequency observations taken from satellites which can extend the range of heights right through interplanetary space. And through some quite brilliant observations by Joe Fainberg, Bob Stone and others at NASA's Goddard Space Flight Centre, as well as at the University of Michigan, one can trace the electron trajectories right along the Archimedes spiral to the Earth's orbit.

That is all I am going to say about the flash except to point out that the bursts of electrons that we pick up not only cause type III bursts and produce detectable electrons in space, but similar electrons traveling inwards towards the photosphere produce microwave bursts and X-rays which in turn cause the sudden ionospheric disturbance effects on the Earth. And so a whole range of phenomena can be explained simply from the one basic cause.

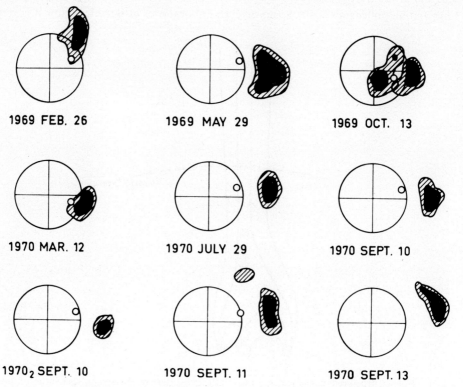

Fig. 10. 80 MHz heliograms of nine type III bursts which accompanied electrons detected directly from satellites. (After I. D. Palmer and R. P. Lin, to be published).

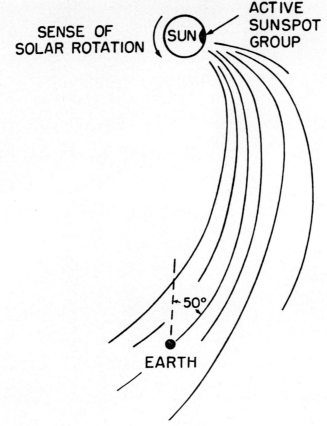

Fig. 11. Archimedes spiral traced out by the interplanetary magnetic field.

4. The Blast Wave

We now pass onto the second phase of a solar flare. At the same time as the explosive instability causing the type III burst takes place, a blast wave moves out at a speed of the order of a 100 km s^{-1} and forms into a sharply defined shock wave. As it advances we can see on the radio spectrum the spectacle of a type II solar burst, a quite beautiful

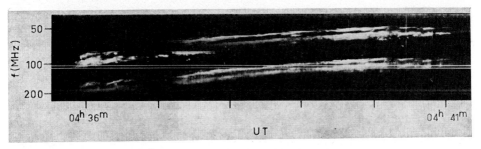

Fig. 12. Dynamic spectrum of a type II burst.

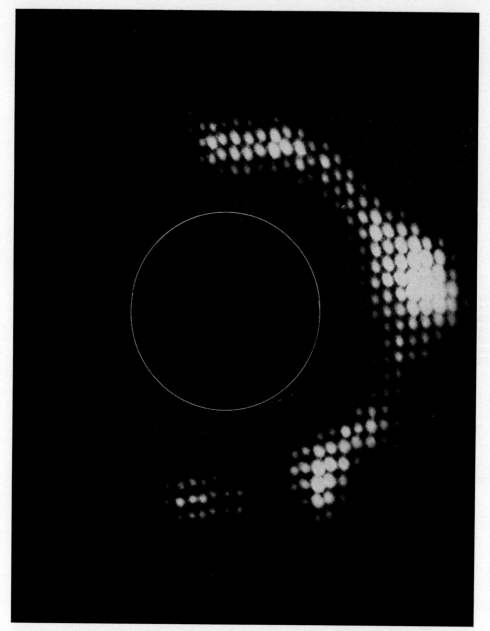

Fig. 13. 80 MHz heliogram taken during the early stage of a great outburst on 30th. March 1969 consisting mainly of type II bursts. The flare occurred beyond the west limb. (Smerd, 1970.)

phenomenon. An example is shown in Figure 12 and we note once again the presence of fundamental and harmonic bands and of frequency drift just as with the type III burst, but now with a different time scale about 200 times slower. This difference corresponds to that between velocities of 1000 km s^{-1} and one-third the velocity of light. This is really the definitive phenomenon of the second phase. When the shock front reaches the 80 MHz level we can expect it to register on our radioheliograph. Figure 13 is a dramatic example of a radioheliogram of an outburst in which two type II bursts occurred. The shock waves travelled out from the source near the limb – actually behind the limb in this case; the event, recorded on 1969 March 30, was accompanied by intense proton emission.

With radio observations we can detect the shock wave from a flare by the type II burst quite as a matter of routine. It is only rather rarely that one can pick up the shock wave by optical means; some fine examples have been recorded at the Lockheed Observatory by Moreton and Ramsey and others, and a clear association with type II bursts has been established.

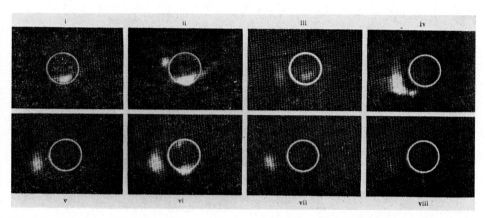

Fig. 14. Sequence of 80 MHz heliograms showing a type II burst (i–iii) followed by a synchrotron burst above an eruptive prominence (iii–viii) believed to be triggered by the same shock wave. (Wild *et al.*, 1968.)

It is clear from optical records, that a flare can make its presence known across large regions of the solar sphere through the propagation of the shock wave. Radio observations go one step further and show that shock waves can actually trigger the release of high-energy particles at very great distances. A classic example at 80 MHz is shown in the sequence of pictures in Figure 14: type II radio emission first occurs above a flare (located on the 5 o'clock vector, regarding the solar disc as a clock face). Minutes later an optical prominence erupts on the limb (on the 8 o'clock vector) and this in turn is accompanied by 800 MHz synchrotron radiation from a source high above the eruptive prominence. The whole sequence can be explained in terms of the effects of a shock wave emanating from the original flare. Figure 15 illustrates how coronal shock waves may trigger such prominence eruptions and, indeed, other flares. Finally I should not forget to mention that one of the most distant interactions involving the shock wave is when it strikes the Earth's atmosphere and produces aurorae and magnetic storms on the Earth.

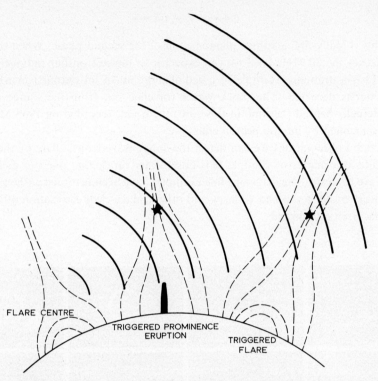

Fig. 15. Diagram showing how advancing shock waves may trigger instabilities (stars) to cause prominence eruptions and other flares.

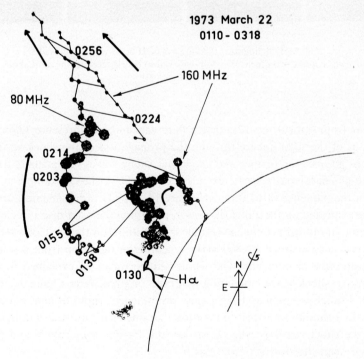

Fig. 16. Diagram showing the evolution of an event in which a slow (100 km s^{-1}) disturbance prompted the start of a radially ejected isolated source (plasmoid?). (McLean, 1974.)

5. A Third Phase to the Solar Flare?

Up till very recently it was widely believed that almost all the important phenomena caused by flares could be traced to the two phases we have just discussed. On the other hand, from time to time one sees peculiar blobs of radiating matter leave the Sun not at supersonic speeds but at rather moderate speeds, a mere 250 km s^{-1}, which is about the Alfvén velocity in the corona. And also one sees optical phenomena such as flare surges which have velocities, typically of the order of 100 km s^{-1}, much slower than coronal shock waves. Now I believe that recent radioheliograph results have begun to make it clear that we should perhaps start thinking of a third basic phase of the flare associated with a slower disturbance. Figure 16 shows what might turn out to be a rather definitive record that depicts a distinctly slower type of disturbance quite different from the disturbances we have previously discussed. Following an Hα surge (shown at 01^{h}30^m) with velocity about 60–100 km s^{-1}, the radio observations at 160 and 80 MHz reveal what might be interpreted as a column of gas

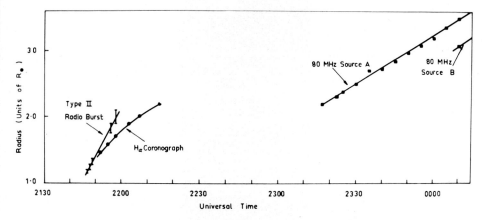

Fig. 17. Height/time analysis of the event of 1–2 March 1969 ('Westward Ho'), showing the 1 hour's time delay between type II shock wave and later ejection. (Riddle, 1970.)

beginning to move out at 01^{h}38^m. The column is channelled round an arc and excites the 160 and the 80 MHz radiation at two different levels, presumably corresponding to the plasma levels. The speed of travel is still about 100 km s^{-1} (very slow) until it reaches what looks like the top of an arch structure. Perhaps you can imagine a helmet structure being drawn around this configuration – it is certainly very suggestive of it. When the disturbance reaches the top of the arch, something else happens; a discrete disturbance, bipolar in structure, moves out radially at about 250 km s^{-1}. Now in the past few years we have witnessed a number of such radially moving disturbances, sometimes reaching solar distances of as much as 5–6 $R_\odot$. We think perhaps these ejections are self-contained plasmoids but it has always been something

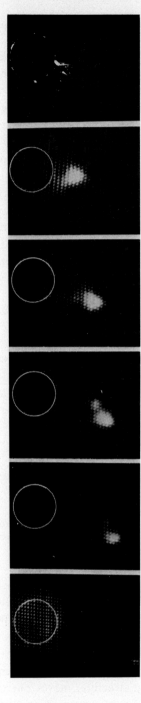

Fig. 18. Sequence of 80 MHz heliograms during the event summarized in Figure 17. The top picture is the associated $H\alpha$ prominance (flare spray) recorded at the University of Hawaii.

of a puzzle that although the ejections come from the general direction of the flare centre the timing is often quite wrong for the ejection to have come from the initial flare explosion. Typically, as illustrated by Figure 17 (which depicts the rather well-known event nicknamed 'Westward Ho'), there is a one-hour delay between the origin of the shock wave and the origin of the slow 'third phase' ejection. The phenomenon in Figure 16 now offers an explanation of the missing hour. It suggests that the plasmoid (if such be the disturbance) is ejected not from the flare centre but rather from an instability located at the point in the helmet structure where the top of the arch meets the base of the radial streamer. And that the instability is triggered by a slow (100 km s^{-1}) column of gas which is first manifest optically as a flare surge and is later seen in radio emission as a source climbing up and around the magnetic arch.

To end this discourse let us take a look at the movie of Westward Ho! itself (Figure 18) travelling out and out at 270 km s^{-1} through more than 2 million kilometres. The accompanying music clearly ought to be the William Tell overture.

Mr Chairman, I think I should finish at this stage and leave you in a position half way between the Sun and the Earth. And although I have taken you through quite a long road of comples and varied phenomena, I hope you will grant that the main contribution of this work is basically one of unification and simplification.

References

Kai, K. and Sheridan, K. V.: 1974, *Solar Phys.*, in press.
McLean, D. J.: 1973, *Proc. Astron. Soc. Australia* **2**, 222.
Morimoto, M., Sheridan, K. V., and Wild, J. P.: 1967, *Proc. IEEE Australia* **28**, No. 9.
Riddle, A. C.: 1970, *Solar Phys.* **13**, 448.
Smerd, S. F.: 1970, *Proc. Astron. Soc. Australia* **1**, 305.
Wild, J. P. (ed.): 1967, *Proc. IEEE Australia* **28**, No. 9.
Wild, J. P.: 1969, *Solar Phys.* **9**, 260.
Wild, J. P.: 1970, *Proc. Astron. Soc. Australia* **1**, 365.
Wild, J. P. and Smerd, S. F.: 1972, *Ann. Rev. Astron. Astrophys.* **10**, 159.
Wild, J. P., Sheridan, K. V., and Kai, K.: 1968, *Nature* **218**, 536.

EARLY STAGES OF THE UNIVERSE

D. W. SCIAMA

Dept. of Astrophysics, Oxford University, Oxford, England

1. Introduction

When I was young theoretical cosmology was more a series of exercises in geometry than a branch of astrophysics. Moreover these exercises were based on little more than the Hubble law of red shifts and a general impression that on a large scale the Universe was roughly homogeneous and isotropic. Now all that has changed. While geometrical considerations retain their importance through the dominant role played by general relativity, cosmology has also become highly astrophysical. As a result we can now say a great deal about the early stages of the Universe. My aim is to give a brief account of our present understanding of this fascinating subject, at the same time highlighting the most important problems that remain.

Let us first recall the situation as it stood before these recent developments took place, say in the early nineteen fifties. The Hubble law then told us that the red shift z of a galaxy (defined by $\delta\lambda/\lambda_{\text{rest}}$) was proportional to the distance of the galaxy. This law had two immediate consequences:

(i) All galaxies would see the same linear law of red shifts relative to themselves and with the same Hubble constant. This strengthened the inference from the observed rough homogeneity of the large-scale distribution of galaxies, that our own Galaxy was in a *typical* place in the Universe and not a special one.

(ii) Taken together with the observed homogeneity and isotropy, and assuming that the red shift was a Doppler shift, it implied that a finite time ago the Universe had a high density (strictly speaking an infinite density, if the symmetries were taken as exact, and Einstein's cosmical constant ignored). The corresponding 'age' of the Universe depended upon the value of the Hubble constant and so upon the distance scale for galaxies, which was being substantially revised in the 1950's and is even now uncertain to at least a factor 2. This important astronomical problem lies outside our present subject; for our purposes it will suffice to use the round figure of 10^{10} yr for the Hubble constant. The time since the initial singularity depended also on the present mean density of matter in the Universe, through its gravitational influence on the deceleration of the expansion. Here the possibilities ranged from a low density universe, in which galaxies made the main contribution to the mean density ($\sim 10^{-30}$ gm cm^{-3}), to a high density universe, in which intergalactic gas (or faint stars, bricks, neutrinos, gravitational waves or black holes) contributed more substantially than galaxies. An upper limit on the possible density came from the requirement that the age of the Universe should not be less than the reasonably well-determined ages of its contents (for instance, the Earth, the Sun, the Milky Way). This upper limit was about 10^{-28} gm cm^{-3}. A low density universe would be destined to expand forever, while a high density one (with $\varrho > 2 \times 10^{-29}$ gm cm^{-3}) would eventually re-contract into a singularity. (Some of these statements would need modifying if a cosmical constant were retained in Einstein's field equations of general relativity. The resulting new possibilities are discussed in e.g. Bondi (1960).)

An important speculation that had recently been made (Gamow, 1948, 1949) was that in the early stages of the Universe there might have been enough heat present to promote the build-up of heavier elements out of hydrogen by means of thermonuclear reactions, a proposal now conjured up in the phrase 'hot big bang'. According to the α-β-γ theory (Alpher et al., 1948) essentially all the elements heavier than hydrogen now observed were built up in this way. In this picture the early heat, present as an initial condition at $t=0$, would have rapidly attained a black body spectrum by interacting with matter, and would thereafter retain such a spectrum to a good approximation even when, after considerable expansion of the Universe, the interaction with matter became much reduced (for zero interaction the spectrum would remain exactly that of a black body, the temperature dropping in proportion as the scale factor of the Universe expanded). The α-β-γ calculations suggested that the present temperature of the black body radiation field would be measured in tens of degrees absolute, a prediction which it was not then possible to verify (and later became forgotten).

On the theoretical side, the homogeneous and isotropic solutions of Einstein's field equations had been codified into the Robertson-Walker models (also called the Friedman models). This phase of the theory is fully discussed in Bondi (1960). One property of these models, which was first fully elucidated by Rindler (1956), was the occurrence of various kinds of *horizons*. The one which will be important for our later considerations is the *particle horizon*. In a cosmological model with such a horizon a fundamental observer at time t cannot see all the particles (galaxies) in the Universe, but only those within his particle horizon, the size of which increases with t. This means that there has not been time enough for radiation from the more distant galaxies to reach the fundamental observer by the time t. This may seem odd at first sight, and it does depend on the Universe having had a singular origin. Roughly speaking we may say that the initial rate of expansion of the Universe was infinite, so that some fundamental observers were thrown apart to distances such that their light time away exceeded the age of the system. (Rindler shows how this remark can be made rigorous).

So far we have been recalling the conventional picture of the early 1950's, based on an acceptance of general relativity and of the assumption that the matter and radiation content of the Universe obeyed familiar kinds of equation of state. There were also extant a number of unconventional ideas involving deviations from standard theory, such as having a time-varying gravitational constant (Milne, 1948; Dirac, 1938, 1973a, 1973b; Jordan, 1949, 1973; Hoyle, 1973; see also Brans and Dicke (1961) and Dicke (1962)). Of these the most influential and the most closely related to astrophysical questions was undoubtedly the steady state theory of Bondi and Gold (1948) and Hoyle (1948). This theory still has its adherents despite the hostile evidence which has accumulated during the recent astrophysical phase of cosmology. Indeed one possible lecture I could have given would have involved presenting a critical account of all the modern data relevant to this question. However, I believe that this would not be the appropriate occasion for such an account. I prefer therefore to accept for this lecture what I think is generally agreed to be the most plausible outcome of such a discussion, and to assume that conventional theory is correct at least as far as saying that the Universe *did* have early stages different from its present stage.

This is the essential point because the steady state theory, as its name implies, took the radical view that on a large scale the Universe does not change with time at all. This view was reconciled with the observed expansion by the postulate that matter was being continually created at the appropriate rate, a rate far too small to be subject to direct observational check. The basic question of whether the Universe evolves with time or not is then best answered by seeking clear-cut evidence as to whether the universe *was* different in the past. The nearest approach to such evidence is provided by the 3 K microwave background, whose observational existence was hardly suspected before 1965, although, as we have seen, it was in fact predicted by Gamow in 1948. (The counts of radio sources and quasi-stellar objects also probably convince most astronomers that the Universe is evolving, but since they are subject to greater controversy than the microwave background, I shall not consider them here.)

The story of the great discovery by Penzias and Wilson (1965) of excess radiation at a wavelength of 3 centimeters, of its interpretation as a cosmological black body radiation field at 3 K by Dicke *et al.* (1965), and of the gradual strengthening of our conviction that the microwave background does have an accurately black body spectrum, is too well known to be worth repeating here. (Accounts can be found in Field (1969), Partridge (1969), Peebles (1971), Sciama (1971a, b), and Weinberg (1972)). Suffice it to say that recent claims to have discovered excess radiation with $T > 3$ K which were in conflict with other observations in the submillimetre range have now been withdrawn, but that we still eagerly await (satellite) observations in this decisively important region, which lies beyond the predicted peak at ~ 1 mm. For the purposes of this lecture I shall simply assume that the microwave background does have an accurately thermal spectrum, and shall follow the conventional view that this implies that the Universe was denser in the past than it is to-day. We are then led to a more general question, namely how far back and with what confidence, can we extrapolate into the past? What does observation tell us directly about the early stages of the Universe?

2. Extrapolating into the Past

2.1. Thermalisation of the Microwave Background

We are making the assumptions that the microwave background has interacted sufficiently with matter at some stage in the development of the Universe to have been brought into thermal equilibrium and that thereafter this equilibrium has not been significantly disturbed. These assumptions impose a number of important constraints on the thermal history of the Universe. To understand these constraints it is helpful to consider first the mean free path of a microwave photon as it propagates through the Universe. The most important process determining this mean free path is probably Thomson scattering by intergalactic (or pregalactic) free electrons. We denote by z_0 the red shift of an emission process which occurred one mean free path away. A straightforward calculation shows that this red shift z_0 ranges from about 7 in a high density universe consisting mainly of ionised intergalactic gas, to about 1000 in a low density universe. The corresponding age of the Universe when the photon was emitted ranges from about 10^9 yr to about 10^7 yr.

Now the immediate source of the observed microwave background is the free electrons located on the 'last-scattering surface' at red shift z_0. We may therefore regard ourselves as completely surrounded by such a surface, whose radiation temperature was $3(1+z_0)$ K at the moment of emission, that is, about 24 K in a high density universe and 3000 K in a low density universe. However, to achieve adequate thermalisation requires many scatterings (and also transitions in which photons are absorbed or emitted and not simply scattered, since if photon number were conserved an arbitrary radiation field could not reach thermal equilibrium with matter). Thus if these are the responsible processes thermalisation must have occurred at red shifts considerably greater than z_0 and at correspondingly further times into the past. In fact the most recent thermalisation epoch must have occurred at a red shift of about 4×10^5 in a high density universe, when the age of the system was about 30 years and its density nearly 10^{17} times the present density. In a low density universe the corresponding red shift was greater (up to 4×10^7), the associated density also greater (up to nearly 10^{23} times the present density) and the age of the Universe less (down to 5×10^5 s). Hence if thermalisation is due to the coupling between radiation and ionised material then the present observation that the microwave background has a black body spectrum gives us direct access to processes occurring when the Universe was at most 30 years old. (An alternative possibility is thermalisation by interaction with dust particles (Layzer and Hirely, 1973). In this case thermalisation would have occurred more recently.)

Once thermalisation ceases the black body character of the microwave background is preserved throughout the expansion (with temperature $T \propto 1+z$), unless further interaction with free electrons then distorts the spectrum. This distortion is a real possibility for two reasons. Firstly, after thermalisation ceases the free electrons tend to cool more rapidly than the radiation, and secondly there may be a later phase when the free electrons are heated to a much higher kinetic temperature as a result of interaction with cosmic rays and electromagnetic radiation emitted by galaxies, radio galaxies and quasi-stellar objects. Insofar as distortions from a black body spectrum are not observed we must impose the appropriate constraints on the possible thermal history of the Universe. This is a very interesting and important question, but since it concerns stages later than 30 years from the big bang we shall not consider it here, (see Peebles, 1971; Zel'dovich and Novikov, 1974.)

2.2. THE HELIUM PROBLEM

For possible direct evidence concerning a Universe younger than 30 years old we must turn to the helium problem. As is well known, there is good, but not decisive, evidence that the abundance of helium relative to hydrogen is approximately constant throughout the Galaxy and is about 10% by particle number (Searle and Sargent, 1972; Peimbert, 1974). Attempts to account for this helium abundance in terms of nuclear processes occurring in the Galaxy since its formation have not yet been successful. On the other hand, such processes occurring in the early stages of the Universe do yield a relative abundance of about 10%, the actual value depending slightly on the density of the Universe, being rather more in a high density than in a low density universe (Wagoner, 1973). Elements heavier than lithium, however, are not made appreciably in the hot big bang, so that the α-β-γ theory would not be valid for such elements.

If the helium we now observe is primordial, we thereby have access to processes occurring earlier than the most recent thermalisation epoch. The limiting factor here is that the radiation field must not be so intense that nuclei intermediate between hydrogen and helium are photo-disintegrated before they can be built up into helium. The maximum temperature permitted by this consideration is about 10^9 K, corresponding to a red shift of 3×10^8, and an age for the Universe of about 100 s. Thus if the helium argument could be made firmer, we would have direct access to a very early stage of the Universe indeed.

A similar argument has recently been made in connection with the formation of deuterium (Reeves *et al.*, 1973). This would be particularly interesting because the amount of deuterium formed in the early stages depends more sensitively on the density of the Universe than does the amount of helium. However, this argument is not a strong one at the moment because it appears that the deuterium may have been formed in the Galaxy (Hoyle and Fowler, 1973; Colgate, 1973). Similar remarks apply to the formation of lithium 7 (Audouze and Truran, 1973).

2.3. THE INITIAL SINGULARITY

To study earlier stages still one must turn to other evidence, namely the *isotropy* of the microwave background. We shall see that according to general relativity this isotropy implies that the Universe was singular a finite time ($\sim 10^{10}$ yr) ago. This is so even if one abandons the artificially strict symmetries of the Robertson-Walker models, and deals with a realistic universe. This singularity is widely regarded as physically unacceptable, but it seems likely that the density of the Universe was extremely high in its earliest stages. It will be convenient to defer discussion of this remarkable result until later.

3. Outstanding Problems

So far I have been considering recent developments in our understanding of the early stages of the Universe which, while not definitive, are at least clear-cut. Now it is

time to discuss the most important completely open problems. Amongst these I select the following:
(a) What is the ultimate origin of the black body radiation field?
(b) Why is the Universe so symmetric?
(c) How can we eliminate the initial singularity?
I shall discuss these in turn.

3.1. THE ORIGIN OF THE BLACK BODY BACKGROUND

What requires explanation is not merely the existence of a black body radiation field, but its amount. It is not convenient to specify this amount in terms of the relative energy densities of the radiation field and of matter, because this relation would be time dependent in the expanding universe. What is time *independent* is the number of black body photons per proton in the Universe. This relative number would be about 10^8 in a high density universe and 10^9 in a low density universe. This number appears as a new kind of dimensionless 'constant of nature', and its observed value has to be explained. I must emphasise that the canonical hot big bang picture of Gamow does *not* offer any explanation. In this picture all the black body radiation was present in the Universe from the initial singularity, and the number of photons per proton is *completely arbitrary*.

Amongst the alternative proposals which have so far been made, there are two which stand out because they are amenable to rather detailed analysis, and appear to offer reasonable prospects of actually leading to the observed number of photons per proton. I shall confine myself to these two.

3.1.1. *The Matter-Antimatter Universe*

It has been argued by Harrison (1968), and in most detail by Omnes (1972) and his colleagues, that initially the Universe might have consisted of black body radiation *and nothing else*. Or rather that it consisted of black body radiation plus the thermally excited particle-antiparticle pairs that would be demanded by standard physics. At sufficiently early times, when the radiation temperature exceeded about 10^{13} K, there would have been roughly as many proton-antiproton pairs (and electron-positron pairs) as photons in the radiation field. If these pairs had remained intimately mixed up with one another, they would have annihilated progressively and completely as the Universe expanded and the radiation cooled. However, this end result is by no means what we observe to-day. In fact, we observe one proton left over for every 10^8 or so photons. This means that in the Gamow picture there were, during the earliest stages, $10^8 + 1$ protons for every 10^8 anti-protons. That extra proton was just what was needed to survive annihilation and to provide the material universe that we know to-day.

It would clearly be more attractive to do without the extra proton, if complete annihilation could be avoided by some separation process which kept matter and anti-matter sufficiently apart from one another. This is the problem which Omnes and his collaborators are currently trying to solve. What they must find is a separation process so efficient that a whole galaxy (and perhaps also a cluster of galaxies) should consist predominantly of matter of one sign, since this is a direct observational requirement. Given a separation process one could calculate, at least in principle, the amount of annihilation that would have taken place, and therefore obtain a definite value for the present ratio of photons to material particles. Thus this theory has within it the potentiality to account for the number 10^8.

This is a beautiful programme, but it suffers at the moment from two main difficulties. The first is that it has not yet been generally accepted that the separation mechanisms so far proposed are efficient enough to guarantee that galaxy-sized objects would separate out. This difficulty might, of course, be resolved by further work (and I personally hope that it will be, in view of the beauty and economy of the theory). The second difficulty is that essentially no helium would have been formed in the early stages (because at that time neutrons would have diffused too rapidly out of regions containing matter of one sign). Thus this theory requires one to find an alternative mechanism for manufacturing helium at a later stage, and would of course benefit from unambiguous evidence that the relative abundance of helium is not uniform throughout the Galaxy.

3.1.2. *Dissipative Processes in the Early Universe*

This explanation is the exact opposite of the previous one in the sense that in it the Universe begins with matter and no radiation, instead of radiation and no (net) matter. However, the potentiality for producing radiation is present in that it is assumed that the matter is undergoing irregular motions whose dissipation would lead to radiation being produced. These irregular motions may also play a decisive role in the formation of galaxies. The possible importance of dissipative processes was first pointed out by Misner (1967, 1968) and by Doroshkevich *et al.* (1967). In its most extreme form it supposes that initially the spectrum of disturbances corresponded in some sense to 'white noise'. In other words 'all' disturbances are assumed to be present initially with comparable amplitudes, and those disturbances not dissipated away should have survived to the present day. According to this approach, sometimes described as chaotic cosmology, any other assumption about the initial conditions would involve special restrictions which would require explicit justification.

Whatever view one might take about this methodological point, the chaotic cosmology programme has the advantage that, like the matter-antimatter programme, it has calculable consequences which can be compared with observation. The most attractive variant proposed so far is that due to Rees (1972) who assumes that the irregular motions to be dissipated involve velocities which are substantial compared with the velocity of light (mildly relativistic turbulence). If dissipation is efficient this would mean that the resulting radiation field would have an energy density ϱ_γ comparable with the rest density of matter ϱ_m. More precisely, as the Universe expands and the size of the particle horizon steadily increases, larger scale irregularities would become available for dissipation, and so the approximate equality $\varrho_\gamma \sim \varrho_m$ would be maintained until the largest-scale irregularity is reached. However, this process must cut off at the latest when the matter is still capable of thermalising the heat produced by the dissipation; otherwise the microwave background would not now have a black body spectrum. If the cut-off does occur close to the last possible moment then, as Rees points out, the final photon/proton ratio would indeed be about 10^8, and also the largest-scale irregularities would correspond in mass to clusters of galaxies. These two beautiful consequences of the chaotic cosmology programme suggest that it should be investigated in much more detail. In particular its implications for the helium problem need elucidation. At first sight the predicted helium abundance appears to be on the high side, since the photon/proton ratio would have been less than 10^8 when the relevant nuclear reactions occurred. However, the assumed irregularities are likely to change the expansion time-scale in the early stages and this would affect the amount of helium produced. This question deserves further study.

3.2. Why is the universe so symmetric?

The Robertson-Walker models are based on the assumptions that the Universe is exactly isotropic and homogeneous. These assumptions are certainly not true on a scale of, say, 100 Mpc (de Vaucouleurs, 1971), but the important question for cosmology is how true they are on a scale closer to, say, 1000 Mpc, that is on a scale somewhat smaller than the radius of the Universe itself (3000 Mpc). Here the best evidence comes from the observed isotropy of the microwave background, which is good to about 0.1% on a variety of angular scales (Partridge and Wilkinson, 1967; Conklin and Bracewell, 1967; Partridge, 1969; Parijskij, 1973; Carpenter et al., 1973). This is by far the most accurate observation of importance to cosmology. Regarding the immediate source of the background as the last-scattering surface, we can conclude that the radiation temperature is constant over the surface to a precision of 0.1%, and that the redshift of the surface, and so the rate of expansion of the Universe, is the same in all directions to this precision.

We can also draw conclusions about the large-scale homogeneity of the Universe (Sachs and Wolfe, 1967; Rees and Sciama, 1968). Consider for instance an irregularity on a scale ~ 1000 Mpc, in which the density were twice the mean density of the Universe. The presence of such an irregularity would influence the temperature of the black body radiation moving through it, and so would impose an anisotropy on the background, which would be of the order of 0.1%. The observed upper limit therefore places constraints on possible large-scale inhomogeneities in the Universe.

We conclude from this that on a large scale the universe is indeed both isotropic and homogeneous to high precision, and that the Robertson-Walker models are adequate representations at least back to the last scattering surface. Indeed they are better representations than mathematical cosmologists had previously had the right to expect. However, what is convenient mathematically may create serious physical problems. Here the problem is, why should the Universe be so symmetric? The urgency of this question derives from the fact that if the Robertson-Walker models are good representations at all epochs, then two points on the last-scattering surface more than, say, 30° apart would have been outside each other's particle horizon up to the moment that the radiation we now observe was emitted. In other words, no causal influence, even moving with the speed of light, could have linked together the physical processes occurring in the neighbourhood of the two points. Yet their radiation temperatures are the same to within 1 part in a thousand!

This equality represents an unsolved problem. I would like, however, to mention briefly some of the solutions that have so far been proposed.

3.2.1. *Initial Conditions*

According to this proposal, the initial conditions governing the Big Bang simply had the necessary properties. While this may be the correct answer, no self-respecting cosmologist would adopt it while any alternative is conceivable.

3.2.2. *The Mix-Master Universe*

This proposal has the paradoxical property of trying to explain the observed isotropy by assuming that initially the Universe was highly anisotropic (though possibly homogeneous). The idea is that in the presence of a large early deviation from the Robertson-Walker symmetry particle horizons would no longer exist, so that dissipative processes might be able to remove the anisotropies in the time available. The detailed discussion of this idea is rather complicated because in these anisotropic models particle horizons are abolished first in one direction, then in another and so on, so that eventually the whole universe can become smoothed out; hence the name mixmaster universe (Misner, 1969). Unfortunately it now appears that models possessing this property form such a special class amongst all possible anisotropic models, that one is back to appealing to special initial conditions (MacCallum, 1971; Matzer and Misner, 1972).

3.2.3. *The Parabolic Universe*

Amongst the homogeneous (but possibly anisotropic) models of the Universe there are those which expand for ever with energy to spare (hyperbolic cases), those which just have the escape velocity (parabolic), and those which re-contract (elliptic). It has recently been discovered (Collins and Hawking, 1973a) that the hyperbolic models are unstable to the growth of a perturbation in which the anisotropy eventually increases with time. Unless therefore we live at a special time in such a model when the anisotropy happens to be very low we can conclude that the Universe is not hyperbolic. On the other hand in an elliptic universe an arbitrary initial anisotropy would not have time to die away to a low level before the universe re-collapses.

There remain the parabolic homogeneous universes, which do in general tend towards isotropy. However, as Collins and Hawking point out, if we assume that the Universe is parabolic, we are again appealing to a very special initial condition. Is there any independent way of understanding why the universe should have this special property? Collins and Hawking propose that there might be, following a suggestion made by Dicke (1961), Carter (1968) and others. According to this suggestion, one should contemplate the existence of a variety of universes in which the initial conditions, and perhaps also the laws and constants of nature, take on all conceivable properties. Only in a very restricted subclass of cases would galaxies and stars form, and intelligent life develop. Thus our very presence would impose strict limitations on the laws and constants of nature, and on the initial conditions of the Universe. In particular Collins and Hawking suggest that it might be difficult for galaxies to form in hyperbolic universes, since self-gravitation may not be important during the later stages when the black body radiation itself has ceased to prevent galaxy formation. Their discussion of elliptic universes seems less satisfying since they argue that in this case there is no infinite future available in which an arbitrarily large initial anisotropy would die away, and that small density perturbations would not have time to develop into galaxies before the Universe re-collapses. However, the first of these arguments overlooks the likely fact that intelligent life is only possible for a finite time in a parabolic or hyperbolic universe also. In any case we certainly are observing the Universe at a finite time, and one whose value depends very little on the dynamical character of the Universe. Nevertheless it is an interesting point that a parabolic universe does not have this particular instability to an anisotropic perturbation, and this fact may be relevant to an understanding of the observed isotropy of the Universe.

3.2.4. *The Machian Universe*

According to Mach's principle local inertial frames are determined by the large-scale distribution of matter in the Universe and not by absolute space. One consequence of this principle would be that the Universe could not be in a state of absolute rotation. Astronomers usually take this for granted when they attempt to use extragalactic objects such as galaxies or quasi-stellar objects to define the best 'fixed' reference frame. However, general relativity gives us no grounds for believing it. There are cosmological solutions of Einstein's field equations in which the Universe is homogeneous at any one time, but is rotating relative to absolute space and so is anisotropic. In more technical language we would say that such a Universe possesses vorticity. This would mean that the whole Universe would appear to rotate relative to a local inertial frame of reference (e.g. relative to one in which a Foucault pendulum on the Earth oscillated in a fixed plane, or more correctly in view of the 'dragging of inertial frames' by localised rotating objects, relative to one which was appropriate for discussing dynamics on a scale of, say, 100 Mpc).

We must add a further condition to general relativity if we wish to eliminate this absolute rotation on the grounds that it is unphysical. This is where Mach's principle comes in (Sciama, 1971). Of course one should first ask the purely empirical question, with what precision is the Universe observed not to rotate relative to a local inertial frame? If the observed lower limit on the rotation period of the Universe were less than or of the order of 10^{10} yr, this would hardly constitute good evidence in favour of Mach's principle since 10^{10} yr is also the characteristic time associated with the size of the Universe. This was precisely the situation before the microwave background was discovered. At that time the best limit that could be placed on the rotation period of the Universe derived from the absence of *transverse* Doppler shifts in the spectra of distant galaxies. This led to a lower limit for the rotation period of 10^{10} yr (Kristian and Sachs, 1966), which was too weak to support Mach's principle.

The situation is quite different now that the isotropy of the microwave background has been established to 0.1%. This isotropy imposes a constraint on the transverse Doppler shift associated with any rotation of the last-scattering surface, and so on the present rotation of the universe. This question has been analysed by Hawking (1969). His detailed conclusions are somewhat model-dependent, but in all cases the rotation period has a lower limit of at least 10^{14} yr. This result appears to give strong support for Mach's principle. It could be argued, however, that the comparison between rotation period and expansion time-scale is itself time-dependent, and that even if the rotation period is relatively long now, it need not have been in the past. This question has been studied by Collins and Hawking (1973b), who concluded that in any reasonable model of the universe the rotation period must at all times have been long compared with the expansion time-scale. They also succeeded in placing much sharper limits on the period than did Hawking, at least for the case of closed universes.

We conclude from this that the evidence in favour of Mach's principle is strong enough to justify trying to incorporate it into general relativity. We might expect this to give us a coherent reason for excluding those anisotropies of the universe which are due to vorticity. However, it appears that we can in this way also exclude shear, which would mean that (homogeneous) Machian universes would have to be completely isotropic. The detailed incorporation of Mach's principle into general relativity is a technical question which I cannot go into here. Suffice it to say that, starting out from work by Alt'shuler (1969), Lynden-Bell (1967), Sciama *et al.* (1969), and Gilman (1970), Raine (1974) has gone far towards formulating a rigourous criterion for Mach's Principle in general relativity. In terms of this criterion he is indeed able to show that a homogeneous Machian universe must be isotropic (at least when the matter in the Universe has a perfect gas equation of state, although this restriction is probably inessential). What must now be done is to show, as seems plausible, that a nearly-homogeneous Machian universe is nearly isotropic. If this statement could be made quantitative it would enable one to correlate different aspects of the angular distribution of the microwave background.

If this programme were successful there would still remain the problem of understanding why the Universe is so homogeneous. Mach's principle seems to shed no light on this. So long as, roughly speaking, there is sufficient matter everywhere in the Universe, so that one cannot get arbitrarily far from all the matter that there is, it would seem that Mach's principle could be satisfied whatever the length scale or the amplitude of the inhomogeneities. Their observed limitation to less than 1000 Mpc for amplitudes $\gtrsim 1$ remains a mystery.

3.3. How can we eliminate the initial singularity?

The Robertson-Walker models (with zero cosmical constant) contain a point singularity a finite time ago in which the density of the Universe was infinite. Most cosmologists (though not all) regard this singularity as physically unacceptable. Until recently it was thought that the singularity was an artefact of the assumption that the Universe is exactly homogeneous and isotropic. In the presence of some irregular motions one might have expected that, going backwards in time, the material of the Universe would not be exactly focussed onto one point, but would either bounce or interpenetrate without a singularity. It turns out that this is not so. In the most important theoretical development in cosmology since the Robertson-Walker models were discovered, it has been found that, according to general relativity, self-gravitation was so strong in the early stages of the Universe that the existence of a singularity is guaranteed. The argument does, however, involve a number of (reasonable) assumptions which I shall discuss in a moment. We owe this development mainly to work by Penrose, Hawking and Ellis, and a full technical account of it may be found in a comprehensive monograph by Hawking and Ellis (1973).

There is in fact a variety of singularity theorems, both for collapsing stars and for expanding universes. Some of these theorems are based on assumptions which, though plausible, could not actually be checked in practice. Fortunately there is one theorem in the cosmological case where the basic assumptions can indeed be checked. Remarkably enough, the existing observations of the microwave background can be used for this purpose (Hawking and Ellis, 1968). In fact the background plays two roles in the argument, namely through (i) its isotropy and (ii) its energy density.

The argument is based on the following theorem, which I first state precisely and then comment on.

Space-time is not singularity-free if the following conditions hold:

(a) Einstein's field equations,

(b) The energy condition ($T_{ab}W^a W^b \geqslant \frac{1}{2}T$ for all time-like vectors W^a),

(c) Strong causality (every neighbourhood of a point contains a neighbourhood of that point which no non-spacelike curve intersects more than once),

(d) There exists a point P such that all past-directed time-like geodesics through P start converging again within a compact region in the past of P.

Condition (b) is a very weak and plausible restriction on the energy-momentum tensor, while a violation of (c) would perhaps be worse than a singularity, being a global rather than a local breakdown of our ordinary physical concepts. The really important condition is (d), which is a precise statement of the idea that the gravitation due to the material in the Universe is sufficiently attractive to produce a singularity. The behaviour of time-like geodesics required is reminiscent of that of light-rays in a curved space-time. In the latter case the re-convergence would correspond to the angular diameter of an object of fixed intrinsic size passing through a minimum as the distance of the object increased from the point P (which we may take to represent the Earth). This lens effect certainly occurs in Robertson-Walker models. In fact in a high density universe the minimum occurs at a red shift of less than 2. Since many quasi-stellar objects have red shifts exceeding 2 this behaviour is of importance in the interpretation of measurements of their (radio) angular diameters.

The argument now proceeds by showing that the actual Universe is sufficiently like a Robertson-Walker one for the re-convergence to be preserved. In fact the observed isotropy of the microwave background shows that the universe is Robertson-Walker-like to high precision at least back to the last-scattering surface. The aim of the argument is therefore to show that the re-convergence begins before the last-scattering surface is reached. To simplify the argument we consider only the two extreme cases of a high density and a low density universe. In the former case the last-scattering surface occurs at a red shift of about 7. A straightforward calculation then shows that there is enough matter in this model for its gravitation to start producing re-convergence at a red shift less than 7. In the low density case we can be sure that the Universe is Robertson-Walker-like at least out to the much larger red shift of 1000. However, in this case there may not be enough matter to produce reconvergence at red shifts less than this. Nevertheless the argument is again saved by the microwave background, but now in its role as a contributor to the energy density of the Universe. For this contribution alone is sufficient to guarantee re-convergence before a red shift of 1000 is reached. A detailed investigation shows that models with intermediate density also exhibit re-convergence before the last-scattering surface. (The introduction of a non-zero cosmical constant with a value permitted by observation would not affect the argument, since its contribution to Einstein's field equations would be swamped by that of matter and radiation at and prior to the last-scattering surface).

We conclude that, according to standard theory, the actual Universe was singular at some time or times in the past. We here face a major crisis of fundamental science. The most conservative method of dealing with it would be to argue that when extremely high but finite densities are reached quantum mechanical and elementary particle effects would become important. In particular, it may become necessary to quantise the gravitational field. It has been argued that this might suffice to eliminate the singularity, and certain preliminary calculations do support this contention. However, other calculations suggest that the singularity may survive the quantisation process and therefore that new laws of physics would be needed to resolve the problem. The matter is a highly technical one which is unsettled at the present time. Whatever the outcome, it is clear that radio astronomy, general relativity, quantum field theory and elementary particle physics have combined together to present us with one of the greatest scientific problems of our time. It is a very difficult problem indeed, but I hope one day to listen to an Invited Discourse in which the solution is presented.

Acknowledgements

I am very grateful to C. B. Collins, D. J. Raine and M. J. Rees for their critical reading of the manuscript.

References

Alpher, R. A., Bethe, H. A., and Gamow, G.: 1948, *Phys. Rev.* **73**, 803.
Alt'shuler, B. L.: 1969, *Soviet Phys. JETP* **28**, 687.

Audouze, J. and Truran, J. W.: 1973, *Astrophys. J.* **182**, 839.
Bondi, H.: 1960, *Cosmology* (2nd ed.), Cambridge.
Bondi, H. and Gold, T.: 1948, *Monthly Notices Roy. Astron. Soc.* **108**, 252.
Brans, C. and Dicke, R. H.: 1961, *Phys. Rev.* **124**, 925.
Carter, B.: 1968, Unpublished Cambridge preprint.
Carpenter, R. L., Gulkis, S., and Sato, T.: 1973, *Astrophys. J.* **182**, L61.
Colgate, S.: 1973, *Astrophys. J.* **181**, L53.
Collins, C. B. and Hawking, S. W.: 1973a, *Astrophys. J.* **180**, 317.
Collins, C. B. and Hawking, S. W.: 1973b, *Monthly Notices Roy. Astron. Soc.* **162**, 307.
Conklin, E. K. and Bracewell, R. N.: 1967, *Phys. Rev. Letters* **18**, 614.
Dicke, R. H.: 1961, *Nature* **192**, 440.
Dicke, R. H.: 1962, *Rev. Mod. Phys.* **34**, 110.
Dicke, R. H., Peebles, P. J. E., Roll, P. G., and Wilkinson, D. T.: 1965, *Astrophys. J.* **142**, 414.
Dirac, P. A. M.: 1938, *Proc. Roy. Soc.* **A165**, 199.
Dirac, P. A. M.: 1973a, in *The Physicists' Conception of Nature*, North-Holland, in press.
Dirac, P. A. M.: 1973b, *Proc. Roy. Soc.* **A333**, 403.
Doroshkevich, A. G., Zel'dovich, Y. B., and Novikov, I. D.: 1967, *Soviet Phys. JETP* **53**, 644.
Field, G. B.: 1969, 'The Growth Points of Physics', *Rivista Nuovo Cimento* **87**.
Gamow, G.: 1948, *Phys. Rev.* **70**, 572.
Gamow, G.: 1949, *Rev. Mod. Phys.* **21**, 367.
Gilman, R. C.: 1970, *Phys. Rev.* **D2**, 1400.
Harrison, E. R.: 1968, *Phys. Rev.* **167**, 1170.
Hawking, S. W.: 1969, *Monthly Notices Roy. Astron. Soc.* **142**, 129.
Hawking, S. W. and Ellis, G. F. R.: 1968, *Astrophys. J.* **152**, 25.
Hawking, S. W. and Ellis, G. F. R.: 1973, *The Large Scale Structure of Space-Time*, Cambridge.
Hawking, S. W. and Penrose, R.: 1970, *Proc. Roy. Soc.* **A314**, 529.
Hoyle, F.: 1948, *Monthly Notices Roy. Astron. Soc.* **108**, 372.
Hoyle, F.: 1972, *Quart. J. Roy. Astron. Soc.* **13**, 328.
Hoyle, F. and Fowler, W. A.: 1973, *Nature* **241**, 384.
Jordan, P.: 1949, *Nature* **164**, 637.
Jordan, P.: 1973, in *The Physicists' Conception of Nature*, North-Holland, in press.
Kristian, J. and Sachs, R. K.: 1966, *Astrophys. J.* **143**, 379.
Lynden-Bell, D.: 1969, *Monthly Notices Roy. Astron. Soc.* **135**, 413.
Layzer, D. and Hively, R.: 1973, *Astrophys. J.* **179**, 361.
MacCallum, M. A. H.: 1971, *Nature Phys. Sci.* **230**, 112.
Matzner, R. A. and Misner, C. W.: 1972, *Astrophys. J.* **171**, 415.
Milne, E. A.: 1948, *Kinematic Relativity*, Oxford.
Misner, C. W.: 1967, *Nature* **214**, 40.
Misner, C. W.: 1968, *Astrophys. J.* **151**, 431.
Misner, C. W.: 1969, *Phys. Rev. Letters* **22**, 1071.
Omnes, R.: 1972, *Physics Reports* **3c**, 1.
Parijsky, Y. N.: 1973, *Astrophys. J.* **180**, L47.
Partridge, R. B.: 1969, *Am. Scientist* **57**, 37.
Partridge, R. B. and Wilkinson, D. T.: 1967, *Phys. Rev. Letters* **18**, 557.
Peebles, P. J. E.: 1971, *Physical Cosmology*, Princeton.
Peimbert, M.: 1974, in J. Shakeshaft (ed.), 'Formation and Dynamics of Galaxies', *IAU Symp.* **58**, in press.
Penzias, A. A. and Wilson, R. W.: 1965, *Astrophys. J.* **142**, 419.
Raine, D. J.: 1974, to be published.
Rees, M. J.: 1972, *Phys. Rev. Letters* **28**, 1669.
Rees, M. J. and Sciama, D. W.: 1968, *Nature* **217**, 511.
Reeves, H., Audouze, J., Fowler, W. A., and Schramm, D. N.: 1973, *Astrophys. J.* **179**, 909.
Rindler, W.: 1956, *Monthly Notices Roy. Astron. Soc.* **116**, 662.
Sachs, R. K. and Wolfe, A. M.: 1967, *Astrophys. J.* **147**, 73.
Sciama, D. W.: 1969, *The Physical Foundations of General Relativity*, Science Study Series.
Sciama, D. W.: 1971a, *Modern Cosmology*, Cambridge.
Sciama, D. W.: 1971b, in R. K. Sachs (ed.), *General Relativity and Cosmology*, Proceedings of the

International School of Physics 'Enrico Fermi'. Course XLVII, Academic Press, New York.
Sciama, D. W., Waylen, P. C., and Gilman, R. C.: 1969, *Phys. Rev.* **187**, 1762.
Searle, L. and Sargent, W. L.: 1972, *Comm. Astrophys. Space Sci.* **14**, 179.
Vaucouleurs, G. de: 1971, *Publ. Astron. Soc. Pacific* **83**, 113.
Wagoner, R. V.: 1973, *Astrophys. J.* **179**, 343.
Weinberg, S.: 1972, *Gravitation and Cosmology*, Wiley.
Zel'dovich, Y. B. and Novikov, I. D.: 1974, *Relativistic Astrophysics*. Vol. II: *Cosmology*, Chicago.

THE PHYSICS OF THE INTERSTELLAR MATTER

G. B. FIELD

*Center for Astrophysics, Harvard College Observatory and
Smithsonian Astrophysical Observatory, Cambridge, Mass., U.S.A.*

1. Introduction

Some of the most striking recent discoveries about interstellar matter involve molecules. It has been known for a long time that there are atoms and ions in space – mainly hydrogen and helium, of course – but also heavier elements like sodium and calcium. In addition, there are solid particles of dust, about 10^{-5} cm across, which must be composed of heavier elements, as hydrogen and helium cannot condense under interstellar conditions.

In 1972, the Orbiting Astronomical Observatory-3, which employs an 80-cm telescope at wavelengths between 1000Å and 3000Å, was launched in the United States and put into operation. In the ensuing year, it has demonstrated that much of the interstellar medium is composed of hydrogen molecules. This result, based upon the observation of Lyman-band absorption in the spectrum of early-type stars, had been anticipated by a rocket observation of H_2 by Carruthers in 1970.

The same OAO-3 instrument observed resonance lines of many cosmically abundant elements, and found that these elements often appear to be less abundant in interstellar space than in the solar system, relative to hydrogen. As young stars born recently from the interstellar medium do not show this effect, the heavy elements must in fact be present in some other form. Here I will argue that the heavy elements are largely locked up in the form of the dust and, further, that dust is critical for the formation of the molecules in interstellar space.

It is appropriate that the Orbiting Astronomical Observatory-3 which made these discoveries has been named in honor of Copernicus, the Polish astronomer we honor here on the 500th anniversary of his birth. Just as his discoveries were revolutionary for the understanding of the solar system, those made using the Observatory named in his honor have been revolutionary for the understanding of the Galaxy.

2. Interstellar Dust

The chemical composition of interstellar dust has been hotly debated for many years. A resolution of this question was not possible as long as astronomers were limited by the information available in the visible region, because there the extinction by dust exhibits only a slow variation with wavelength. This can be matched with several different materials which might be expected to be present, such as ice, silicates, graphite, metallic iron, etc. Recent exploration of the infrared and ultraviolet bands has provided critical new information, however.

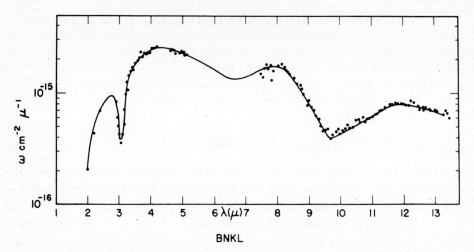

Fig. 1. Spectrum of the Becklin-Neugebauer point infrared source in Orion, due to Gillett et al. (1972). Note the two absorption bands, at 3.1 μ and 9.7 μ, characteristic of water ice and of silicates.

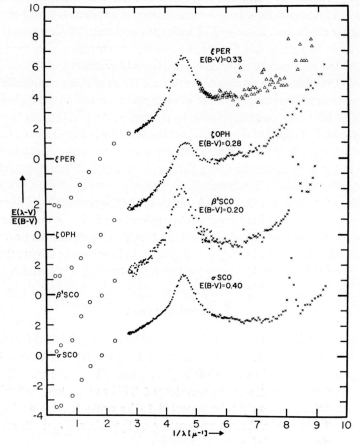

Fig. 2. Interstellar extinction curves for four stars, with normalized absorption increasing upwards (Bless and Savage, 1972). The absorption feature at 2200 Å ($\lambda^{-1} = 4.6 \mu^{-1}$) has been identified as graphite.

Figure 1 shows the infrared spectrum of the infrared point source in Orion. Two absorption features are seen: that at 3.1-μ wavelength is characteristic of water ice, while that at 9.7-μ wavelength is characteristic of silicates such as $MgSiO_3$ or Mg_2SiO_4. Figure 2 shows the interstellar extinction curve in the ultraviolet for several early-type stars. One notes an absorption feature at 2200Å, which is seen in each of the about 200 stars observed. This feature is characteristic of small particles of graphite. Recently, a feature of silicon carbide has been observed at 11.5 μ. Thus, there is direct evidence for particles containing ice, silicates, graphite and silicon carbide in interstellar space.

How do such particles originate? Attempts to explain their origin in interstellar space have failed; the time scales for atoms to collide and stick are too large at the very low gas densities involved, 10–100 cm^{-3}. It was suggested, rather, that dust grains condense in the relatively high-density conditions prevailing in the outer envelopes of stars, provided such envelopes are cool enough. The condensation process has been studied theoretically, with the results in Table I. One sees that in an oxygen-rich atmosphere the most abundant elements (C, N, O, Mg, Si, and Fe) condense as methane, water, ammonia, magnesium silicates, and metallic iron, while in a carbon-rich atmosphere, one expects graphite, silicon carbide, and iron carbide. Evidence that condensation actually does occur in some stars is provided by infrared spectra in which the 9.7-μ feature appears in emission, the interpretation being that the circumstellar dust is heated by the star and that it emits in the 9.7-μ band, where its absorptivity is high. For this to work, the dust must be near the star.

Independent confirmation of circumstellar dust has been obtained through polarization studies. While most stars show a fixed polarization due to aligned dust particles in interstellar space, many late-type stars also show variable polarization which can be explained only if the responsible dust is very close to the star. Furthermore,

TABLE I

Condensation temperatures[a]

Stage	Temperature (K)	Condensates	Elements removed
1	1400–1600	$CaTiO_3$	Ti
		$Mg_2Al_2O_4$ Al_2SiO_3 $CaAl_2Si_2O_8$	Al
		$CaMgSi_2O_6$ Ca_2SiO_4 $CaSiO_3$	Ca
2	1220–1320	$MgSiO_3$ Mg_2SiO_4 $BeAl_2O_4$	Si, Mg, Be
3	1280	Metallic Fe Ni	Fe, Ni
4	1210	$MnSiO_3$	Mn
5	970–1070	Alkali Silicates	Na, K, Rb
6	600–700	FeS $NaBO_2$	S, B
7	180	H_2O	O
8	120	$NH_3.H_2O$	N
9	75	$CH_4.XH_2O$	C
10	25	Ar(solid)	Ar

[a] See, e.g., Lewis (1972).

many late-type stars are known from spectroscopic studies to be losing gaseous material at substantial rates. Hence, one concludes that some late-type stars are losing material and that much of the associated heavy elements has condensed to form dust. Quantitatively, it appears that the rate at which dust is being ejected into space by this mechanism is adequate to account for the amount of interstellar dust which is observed.

3. Depletion of Gas-Phase Elements

If it is really true that heavy elements which are injected into space by stars have condensed partly into dust grains, one would expect this effect to show up when atoms and ions are sought spectroscopically in the gas phase. The Copernicus satellite telescope observed resonance lines of interstellar H, B, C, N, O, Mg, Si, P, S, Cl, Ar, Mn, and Fe, while previous ground-based measurements gave data on Li, Be, Na, Al, K, Ca, and Ti. In the direction of the 09.5 star ζ Oph, there is a wealth of information, including a measurement of the ratio of atomic calcium to ionized calcium in the cloud between us and ζ Oph. This ratio permits one to estimate the electron density in the cloud and, therefore, to calculate the abundances of certain other ions which are not directly observed.

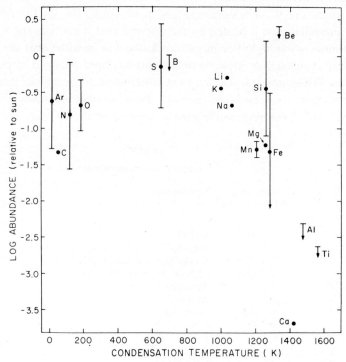

Fig. 3. Abundances toward ζ Oph on a logarithmic scale, relative to solar-system values (Field, 1974). The horizontal axis is condensation temperature, defined as that temperature at which the element condenses into solid particles, according to the sequence of Table I. Note the tendency for the most condensible elements to be most depleted, suggesting that they are condensed into dust grains.

The results, expressed as column densities of various gas-phase elements relative to those expected if the abundances are normal, are shown in Figure 3. It is instructive to plot the observed abundances against the theoretical condensation temperatures given in Table I. The data indicate that those elements which should condense at the highest temperatures (notably Al, Ca, and Ti) are most underabundant. There is a general tendency for the underabundance to correlate with condensation temperature, suggesting that the heavy elements have actually condensed into dust grains, this process being most effective for the highest temperature condensates. This would be expected if, for example, the condensation process occurs while gas is flowing away from a star, for high-temperature materials would condense near the star, where the gas density is high. Low-temperature materials would condense only far away from the star, where the gas density is so low that the process would not be completed.

From Figure 3 one sees that the abundant elements C, N, O, Mg, Si, and Fe are all depleted in the gas phase by substantial factors. One may wonder whether these elements are really locked up in dust or, rather, are present in some other form, such as gas molecules. One finds spectroscopically that while gas molecules are indeed present in the ζ Oph cloud, they have much too low an abundance to account for the observed depletion of atoms and ions. We conclude that these elements are in the dust.

One is, therefore, led to consider models of grains which are composed of these elements, in those forms predicted by the condensation theory, notably graphite, methane, ammonia, water, magnesium silicates, silicon, carbide, and metallic iron. If one confines attention to spherical grains (perhaps an unreasonable assumption), one finds that one can just explain the observed visual extinction if all the missing abundant elements are in the form of these materials. Even more significant, quantitative agreement is obtained with the observed silicate absorption in the Orion infrared source (Figure 1) and with the observed graphite absorption in ζ Oph (Figure 2).

Difficulty is encountered, however, with water ice. If all the missing oxygen which is not chemically combined in silicates is assumed to be in water, the predicted 3.1-μ band is about ten times stronger than the observed one (Figure 1). This problem is discussed further below.

4. Interaction of Gas and Dust

Figure 3 reveals an apparent anomaly. Depletion factors decrease steadily with decreasing condensation temperature down to 650 K, where S is depleted by a factor less than 5. Yet the very low-temperature condensates C, N, O, and Ar are depleted by even greater factors. This suggests that a different process may be involved for the low-temperature condensates; we tentatively identify this to be the slow accumulation of ice mantles on the dust grains after they have become part of the interstellar cloud.

Does this actually occur?

To answer this question requires an understanding of the surface physics of graphite, iron, and silicate dust grains, which is not available at present. One can make some general points, however.

In a typical interstellar cloud, a hydrogen atom collides with a dust grain about once every 10 million years. Very likely it will stick there, as it is bound by van der Waals forces. Because hydrogen is the dominant constituent of interstellar gas, a layer of H atoms builds up on the grain. Hence the H atoms on the grain are closely surrounded by other H atoms, and there is a fair probability that they will react to form H_2 via

$$H + H \rightarrow H_2 + 4.5 \text{ eV}. \tag{1}$$

The 4.5-eV energy released goes either into heating the grain or into kinetic energy of the molecule. As it is much greater than the binding energy to the grain, the newly-formed H_2 probably is ejected at once. We discuss the fate of the resulting H_2 molecules further below.

Other atoms, particularly the abundant ones C, N, and O, also collide with dust grains about every 10 million years, and it is believed that they, too, will stick. But these atoms are surrounded not by others of the same type, but by the H atoms in the layer described above. Hence, they are likely to react with H to form the radicals CH, OH, and NH.

At this point, it is uncertain what happens. On the one hand, the radicals may be ejected at once, like H_2, and, in fact, gas-phase CH and OH have been observed. On the other hand, their much greater mass may reduce the ejection probability, so that they stay on the grain. In the latter event, they will react further, ultimately forming the saturated molecules CH_4, NH_3, and H_2O. The latter molecules, accumulated on the grain, would constitute the ice mantle I spoke of earlier and would explain the observed depletion of C, N, and O.

Does mantle formation actually occur in interstellar space?

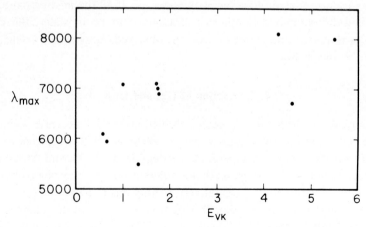

Fig. 4. The wavelength of maximum polarization, plotted against the extinction $A_V - A_K \simeq A_V$, for various stars shining through the dark cloud near ϱ Ophiuchi (Carrasco et al., 1973).

The evidence on this point is often ambiguous, but recently, observations of the well-known dark cloud near ϱ Oph indicate strongly that the dust in this cloud is accreting mantles. The density in the cloud varies from about 10 cm^{-3} 6 pc from the center to about 1000 cm^{-3} 0.4 pc from the center. The extinction of stars seen through the cloud also increases inward. The light from stars shining through the cloud is polarized and the degree of polarization varies with wavelength in the usual way, reaching a maximum at λ_{max} located in the observable range. Normally λ_{max} is about 5400 Å, but in this cloud, it rises from about that value in the outer parts of the cloud to about 8000 Å in the inner part of the cloud (Figure 4). Theory indicates that λ_{max} should be roughly proportional to particle size, so this indicates that the size has increased 60% in going from the outer to the inner part of the cloud. This can be explained if substantial amounts of C, N, and O have accreted on to the dust grains in the cloud. The rate at which this happens is proportional to density, so it is natural that the dense inner parts of the clouds, where A_V is larger, should show the larger effect. I conclude that accretion of mantles occurs in at least some clouds.

However, one is left with the problem of the weak 3.1-μ band of H$_2$O ice. One may speculate here that absorption of stellar ultraviolet light may lead to further chemical changes in the mantle. Each absorption dissociates a molecule, momentarily forming a radical which reacts rapidly with neighboring molecules. Such processes have been observed in the laboratory to yield complex molecules at a rate which would give a time scale of a few million years if translated to interstellar conditions. Thus, most of the oxygen may be combined into complex organic molecules and the fraction of oxygen combined in water may be small, as observed.

We referred above to H$_2$ molecules forming on grains and being ejected into the interstellar medium. This process is very important in the Galaxy, for it is the only known way to efficiently catalyze the formation of H$_2$ molecules from H atoms in space. The reason for this is that if two H atoms approach each other in the gas phase, there is no third body to take up the energy of formation, so that the atoms may form a bound system. Consequently, the only way that a molecule can form is by the emission of radiation, which for a homonuclear molecule like H$_2$ requires quadrupole emission. This is so weak that H$_2$ molecules cannot form in appreciable amounts in this way. Thus, the mere fact that H$_2$ is observed in interstellar space indicates that the H$_2$ formation is being catalyzed by dust grains.

Of course, H$_2$ can also be destroyed in interstellar space. The most important process is

$$\begin{aligned} \text{H}_2(X'\,\Sigma_g^+) + h\nu &\to \text{H}_2(B'\,\Sigma_u^+) \\ &\to \text{H}_2^*(X'\,\Sigma_g^+) \to \text{H} + \text{H}, \end{aligned} \quad (2)$$

where H$_2^*$ indicates vibrational continuum and where $h\nu$ is a photon of the Lyman band ($\lambda < 1108$ Å). In low-density regions, this occurs every 300 years, a much smaller time scale than the millions of years it takes to form H$_2$. Hence, in unshielded space, one expects that H$_2$/H $\ll 1$.

However, in a denser cloud, the newly formed H_2 molecules on the outside of the cloud absorb the Lyman-band photons before they reach the interior and the molecules inside are thereby shielded. The theory of this effect yields the results in Figure 5, where I have also plotted some of the observational data from Copernicus. There is a qualitative agreement, particularly if one includes slightly lower cloud masses. The theory predicts that in unreddened stars, presumably seen through the intercloud medium, the fractional H_2 abundance should be 4×10^{-8}. The observations give upper limits of about this value. Therefore, the observations agree quite well with the theory that H_2 is catalyzed by grains and destroyed by ultraviolet light.

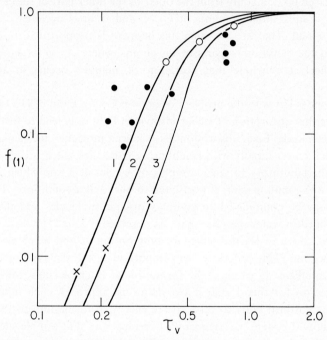

Fig. 5. Fractional abundance of H_2 in interstellar clouds, plotted against optical depth to the cloud center in the visible, τ_V, calculated by Hollenbach *et al.* (1971). One can show that on the average $\tau_V = \frac{3}{4} A_V$, where A_V is the extinction through the cloud at a random point. Since $A_V = 3E_{B-V}$, $\tau_V = 2.25\ E_{B-V}$ and the observations may be plotted as shown by the filled circles. The observations are from Copernicus (Spitzer *et al.*, 1973).

This has immediate consequences for the interstellar medium. It has been estimated that up to 40% of all interstellar gas may be in clouds which are thick enough that they are mostly H_2. Hence, previous estimates of the mean density of the interstellar medium may be increased as much as 75%.

To summarize this section, I have described how molecules can form upon grains and then either enter the gas phase or accumulate as part of a mantle. That the first process occurs is demonstrated by the discovery of large amounts of H_2 by

Copernicus. That the second process occurs is indicated by the depletion of C, N, and O found by Copernicus and by the systematic variations with radius in the dark cloud near ϱ Oph.

5. Molecules

We have seen that H_2 is catalyzed by dust grains. But radio astronomers have observed the spectral lines of a large variety of molecules (Table II), ranging from diatomics like OH, to seven-atom molecules like acetaldehyde (CH_3CHO). Several diatomics were discovered optically many years ago by their absorption lines in stars, but all the complex molecules have been discovered quite recently by radio astronomers.

TABLE II

Molecules observed in the interstellar medium 1973

H_2	Molecular hydrogen
CH^+	Methylidyne (ion)
CH	Methylidyne
CN	Cyanogen radical
OH	Hydroxyl radical
NH_3	Ammonia
H_2O	Water
H_2CO	Formaldehyde
CO	Carbon monoxide
HCN	Hydrogen cyanide
HCO^+ (?)	Formyl radical (ion)
HC_3N	Cyanoacetylene
CH_3OH	Methyl alcohol
HCOOH	Formic acid
CS	Carbon monosulfide
NH_2CHO	Formamide
OCS	Carbonyl sulfide
SiO	Silicon monoxide
CH_3CN	Methyl cyanide
HNCO	Isocyanic acid
HNC (?)	Hydrogen isocyanide
CH_3CCH	Methyl acetylene
CH_3CHO	Acetaldehyde
H_2CS	Thioformaldehyde
H_2S	Hydrogen sulfide
CH_2NH	Methanimine

The fact that some of the complex interstellar molecules resemble fragments of organic molecules which are important in biological processes – such as amino acids – has led to speculation that they are produced by living organisms beyond the Earth. As a Copernican, I am open to the idea that the Earth is not the unique locus of life in the universe. But there is no need to postulate life in space to understand the molecules observed.

Dust grains can also catalyze other molecules – at least simple ones like OH.

However, it is not clear that they can also catalyze more complex ones. Hence there is interest in a recent theory which demonstrates that it is possible to build up more complex molecules by gas-phase reactions. The key to this theory is the ionization of H_2 molecules by cosmic rays, which are believed to penetrate all space in the Galaxy. The important ionization reaction is

$$H_2 + CR \rightarrow H_2^+ + e^- + CR. \tag{3}$$

This process occurs about once every 3×10^9 yr to each H_2 molecule. Neither the H_2^+ nor e^- density ever becomes large, so recombination is very slow. Instead, the H_2^+ reacts rapidly with H_2:

$$H_2^+ + H_2 \rightarrow H_3^+ + H. \tag{4}$$

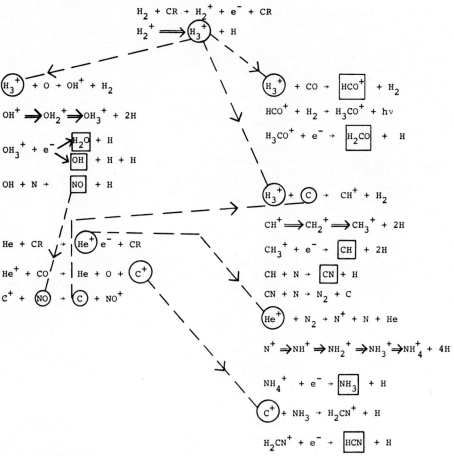

Fig. 6. Ion-molecule reactions in dark clouds (Herbst and Klemperer, 1973). The reactions are initiated by ionization of H₂ (top) and He (bottom left) and result in the species denoted by a box. The ingredients are H₂, He, CO, N, O, and 100-MeV cosmic rays (CR). The symbol ⇒ refers to reaction with H₂, e.g., A⁺ + H₂ → AH⁺ + H. The species produced include CH, OH, CN, NO, H₂O, NH₃, HCO⁺, H₂CO, and HCN, of which all but HCO⁺ and NO have definitely been identified in interstellar space. (HCO⁺ has been tentatively identified.)

TABLE III

Comparison of observed abundances with a theory of molecule formation[a]

Molecule	$\log N_{obs} - \log N_{theor}$		
	Dark clouds	Orion A	Sgr B2[b]
OH	+1.1	−0.1	+0.4
CN	−	0.0	−
NO	<+0.5	<+1.2	<+0.5
NH_3	−	−	− 0.3
HCO^+	−	−0.3	−
H_2CO	−0.5	+0.5	−0.4
HCN	−	+0.5	−

[a] Based on Herbst and Klemperer (1973).
[b] 10 × local cosmic-ray flux.

Now H_3^+ reacts rapidly with O to form OH^+, and a further chain of reactions with H_2 makes H_2O and OH. None of these reactions requires emission of radiation. Furthermore, because the ions involved exert a strong attractive force, there is no activation energy. Hence, these 'ion-molecule' reactions are much more rapid than reactions between neutrals.

It has been found observationally that in dark clouds, where hydrogen is mostly H_2, much of the carbon is in the form of CO. Hence the theory has been applied to dark clouds, where the principal ingredients available are H_2, He, CO, N, and O. Figure 6 summarizes the key reactions which occur. The products include CH, OH, CN, NO, H_2O, NH_3, HCO^+, H_2CO, and HCN. Seven of these molecules, and perhaps one more (HCO^+) have already been observed, and the other (NO) has been sought but not yet found. Table III presents a comparison between the observed and predicted column densities in a number of molecular clouds. The agreement is quite striking, lending support to the view that gas-phase reactions play a role in making interstellar molecules.

Clouds near HII regions are often particularly rich in molecules; the radio object Orion A is an example of such a cloud. Its optical counterpart, the Orion Nebula, appears to be on the near side of a giant dark cloud which is not very apparent in photographs but which is easily detected in the radio region. It is not understood why dark clouds near HII regions should be so rich in molecules. Perhaps this is an effect of observational selection, because HII regions are caused by the interaction of young stars and interstellar gas, and the formation of young stars presupposes the collapse of interstellar gas to high densities, which favor molecule formation.

6. Summary

Our theme is that cores of interstellar dust grains, which are made in and ejected from cool stars, determine the chemical nature of the interstellar medium. Figure 7, our last, is a sketch of a hypothetical dust grain, showing a core of silicates and iron, a mantle of ice and other molecules, and a thin layer of adsorbed H atoms. Hydrogen atoms hitting the grain react with those on the surface to form H_2. By the ion-molecule reactions in Figure 6, some of the H_2 reacts with O to form OH and H_2O.

Oxygen atoms hitting the grain react with H in the adsorbed layer to become OH. If the OH is ejected on formation, gaseous OH is produced. If not, the OH reacts further with H on the grain to form H_2O, which becomes part of the mantle. Along with other molecules in the mantle, these are processed by ultraviolet light to become more complex molecules.

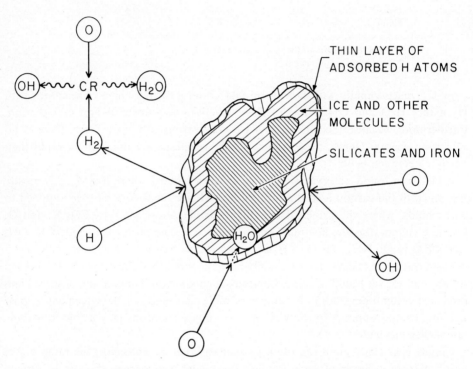

Fig. 7. Sketch of a hypothetical interstellar dust grain, magnified one million times. Surface reactions with H and O are shown. Aided by cosmic-ray ionization, the H_2 reacts further with O to form OH and H_2O. Other elements, such as C and N, produce other molecules by similar processes.

The net result of these processes is a core-mantle grain which fits the interstellar extinction observations, and gas-phase molecules which fit the radio observations. That atoms are indeed lost to the grains and to molecules is proved by the deficiencies of gas-phase atoms discovered by Copernicus.

In conclusion, I would like to point out an amusing analogy. The dust grain in Figure 7 bears qualitative resemblance to a planet! The interior of the grain, like that of the Earth, is composed of iron and silicates. Its outer envelope, like the oceans of Earth, is water. The whole is immersed in a gaseous atmosphere, and is bathed, like the Earth, in ultraviolet light and cosmic radiation. Radio astronomy has shown that the chemical reactions in and around interstellar dust grains produce molecules of surprising complexity. Perhaps there is a lesson here for the chemists who are trying to reconstruct the synthesis of organic molecules which must have preceded life on Earth. While we do not need to postulate life in space to explain interstellar molecules, the processes which are producing them in space may be very similar to those which occurred three billion years ago on primitive Earth.

References

Bless, R. C. and Savage, B. D.: 1972, *Astrophys. J.* **171**, 293.
Carrasco, L., Strom, S. E., and Strom, K. M.: 1973, *Astrophys. J.* **182**, 95.
Field, G. B.: 1974, *Astrophys. J.* **187**, 453.
Gillett, F. C., Forrest, W. J., and Cohen, M.: 1972, in preparation.
Hackwell, J. A.: 1971, Thesis, University College, London.
Herbst, E. and Klemperer, W.: 1973, *Astrophys. J.* **185**, 505.
Hollenbach, D. J., Werner, M. W., and Salpeter, E. E.: 1971, *Astrophys. J.* **163**, 165.
Lewis, J. S.: 1972, *Icarus* **16**, 241.
Spitzer, L., Drake, J. F., Jenkins, E. B., Morton, D. C., Rogerson, J. B., and York, D. G.: 1973, *Astrophys. J.* **181**, L116.

GALAXIES AND THEIR NUCLEI

V. A. AMBARTSUMIAN
Academy of Sciences, Armenia, U.S.S.R.

Our Extraordinary General Assembly is devoted to the memory of one of the greatest men of science to the genial Polish astronomer – Nicolaus Copernicus. The main service of Copernicus which has made his name immortal was in finding the correct interpretation of the planetary motions we observe. Instead of geocentric notions, which proved unable to explain the accumulated bulk of empirical data on the apparent motions of planets he has put forward and advocated the notion of a *solar system* thus presenting the true picture of the part of the Universe we live in. The scientific revolution started by him was continued by Galileo and Kepler and was crowned with the great theoretical generalizations of Newton. As a result a foundation has been created for the most exact theories of motions in the solar system which were developed during the next centuries. These theories in their modern form give also the possibility to solve all the problems concerning the orbital motions of spaceships.

At this stage we can not yet boast that in the study of nuclei of galaxies and their activity we have reached the level which existed in planetary astronomy even before the works of Newton. Only 15 years elapsed from the moment when the idea of activity of nuclei of galaxies was clearly formulated (Ambartsumian, 1958). During these years discoveries of the greatest importance have been made. New unexpected discoveries occur almost each year. These discoveries influence decisively our notions on the diversity of objects and phenomena in the distant space, but they are still insufficient for the construction of adequate theories. In order to penetrate into the very nature of nuclear phenomena we require new observations, new measurements and new data. And if some optimists imagine that the time has already ripen to build a general theory of these phenomena, the more cautious astronomers would like to consider a more or less satisfactory systematization of observational data concerning the activity of nuclei and the understanding of external physical processes accompanying it as a tremendous success.

These external processes reach often such a large scale that they influence the appearance and integral parameters characterizing the galaxies. Therefore the study of the nature and the activity of nuclei sometimes means to investigate the problems of the structure of galaxies and of their evolutionary changes.

In this report we consider some properties of galaxies which are immediately connected with the activity of nuclei and ultimately with the nature of nuclei themselves.

1. Soon after the introduction of the concept of the activity of nuclei the observations have revealed some new forms of that activity. Therefore we can speak now about the considerable diversity of the external forms of activity of the nuclei of galaxies. Let us mention here some important forms:

(a) The ejection from the nucleus, and from the volume of the galaxy itself, of giant masses which transform into large clouds of relativistic plasma. Owing to this a galaxy transforms into a *radiogalaxy*.

(b) The enhancement of the optical luminosity of the nucleus. Owing to this form of activity a galaxy passes into class 5 of the Byurakan classification (Ambartsumian, 1966) or into a Seyfert galaxy. In the case of a stronger increase of the luminosity we have an *N-galaxy*. The extreme form of the same kind of activity are the *quasars*, where the nuclei reach the absolute magnitude of about -25 and even higher.

(c) The ejections and motions of gaseous clouds in and from the nuclei of Seyfert galaxies and from quasars.

(d) Great explosions which lead to the ejection of large gaseous masses of the order of $10^6 \, M_\odot$, like the ejection that occurred in M 82.

(e) Relatively small but recurrent explosions in nuclei which manifest themselves as increases in radioluminosity at the centimeter range of wavelengths.

2. There are many indications that alongside with the forms of activity we have mentioned above, which we observe immediately, there are also the following *supposed* forms:

(a) The ejection of condensations from the nuclei of supergiant galaxies which are capable to transform into new galaxies (mostly into a satellite galaxy or a member of the group which surrounds the supergiant galaxy under consideration).

(b) The outflow of matter, from the nucleus which can produce the spiral arms.

Of course in studying and classifying the manifestations of nuclear activity one must strictly differentiate the well established forms from the supposed forms we have just mentioned. However in the course of the systematization of the known data on galaxies and phenomena occurring in them it seems appropriate to assume that these forms also exist in reality. The possible (though at the moment improbable) fallacy of such an assumption cannot discredit the results of such systematization work since the concepts on the forms of nuclear activity serve only as suggestion for such work. Evidently we can choose as a basis for the work of systematization and classification of facts any consideration or idea about the nature of galaxies. Only the result of the work can show how fruitful the considerations chosen as a basis proved to be.

3. The manifestations of the nuclear activity are very unusual physical processes and we are still far from understanding their real nature. Therefore it is too early to build models explaining them. Owing to the fact that the majority of known galaxies having active nuclei are very distant, not only the nuclei themselves but the much larger central volumes in which the most important processes take place remain usually unresolved with our instruments. In the majority of cases we are not convinced, that the radiation we receive is coming immediately from the central body which is the main source of activity. Therefore, as a rule it is difficult to clear up *even the geometrical picture of the external processes*, not to speak about the mechanism of the active source or structure of the true nucleus.

It seems that before proceeding to the construction of models a considerable amount of work must be done in order to find the empirical regularities. For such a work the correct classification of objects and phenomena should serve a basis.

Evidently such a classification should be founded on direct observations. However, since only the external manifestations of nuclear activity are accessible for observations, the classification of processes and objects connected with the activity of nuclei is to be established from the observations of these external manifestations. Only after the study of external manifestations we may find the way to the very essence and true causes of phenomena taking place in the nuclei of galaxies.

4. Since at this stage the *systematization* of data on external manifestations of nuclear activity must be the most important aim we would like to emphasize the broad character of this problem. There is no doubt now that the nuclei sometimes can cause fundamental changes in the properties of the galaxy.

During the last twenty years the present reporter has defended the opinion that each galaxy including all subsystems of which it consists (spherical system, disk, spiral arms) is the result of nuclear activity. If so the systematization of data related to the external manifestations of nuclear activity means the systematization of all accessible data on galaxies. However we have in our view such systematization which takes into account *more direct* external manifestations of nuclear activity. The first step in this direction must be the classification based on the existing notions on different forms of nuclear activity.

Of course such a work is made easier by the fact that during our century several systems of classifications of galaxies have been worked out and practically applied. Two of these systems (classifications of Morgan and of Byurakan Observatory) put considerable emphasis on the properties of the structure of central circumnuclear regions of galaxies. As regards to Hubble's system – it takes into account only one of the parameters connected with the activity of the nuclei – the presence and strength of spiral arms. However the building of a classification system which takes into account different known kinds of nuclear activity is a difficult task. In the present report we are going to consider only a partial question related to the classification of a special category of galaxies. Thus our aim is rather *to consider the diversity of parameters and of properties which are connected with nuclear activity and by which galaxies of that category differ between themselves.*

5. Let us concentrate our attention on galaxies of which the spherical component only (Population II) has the total absolute magnitude $M_V < -21.0$ independently of the presence of other components (disc, spiral arms). If the stars of Population I are present the integral luminosity of such galaxy as a whole will exceed the given limit $M_V = -21.0$ even more. Thus we have chosen the galaxies of highest luminosity and mass.

Having concentrated on those supergiant stellar systems we see that they are different in many respects and particularly in properties which depend on the activity of their nuclei. Our aim is to specify here different parameters which are essential characteristics of these systems and to consider the possibility of classification based on such parameters.

We shall have in mind that though the supergiant galaxies form only a small minority among all galaxies, their total mass in a given volume represents a considerable part of the total mass of all galaxies as it was indicated by us already in our report at Berkeley (Ambartsumian, 1962).

6. Among the parameters decribing the galaxies under considerations we shall consider as important:

(a) *The Radioluminosity L_R*. As we know radiogalaxies are defined as systems emitting strong radioemission with radioluminosity $L_R > 10^{41}$ erg s^{-1}. They have at the same time high optical luminosities. Therefore all radiogalaxies belong to the category of galaxies under consideration. Curiously enough when choosing the galaxies according to the only criterium $L_R > 10^{41}$ erg s^{-1}, we obtain a sample of systems (radiogalaxies) for which the dispersion of optical luminosites is much smaller than the dispersion of radio luminosities.

It is known that spiral galaxies also are often the sources of radioemission, however their radioluminosity is always lower than the indicated lower limit. Therefore the radiogalaxies are always *E*-systems, which however sometime contain a significant quantity of dust and some number of Population I stars.

(b) *The optical luminosity of the nucleus*. This is the second important parameter. Alongside with optically very weak nuclei (NGC 4486 where $M_n > -15$) the galaxies of the category under consideration sometimes have intense nuclei (for example the nuclei of *N* galaxies, which were wittily called mini-quasars) and even contain quasars with absolute magnitude reaching $M_n = -27.5$. Thus the range of optical luminosities of nuclei of systems under consideration is of the order of 10^5.

(c) *The presence, strength and the degree of development of spiral arms* and generally the relative strength of stellar Population I. It is clear that in this case it is very disirable for classification-purposes to choose again one definite numerical parameter. Perhaps as such a parameter one can use the ratio of the mass of neutral hydrogen in the given galaxy to its total mass. The use of another important parameter – of the ratio L/M which according to the work of the Meudon group (Balkowski *et al.*, 1973) changes abruptly when passing from spiral and lenticular galaxies to ellipticals seems not very practical. Probably as a substitute of the parameter under consideration the value of $B-V$ of the galaxy can be of some use.

Considering the question more precisely one must also take into account that apparently for the description of the spiral structure several independent parameters are necessary. Here we limit ourselves with one parameter since as a supplementary parameter we consider below the strength and the degree of the development of the bar.

(d) *The presence, intensity and the degree of development of the bar*. According to the surface photometry of giant SB galaxies (Kalloghlian, 1971) the surface brightness of the bar has some prefered value (the mean photographic surface brightness along the axis of the bar is about 21 magnitudes per square second of arc). Therefore very roughly we can consider this parameter as having only two possible values (0 or 1 depending on the absence or the presence of a bar).

(e) Finally the spherical component of population (which in the case of E systems coincides with the galaxy as a whole) can have a great or a small diameter. Since we have confined ourselves to the high luminosity systems only, this means that they can have different surface brightnesses. This corresponds to the division of supergiants according to the *degree of compactness*. At this stage it will be convenient to consider three species of galaxies: extended systems (with a diameter larger than 40000 pc), normal systems (with a diameter between 15000 and 40000 pc), and compact galaxies (diameter less than 15000 pc). Of course to make the division more exact one must add to this some definition of the diameter of a galaxy.

When elaborating the question it will be expedient to introduce some more definite numerical parameter describing the degree of compactness. Zwicky has suggested that the mean surface brightness can serve as such. However it is evident that the simple average:

$$\langle i \rangle_0 = \frac{\int i \, ds}{\int ds},$$

where i is the surface brightness is not very suitable since increasing the domain of integration we can obtain as low value of $\langle i \rangle_0$ as we wish. One can propose instead a weighted average of the surface brightness for example

$$\langle i \rangle_1 = \frac{\int i^2 \, ds}{\int i \, ds}.$$

However in this case the relative role of the central region is very large and therefore it is necessary to know sufficiently well the exact behaviour of i near the centre of the galaxy, which is difficult since the angular resolution by photometric measurements is low. Probably the best practical alternative is to adopt an average of the type

$$\langle i \rangle_2 = \frac{\int i f(i) \, ds}{\int f(i) \, ds},$$

where $f(i) = i_0$ when $i > i_0$ and $f(i) = i$ when $i < i_0$. Here i_0 is some conventional, fairly high surface brightness corresponding for example to 22th mag. from a square second of arc.

Since the degree of compactness must be closely related to the ratio of the absolute value of gravitational energy of the system to the square of its total mass (since this ratio is according to definition proportional to the radius of the system determined in some specific way) it is clear that this degree of compactness must depend on the mechanism and conditions of formation of the spherical component of the galaxy and particularly from the mean kinetic energy of the member-stars. If one assumes that the nucleus plays the fundamental part in the formation of the spherical component of the galaxy the degree of compactness may serve as one of external manifestations of the activity of the nucleus.

7. The importance of the degree of compactness as one of the properties of galaxies became clear when Zwicky had given attention to the existence of many compact galaxies, and having accomplished his valuable catalogue has even discovered several clusters consisting of compact galaxies (Zwicky, 1971).

At the start of this year Robinson and Wampler of the Lick Observatory have published a paper in which they have shown that the cluster Shakhbazian 1 is a *compact group of compact galaxies*. Immediately after that Shakhbazian (1973) has presented a list of similar groups consisting of compact galaxies. Owing to these studies the compact galaxies recently have attracted considerable attention from investigators.

However, commenting his studies on compact extragalactic objects Zwicky has expressed the view that quasars are the extreme cases of highly compact systems. There are now fairly good evidences that quasars generally have underlying galaxies. These underlying galaxies often are extended objects. Therefore using the terminology used above we can assert than the fifth (compactness) and the second (the luminosity of the nucleus) parameters we have introduced are to be considered independently. The question of statistical correlation or anticorrelation between them must be solved from observations. *One of these characteristics describes the distribution of stellar population, the other – the state of the nucleus.*

8. Since within the category of systems we are considering among others enter spirals with sufficiently luminous spherical subsystems it is appropriate to make the following reservation. It is known that the ellipticals, which consist only of spherical subsystems have an L/M ratio four times smaller than in the case of spirals. This means that in order to have a spherical subsystem of absolute magnitude $M_V = -21.0$ a spiral must have a total visual magnitude not lower than -22.5.

There are many ellipticals of such high luminosity but the spiral systems with $M_V < -22.5$ are very few indeed. Among the possible candidates for this category is Markarian 10, for which $M_V \sim -23.0$. However in order to introduce it into the category of systems under consideration it is necessary to show that its spherical subsystem has in fact a partial luminosity of $M_V < -21.0$. It may happen that the real number of such spirals is very small or they don't exist at all. In the last case our conclusions are liable to some changes.

9. Thus we see that considering only supergiant systems, i.e. fixing the value of one of the important parameters (the luminosity) we see that the totality of states of galaxies depends on at least five different parameters.

The question arises wether all possible combinations of different values of these parameters are represented among the real galaxies or some ot them depend on the values of others thus reducing the number of truly independent variables.

We shall simplify this complex problem by means of the very rough discretization of the values of parametes describing each galaxy. Namely:

(a) If the radioluminosity of a galaxy $L_R > 10^{41}$ erg s^{-1} we shall say that the corresponding discrete parameter $\alpha = 1$. If $L_R < 10^{41}$ erg s^{-1} we shall write $\alpha = 0$. For all radiogalaxies $\alpha = 1$. For all other galaxies $\alpha = 0$.

(b) If the nucleus of a galaxy has an absolute photographic magnitude $M_{pg} < -21$ we shall say that another discret parameter $\beta = 1$. If the nucleus is fainter than $M_{pg} = -21$ we shall write $\beta = 0$. For all other galaxies $\beta = 0$.

(c) If on the plates of sufficient resolution and density in photographic light a galaxy has noticeable spiral arms we shall write that the corresponding discrete parameter $\gamma = 1$. If they are unnoticeable $\gamma = 0$.

(d) If on the plates of sufficient resolution in photographic light a galaxy has a discernible bar we shall write $\delta = 1$. In the opposite case we shall adopt $\delta = 0$.

(e) If a galaxy is compact, i.e. on the plates which are as effective in showing the faint peripheries as the maps of the Palomar Sky Survey, its radius is less than 15000 pc, we shall then write $\varepsilon = 1$. In the opposite case $\varepsilon = 0$. Incidentally this definition of compactness based on the value of the diameter is correct only for the high luminosity galaxies under our consideration. For the less luminous galaxies this limiting value of diameter must be smaller.

10. The five-digit binary number $S = \alpha\beta\gamma\delta\varepsilon$ determines a particular value of each of the parameters introduced above and describes roughly the state of the given galaxy. Correspondingly in decimal numeration the state of a galaxy can be given by one of the numbers from $S = 0$ to $S = 31$. For example – the case $S = 0$ means a radioquiet galaxy, without a quasar in its centre, without spiral arms and bar which is not compact. *It is simply a normal elliptical galaxy.*

Now we can discuss the problem of the independence of the introduced parameters in two different ways.

(A) Do all 32 values of $S = \alpha\beta\gamma\delta\varepsilon$ (in decimal system the numbers from 0 to 31) have their counterparts among the galaxies? In other words, are all combinations of discrete quantities α, β, γ, δ, ε realized in the Universe? We have seen above that the combination 00000 corresponds to a normal elliptical and therefore is very frequent. On the other hand some combinations for example 10100 (in the decimal notation $S = 20$) are not realised. Unfortunately we don't know whether the combination 11111 ($S = 31$) is realized anywhere, i.e., are there quasistellar radiosources for which the underlying galaxy is of SB type and the spherical subsystem is compact.

(B) Can we represent the distribution function P_S of the values of S for the set of galaxies in a unit volume as a simple product

$$P_S = \varphi_1(\alpha)\varphi_2(\beta)\varphi_3(\gamma)\varphi_4(\delta)\varphi_5(\varepsilon),$$

where $\varphi_i(\alpha)$ is the probability of the given value of α, and $\varphi_k - S$ have a similar meaning.

Evidently the answer is negative. This follows from the fact that for some values of S we have $P_S = 0$. But this means that at some value of its argument at least one of the functions φ_i must be equal to zero. But this cannot be the case, since this means that one of two values of that argument is not realized in the Universe at all while we have introduced our parameters on the ground that both values have been observed somewhere in nature.

11. Returning to the first problem (A) we can confine ourselves with the simple question of the compatiblity of values of some pairs of parameters (for example of α and β). Thus from Table I containing the data on four extragalactic objects we can see that all four combinations of values of α and β are realized in the Universe.

TABLE I

Object	Parameter	
	α	β
NGC 4889	0	0
Ton 256	0	1
NGC 4486	1	0
3C 371	1	1

This statement means that both the presence or absence of strong radioemission are equally compatible with the presence or absence of a quasar (or of miniquasar) in the centre of a galaxy.

Apparently we have a similar situation when comparing the parameters β and ε, i.e. the presence of a quasistellar source and the compactness. This may be seen from Table II

TABLE II

Object	Parameter	
	β	ε
NGC 4889	0	0
I Zw 94	0	1
Ton 256	1	0
Zw 0039 + 4003	1	1

However, statistically the compactness of a galaxy is rather anticorrelated with the presence of quasistellar objects in its central region.

The situation is more complicated when we consider another pair of parameters: α and ε. Among the radiogalaxies there are both the extended stellar systems and the systems of normal size. But we don't know any compact radiogalaxy (with the diameter less than 15000 pc) nearer than 500 Mpc. Thus if $\alpha=1$ we have $\varepsilon=0$. But if $\alpha=0$ the quantity ε can be equal either to 1 or to 0. Thus between the values of our two parameters there is no one to one correspondence.

We cannot exclude also the possibility that among the very distant (more than

500 Mpc) radiogalaxies there are compact systems which will have the external appearance similar to quasars. Therefore before making any final conclusions we must wait for more refined data about the sizes of the optical images of quasistellar radiosources.

The survey of external forms of radiogalaxies shows that none of them has developed and regular spiral arms. This means that strictly speaking the values $\alpha=1$, $\gamma=1$ are incompatible. On the other hand the radiogalaxies NGC 5128 and 2175 show the presence of dust, gas and of stellar Population I. This is not equivalent to the presence of developed and regular spiral arms. Therefore in this case also we cannot write $\gamma=1$ or $\delta=1$. But probably these cases indicate that the phase of radiogalaxy precedes the phase of evolution of supergiant galaxies at which the developed spiral arms are formed.

12. But also in this case form the fact that the combination $\alpha=1$, $\gamma=1$ never occurs we cannot conclude that the value of γ is determined by the value of α. In fact, when $\alpha=0$ both cases $\gamma=0$ and $\gamma=1$ can happen. Thus here also we do not have one to one correspondence.

Continuing these consideration we may show that all five parameters introduced are physically independent. But they may have correlations and statistical dependences which we shall not discuss further.

The question arises on the evolutionary interpretation of the chosen parameters and their mutual relations.

13. Evidently the diversity of forms and states of the galaxies we observe one should explain by (a) differences of age and (b) differences of initial conditions. Among the initial conditions such quantities as the mass of the system, its total internal energy and the rotational momentum play an important part. Some significance can be attached also to the differences in the initial chemical composition. However there is no doubt that during the life of a galaxy the chemical elements in it undergo essential evolutionary changes. Therefore, if the dominant state of matter from the beginning was of atomic type (and not of the type of nuclei or particles having masses of stellar or larger order), the simplest assumption would be the similarity of initial chemical composition. If however in the beginning the nuclear phase was predominant (of the type of the baryon star structures), it is probable that after the transition to atomic structure of matter approximately the same chemical composition emerged. Therefore it seems possible to disregard the possibility of differences of the initial chemical composition.

Since we are considering such systems (supergiant galaxies) which have masses of about the same order of magnitude, there remain only three parameters: (a) the age, (b) the total energy, and (c) rotational momentum.

Thus we have the situation when the number of empirically determined parameters, which specify the different states of the systems, is larger than the number of parameters we can foresee from physical considerations related to the diversity of initial conditions.

However one must take into account that the activity of a nucleus takes sometime such intense and cataclysmic forms that in a short time it can not only cause essential changes in the properties of the galaxy, but even originate new temporary properties which should be described by the values of new parameters.

This can happen for example in the case of a sudden appearance of strong radioemission (formation of large clouds of relativistic charged particles), or X-ray emission or of strong nonstellar optical emission (quasars). As we have seen the appearance of new properties in each case means that some parameter normally having a constant value (in the example just mentioned not much differing from zero), for some interval of time acquires new, sometime variable value.

The intervals of time during which different new properties are maintained can overlap. If for two given properties owing to the regularities of evolutionary processes there is partial overlap of intervals τ (for example the interval τ_B begins somewhere in the middle of τ_A), we shall observe (1) the cases when both properties are present in a galaxy, (2) the cases when galaxies have only one of them, of (3) the cases when a galaxy have none of them. This exactly happens in the case of the pair α and β or in the case of properties β and ε.

If the intervals don't overlap both properties never meet in the same system. However from the absence of one of them we cannot conclude about the presence of the second. Exactly such a situation is present in the case of the parameters α and γ, i.e. the presence of spiral arms is incompatible with the strong radioemission, but the absence of radioemission does not mean necessarily the presence of the arms.

We can hope that further work on the classification of galaxies and measurements of the essential parameters will allow us to determine the length of intervals during which the properties under consideration are maintained and will ascertain the circumstances that precede them.

14. If we will try to extend the above considerations to the galaxies for which the spherical subsystem is considerably fainter than $M = -21.0$ we will meet some important circumstances.

(a) Such galaxies are never strong sources of radioemission. But they can emit weak radioemission (normal spiral) or moderate as some Seyfert galaxies (for example NGC 1068) or the irregular M 82 do. But for them always $L_R < 10^{41}$ erg s^{-1}.

(b) Such galaxies have spiral arms more frequently.

(c) In this category the moderate radio-emission and the presence of spiral arms are compatible.

(d) Many examples of Seyfert and Markarian galaxies show that these objects of low luminosity can have nuclei of relatively high luminosity. Moreover such objects are sometimes the galaxies that underlie the radio-quiet quasars. This does not exclude the fact that many radio-quiet quasistellar objects can have as their underlying galaxies high luminosity systems, i.e., supergiant galaxies.

(c) There is some definite lower limit for the integral luminosity of the spherical component of galaxies capable to form the spiral arms of more or less regular form. The exact value of this limit is not known but probably it is near $M = -14$.

The galaxies with still fainter spherical subsystems can have some stellar population I and interstellar material of appreciable density, however in such cases they have irregular forms.

15. If one takes as starting point the assumption that the formation of spiral arms is the result of nuclear activity one can resume these facts in the following way:

(a) If the spherical component has an integral absolute magnitude $M_V < -21.0$ its nucleus is capable to form large radioemitting clouds but seldom is able to form the regular and bright spiral arms.

(b) If the luminosity of the spherical subsystem is confined within limits

$$-21.0 < M < -14.0$$

the nucleus of such galaxy is unable to form strong radio emitting clouds but frequently forms regular spiral arms.

(c) If $M > -14$ the nucleus cannot form regular spiral arms but still is able to produce relatively abundant population I.

16. Thus according to observations the kind of nuclear activity depends on the absolute magnitude and therefore on the mass of the spherical subsystem. On the other hand it is clear that the spherical component of the galaxy hardly can have any direct influence on the properties of the nucleus. Therefore there remain two possibilities:

(a) The spherical component itself is the result of the nuclear activity. Therefore it is strongly correlated with the other external manifestations of the same activity.

(b) The nucleus and the spherical subsystem have been formed together. The properties of the nucleus and the mass of the spherical component are determined by the integral mass of the galaxy.

It seems that at this stage of our knowledge it is difficult to decide which of these alternatives corresponds to reality. Only the more general considerations concerning the universal role of nuclear activity make the first possibility more likely.

17. We have stated above that compact galaxies displaying strong radioemission are not known. But there is no doubt that the compact galaxies sometime are able to produce in themselves a considerable population of type I and even form the ejections and plumes which in some degree are similar to the spiral arms. As an example we have the galaxy NGC 1614. As a result of a very preliminary survey of *compact groups of compact galaxies* carried out in Byurakan it has been concluded that some of them contain a blue compact galaxy. Usually these blue members have almost elliptical appearances somewhat disturbed by the presence of absorbing matter. There is no doubt that the study of the color distribution in such galaxies will bring interesting results.

18. The study of clusters and groups of galaxies is extremely important for the understanding of processes of the formation and evolution of galaxies. Such studies inevitably bring the conclusion that *the supergiant galaxies play a particularly important role in the Universe*. For example let us quote one of the conclusions reached in the recent paper of Sandage (1972) 'The luminosity of the brightest cluster member does not depend strongly, if at all, on the luminosities of the fainter cluster members'. The dispersion of absolute magnitudes of the brightest cluster members is of the order of 0.25 mag. The deviations from these rules happen in the specific cases of compact groups of galaxies.

Excluding for a moment such compact groups from our consideration we can say that each cluster contains at least one member having a mass of the order of 5×10^{12} $M_\odot$. If one adheres to the theory of formation of the clusters of galaxies from a large cloud of diffuse matter, the existence of definite upper limit for the masses of the parts into which the large cloud splits and at the same time the necessary formation of at least one part which has mass of the order of that limit is difficult to understand.

If in order to explain the origin of clusters of galaxies we consider the alternative hypothesis of fragmentation of an initial dense and massive body, it is quite natural to suppose that during each step of such fragmentation a body divides into several pieces having masses of equal order. In this way at some stage the dense bodies with masses of the order of 5×10^{12} $M_\odot$ will be formed inevitably. Then by some reason the division into masses of equal order of magnitude stops and each part behaves as an active nucleus. Perhaps at this stage it is better to say 'protonucleus'. This means that each such part forms around itself a galaxy, consisting of stellar populations of different kinds. Moreover the ejection of secondary nuclei of smaller masses (10^{11} $M_\odot$) is possible. Thus the nucleus of a supergiant galaxy contributes to the formation of the less massive population of the cluster of galaxies.

19. Zwicky has established the existence of several large *clusters consisting of compact galaxies*. Since however it is difficult to judge the compactness of faint members it is more correct to say that a number of bright galaxies in each such cluster are compact.

Among these clusters is Zw Cl 0152+33 which has the angular diameter of about one degree. Since the distance must be of the order of 5×10^8 pc (this corresponds to the radial velocity $V_r = 26\,300$ km s^{-1} determined by Sargent (1972)) the linear diameter is of the order of 10^7 pc. Since the dispersion of radial velocities is of the order of 1000 km s^{-1} we arrive to the conclusion that galaxies cross the whole cluster during the time interval of the order of 10^{10} yr. Therefore one can suppose that the differences in the ages of galaxies in the cluster are of this same order of magnitude. This means that compactness is not a quickly passing property of a galaxy and lasts at least hundred of millions, perhaps billions of years. This requires that these galaxies are in a steady state. But from this follows that these systems *will remain compact* also in the future, during the life of stars which enter in these systems.

Perhaps somewhat extrapolating we can suppose that the compact galaxies as a rule are born as such and remain compact during the length of their life. In any case they are systems *sui generis* and not some stages of evolution of normal galaxies.

The division of clusters and groups of galaxies into systems consisting of normal galaxies on one hand and of compact galaxies on other hand has therefore fundamental significance. This division must be intimately connected with the mechanism of the formation of galaxies in clusters. It is very difficult to imagine that one can explain such a division on the basis of hypothesis of formation of galaxies from diffuse matter.

20. Of great interest are *compact groups of compact galaxies*. Such systems usually have from half a dozen to two dozen members, though there are richer groups. Typical representatives of such groups are the No. 1 and No. 4 of Shakhbazian's list which will be published shortly. The first of these groups consist of 17 members, the second of 7 members. The linear size of these groups are of the order of 2×10^5 pc.

The first of these groups has been found at Byarakan in 1957 (Shakhbazian, 1957) during the study of the Palomar Sky Survey maps. Owing to compactness of its members and of the group itself it looks very different from other groups of galaxies. This was the reason that with some hesitation we first supposed that it was a stellar cluster situated at some distance from our Galaxy. Later Kinman and Rosino (1962) found on large scale plates that some members of the group are galaxies. But since the other members were seemingly stars they concluded that the group is a chance agglomeration of galaxies and stars on the sky. Only recently Robinson and Wampler (1973) have found from spectral observations that it is a definite physical group of compact galaxies. Meanwhile new groups of similar type have been found at Byurakan.

The group Shakhbazian 1 has the redshift $z=0.1$, i.e. it is at distance of six hundred million parsecs from us. The brightest centrally located member of the group has an absolute magnitude of the order of $M_V = -23$. It is interesting that the brightest member of the Zwicky cluster 0152+33 mentioned above according to rough estimates has the same luminosity.

The determination of dispersion of radial velocities of the members of Shakhbazian 1 from redshifts and the application of the virial theorem has shown that the M/L ratio expressed in solar units is of the order of unity.

Thus in this case the virial theorem gives too small masses. In this sence the situation is opposite to what we have in usual clusters of galaxies.

However not in all similar compact groups of compact galaxies the dispersion of radial velocities is as small. Thus according to unpublished observations by Khachikian the dispersion of radial velocities in the remarkable compact group Shakhbazian 4 is of the same order of magnitude as in usual clusters. Probably in this case we have again an expanding group.

The compact clusters of compact galaxies differ from usual clusters (as catalogued by Zwicky and Abel) in that the integral magnitudes of member-galaxies are contained in a narrow interval of stellar magnitudes and the difference of magnitudes of first ranked and the second galaxy is relatively small.

The search for new groups of Shakhbazian 1 type is now in progress at Byurakan. Already the number of groups found reached several dozens. The very preliminary statistics shows that the number of such clusters till 18.5 red mag. for the brightest member must exceed one thousand.

Thus the compact groups of compact galacies (CGCG) represent *one of the important constituents of the Metagalaxy.*

21. From what has been said above one can conclude that the study of compact galaxies and their clusters will bring new conclusions, which may have close bearing to the problem of the origin and evolution of galaxies and the nature of the activity of their nuclei.

At the same time we shall not think that the phenomena connected with compact galaxies are strictly isolated from the world of normal galaxies. The opposite is true, and there are cases when it is difficult to know whether to relate a given galaxy to the compacts or normals. It seems that the attention given to such intermediate cases will be rewarding.

Though the groups of galaxies included by Shakbazian in her first list contain almost exclusively the compact galaxies this is the result of intentional selection. It may happen that there exist mixed systems and their study will help us understand the connection between opposite phenomena in the extragalactic world.

22. There is nothing astonishing in the existence of the compact galaxies and in their properties. It is natural to suppose, that any mechanism of the origin of galaxies must provide the possibility for formation of systems for which the ratio M^2/H, where H is the total internal energy, is smaller than for others. Such systems will appear as less extended, compact galaxies. However it is remarkable that:

(a) There are some rich clusters of galaxies each of which contains dozens of high luminosity compact systems and do not have any extended system of the same luminosity.

(b) In the observable part of the Metagalaxy there are thousands compact groups of compact galaxies containing from five to ten compact systems but don't contain normal or extended galaxies.

(c) Inspite of differences in the nature of compact and normal galaxies and of probable differences in the values of M/L among them (perhaps more than ten times) the upper limit of luminosities for the normal galaxies ($M_V = -23.7$) is apparently a sufficiently exact upper limit also for compact galaxies.

(d) The colours of compact galaxies apparently are not very different from the colours of some normal galaxies. However there was until now no extensive and precise study of the colours of compacts.

23. Aiming to the detailed study of compact galaxies and their clusters we shall keep in mind the difficulty of this problem. The compact galaxies apparently comprise only a small percentage of all galaxies. The nearest compact galaxies of high luminosity are at distances not less than 50 million parsecs from us. I am not quite sure that in the Shapley-Ames catalogue there is even one high luminosity compact. However, beginning with $m=13.0$ they appear. But owing to their high surface brightness the compact galaxies of 13th or 14th apparent magnitude must have diameters smaller than 20". Therefore their detailed morphological study is difficult.

At the same time I should like to warn you against the unfounded pessimism. During the time after Copernicus astronomy has overcome the distance barrier from 10^{-4} pc to some billions in parsecs. In any case at the distances reaching 800 million parsecs the contemporary extragalactic astronomer feels himself almost at home.

Now it is necessary to overcome the barrier of low angular resolution in optical observations.

Apparently this will be achieved by combination of optical interferometry with the use of observations from outer space. If these barriers will be conquered new prospects of extragalactic research and specially for investigations of compact galaxies will open.

24. Finishing this discourse I should like to pay tribute to the astronomers who have mostly contributed to the solution of problems considered above.

Owing to tremendous observational work *Sandage* has reached the final conclusion that quasars are nuclei of supergiant elliptical galaxies. Having understood the significance of high luminosity E-systems in the Metagalaxy he has established the regularities concerning the brightest objects in cluster of galaxies.

Essentially I agree with his important conclusions and they have been used above.

By his studies of compact galaxies *Professor Zwicky* has opened a new page in extragalactic astronomy. Every compact galaxy which has emission lines in its spectrum arouses great interest individually as it happens in the case of quasars and Markarian's galaxies. But we insist that the totality of compact galaxies (the majority of which have no emission lines) presents much deeper interest and significance.

Attracting the attention of astronomers to the compact galaxies Zwicky has shown once again how far from reality are those who think that we already know the composition and regularities of the structure of Universe, and that it is up to the theoreticians to put the last touches to their models. Nature showing us new types of objects in the Universe, and demonstrating new kinds of processes, literally compels us not to follow such oversimplified views.

The conviction of the inexhaustibility of the Universe has led modern astronomy to its great discoveries. And if we modestly recognize this inexhaustibility we shall continue to inspire ourselves with the increasing difficulty and deepness of the arising problems, and we can hope that astronomers of 2473 celebrating the thousandth anniversary of Copernicus will mention that the generation which lived halfway was not always sitting idle, but was sometimes unrestrained and fearless in the search of the unknown properties of the Universe.

References

Amburtsumian, V.: 1958, *XI Solvay Conference Report*, Bruxelles, p. 241.
Amburtsumian, V.: 1962, *Trans. IAU* **XIB**, 145.

Amburtsumian, V.: 1966, *Trans. IAU* **XIIB**, 578.
Balkowski, V., Botinelli, L., Gougenheim, L., and Heidmann, J.: 1973, '21 cm Neutral Hydrogen Line Study of Early Type Galaxies', in press.
Kalloghlian, A.: 1971, *Astrofizika* **7**, 189.
Kinman, T. D. and Rosino, L.: 1962, *Astron. J.* **67**, 644.
Robinson, L. B. and Wampler, E. J.: 1973, *Astrophys. J. Letters* **179**, 135.
Sandage, A.: 1972, *Astrophys. J.* **178**, 1.
Sargent, W. L.: 1972, *Astrophys. J.* **176**, 581.
Shakhbazian, R.: 1957, *Astron. Circ.*, No. 177, 11.
Shakhbazian, R.: 1973, *Astrofizika*, in press.
Zwicky, F.: 1971, *Catalogue of Selected Compact Galaxies and of Post Eruptive Galaxies*, Offsetdruck L. Speich, Zürich, Switzerland.

THE ASTRONOMY AND COSMOLOGY OF COPERNICUS

OWEN GINGERICH

Center for Astrophysics, Harvard College Observatory and Smithsonian Astrophysical Observatory, Cambridge, Mass. 02138, U.S.A.

It was close to the northernmost coast of Europe, in the city of Toruń, that the King of Poland and the Teutonic Knights signed and sealed the peace of 1466, which made West Prussia part of Polish territory. And it was in that city, just seven years later and precisely 500 years ago, in 1473, that Nicholas Copernicus was born. We know relatively few biographical facts about Copernicus and virtually nothing of his childhood. He grew up far from the centers of Renaissance innovation, in a world still largely dominated by medieval patterns of thought. But Copernicus and his contemporaries lived in an age of exploration and of change, and in their lifetimes they put together a renewed picture of astronomy and geography, of mathematics and perspective, of anatomy, and of theology.

When Copernicus was ten years old, his father died, but fortunately his maternal uncle stepped into the breach. Uncle Lucas Watzenrode was then pursuing a successful career in ecclesiastical politics and in 1489 he became Bishop of Varmia. Thus Uncle Lucas could easily send Copernicus and his younger brother to the old and distinguished University of Krakow. The Collegium Maius was then richly and unusually endowed with specialists in mathematics and astronomy; Hartmann Schedel, in his *Nuremberg Chronicle* of 1493, remarked that "Next to St. Anne's church stands a university, which boasts many eminent and learned men, and where numerous arts are taught; the study of astronomy stands highest there. In all Germany there is no university more renowned in this, as I know from many reports." At the university the young Nicholas embraced the study of astronomy with a passion found only in the most exceptional of undergraduates. There he learned about the works of Sacrobosco, Regiomontanus, Ptolemy, and Euclid.

After leaving the Collegium Maius, Copernicus journeyed to the great university cities of Bologna, where he studied canon law, and Padua, where he studied medicine. Italy, then as now, bore the visible imprint of ancient Rome. It had become the recent home of Greek scholars, refugees from Byzantium, and in Italy Copernicus seized the opportunity to learn Greek. Italy was then in the high Renaissance, with Leonardo, Michaelangelo, and Raphael creating their great masterpieces. But Copernicus, like many before him, had been drawn to Italy not for art but in search of a degree, and before he went home, he picked up a doctorate in canon law at the University of Ferrara. He thus became a lawyer by profession, with astronomy remaining an avid avocation.

In 1503, the 30-year-old Copernicus returned to Poland to take up a lifetime post as a canon of the Cathedral of Frombork, an appointment arranged through the benevolent nepotism of his uncle Lucas. Bishop Lucas was the head of the local government in Varmia, and the sixteen canons of the Cathedral Chapter constituted the next highest level of administration. In this northernmost diocese of Poland, Copernicus led an active and fruitful life for 40 years.

It was here that Copernicus served as an administrator of the Cathedral estates, collecting rents, resettling peasants, and writing an essay on currency reform. He served for a while as private secretary, personal physician, and diplomatic envoy for his uncle. And here in northern Poland, imbued with the spirit of Italian humanism, he made a Latin translation of a Greek work by Theophylactus Simocatta, a 7th-century Byzantine epistolographer, and perhaps he even painted his own self portrait. Each of the Cathedral canons received an ample income derived from the peasants working the farmlands administered by the Chapter, and with such a tenured position Copernicus had the financial security to pursue his sideline of astronomical researches.

It was in Frombork that he wrote "For a long time I reflected on the confusion in the astronomical traditions concerning the derivation of the motion of the spheres of the Universe. I began to be annoyed that the philosophers had discovered no sure scheme for the movements of the machinery of the world, created for our sake by the best and most systematic Artist of all. Therefore, I began to consider the mobility of the Earth and even though the idea seemed absurd, nevertheless I knew that others before me had been granted the freedom to imagine any circles whatsoever for explaining the heavenly phenomena."

We do not know precisely when Copernicus began to meditate on the mobility of the Earth. He first announced his assumptions in an anonymous tract, today called the *Commentariolus*, that is, the Little Commentary. The *Commentariolus* was written before 1514, because in that year Matthew of Miechow, a Krakow University professor, cataloged his books and noted that he had "a manuscript of six leaves expounding the theory that the Earth moves while the Sun stands still." This brief document represents a first account of planetary motion, which was considerably extended and elaborated by Copernicus in later years. We do not know if the *Commentariolus* was widely distributed. In any event, it dropped completely out of sight until around 1880, when an example was found in Vienna and another in Stockholm. More recently a third copy has been found in Aberdeen, Scotland.

In Copernicus' day the sciences, and astronomy not least, were beginning to respond to the new opportunities offered by the printing press. It is interesting to notice that his lifetime of astronomical studies was to a large part made possible by his access to printed sources. During the Thirty Years' War, the Frombork Cathedral library was carried off to Sweden, and as a result most of his books are now found in the Uppsala University library. They include the beautiful Ptolemaic atlas printed in Ulm in 1486, Argellata's book on surgery, two editions of Pliny the Younger, plus works by Cicero, Herodotus, Hesiod, and Plato.

One of the earliest books he bought, presumably while he was still a student at the Collegium Maius, was the 1492 edition of the *Alfonsine Tables*. His personal copy is still preserved in its Krakow binding. These tables, originally constructed in 1273, represented the state of the art when Copernicus was a young man. They enabled him to calculate solar, lunar, and planetary positions for any date according to the Ptolemaic theory. Among the other scientific volumes remaining from Copernicus' personal library is the beautiful first edition of Euclid's *Elements*, printed by Ratdolt in 1483, and Stoeffler's *Calendarium Romanum Magnum* of 1518. The annotations in this latter book show that Copernicus witnessed celestial phenomena on numerous occasions not mentioned in his published work.

A book that must have been enormously important during Copernicus' formative years was the Regiomontanus *Epitome of Ptolemy's Almagest*. His personal copy of this book is lost, but perhaps it is still waiting to be recognized by some sharp-eyed scholar. Our astronomer's principal access to the Ptolemaic theory must have come at first through the *Epitome*. It was not until after he had written the *Commentariolus* that the full text of Ptolemy's *Almagest* became available, in the edition printed in Venice in 1515. Copernicus studied the work carefully, as the manuscript notes and diagrams in the margins clearly show. Through this work he must have become more fully aware of the tremendous task facing any astronomer with the courage to construct a complete celestial mechanism.

During the 1520's, Copernicus worked extensively to elaborate his ideas, especially the planetary theory, if we are to judge by the scattered planetary observations recorded in his work. The *Commentariolus* had already hinted at a larger work, which Copernicus composed and continually revised during these years. By heroic good fortune, which we could scarcely have expected, his original manuscript has survived all these years. Perhaps the most priceless artifact of the entire scientific renaissance, it is now preserved in the Jagiellonian Library of Krakow University. The skilled draftsmanship, the precise hand, and, above all, the way in which he has elegantly written his text around the famous diagram of the heliocentric system (see Figure 1) convey the impression that this was a piece of calligraphy for its own sake, not a manuscript to be destroyed in the printing office, but an opus destined for the library shelf in the quiet cloisters of Frombork.

It is quite possible that his manuscript would have gathered dust, unpublished and virtually unknown, had it not been for the intervention of a young professor of astronomy from Wittenberg, Georg Joachim Rheticus. Exactly how Rheticus heard about Copernicus' work is still a mystery, although he may have seen a copy of the *Commentariolus*. In any event, he decided that only a personal visit to the source would satisfy his curiousity about the new heliocentric cosmology. Thus, in 1539, the 25-year-old Rheticus set out to that "most remote corner of the Earth," as Copernicus himself described it. Although he came from the central bastion of Lutheranism, the Catholic Copernicus received him with courage and cordiality.

Swept along by the enthusiasm of his young disciple, Copernicus allowed him to publish a first printed report about the heliocentric system. In a particularly beautiful passage of the *Narratio Prima*, Rheticus wrote:

"With regard to the apparent motions of the Sun and Moon, it is perhaps possible to deny what is said about the motion of the Earth.... But if anyone desires to look

Fig. 1. Autograph manuscript of Copernicus' *De Revolutionibus*, folio 9v, showing the heliocentric system. Photograph by Charles Eames, courtesy of the Jagiellonian Library, Krakow.

either to the order and harmony of the system of the spheres, or to ease and elegance and a complete explanation of the causes of the phenomena, by no other hypotheses will he demonstrate more neatly and correctly the apparent motions of the remaining planets. For all these phenomena appear to be linked most nobly together, as by a golden chain; and each of the planets, by its position and order and very inequality of its motion, bears witness that the Earth moves."

Rheticus had not come to Polish Prussia empty-handed. He brought with him three volumes, the latest in scientific publishing, each handsomely bound in stamped pigskin. These he inscribed and presented to his distinguished teacher. Included were Greek texts of Euclid and Ptolemy, as well as three books published by Johannes Petreius, the leading printer of Nuremberg. By the time Rheticus returned to Wittenberg in September of 1541, he had persuaded Copernicus to send along a copy of his work, destined for Petreius' printing office.

Tantalizingly little information survives concerning the actual publishing of Copernicus' book. We do not know the time required for the printing, the size of the edition, the methods of distribution, or the price. A few things can be conjectured from the standard practices of the day. Thus we can deduce that if a single press were used for the folio sheets, the printing of the 404-page treatise would have taken about four months. It is likely that the type would have been redistributed and continually reused, so that a competent technical proofreader would have been required on the scene.

Wildly diverse guesses about the size of the first edition have appeared in the literature. At the present time, I have located approximately 200 copies; perhaps an additional hundred exist that I have not found, and I would appreciate help in locating other copies. These numbers suggest an edition of at least 400, and perhaps five or six hundred. If many more were sold, it seems improbable that a second edition of about the same size would have been required 23 years later. In any event, enough copies were issued so that its ideas could not easily be suppressed or forgotten.

By the time the printing had got under way, Rheticus had taken a professorship at Leipzig, too far from Nuremberg to assist directly with the proofreading. Thus the printer, Petreius, turned to a local scholar and theologian, Andreas Osiander, who had helped him on at least one previous occasion.

In order to disarm criticism of the unorthodox cosmology in the book, Osiander added an unsigned introduction on the nature of hypotheses. He wrote: "It is the duty of an astronomer to record celestial motions through careful observation. Then, turning to the causes of these motions he must conceive and devise hypotheses about them, since he cannot in any way attain to the true cause.... The present author has performed both these duties excellently. For these hypotheses need not be true nor even probable; if they provide a calculus consistent with the observations, that alone is sufficient.... So far as hypotheses are concerned, let no one expect anything certain from astronomy, which cannot furnish it, lest he accept as true ideas conceived for another purpose, and depart from this study a greater fool than when he entered it."

I doubt that Osiander's anonymity stemmed from any malicious mischievousness, but rather simply from a Lutheran reluctance to be associated with a book dedicated to the Pope. In any event, Kepler and the other leading astronomers of that century were fully aware of the authorship; in Kepler's copy, preserved at the University of Leipzig, Osiander's name has been written above the introduction. There exists a presentation copy given by Rheticus to Andreas Aurifaber, who was then Dean of the University of Wittenberg. The inscription is dated April 20, 1543, and Rheticus probably had the book a little while before he gave it away, since he had started to annotate it. Thus a copy of the book could have easily reached Copernicus a few weeks before he died on May 24, 1543, but because he had been incapacitated by a stroke, he was probably unaware of Osiander's introduction.

Rheticus himself was so offended by the added introduction that he struck it out in the copies he distributed. He also deleted the last two words of the printed title *De Revolutionibus Orbium Coelestium*. There is an old tradition that Osiander assisted the printer in changing the title from 'Concerning the Revolutions' to 'Concerning the Revolutions of the Heavenly Spheres.' It is difficult to see precisely what Rheticus thought was offensive about the additional words except that, like the introduction, the expression 'Heavenly Spheres' perhaps suggests too much the idea of model building. As I shall explain, the idea that astronomers were merely playing some kind of geometrical game had a widespread currency in the 16th century, and Osiander's preface simply served to reinforce what astronomers thought they saw in the major part of *De Revolutionibus*. When we notice that Copernicus used an entirely different arrangement of circles for predicting latitudes than for predicting longitudes, we realize that any reader who studied the great bulk of the book carefully would necessarily have seen Copernicus as a builder of hypothetical geometrical models.

Despite the existence of the manuscript with its many layers of revisions, and even the *Commentariolus*, which provides a glimpse of an earlier formulation, we have no definite idea of the circumstances that caused Copernicus to adopt a Sun-centered cosmology. Attempting to answer this question is one of the intriguing problems that face Copernican scholars today.

If we, as 20th-century astronomers, were to speculate freely, we might well invent some quite convincing causes. First, we might suppose that the *Alfonsine Tables* were no longer in accord with the actual observations. This is true, but mostly irrelevant. Second, we might imagine that successive generations of theory-patching had left the Ptolemaic system too cumbersome for practical use, so that a massive simplification was in order. This second supposition is entirely false.

Let us first consider the matter of predictions vs observations. Was Copernicus motivated to reform astronomy because the current almanacs were bad? Because we can compare 15th-century ephemerides with the far more accurate calculations carried out recently by Dr Tuckerman at the IBM Corporation, we know nowadays that they often had errors of several degrees. But did Copernicus know this?

Soon after Copernicus had returned to Poland from Italy, the planets put on a particularly spectacular celestial show. Saturn and Jupiter, the slowest moving planets, moved into the constellation Cancer for one of their scarce conjunctions, once in twenty years. In addition, Mars, Venus, and Mercury, and eventually the Sun and the Moon, all congregated within this single astrological sign. In the winter of 1503–1504, Mars went into its retrograde motion, making repeated close approaches to Jupiter and Saturn.

My assistant, Barbara Welther, has charted for us the geocentric longitudes of the superior planets as a function of time (Figure 2). You can see how Mars bypasses Jupiter and Saturn in October 1503, and then, as all the planets go into retrograde,

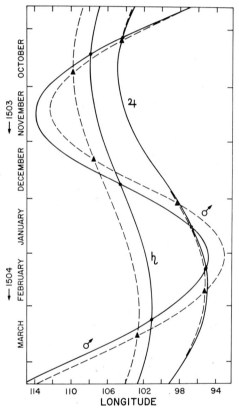

Fig. 2. Apparent motions of the superior planets just before the great conjunction of 1504. The solid lines and circles show the actual positions and conjunctions. The broken lines and triangles show the predicted positions and conjunctions.

Mars backs up past Saturn and Jupiter, and then passes them directly once more in the winter of 1504. We have not shown the great conjunction of Jupiter and Saturn at the end of May, because by that time they were too close to the Sun. We have marked with dashed lines the predicted positions of the planets according to the *Alfonsine Tables*. Notice particularly that in February and March the Mars predictions erred by 2° and Saturn by 1.5°, whereas Jupiter was predicted rather accurately. The

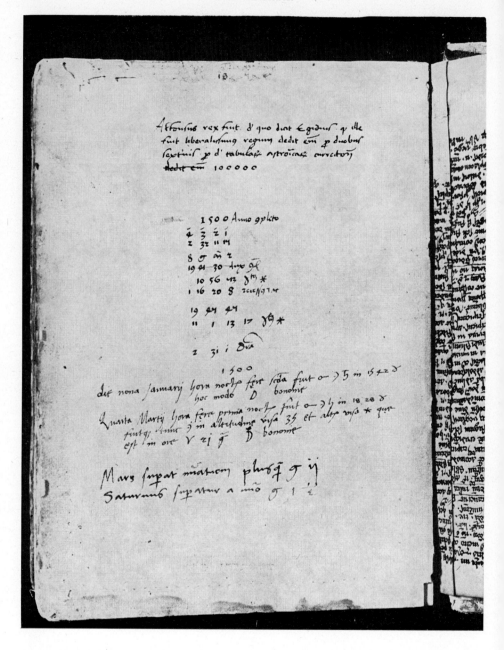

Fig. 3. Copernicus wrote these notes on observations at the end of his printed copy of the *Alfonsine Tables* (1492). By permission of the Uppsala University Library.

predicted times of the conjunctions differ by about one or two weeks from the actual times shown by the intersections of the curves.

Anyone as interested in astronomy as Copernicus could scarcely have failed to observe these phenomena, but I was curious to know whether he had noticed these deficiencies in the *Alfonsine Tables*. Although there is no direct record that Copernicus made these observations, Dr Jerzy Dobrzycki suggested to me a way whereby we can be certain that our astronomer followed the planetary motions in the year of the great conjunction. Bound in the back of his copy of the *Alfonsine Tables* are sixteen extra leaves on which Copernicus added carefully written tables and miscellaneous notes. Below the record of two observations made in Bologna in 1500, there is, in another ink, a cryptic undated remark in highly abbreviated Latin (Figure 3):

"Mars surpasses the numbers by more than two degrees.
Saturn is surpassed by the numbers by one and a half degrees."

If we examine carefully the error pattern between the positions predicted for the superior planets by the *Alfonsine Tables* and the calculations made by Tuckerman, we find a virtually unique error pattern for February and March of 1504 corresponding to the note. Thus, our astronomer must have been fully aware of the discrepancies between the tables and the heavens.

Why, then, are such glaring inadequacies never mentioned by Copernicus as a reason for introducing a new astronomy? I believe the answer is quite simple. Copernicus knew very well that discrepancies of this sort could be corrected merely by changing the parameters of the old system. A new Sun-centered cosmology was hardly required for patching up these difficulties with the tables.

But furthermore, if we turn once more to the analysis made possible by modern computers and if we examine the old ephemerides, we are shocked to discover that there is relatively little difference in the average errors before and after Copernicus. His work has scarcely improved the predictions.

Rather than condemn Copernicus, we should remember that he had no procedure for handling errors in a multiplicity of data. He had only a few score ancient observations, those recorded by Ptolemy in the *Almagest*. Since these were the minimum number required to establish the parameters, he was obliged to assume that they were perfect and to force his own parameters to fit them. From his own planetary observations he only slightly modified Ptolemy's eccentricities and apsidal lines, and he reset the mean longitude, somewhat akin to resetting the hands of a clock whose mechanism is still basically faulty. Copernicus himself must have realized that he had not achieved as much in this direction as he might have hoped, and perhaps this partly explains his reluctance to send his great work to the printer.

After *De Revolutionibus* was published, Erasmus Reinhold reworked the planetary tables into a far handier form. His *Prutenic Tables* superseded the *Alfonsine Tables* remarkably quickly. This is actually very curious because, in the absence of systematic observations, nobody really knew how good or bad any of the tables were. In fact, it was not until Tycho Brahe that a regular series of observations established the inadequacies of all the tables.

Tycho himself was something of a child prodigy; when he saw an eclipse at age 13 it struck him as "something divine that men could know the motions of stars so accurately that they could long before foretell their places in relative positions." But three years later, at the great conjunction of Saturn and Jupiter in 1563, he was astonished and offended to discover that even Prutenic-based ephemerides foretold the event on the wrong day. From the time of that great conjunction onward, he kept regular observations of increasing precision that eventually became the basis for another sweeping reform of astronomy.

Let us now turn quickly to a second imagined defect in the ancient geocentric astronomy, which, if true, would give more than adequate grounds for introducing a new system. This is the story, widely repeated in the secondary literature, that by the Middle Ages the Ptolemaic theory had been hopelessly embroidered with epicycles-on-epicycles. I fear that we modern astronomers have been particularly fond of this legend because it reminds us of a Fourier series. In Ptolemy's original scheme, the Earth is placed near but not exactly at the center of a large orbital circle called the deferent. Each planet moves in a secondary circle of epicycle, which produces the retrograde motions of the sort that we have noted at the time of the conjunctions in 1504. From a modern heliocentric viewpoint we would say that the planetary epicycles are reflections of the Earth's own orbit.

About a century ago, the story began to propagate that Ptolemy's rather simple system had been overlaid with dozens of additional secondary circles. The seed for this mythology was planted by Copernicus himself when, at the end of his *Commentariolus*, he concluded: "All together, therefore, 34 circles suffice to explain the entire structure of the universe and the entire ballet of the planets." Nineteenth-century commentators used their imaginations to embellish Copernicus' simple claim. Without checking the facts, they created a fictitious pre-Copernican planetary theory hovering on the brink of collapse under the burden of incredibly complex wheels upon wheels.

I suspect that at the end of the 13th century, Alfonso the Great may have contributed to the legend, because he supposedly told his astronomers that if he had been present at creation, he could have given the Good Lord some hints. Again, modern electronic computers have helped us put this legend to rest. I have recomputed his planetary tables in their entirety to show that they are based on the classical and simple form of the Ptolemaic theory with only two or three minor changes of parameter in the whole set.

Next, I used these 13th-century tables to compute a daily ephemeris for 300 years, and this I compared with the best almanacs of the 15th and early 16th centuries. The comparison showed, without any question, that the leading almanac makers, such as Regiomontanus, were using the unembellished Ptolemaic theory as found in the *Alfonsine Tables*.

Is it possible that the epicycles-on-epicycles existed but simply did not get to the level of almanac making? The answer is both no and yes. From antiquity there were actually two competing cosmological views. First was the system of concentrically nested spheres, espoused by Aristotle because it made such a tidy, compact, mechanical universe. In contrast, the Ptolemaic system had large clumsy epicycles that were difficult to place in concentric nests.

Peurbach's *New Theory of the Planets*, the most important work on astronomy written in the generation immediately preceding the birth of Copernicus, added no new epicycles, but instead attempted to resolve the cosmological competition by incorporating large eccentric zones of crystalline ether. By providing something of an off-center tunnel for the epicycle, the mechanism for each planet could be contained within two concentric bounds. Thus in principle the entire planetary system of Ptolemy could be nested together within the homocentric aethereal spheres of Aristotle. Such was the *New Theory of the Planets*, and I hasten to say that this idea was not really new, as it had already been described by Islamic scientists, and proposed even earlier by Ptolemy himself.

In recent years, the historians of science have discovered that, interestingly enough, 13th- and 14th-century Islamic astronomers discussed one important case of an epicycle-on-epicycle, designed not to improve the fit to observations, but to satisfy a philosophical principle. Because this same philosophical point played a major role in the motivation of Copernicus, let me now return to his work and present the two major reasons that Copernicus himself gives as primary motivations for his astronomical work.

In the *Commentariolus*, our astronomer wrote concerning the planetary motions that "Eventually it came to me how this very difficult problem could be solved with fewer and much simpler instructions than were formerly used, if some assumptions were granted me." If we put aside the spurious relevance of counting circles, the

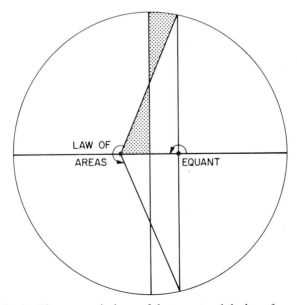

Fig. 4. The near-equivalence of the equant and the law of areas.

heliocentric system does provide a profound simplification, and I must necessarily return to this point before the end of the paper. However, Copernicus awarded virtually equal weight to a second philosophical principle, the Platonic-Pythagorean concept of uniform circular motion. Copernicus opened his *Commentariolus* with an attack on the Ptolemaic equant, which appeared to violate this principle of uniform circular motion. The equant is a seat of uniform circular motion placed equal and opposite to the Earth within the deferent circle; it drives the epicycle around on the deferent more swiftly at the perigee than at the apogee.

Figure 4 illustrates the relation between Kepler's law of areas and the equant; because the equant turns uniformly, the planet will move in equal time in each of the four quadrants. The law of areas tells us that the planet will move through these same arcs in equal times provided that the areas swept out from the primary focus are equal. Because the equant is at the empty focus, the shaded triangles are virtually equal except that the upper one has a curved side; to this extent, the equant is a good approximation to the true motion, especially at the quadratures. As Kepler was later to show, the major discrepancy occurs in the octants.

In any event, Copernicus despised the equant and he felt that Ptolemy had cheated by introducing it. Figure 5 shows how Copernicus replaced the equant with an eccentric circle and a small epicyclet. In the *Commentariolus* he preferred to use a concentric circle with a double epicyclet, which was precisely the same mechanism suggested two centuries earlier by Ibn ash-Shatir in Damascus; whether there was any transmission from those Islamic astronomers to Copernicus is still debatable. After Copernicus discovered the motion of the planetary apsidal lines, it became more convenient to use the eccentric circle and single epicyclet shown here. I shall not take the time here to explain the equivalence between this mechanism and the equant, but

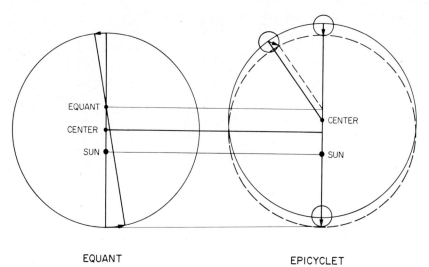

Fig. 5. Copernicus' replacement of the equant by a pair of uniform circular motions. The epicycle has a radius of $e/2$ and always moves to form the isosceles trapezoid shown above.

I shall simply say that the great bulk of the *De Revolutionibus* involves the use of this mechanism.

Nowadays the epicyclet seems esoteric and forgettable. When we commemorate Copernicus, we praise his profound insight in seeing the philosophical and esthetic simplicity of his system, but we try to ignore his infatuation with the second, very deceptive principle of uniform circular motion. I should now like to demonstrate that Copernicus' 16th-century successors, living in an age long before Newtonian dynamics, evaluated these philosophical principles in precisely the opposite way, rejecting the simplicity of the heliocentric cosmology but admiring the epicyclets.

About three years ago I had an interesting discussion with another Copernican scholar, Dr Jerome Ravetz, and we asked ourselves if *De Revolutionibus* actually had any careful readers. We speculated that there are probably more people alive today who have read this book carefully than in the entire 16th century. I have already introduced some of the candidates for that early era: Georg Rheticus, the Wittenberg scholar who persuaded Copernicus to publish his book; Erasmus Reinhold, the Wittenberg professor who stayed home but who later composed the *Prutenic Tables*; and Tycho Brahe, the great Danish observer. Others would include Johann Schöner, the Nuremberg scholar to whom Rheticus addressed the *Narratio Prima*; Christopher Clavius, the Jesuit who engineered the Gregorian calendar reform; Michael Maestlin, Kepler's astronomy teacher; and Johannes Kepler himself.

At that time I was on a sabbatical leave from the Smithsonian Astrophysical Observatory, and two days after talking with Dr Ravetz I happened to visit the remarkable Crawford collection of rare astronomical books at the Royal Observatory in Edinburgh. There I admired one of their prize possessions, a copy of the first edition of *De Revolutionibus*, legibly annotated in inks of several colors. As I examined the book, I deduced that the intelligent and thorough notations were undoubtedly made before 1551, that is, within eight years after its publication. Our speculation from two days earlier seemed completely demolished, because it appeared that if intelligent readers were so rare, it would be unlikely that the very next copy of the book that I saw could be so carefully annotated. But then a second thought crossed my mind: Perhaps the Crawford copy had been annotated by one of the handful of astronomers we had mentioned. The list quickly narrowed to Rheticus, Reinhold, and Schöner, the only ones active before 1550. Internal evidence suggested Erasmus Reinhold, and although his name is not in the book, I soon found his initials stamped into the decorated original binding. Ultimately I was able to obtain additional specimens of Reinhold's distinctive handwriting, which settled the matter beyond all doubt.

One of the most interesting annotations in Reinhold's copy appears on the title page, where he has written in Latin "The axiom of astronomy: Celestial motion is uniform and circular or composed of circular and uniform parts." Reinhold was clearly fascinated by Copernicus' epicyclets and his adherence to the principle of circular motion. The paucity of annotations in the first twenty pages, which Copernicus devoted to the new cosmology, shows that Reinhold was not particularly interested in heliocentrism. Accepting Osiander's statement that astronomy was based on hypotheses, Reinhold was apparently intrigued by the model-building aspects. Whenever alternative mechanisms for expressing the motions appeared in the book, he made conspicious enumerations with Roman numerals in the margins.

Because Reinhold published the *Prutenic Tables*, naming them in part for Copernicus, he is sometimes listed as an early adherent of the heliocentric cosmology. However, the nature of the tables makes them independent of any particular cosmological system, and although his introduction is full of praise for Copernicus, he nowhere mentions the heliocentric cosmology. With his great interest in hypothetical model building, there is reason to suspect that Reinhold was on the verge of an independent discovery of the Tychonic system; unfortunately, he died of the plague at an early age before he could consolidate any cosmological speculations of his own.

Flushed with the success of identifying Reinhold's copy, I resolved to examine as many other copies of the book as possible in order to establish patterns of readership and ownership, always hoping to find further interesting annotations. For three years I have systematically examined copies in such far-flung places as Budapest and Basel, Leningrad and Louisville, Copenhagen and Cambridge. In the process I saw and photographed several particularly interesting copies, including the *De Revolutionibus* owned by Michael Maestlin, preserved in Schaffhausen, Switzerland; this is one of the most thoroughly annotated copies in existence. I also examined copies once owned by Rheticus, by Kepler, and by Tycho Brahe – the last being a heavily annotated second edition in Prague. In all, I had managed to see 101 copies by the spring of 1973. The investigation confirmed that the book had rather few perceptive readers, at least among those who read pen in hand. Despite this, however, the book seems to have had a fairly wide circle of casual readers, much larger than generally supposed.

In May of 1973, I had the opportunity to visit Rome, where there were seven copies of the first edition that I had not examined. My quest took me first to the Vatican Library, where I went armed with shelf mark numbers provided by Dr Dobrzycki. Some of the books in the Vatican Library came there with the eccentric Queen Christina of Sweden, who abdicated her throne in 1654, abandoning her Protestant kingdom for Rome. Her father, Gustavus Aldolphus, had ransacked northern Europe during the Thirty Years' War and among other things had captured most of Copernicus' personal library. Dr Dobrzycki had gone to Rome in search of Copernican materials that Queen Christina might have taken along. In the Vatican, he found an unlisted copy of Copernicus' book among the manuscripts, that is, a third copy beyond the two examples cataloged among their printed books. Fortunately, Dr Dobrzycki gave me the number for the volume, which could not have been found in any of the regular Vatican catalogs.

When I examined this copy, I recognized that the extensive marginal annotations must have been made by a highly skilled astronomer. At the end were thirty interesting manuscript pages, full of diagrams made by someone working along the same lines as Tycho Brahe, and dated 1578. Although there was no name any place on the volume, I quickly conjectured that the annotations had been made by the Jesuit astronomer

Christopher Clavius. In the first edition of his learned *Commentary on the Sphere of Sacrobosco*, published in 1570, he failed to mention Copernicus. But in the third edition, published in 1581 – after the time these manuscript notes were written – he commented rather extensively and wrote "All that can be concluded from Copernicus' assumption is that it is not absolutely certain that the eccentrics and epicycles are arranged as Ptolemy thought, since a large number of phenomena can be defended by a different method."

In a state of considerable excitement, I contacted Dr D. J. K. O'Connell, former Director of the Vatican Observatory, and with his help I obtained Xerox copies of two Clavius letters from the Jesuit Archives. I eagerly returned to the Vatican Library, only to have my hypothesis smashed within a few minutes. There was no possibility that the handwriting in the *De Revolutionibus* could be that of Christopher Clavius.

I left Rome in a baffled and troubled state for a Copernicus conference in Paris. There, by a fantastic stroke of luck, I received the new Prague facsimile of the second-edition *De Revolutionibus* with the annotations by Tycho Brahe. I think my heart must have skipped a beat when I saw the handwriting in the fascimile, because I then realized that the first edition in Rome was probably also in Tycho's hand. What I had discovered was the original working copy, probably the most important Tycho manuscript in existence. The example in Prague was a derivative copy, being annotated by Tycho for possible publication. I rebooked my flights, went back to Rome, and after I put the Prague facsimile side by side with the Vatican copy, it took only a few minutes to prove my conjecture. Afterward, the Vatican librarians traced the book to Queen Christina, who must have gained possession of it in 1648 when her troops captured the collections founded by Rudolph II in Prague.

Of many remarkable things about this copy, the first appears on the title page itself. We find the very same words that Reinhold had inscribed on the title page of his copy, "The axiom of astronomy: Celestial motion is uniform and circular, or composed of uniform and circular parts." I had already known that Tycho Brahe had visited Wittenberg on at least four occasions, and that in 1575, three years before the dated annotations in this book, he had visited Reinhold's son and had seen Reinhold's manuscripts. In an article that I had written earlier for the Copernicus celebrations in Torun, I had stated "We are tempted to imagine that Tycho's own cosmological views grew from seeds planted at Wittenberg by a tradition that honored Copernicus, but which followed Osiander's admonition that it is the duty of the astronomer to 'Conceive and devise hypotheses, since he cannot in any way attain the true causes'." The newly found Tycho copy dramatically confirms this intellectual heritage, not only through this motto on the title, but within the book, where numerous annotations are copied word for word from Reinhold's copy. In particular, Tycho like Reinhold specifically numbered any alternative arrangements of circles indicated by Copernicus.

In the Tycho Brahe manuscript bound at the end of the Vatican *De Revolutionibus*, the first opening is dated January 27, 1578, the day after the spectacular comet of 1577 had been seen for the last time. The diagrams on those two pages are heliocentric, and a note in the corner indicates that it was drawn according to the third hypothesis of Copernicus. In the next two weeks, Tycho explored additional heliocentric arrangements for the planets and geocentric models for the Moon. On February 14 and 15, he began to investigate *geocentric* constructions for Venus and Mercury, especially alternate positions of the single epicyclet for Venus and the pair of epicyclets for Mercury. He specifically noted that "This new idea occurred to me on February 13, 1578."

Three days later Tycho drew the most interesting diagram of the entire sequence, a proto-Tychonic system with the Earth at the center circled by the Moon and the Sun (Figure 6). Around the Sun are the orbits of Mercury and Venus. The three superior planets are still arranged in circles about the Earth, but each epicycle has been drawn the same size as the Sun's orbit. To finish the construction of the Tychonic system, it is necessary only to complete the parallelograms for Mars, for Jupiter, and for Saturn. Tycho was now surely within grasp of his final system. But notice the caption: "The spheres of revolution accommodated to an immobile Earth from the Copernican hypotheses." Here we see Tycho playing the astronomical geometry game, greatly under the influence of Copernicus, and somehow supposing that a geocentric system is compatible with the teachings of the master.

It is very curious that Tycho did not publish his new system until a decade later. Tycho was a dynamic young man of 31 when he wrote this manuscript, already well-established on the island of Hven, but perhaps still uncertain where his observations for the reform of astronomy would lead him. A passage in his book implies that he did not establish the Tychonic system until around 1583, five years after he drew these diagrams. I can only suppose that these five years were an important time of maturing. In that interval, Tycho must have speculated on the movement of the great comet of 1577, realizing that it would have smashed the crystalline spheres of the ancient astronomy, had they existed. Perhaps he began to look for greater certainty in astronomy and to suppose that, after all, the observations made with his giant instruments at his Uraniborg Observatory could lead beyond hypothesis to physical reality. If so, like this contemporaries in that pre-Newtonian, predynamical age, he must have viewed the physics of the sluggish, heavy Earth as a most important phenomenon to be preserved. Concerning the Copernican system, Tycho Brahe wrote: "This innovation expertly and completely circumvents all that is superfluous or discordant in the system of Ptolemy. On no point does it offend the principle of mathematics. Yet it ascribes to the Earth, that hulking, lazy body, unfit for motion, a motion as quick as that of the aethereal torches, and a triple motion at that." I can well imagine that Tycho believed he was making a great step forward toward understanding the physical reality of the universe when he adopted his own geocentric system.

To us, the Tychonic system looks clumsy and wrong. To us, there is something more neat and orderly about the heliocentric system. Indeed, it is precisely this elegant organization that Copernicus found pleasing to the mind, and that led to his cosmology. In a powerful plea for the heliocentric world view near the beginning of *De Revolutionibus*, Copernicus wrote: "At rest in the middle of everything is the Sun.

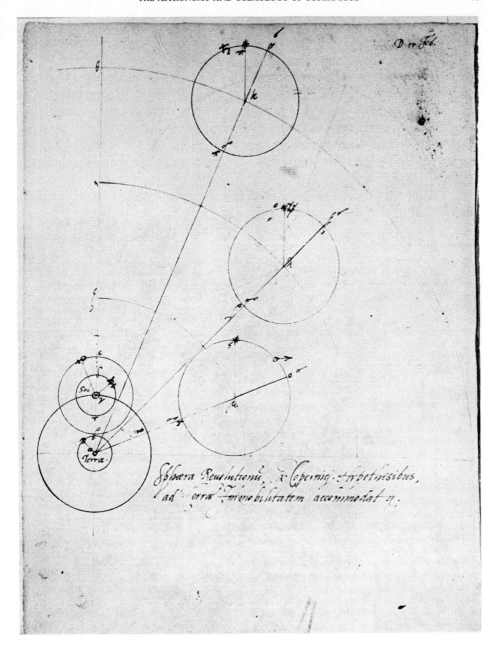

Fig. 6. Tycho's sketch of a geocentric planetary system, folio 210v in the manuscript notes bound at the end of his annotated copy of *De Revolutionibus* (1543), Vatican Library Ottob. 1901.

For in this most beautiful temple, who would place this lamp in another or better position? From here it can light up the whole thing at the same time. Thus as though seated on a royal throne, the Sun governs the family of planets revolving around it.

"In this arrangement, therefore, we discover a marvelous symmetry of the Universe, and an established harmonious linkage between the motion of the spheres and their size, such as can be found in no other way. Thus we perceive why the direct and retrograde arcs appear greater in Jupiter than in Saturn and smaller than in Mars, and why this reversal in direction appears more frequently in Saturn than in Jupiter, and more rarely in Mars and Venus than in Mercury. All these phenomena proceed from the same cause, which is the Earth's motion. Yet none of these phenomena appears in the fixed stars. This proves their immense height, which makes the annual parallax vanish from before our eyes."

There is a whiff of reality here, especially in the resounding conclusion, "So vast without any question is this divine handiwork of the Almighty Creator." Yet very few people in the 16th century grasped the harmonious, esthetic unity that Copernicus saw in the cosmos. And that is why we must also salute another perceptive genius, born almost a century later than Copernicus. Like Copernicus, Johannes Kepler saw the Sun seated upon its royal throne as the governor of the planetary system, and he tried mathematically to find the harmonious linkage between the motions of the spheres and their sizes. To us Kepler's neo-Platonic attempts to find an archtypal geometrical structure in the planetary arrangement smack of mystical numerology – yet this is hardly a criticism, considering that numerology has not been banished from modern cosmology. But more important, Kepler saw in the Copernican arrangement of the planets about the Sun the real possibility of a celestial physics, and he made the first groping steps toward a dynamics of the heavens – a dynamics that, reshaped and powerfully formulated by Isaac Newton, ultimately proved to be the primary justification for the heliocentric universe.

Although Copernicus is not celebrated for his observations, yet it was in the Copernican tradition that Kepler and Galileo taught us to use our senses to distinguish between the various hypothetical world views, leaving only those consistent with the observations. In a way we are still model builders, as Osiander suggested, but unlike Andreas Osiander and Erasmus Reinhold, we are no longer content to entertain alternatives without trying to choose one as physically most acceptable. Modern science still plays its games, but in an entirely different way than did the ancient and medieval astronomers. Certainly Copernicus, Tycho, Kepler, Galileo, and Newton are heroes in this epic reformation in our understanding of what nature is and what learning and observation should be.

Although I have said perhaps too much about the technical astronomy of Copernicus and rather little about his cosmology, I hope that within this broader context you have been able to appreciate all the more how unique was Copernicus' own intellectual adventure. Only in our own generation have we been able to break the terrestrial bonds; men flung out toward the Moon have seen the spinning Earth, a blue planet, sailing through space. Although rejected by the astronomers of his day, the Copernican idea became the point of departure for the law of universal gravitation. In reality, the Copernican quinquecentennial celebrates the origins of modern science and our contemporary understanding of the universe. In setting the Earth into motion, Copernicus was right: his daring idea still guides the unfinished journey of modern science.

Acknowledgements

Professor Edward Rosen's Copernican biography in his *Three Copernican Treatises* (New York, 1971) has provided an authoritative source for details of Copernicus' life, and I have borrowed from him several felicitous turns-of-phrase as well as English translations of some of the Latin texts (which I have generally abridged). Other writings that have been particularly stimulating include J. Ravetz, *Astronomy and Cosmology in the Achievement of Nicolaus Copernicus* (Wroclaw, 1965), P. Duhem, *To Save the Phenomena* (Chicago, 1969), and L. A. Birkenmajer, *Mikolaj Kopernik* (Krakow, 1900) (English translation under joint preparation by J. Dobrzycki and myself). Also useful are A. Koyré, *La révolution astronomique* (Paris, 1961) and *The Great Books of the Western World* (Chicago, 1952) volume 16, which contains an English translation of *The Revolutions* as well as Ptolemy's *Almagest*.

Many persons and institutions in Europe and America made possible the triple-screen presentation of this lecture in Warsaw. I should particularly like to thank the design office of Charles and Ray Eames for many of the photographs. I deeply appreciate the gracious cooperation we received from libraries throughout the world, especially the manuscript department of the Jagiellonian University in Krakow, the Uppsala University and Uppsala Observatory Libraries, the Crawford Collection of the Royal Observatory in Edinburgh, Harvard College Library, and, of course, the Biblioteca Apostolica Vaticana.

SELECTED PAPERS

X-RAY SOURCES IN CLOSE BINARY SYSTEMS

M. J. REES

Institute of Astronomy, Cambridge, England

1. Introduction

The discovery by Giacconi and his colleagues of variable X-ray sources in close binary systems certainly ranks as one of the highlights of astronomical research during the last 3 years. These remarkable objects have already been extensively studied, by optical and radio observations as well as in the X-ray band; and they seem likely to prove as significant and far-reaching in their implications as pulsars.

The 'Third Uhuru Catalogue' (Giacconi *et al.*, 1973a) contains about 160 sources, of which about 100 lie in our Galaxy. Their distribution over the sky (together with other arguments) suggests that these sources have luminosities of the general order 10^{36}–10^{38} erg s^{-1}, and that their typical distances are ~ 10 kpc. These galactic sources generally display rapid variability. Little else is known about most of them, but they are probably of the same general class as systems such as Her X1, Cen X3, Cyg X1 and Cyg X3. These sources have been investigated in detail, and in all cases one infers a system where the X-ray source is orbiting around a relatively ordinary star. Six sources have been optically identified, and there are some others whose binary nature is established by the occurrence of an X-ray eclipse. Orbital periods range from 4.8 h (Cyg X3) up to ~ 10 days.

In this talk I shall review some of the theoretical implications of these systems, summarising the relevant observations where necessary. A fuller account of the data may be found in Giacconi (1973a, b), and in the many recent papers – mainly published in *Astrophys. J.* – referred to in these articles.

The idea of X-ray sources being associated with close binary systems dates back to the earliest days of X-ray astronomy. It was Hayakawa and Matsouko (1964) and Zel'dovich and Guseynov (1965) who first made the suggestion that binary stars might be X-ray sources. After Sco X1, the brightest object in the X-ray sky, had been identified with an object reminiscent of an old nova, many astrophysicists proposed that X-ray sources involved transfer of matter from one star onto a compact companion (see Burbidge, (1972) for an account of these developments). It is still unclear whether this is actually happening in Sco X1, and many other interpretations of this object have been proposed; but there now seems little doubt that it *is* the case for a major class of X-ray sources in the Galaxy.

2. General Theoretical Remarks

What, then, can be said concerning the general nature of these system? The first general point is that the rapid variability suggests, though of course it does not prove, that a very small object – probably even smaller than a white dwarf – is involved. The gravitational potential well associated with such an object is very deep indeed, and accretion therefore provides an efficient energy source. If, as seems to be the case in the observed X-ray binaries, the compact object is in a close orbit – almost a grazing orbit – around another star, then a copious supply of material is available from the companion. If the compact object is a neutron star or black hole, then $\sim 10\%$ of the rest mass energy of the accreted material (10^{20} erg gm^{-1}, or ~ 100 MeV per nucleon) can be liberated in the form of radiation (~ 10 times as much as can be obtained from nuclear fusion – in contrast to the 'classical' case of accretion onto ordinary stars, where the gravitational energy is generally insignificant (Bondi, 1952; Mestel, 1954)): for accretion onto a white dwarf the efficiency is 10^{-2}–$10^{-1}\%$ (0.1–1 MeV per nucleon). This means that, for accretion onto a neutron star or collapsed object, the accretion rates need only be in the range 10^{16}–10^{18} gm s^{-1} (10^{10}–10^{-8} $M_\odot$/yr^{-1}) in order to produce the observed luminosities. These are modest compared to the inferred transfer rates in other binary systems, and could be supplied by a stellar wind even if the companion star did not overflow its Roche surface.

A second general point is that most of the gravitational energy is liberated deep in the potential well – at or near the surface of the compact object if it is a neutron star; within a few Schwarzschild radii if it is a black hole. Thus the effective dimensions of the source (assuming that the compact object is in the stellar mass range) are only $\sim 10^6$ cm. If 10^{36}–10^{38} erg s^{-1} are radiated thermally from such a small region, a temperature high enough that the energy emerges predominantly in the X-ray band is therefore guaranteed.

The most popular interpretation is one in which the X-ray source is regarded as being associated with either a neutron star or a black hole (Pringle and Rees, 1972). The evidence favouring this model now seems fairly compelling, but the case is certainly not completely watertight; and some quite different interpretations for various aspects of these phenomena still remain tenable. These include models involving pulsating or rapidly rotating white dwarfs (Mock, 1968; Brecher and Morrison, 1972), analogies with pulsars, or particle acceleration at reversing layers of strong magnetic fields in a binary star system which does not contain a compact component at all (for example, Bahcall et al., 1973). My reason for concentrating on this particular model is that it seems more plausible than any specific alternative so far proposed. Also this model has formed the basis for most of the detailed theoretical work carried out so far. Already so much work has been done that it will only be possible for me to sketch most of it; and some interesting aspects of the phenomena will be left out entirely.

The X-ray binaries obviously involve all the problems connected with ordinary close binary systems (see, for example, Paczyński, 1971) – problems which are still ill-understood despite having been with us for many years – together with a whole range of new ones connected with the compact X-ray source itself. I shall first outline the main features of the model, and then comment on some specific consequences as they relate to individual sources. For convenience of exposition, it is appropriate to split the subject into three parts: the mass transfer (relevant length scales $\sim 10^{11}$ cm); the accretion disc (dimensions $\lesssim 10^{10}$ cm); and the compact object itself, which is also the region where the X-rays are presumed to originate (10^6–10^8 cm). Unfortunately these three areas cannot be regarded as entirely disjoint, despite the very different length scales involved. For example, the X-ray intensity and spectrum is probably determined by processes occurring close to the compact object, but it may nevertheless have an important influence on the flow of matter from the companion because of heating and radiation pressure effects. (It has even been proposed that X-ray heating of the companion star's atmosphere may excite a stellar wind which maintains the accretion flow which generates the X-rays which....)

2.1. The mass transfer ($\sim \lambda 10^{11}$ cm)

Much theoretical work has been based on the hypothesis that the companion star fills its Roche lobe, and that material flows across the Lagrangian point. It is important, however, to remember that these analyses are only strictly valid if the star corotates with the orbital period. This is probably quite a good assumption in these close systems, unless they were perturbed so recently that tidal effects have not yet re-established synchronous rotation. Some calculations – for example, estimated limits on the masses of the X-ray sources – depend rather heavily on this postulate. It is also possible that the star does *not* fill its Roche lobe, but has a strong stellar wind. Gas streams may cause the optical emission lines observed in these systems (and also, incidentally, confuse attempts at radial velocity determinations).

2.2. The accretion disc ($\gtrsim 10^{10}$ cm)

By whatever process material is captured from the companion star, it is likely to have so much angular momentum that it cannot fall directly onto the compact object. The matter will instead dissipate its motions perpendicular to the plane of symmetry and form a differentially rotating disc, the rotational velocity at each point being approximately Keplerian, and then gradually spiral inwards as viscosity transports its angular momentum outwards. If the companion star is overflowing its Roche lobe, it is conventionally assumed that the matter joins the disc at the radius where its angular momentum relative to the compact object equals that of a Keplerian orbit. This argument suggests that the so-called 'hot spot' appears at a radius which is $\sim 20\%$ that of the Roche lobe around the compact star. The structure of the outer part of the disc is not well understood. The disc must extend further out than the hot spot, because *some* of the material transferred from the companion star has to carry away the angular momentum – it cannot all be accreted by the compact object. A further complication is that the gravitational field of the companion star probably cannot be ignored in the outermost part of the disc, so the gas will not circulate in simple Keplerian orbits.

If the accreted matter is captured from a strong stellar wind, it will tend to have less net angular momentum; but the disc would still extend out to a radius $\sim 10^9$ cm in general.

The structure of accretion discs has been discussed by many authors – for example Prendergast and Burbidge (1968), who considered a disc surrounding a white dwarf; Lynden-Bell (1969), Pringle and Rees (1972), Shakura and Sunyaev (1973), and Novikov and Thorne (1973).

An obvious prerequisite for the existence of a disc (whose thickness must, by definition, be only a small fraction of its radius) is that radiative cooling should be efficient enough to remove most of the energy liberated by viscous friction, so that the internal energy is small compared with the gravitational binding energy – i.e.

$$kT\left(1 + \frac{p_r}{p_g}\right) \ll \frac{GMm_p}{r}, \tag{1}$$

where p_r/p_g is the ratio of radiation pressure to gas pressure, and m_p is the proton mass. For accretion flows with the parameters appropriate to X-ray sources the densities are high enough, and the timescales long enough, to ensure that (1) is almost certainly fulfilled. Also, the mass in the disc is gravitationally negligible compared to that of the central object.

If a steady state has been set up, the structure of the disc is governed by the following system of equations. First, the same mass flux $\dot{M}$ must flow across any radius r, so that

$$\dot{M} = 2\pi r \int \varrho(r, z) v_r(z) \, dz \tag{2}$$

for all r, when z is the coordinate perpendicular to the disc measured from the plane of symmetry.

A second, and somewhat less trivial, requirement is that in a steady state the *flux of angular momentum* should be the same at all r. Angular momentum is transported inward by the accreted matter, but transported outward by the viscous stresses. The difference between these quantities represents the rate at which the central compact object is gaining angular momentum. Following Novikov and Thorne (1973) we assume that angular momentum is being accreted at a rate $\beta\dot{M}(GMr_1)^{1/2}$, where r_1 is the radius of the inner boundary of the disc. Since the specific angular momentum deposited on the compact object cannot exceed the Keplerian value at r_1, we have $\beta \leq 1$. One then finds that the heat dissipated per unit surface area of the disc at a radius $r > r_1$ is

$$p(r) = \frac{3\dot{M}}{4\pi r^2} \frac{GM}{r}\left(1 - \beta\left(\frac{r_1}{r}\right)^{1/2}\right). \tag{3}$$

It is important to note that β is a second parameter which is not completely determined by $\dot{M}$ – one can imagine situations with the same $\dot{M}$ but different torques in the disc, and therefore different values of $p(r)$. When $r \gg r_1$, however, one finds, independently of β, that the energy radiated at radii $\geq r$ is *3 times larger* than the energy lost by the accreted material while spiralling inward to that radius. The extra contribution arises because the viscous stresses transport *energy* outward as well as momentum, so that the energy liberated by gravitation is actually radiated at a somewhat larger radius.

One might at first sight worry about the energy budget for the disc as a whole. However, when $\beta = 1$ one finds that the total energy radiated, integrating over all $r \geqslant r_1$, is precisely equal to $\dot{M}$ multiplied by the binding energy of Keplerian orbit of radius r_1; when $\beta = 0$, the factor of 3 enhancement applies right in to $r = r_1$, but in this case the extra energy comes from viscous torques which apply a drag to the compact object – i.e. twice as much energy in this case is supplied by the central spinning object as comes from the infalling material itself. (The discussion leading to Equation (3) is strictly Newtonian. When one considers an accretion disc surrounding a black hole, then one finds that the total energy radiated by the disc *equals* the energy lost by infalling matter when the black hole accretes a specific angular momentum appropriate to the circular orbits at the inner edge of the disc (see Novikov and Thorne (1973) for the details of the relativistic case). The appropriate inner boundary condition in this case is that the *viscous stresses* should be zero at $r = r_1$.)

These deductions do not depend on the magnitude of the viscosity – if this is low, then the radial velocity v_r is small, so the equilibrium value of ϱ needed in order to give a given $\dot{M}$ must be high; and conversely. But to analyse the structure of the disc in any further detail one *must* know something about the viscosity, and this is the stumbling-block to further progress. Possible causes of viscosity include turbulence induced by the differential rotation, convective motions, or sheared magnetic fields. Pringle and Rees (1972) and Shakura and Sunyaev (1973) made specific simplifying assumptions about the viscosity, which enabled them to discuss the vertical structure of the disc (i.e. the balance between the pressure gradient perpendicular to the disc and the component of gravity in that direction), and the spectrum of the emergent radiation. However one has little confidence that one knows even the appropriate order of magnitude for the viscosity, and it therefore seems premature to discuss the spectrum of the disc in great detail. The dominant emission mechanism is probably thermal bremsstrahlung, though the spectrum may be appreciably distorted as a result of scatterings by the hot thermal electrons (Felten and Rees, 1972). All that can be said is that the effective temperature must be at least as high as the black body temperature needed to radiate a power $p(r)$. Any line emission would be broadened and distorted by electron scattering, and by the Doppler effect associated with the Keplerian rotation.

2.3. The compact object and the x-ray emission (10^6–10^8 cm)

When the compact object is a black hole, the emission is concentrated within a few Schwarzschild radii. It is normally assumed that the amount of energy radiated per unit mass accreted equals the binding energy of the innermost stable circular orbit. This implies an efficiency of $\sim 6\%$ if the black hole is described by a Schwarzschild metric, and up to 42% for accretion discs around Kerr black holes, the precise upper limit depending on how much of the emitted radiation is captured by the hole. Once material is closer than the innermost stable orbit, it can be swallowed by the hole without any further loss of angular momentum. However if the viscosity is high enough there would still be emission from this part of the disc, resulting perhaps in even higher efficiencies than those just quoted. (Note that the efficiency would be very much lower if the accreted matter had so little angular momentum that it could fall almost radially inward. This situation, which is relevant to isolated black holes accreting interstellar matter, has been discussed by Schwartzman (1971) and Shapiro (1973a, b). In general, only a small fraction of the mass-energy is radiated away before the infalling matter is swallowed by the hole.)

Attempts to determine the expected radiation spectrum from accretion discs are impeded by our ignorance about the viscosity, which introduces far larger uncertainties than those corresponding to the difference between a Schwarzschild and 'extreme Kerr' black hole. In general, the temperature decreased outwards and, even though the emission is thermal, the integrated spectrum may resemble a power law. Radiation from the outer parts of the disc would not be energetically significant unless, as discussed by Shakura and Sunyaev (1973) the disc were so thick in relation to its radius that X-rays from the inner regions were intercepted by the disc and re-radiated at softer energies. Some further aspects of this model, as it may apply specifically to Cyg X1, are discussed later.

When the central object is a spinning, magnetised neutron star, a far more complex situation ensures, which has been discussed extensively by Pringle and Rees (1972), Davidson and Ostriker (1973), and Lamb et al. (1973). If the neutron star were unmagnetised, then the disc would extend inwards until the accreted material grazed the star's surface. If, however, the neutron star has a surface magnetic field of the same strength as is inferred for pulsars ($\sim 10^{12}$ G) then the magnetic stresses will influence the dynamics out far beyond the surface of the star. We define the 'Alfvén radius' to be that distance at which the magnetic stresses are comparable with the viscous stresses in the disc – i.e.

$$\frac{(H(r_A))^2}{4\pi} \simeq \varrho(r_A) v_r(r_A) v_\theta(r_A).$$

The Alfvén radius depends on $\dot{M}$, but somewhat insensitively because H^2 depends on r at least as steeply as r^{-6}, and for typical parameters is 10–100 times larger than $r*$. r_A, defined as above, is fortunately independent of the viscosity except insofar as this affects the scale height. The disc would not be expected to extend inward to radii much less than r_A, which means that the radiation from the disc itself is relatively unimportant. Once matter penetrates within r_A the high field strengths and conductivity ensure that it is constrained to follow the field lines. If the star has an oblique dipole field, the infalling plasma will impact on the surface in the vicinity of the magnetic polar caps. The situation at $r \simeq r_A$ is analogous to that at the Earth's magnetopause, and is so complicated that one cannot really estimate which of the magnetic field lines can capture matter. These field lines will probably, however, be only a subset of those which would have reached out to radii $\gtrsim r_A$ in the absence of infalling plasma. This guarantees that, when $r_A \gg r*$, the material will be channelled onto only a small fraction of the stellar surface.

The dominant radiation mechanisms would be bremsstrahlung or cyclotron radiation (including emission at the first few harmonics of the basic cyclotron frequency). Lamb et al. (1973), Gnedin and Sunyaev (1973), and Davidson (1973) have discussed the likely beam shape of the emergent radiation. If the dominant opacity were ordinary Thomson scattering, the radiation would tend to leak out of the sides of the accretion column, yielding a fan beam. If the magnetic field is so strong that the cyclotron frequency exceeds the radiation frequency under consideration, then electron scattering is inhibited for radiation propagating along the field direction, and also for radiation travelling across the field which is polarised such that the electric wave vector is at right angles to the magnetic field. Realistic models can yield either pencil beams or fan beams, depending on the strength of the magnetic field and the polarization of the radiation. Modulation of this beam pattern (which is unlikely to possess any especially sharp features) each time the neutron star spins generates the X-ray pulse shape. The spectrum would be broadly thermal; but not exactly a black body, for several reasons – e.g. the temperature may not be the same over all parts of the polar cap where accretion occurs, electron scattering may distort the spectrum, and the surrounding disc may cause absorption below a few keV. The radiation would generally be expected to display a high degree of both linear and circular polarization, especially for the softer X-rays. Detection of such polarization from variable X-ray sources would lend strong support to the accreting neutron star hypothesis.

Some other aspects of this scheme are discussed later in connection with Her X1.

An important role in these models is played by the so-called 'critical luminosity' or 'Eddington limit' at which radiation pressure balances gravity. If Thomson scattering provides the main opacity, and the relevant material is fully ionized, then this luminosity is

$$L_{\text{edd}} = \frac{4\pi G M m_p}{c\sigma_T} \simeq 10^{38} \left(\frac{M}{M_\odot}\right) \text{erg s}^{-1}, \tag{4}$$

σ_T being the Thomson cross-section.

One might therefore expect that the accretion rate $\dot{M}$ could approach, but in no circumstances exceed, the value needed to yield this luminosity. Recently, Margon and Ostriker (1973) have in fact analysed the data on X-ray sources, and find that there does indeed seem to be a luminosity cut-off at around the expected value of L_{edd} for $M \simeq M_\odot$, and that there is a class of sources whose luminosities cluster close to this value. But the 'Eddington limit', as given by (4), is relevant only under relatively restrictive circumstances – circumstances which are *not* generally met by the kinds of X-ray source models usually considered.

The luminosity of a source powered by accretion cannot even approach L_{edd} if the effective cross section per electron is larger than σ_T (see Buff and McCray, 1974). This is quite likely to be the case for a source emitting soft X-rays, because the relevant opacity (unless all the ions are completely stripped) is then primarily due to photo-ionization, for which $\sigma \gg \sigma_T$. If the value of $\dot{M}$ in binary X-ray sources is controlled by processes occurring near the surface of the companion star or the critical Roche surface, as in the 'self-excited wind' hypothesis (Basko and Sunyaev, 1973; Arons,

1973) then one might expect the luminosity to stabilise at a value well below $L_{\rm edd}$. There are, however, several types of situation where luminosities $\gg L_{\rm edd}$ are possible, especially under the extreme conditions prevailing near compact objects. Among these are the following:

(i) The effective opacity may be much *less* than that provided by Thomson scattering. In the context of X-ray source this may, for instance, happen in the accretion column above the magnetic polar caps of neutron stars, where the scattering cross section is $\ll \sigma_{\rm T}$ for photons below the cyclotron frequency travelling along the magnetic field direction.

(ii) Even if the appropriate cross-section *is* $\sigma_{\rm T}$, the Eddington limit can still be violated in a non-spherically-symmetric configuration. Consider again, for example, the accretion column near a magnetised neutron star. If the magnetic field does *not* modify the opacity and make the scattering highly anisotropic, then radiation will tend to escape from the *sides* of the column. This means that the radiation flux along the column, and therefore the pressure opposing gravity, is then less than it would be in an isotropic situation. (An analogous argument may also apply to accretion discs.)

(iii) As has been pointed out by Lamb *et al.* (1973) there are conceivable circumstances when the luminosity may exceed $L_{\rm edd}$ even when the appropriate cross section is $\sigma_{\rm T}$ *and* the acretion is isotropic. This is because $L > L_{\rm edd}$ is merely the condition that infalling matter should be *decelerated*. But unless the total optical depth is sufficiently large, this does not guarantee that radiation pressure can *halt* the accretion. The infalling matter carries momentum across a sphere of radius r at a rate $Mv(r)$, where $v(r)$ is of the order of the free fall speed. If its kinetic energy is converted into radiation at a radius $\sim r_{\rm min}$ the outward momentum flux, ignoring relativistic corrections, is $\sim (M/2c)(v(r_{\rm min}))^2$ (and less, of course, if the conversion efficiency is low). This means that the average photon must undergo more than $2c/v(r_{\rm min})$ scatterings if radiation pressure is to stem the accretion flow (unless the main contribution to the opacity comes from radii $r \gg r_{\rm min}$).

(iv) The Eddington limit is of course irrelevant in an *unsteady* or *explosive* situation: it is, for instance, violated by factors $\sim 10^5$ in supernovae.

It is nevertheless interesting that there are no known objects whose X-ray luminosity greatly exceeds $L_{\rm edd}$ (assuming that the sources have masses of stellar order). If the accretion is $\gtrsim 10\%$ efficient, this implies that the inflow rate is $\lesssim 10^{-7}\ M_\odot$ yr^{-1}. In Cen X3, the *orbital* period is observed to change on a much shorter timescale than 10^7 yr, probably implying rapid mass loss from the companion star. Most of this mass must presumably escape from the system.

3. X-Ray Properties of Her X1 and Cen X3

X-ray observations of Her X1 reveal that it is occulted by a companion for 0.24 days out of every 1.7 days. The X-rays seem to be completely extinguished during the eclipses (at least in the 2–6 keV energy band recorded by Uhuru), and the transition between the eclipse and the high intensity state occupies less than 12 min. The X-rays also display a 1.24 s periodicity. The 'Doppler effect' over the 1.7 day period allows the size of the orbit to be inferred – its diameter is 13.2 cosec i light seconds – and also establishes that this orbit is nearly circular, having an eccentricity $\lesssim 0.05$. This information also determines the mass function to be

$$\frac{M_{opt}^3 \sin^3 i}{(M_x + M_{opt})^2} = 0.85 \, M_\odot, \tag{5}$$

where M_x is the mass of the X-ray source and M_{opt} the mass of the companion star. The companion star of Her X1 has been optically identified, and its light variations are of great interest. Further information is of course required before M_x itself can be deduced, and I shall return to this question later.

Cen X3 has a regular 4.8 s period, and eclipses for ~ 0.49 days out of every 2.087. Its mass function is

$$\frac{M_{opt}^3 \sin^3 i}{(M_x + M_{opt})^2} = 15.4 \, M_\odot. \tag{6}$$

There is, at the time of writing, no firm optical identification for this source.

Her X1 and Cen X3 are clear candidates for systems when the X-ray source may be a neutron star (and it is gratifying that the mass of Her X1 seems to be within the allowable range 0.3–1.6 $M_\odot$ for neutron stars, and that Cen X3 may also have a low mass). The period of Cen X3 is in fact not uncomfortably short for a white dwarf: however, the similarity to Her X1 suggests that the same model is probably applicable in each case. There are several specific observations which can be tentatively explained on the basis of this model. These systems would then resemble pulsars in that a spinning neutron star provides the 'clock'. However the X-ray power radiated cannot derive from rotational kinetic energy – otherwise the rotation would grind to a halt in $\lesssim 10$ yr – but must come instead from accretion. This, as Schwartzman (1971) has pointed out, suggests at least part of the reason why pulsars are not found in binary systems. An isolated spinning neutron star, surrounded only by diffuse interstellar gas, generates the electromagnetically driven relativistic wind which is believed to be a precondition for the coherent pulsed radio emission. When such an object is embedded in a denser environment, the pressure of the relativistic outflow cannot hold the external matter at bay, and we instead get accretion, manifesting itself in the emission of thermal X-rays. One can estimate that Her X1 would have displayed pulsar-like behaviour only if its period were $\lesssim 0.1$ s. (If a rotating white dwarf, whose moment of inertia might be $\gtrsim 10^4$ times larger than that of a typical neutron star, could have a period as short as 1.24 s, its kinetic energy *would* be able to power Her X1 without there being a larger secular change in spin period than is observed.)

3.1. CHANGES IN THE PULSE PERIOD

Since an accreting neutron star is not drawing on its rotational energy as its main power supply, it is not obvious whether its spin rate should slow down or speed up.

An element of gas accreted by the star carries angular momentum corresponding to corotation at the Alfvén radius. This suggests that the spin rate would speed up on a timescale

$$\left|\frac{P}{\dot P}\right| \simeq \frac{M}{\dot M}\left(\frac{r_*}{r_A}\right)^2. \tag{7}$$

It has in fact been found that the period of Cen X3 *decreased* by ~ 3 m s during the time January 1971–September 1972, corresponding to a timescale $P/\dot P$ of only a few thousand years. Even though $M/\dot M \simeq 10^8$ yr, this 'lever-arm' effect certainly allows a speed-up as rapid as that observed in Cen X3. There is, however, a possible opposing effect tending to *brake* the rotation: this is the viscous torque exerted by the accretion disc outside r_A. These two effects can be of the same order of magnitude if

$$\left(\frac{GM}{r_A}\right)^{1/2} \simeq \Omega r_A \tag{8}$$

(and of course if $(2GM/r_A)^{1/2} < \Omega r_A$ it would be energetically possible for material at the inner edge of the disc to be flung out of the system by magnetic forces, leading to a further braking effect). Davidson and Ostriker (1973) suggest that Ω tends asymptotically to a value such that the *net* torque on the neutron star is zero. This value of Ω depends on r_A, which is itself a function of $\dot M$. Therefore, if there were fluctuations in the accretion rate, then Ω would tend to increase (decrease) as $\dot M$ increases (decreases). In the case of Her X1, the period has, on different occasions, been observed both to decrease *and* to increase. The net effect observed over a 15 month interval was a speed-up of ~ 50 μs. If Her X1 were close to this equilibrium state, and the fluctuations in $\dot M$ were small in amplitude, one could perhaps understand why Ω has been observed both to increase and to decrease, and why these changes are slower than in Cen X3.

The effects mentioned above are the dominant ones for causing changes in period. Other effects – for example, the spin-up due to (gradual or sporadic) contraction of the star as it accretes mass – occur on the much slower timescale of $M/\dot M$.

3.2. THE LONG-TERM VARIABILITY OF HER X1

One of the most puzzling properties of Her X1 is that the X-rays are completely extinguished for ~ 24 days out of every 35 (Giacconi *et al.*, 1973b). The source turns on rather abruptly; its mean intensity (when out of eclipse) rises for ~ 4 days; and then the source gradually fades for ~ 7 days. This whole cycle then repeats itself ~ 35 days later. The sharp 'turn-ons' are not strictly periodic. However the data can all be fitted by a model involving an underlying clock with period 34.85 days, plus the additional requirement that the 'turn-ons' always occur around phases 0.2 or 0.7 of the 1.7 day orbital period (which is incommensurable with a 34.85 day cycle). In addition to the eclipses, there is claimed to be a 'dip' which, after the 'turn on' occurs just before the main eclipse, but 'marches' steadily towards earlier orbital phases during the ~ 11 day 'on' period.

There have been many conjectures to explain this peculiar behaviour. In this connection, it is important to bear in mind that optical observations (which will be discussed further in Section 5) impose an important constrain on such suggestions. It appears that the 1.7 day period light variation persist throughouts the 35 day circle with more or less the same amplitude (even though a 35 day periodicity may be discernable in some of the fine details of the light curve (Kurochkin, 1973; Boynton et al., 1973)). Since the thermal inertia of the relevant layers of the companion star is small, this implies that some heating mechanism operates throughout the ~24 days out of ~35 when Uhuru detects no X-rays from Her X1.

3.2.1. *Modulations in Mass Transfer Rate*

One class theory for the 35 day cycle involves supposing that the *mass transfer* is modulated with this period. It seems unlikely that this could be due to some pulsation of the companion star because the expected pulsation periods would be $\ll 35$ days. Another possibility (Pringle, 1973a; Henriksen et al., 1973) is that the spin period of the companion differs by ~5% from the orbital period. If the star displayed some departures from axisymmetry – a 'magnetic spot' associated with an especially vigorous wind for instance – then the transfer rate could vary with a synodic period of 35 days.

Conceivably some kind of feedback process may be operating. McCray (1973) has developed an ingenious model which utilises the fact that the X-ray luminosity is a significant fraction (perhaps ~10%) of L_{edd}. When the X-rays are 'on', the X-ray source behaves with respect to the surrounding gas as though it had a somewhat lower mass. The 'effective' Roche lobe around the companion star might then expand so that material no longer overflowed it. Mass transfer would then cease, and no material would be added to the disc. The disc would then drain away, and the X-ray emission would stop. Mass transfer would then begin again, the disc would be replenished, and so on. McCray speculates that some kind of limit cycle is set up. The time-scale of this cycle would be determined by the length of time for a typical element of gas to spiral inward to the central object. A period of the general order of 35 days would certainly not be unreasonable, but one cannot claim to 'predict' it, because of the wide uncertainty about the efficiency of viscosity in the disc.

A fully developed theory along these lines must also take account of a competing process which might cause *positive* feedback. This arises because the X-rays, by heating the surface layers of the companion star, tend to *raise* the mass transfer rate by increasing the scale height in the atmosphere and/or by stimulating an enhanced stellar wind (Arons, 1973; Basko and Sunyaev, 1973; Alme, 1973). It has in fact been proposed (Lin, 1973) that the 35 day cycle could result from this type of positive feedback if the X-rays stimulate a mass transfer rate which 'overshoots' to such an extent that opacity effects around the compact object quench the X-rays.

At the moment we do not even know whether positive or negative feedback is the more important. A proper theory of the 35 day cycle must also await a fuller understanding of how the X-rays interact with the companion star, and of the factors that determine the residence time of material in the accretion disc.

3.2.2. Processes Occurring in the Accretion Disc

Katz (1973) has suggested that the rim of the accretion disc may not lie in the orbital plane of the system. This might happen if the companion star possessed a component of spin angular momentum which was not aligned with the orbital angular momentum. In this situation, the rim of the disc would precess, and could obscure the X-rays for some fraction of each precession period. To obtain a precession period of 35 days, Katz has to assume that the disc extends outwards to a larger radius than is customarily supposed.

It is also conceivable that the disc might be subject to convective or other instabilities which could cause it to dump material periodically onto the central object. The properties of unsteady accretion discs – in which $\dot{M}$ depends both on r and on t, and (3) no longer holds – have not yet received detailed attention.

3.2.3. Modulation of Inflow from Alfvén Radius

Pines *et al.* (1973) have developed a model according to which the neutron star undergoes free precession in such a way that the angle between the magnetic axis and the plane of the accretion disc varies periodically. When this angle is small, accretion along the 'magnetic funnel' can proceed; but when the magnetic axis points too far out of the plane of the disc accretion is suppressed, and material transferred from the companion accumulates in the disc outside the Alfvén radius. It is not clear how large the precession amplitude would have to be in order for such an 'accretion gate' to operate. However Pines *et al.* list some other reasons why the accretion flow near the Alfvén surface could be sensitive to the orientation of the neutron star's rotation axis, so it is conceivable that a wobble through only a few degrees could suffice. On the basis of this model, Pines *et al.* have attempted to explain the other features of the 35 day cycle. The asymmetry between the sharp rise and the gradual fall in X-ray intensity during the 12 day 'on' period is readily explained. Matter accumulating during the 'off' period will be opaque to the X-rays until it has been photoionized. The fact that the switch-off occurs near orbital phases 0.25 or 0.75 is attributed to the higher density of obscuring matter along the line joining the two stars, which makes it more likely that the first X-rays to be seen will escape perpendicular to this line. The hypothetical 'hot spot' where the gas stream merges with the disc may be thick enough to obscure the X-rays at the phase of the orbit when it lies along our line of sight. The outer radius of the disc would decrease during the 'on' period, and the location of the hot spot would change (it is claimed) in such a way that the dip 'marches' in phase in the matter observed. The apparent tendency of the small ($\lesssim 0.2\%$) amplitude 1.24 s *optical* pulsations (which probably come from gas with cooling time $\lesssim 1.24$ s which is being heated by the pulsed X-rays) to occur at particular orbital phases can also be explained.

3.2.4. Precession of Pencil Beam

Another idea involving precession of the neutron star (Brecher, 1972; Strittmatter *et al.*, 1973) is that the X-rays remain 'on' for the whole 35 day cycle, but that they emerge in a pencil beam which sweeps through our line of sight only for 11 days out of 35. There are some geometrical difficulties associated with this idea. In particular, the broad and relatively smooth observed X-ray pulse profile tells us something about the shape of the beam, and it is hard to reconcile this with the sharp onset of the high state or with the apparent lack of any marked systematic changes in the pulse shape during the 'on' state. A very large wobble amplitude ($\gtrsim 45$ deg) would certainly seem required by this model.

At least in models (a) and (c), the continuous heating of the companion star can only be explained by involving a steady heat source. One possibility (Avni *et al.*, 1973) is that the neutron star emits a steady flux of soft X-rays, powered by the ~ 8 MeV nucleon resulting from nuclear fusion of the accreted matter. This energy is liberated well below the neutron star surface. It is also possible that hear is conducted inwards from the magnetic polar caps, and re-emerges as a steady flux. This emission would not be completely isotropic, because the magnetic field renders the opacity of the crust lower near the magnetic poles. But a serious problem arises with any model in which soft ($\lesssim 0.5$ keV) X-rays play the dominant role in the heating, because these photons (unlike harder X-rays) are absorbed predominantly *above* the photosphere. The associated energy input would then distort the temperature stratification, resulting in the formation of strong emission lines and suppression of the ordinary stellar absorption spectrum (Basko and Sunyaev, 1973; Strittmatter, 1974). It seems more likely that the star HZ Her is heated mainly by *hard* ($\gtrsim 10$ keV) X-rays, though the problem then is the inefficiency resulting from the high albedo (unless one considers photons of $\gtrsim 0.5$ MeV). Heating by fast particles is another possibility. In models (b) and (d), one may suppose that X-rays always hit the companion star even when they cannot propagate along our line of sight (though this requirement places further constraints on the geometry). A more attractive variant of (d) might be to postulate that the star is heated by hard X-rays which are not so strongly beamed as those detected by Uhuru. This is theoretically plausible because the circumstance which might most naturally cause a pencil beam – the reduced scattering cross section for photons travelling along the magnetic field direction – would not be so effective at high photon energies.

35 days is much too short a free precession period for a neutron star with a liquid core. However a neutron star with a *solid* core and the 1.24 s spin period appropriate to Her X1 could plausibly sustain a sufficient deviation from axisymmetry to yield a 35 day precession, and would then automatically be rigid enough to be able to wobble through a large angle. (Mechanisms for exciting this kind of wobble and for sustaining it against damping processes are discussed by Pines (1973).) The question of whether compressed nuclear matter can crystallise at densities $\gtrsim 10^{15}$ gm cm^{-3} is still controversial (see Pandharipande, 1973; Canuto and Cameron, 1973). It will only occur – if at all – in neutron stars with high masses and high central densities, lending added interest to estimates of the mass of Her X1.

In assessing the various models for the 35 day cycle it is of course crucial to know just how regular a phenomenon it really is. It is also relevant that Cen X3 displays extended lows which are apparently *not* strictly periodic. Finally, some explanation is also required for the *very* long ($\gtrsim 10$ yr) time-scale variability inferred from scrutiny of old Harvard plates of Her X1 (Jones *et al.*, 1973) where the 1.7 day optical behaviour changes, implying that whatever agency is responsible for heating the companion star is suppressed. If one were optimistic one might therefore hope that *two* of the possibilities mentioned above might actually be relevant!

4. Cyg X1: A Black Hole?

Cyg X1 is the prime candidate for being an X-ray source involving a black hole. One would expect the accretion disc around a black hole to be subject to various instabilities: thermal instabilities, magnetic instabilities (perhaps analogous to those which Parker has discussed in the context of the interstellar gas in our Galaxy), or perhaps instabilities resulting from irregularities in the mass transfer rate. These could give rise to irregular flickering on all time scales down to the orbital period associated with the most tightly bound stable circular orbits (and perhaps even more rapid fluctuations), but no regular period would be expected. Even if one had no evidence on its mass, one might therefore suspect that the X-rays from Cyg X1 arise from an accretion disc around a black hole. The issue then hinges on the mass of Cyg X1: if this is $\gtrsim 3\ M_\odot$, then it cannot be a stable neutron star or white dwarf, so (assuming that a single compact object is involved) there seems no alternative – at least within the framework of 1973-vintage astrophysical ideas – to the inference that it is a black hole.

The arguments pertaining to the mass have been rehearsed in detail by Giacconi (1973b), and I shall only summarise them here. The first step in this argument depends on the identification of Cyg X1 with the 9th mag., 5.6 day spectroscopic binary HDE 226868. The evidence for this identification is (a) positional agreement to better than 30"; (b) correlations between the X-ray variability and the radio variability (the radio source position agreeing, to better than 1" accuracy, with that of HDE 226868); and (c) recent evidence from the Copernicus satellite for a soft X-ray eclipse near phase zero of the optically-determined 5.6 day period.

The next step concerns the mass of HDE 226868. The mass function determined from optical spectroscopic observations, is

$$\frac{M_x^3 \sin^3 i}{(M_x + M_{opt})^2} = 0.23\ M_\odot \qquad (9)$$

(note that, because there is no regular pulse period, one cannot use X-ray observations to determine a mass function with M_{opt} in the numerator, as was done for Her X1 and Cen X3. On the other hand, the fact that the X-rays from Cyg X1 have a less drastic effect on the companion than seems the case in Her X1 allows the radial velocity of Cyg X1's companion to be determined less ambiguously by optical means). If HDE 226868 were a normal B0 Iab supergiant, its mass would be 15–35 $M_\odot$, and the X-ray source would then be $\gtrsim 6\ M_\odot$. For M_x to be below $2\ M_\odot$ (and $1.7\ M_\odot$ is now the best upper limit to a neutron star mass), one would require HDE 226868 to be below $4\ M_\odot$ even if $i=90°$ (and even lower if $i \lesssim 60°$, as is probably implied by the fact that Uhuru observed no X-ray eclipse). However, as was pointed out by Trimble *et al.* (1973) it is possible for a low mass evolved star (powered by helium burning in a shell) to mimic the spectroscopic appearance of a B0 supergiant. But the star would then have a lower luminosity, and would have to be only ~ 0.5 kpc away. There was, until recently, no very firm evidence on the distance, but Margon *et al.* (1973) have now determined the reddening-vs-distance relation for 50 main sequence stars in the same field. They find a good correlation out to $\gtrsim 2$ kpc. The fact that HDE 226868 displays as much reddening as any of the other stars in the sample thus implies that it is at least ~ 2 kpc away; and this convincingly rules out the possibility of its being an evolved low-mass star.

The natural conclusion seems therefore to be that Cyg X1 involves a black hole of $\gtrsim 6\ M_\odot$. This conclusion could be evaded only if one could devise a plausible model for the source which did not involve a compact object at all. One such possibility (Fabian *et al.*, 1974) involves supposing that the companion of HDE 226868 is itself double: consisting of a neutron star (the source of the X-rays) orbiting a main sequence star of $\gtrsim 6\ M_\odot$. It is the evidence for compactness provided by the rapid X-ray variability which makes Cyg X1 a firmer black hole candidate than the invisible high-mass components of other single-line spectroscopic binaries (ε Aur, for instance). In the latter systems, it is hard to exclude the possibility that, for example, an ordinary star is shrouded by dust.

It is crucially important to determine the shortest timescale on which Cyg X1 varies. (This is an ideal rocket experiment: although a large collecting area is plainly advantageous, 1–2 min of observation should be quite sufficient). Sunyaev (1973) proposed that attempts should be made to search for X-ray pulse trains due to regions of enhanced emissivity orbiting the hole. The typical orbital periods would be $\sim 0.6\ (M/M_\odot)$ ms for the innermost stable orbit around a Schwarzschild black hole, but ~ 8 times faster if the black hole had a maximal Kerr metric whose angular momentum was aligned with the disc, but with the same mass. Sunyaev envisaged this test as a method of diagnosing the metric around the black hole. However it seems quite possible that one would get pulse trains emitted from the region of *unstable* orbits in a Schwarzschild geometry. Thus the discovery of pulses of (say) $0.1 \times (M/M_\odot)$ ms period would not necessarily prove that the metric was close to 'maximal Kerr'. (Further interesting complications involving precession of the disc can occur if the black hole is obliquely oriented relative to the angular momentum vector of accreted material).

More would be learned if an X-ray spectral feature originating in the disc could be discovered and its profile measured, but this seems unlikely to be feasible before 1980. It is important to remember that black holes are a consequence of almost all 'viable' theories of gravity, and much further work is needed before one can diagnose whether the properties of a given black hole agree better with those expected on the basis of general relativity than with the predictions of a rival theory. Nevertheless, the discovery of black holes – objects where gravity is so overwhelmingly strong that it dominates all other effects – opens the way to testing some of the most crucial and remarkable predictions of Einstein's theory, and will surely have a massive impact on gravitational physics.

5. Optical Properties

The interpretation of the optical properties of X-ray binaries raises a whole range of problems. (See Bahcall and Bahcall (1973) for a survey of the observational data.) X-ray heating causes the side of the star facing the compact object to be hotter and brighter than the eclipsed side. A quantitative understanding of this effect involves detailed computations (along the lines of those already done by Arons (1973) and Basko and Sunyaev (1973) of the structure of a stellar atmosphere irradiated by X-rays). A second quite different effect which leads to optical variations with *half* the orbital period arises from the distortion of the companion star by the compact object's gravitational field. Interpretation of actual light curves is complicated by further effects (emission by gas streams, radiation and absorption by the accretion disc itself, etc.) and one suspects that detailed model-building may prove somewhat fruitless unless some very clear-cut correlations between X-ray and optical variability are found.

Her X1 is the system where the X-ray luminosity is highest relative to the intrinsic luminosity of the companion star (which is a late A or early F type main sequence star). It is thus a system where X-ray heating (or heating by some other radiation flux emanating from the compact object) is a dominant effect, being sufficient to make the 'hot side' of the star ~ 10 times more luminous than the unheated side; and the effects of gravitational distortion are relatively minor. The actual light curve, however, is not 'flat-bottomed', implying that the X-ray source is having some observable effect during some of the time when it is eclipsed. This may mean that some of the X-rays are absorbed above the photosphere and reradiated, or else that there is significant emission from an extended disc around the X-ray source. As already mentioned in Section 3, the persistence of the 1.7 day optical variations throughout the 35 day cycle suggests that a 'steady heat source' is operative in addition to the X-rays seen by Uhuru. This conclusion is also supported by simple energetic arguments, which suggest that the excess light emitted from the hot side of the companion star involves more energy than the fraction of 2–6 keV X-rays intercepted by the star.

Because of the drastic perturbing effects of the X-rays a given spectral line is not emitted uniformly over the star's surface, so it is difficult to interpret the radial velocity measurements of the companion star of Her X1. This means that one cannot readily obtain the second equation which, in conjunction with (5), would allow the mass of the X-ray source to be determined. The best estimates, however, suggest $M_x \simeq$ $\simeq 1\ M_\odot$ and $M_{opt} \simeq 2\ M_\odot$. Other methods of determining the mass M_x – which are very uncertain, but yield consistent results – involve using the spectral classification of the companion star, or assuming that the Roche lobe is filled.

In Cyg X1, where the companion star is much more luminous relative to the X-ray source than is the case for Her X1, the heating augments the stellar luminosity by only $\sim 2\%$, and the effects of gravitational distortion are more important. Attempts have been made to derive an independent mass estimate from the theory of this effect, but these are somewhat unreliable. The other known X-ray binaries – for instance, 3U 1700–37, Vela XR1 and SMC X1 – seem to resemble Cyg X1 rather than Her X1, in that the companion star is a highly luminous supergiant, and X-ray heating effects are *not* important.

The only other case where X-ray heating may be important is the peculiar source Cyg X3 (which first attracted attention because of the spectacularly strong radio outburst which it underwent in September 1972). This object has a 4.8 h period, which should probably be interpreted as an orbital period during which the X-ray intensity changes by a factor ~ 2. No optical counterpart has been found, presumably because it is in a strongly obscured region, but synchronous 2.2 variations of $\sim 15\%$ amplitude have been reported (Becklin et al., 1973). The gradual and incomplete character of the X-ray eclipses suggests that in this system the eclipse is caused not by the surface of the companion star, but by scattering and absorption in a strong wind. The observed infrared variations imply that the relevant layer of the heated side of the companion star has a temperature $\gtrsim 10^6$ K. This, however, is quite possible if one is seeing emission from the wind, which is heated to this temperature (Pringle, 1973b).

Although Cyg X3 has a shorter period than the other X-ray binaries, the period is longer than that of systems such as DQ Her. It may differ from such systems merely in having a neutron star (or black hole) as the compact component, instead of this being a white dwarf.

The occurrence of X-ray heating sets a rough *lower limit* to the apparent brightness of the optical counterpart for any eclipsing X-ray source. If there were no interstellar absorption, any X-ray source with an intensity of C Uhuru counts (Her X1 is ~ 100 in these units) which is observed to eclipse for a fraction f of every period should have an optical apparent magnitude

$$m \simeq 15 - 2.5 \left\{ \log\left(\frac{C}{10}\right) + 2 \log (4f) \right\}. \tag{10}$$

Thus any eclipsing source in the Uhuru Catalogue would be optically identifiable were it not for the often severe effects of interstellar extinction.

6. Concluding Remarks

Many interesting and important aspects of binary X-ray sources have not even – for reasons of time – been touched on in the foregoing remarks. In particular, I have said almost nothing about the radio observations, which are certainly not yet adequately explained. However the fact that – even in the extreme case of Cyg X3 at the peak of its radio flare – the radio luminosity is a tiny fraction of the X-ray output, suggests that to concern ourselves with the details of the radio variability may be as premature as it would be to worry about solar flares before understanding the basic elements of stellar structure. Moreover, some binary systems containing two relatively normal stars have similar radio properties and this suggests that the radio behaviour, fascinating and puzzling though it may be, is unlikely to be intimately connected to the compact object itself.

The existence of these close binary systems with compact components raises many astrophysical questions. How do they fit into the general scheme of binary star evolution? How do they evolve to their present state and what role did mass transfer play during their earlier history? How did they avoid disrupting during the catastrophe which formed the collapsed component? Why, nevertheless are there only $\lesssim 100$ such systems in the Galaxy? What will be their eventual fate? – for example, what happens if a neutron star accretes so much material that it comes to exceed the limiting mass; or what happens when, later in its evolution, the companion star swells up and engulfs the compact object? What are their relations to pulsars, and do we really understand why there are no ordinary pulsars observed in binary systems? How does the famous source Sco X1 fit into this pattern?

This is such a new topic that the theorists have a rather better excuse than usual for raising questions instead of providing answers. Let us hope, however, that some of the speculative ideas already proposed are transmuted into 'solid' theoretical models before the theorists are distracted by something even more exciting. We can certainly expect much new data from the next generation of X-ray satellites and from more refined optical observations. It should be feasible to detect X-rays from this type of source in nearby galaxies, thereby acquiring a larger sample for statistical analysis, which will allow us to check the theory of black holes and neutron stars against observations in many key respects.

References

Alme, M.: 1973, paper presented at conference on *Physics and Astrophysics of Compact Objects*. Cambridge, England.
Arons, J.: 1973, *Astrophys. J.* **184**, 539.
Avni, Y., Bahcall, J. N., Joss, P. C., Bahcall, N. A., Lamb, F. K., Pethick, C. J., and Pines, D.: 1973, *Nature Phys. Sci.* **246**, 36.
Bahcall, J. N. and Bahcall, N. A.: 1973, *Proc. 16th Solvay Conf.*, in press.
Bahcall, J. N., Kulsrud, R. M., and Rosenbluth, M. N.: 1973, *Nature* **243**, 27.
Basko, M. M. and Sunyaev, R. A.: 1973, *Astrophys. Space Sci.* **23**, 117.
Becklin, E. E., Neugebauer, G., Hawkins, F. J., Mason, K. O., Sanford, P. W., Matthews, K., and Wynne-Williams, G. C.: 1973, *Nature* **245**, 302.
Bondi, H.: 1952, *Monthly Notices Roy. Astron. Soc.* **112**, 195.
Boynton, P. E., Canterna, R., Crosa, L., Deeter, J., and Gerend, D.: 1973, *Astrophys. J.* **186**, 617.
Brecher, K.: 1972, *Nature* **239**, 325.
Brecher, K and Morrison, P.: 1973, *Astrophys J. Letters* **180**, L107.
Buff, J. and McCray, R. A.: 1974, *Astrophys J.*, in press.
Burbidge, G. R.: 1972 *Comm. Astrophys. Space Phys.* **4**, 105.
Cameron, A. G. W. and Canuto, V.: 1973 *Proc. 16th Solvay Conf.*, in press.
Davidson, K.: 1973, *Nature Phys. Sci.* **246**, 1.
Davidson, K. and Ostriker, J. P.: 1973, *Astrophys J.* **179**, 585.
Fabian, A. C., Pringle, J. E., and Whelan, J. A. J.: 1974, *Nature* **247**, 351.

Faulkner, J.: 1971, *Astrophys J. Letters* **170**, L99.
Felten, J. E. and Rees, M. J.: 1972, *Astron. Astrophys.* **17**, 226.
Giacconi, R.: 1973a, in C. DeWitt (ed.), 'Gravitational Radiation and Gravitational Collapse', *IAU Symp.* **64**, 147.
Giacconi, R.: 1973b, *Proc. 16th Solvay Conf.*, in press.
Giacconi, R., Gursky, H., Kellogg, E., Levinson, R., Schreier, E., and Tananbaum, H.: 1973a, *Astrophys J.* **184**, 227.
Giacconi, R., Murray, S., Gursky, H., Kellog, E., Matilsky, T., Koch, D., and Tananbaum, H.: 1973b, 'The Third Uhuru Catalogue', in press.
Gnedin, Y. N. and Sunyaev, R. A.: 1973, *Astron. Astrophys.* **25**, 233.
Hayakawa, S. and Matsouko, M.: 1964, *Prog. Theor. Phys. Suppl.* **30**, 204.
Henriksen, R. N., Reinhardt, M., and Aschenbach, B.: 1973, *Astron. Astrophys.* **28**, 47.
Jones, C. A., Forman, W., and Liller, W.: 1973, *Bull. Am. Astron. Soc.* **5**, 32.
Katz, J. I.: 1973, *Nature Phys. Sci.* **246**, 87.
Kurochkin, N. E.: 1973, *Inform. Bull. Var. Stars* No. 55.
Lamb, F. K., Pethick, C. J., and Pines, D.: 1973, *Astrophys. J.* **184**, 271.
Lin, D. C.: 1973, *Astron. Astrophys.* **29**, 109.
Lynden-Bell, D.: 1969, *Nature* **223**, 690.
Margon, B. and Ostriker, J. P.: 1973, *Astrophys. J.* **186**, 91.
Margon, B., Bowyer, S., and Stone, R.: 1973, *Astrophys. J. Letters* **185**, L113.
McCray, R. A.: 1973, *Nature Phys. Sci.* **243**, 94.
Mestel, L.: 1954, *Monthly Notices Roy. Astron. Soc.* **114**, 437.
Mock, J.: 1968, Ph.D. Thesis, Columbia University.
Novikov, I. D. and Thorne, K. S.: 1973, in C. De Witt and B. DeWitt (eds.), *Black Holes*, Gordon & Breach, p. 343.
Ostriker, J. P., Rees, M. J., and Silk, J. I.: 1970, *Astrophys. Letters* **6**, 179.
Paczyński, B.: 1971, *Ann. Rev. Astron. Astrophys.* **9**, 183.
Pandharipande, V. R.: 1973, *Proc. 16th Solvay Conf.*, in press.
Pines, D.: 1973, *Proc. 16th Solvay Conf.*, in press.
Pines, D., Lamb, F. K., and Pethick, C. J.: 1973, *Proc. N.Y. Acad. Sci.*, in press.
Prendergast, K. H. and Burbidge, G. R.: 1968, *Astrophys J. Letters* **151**, L83.
Pringle, J. E.: 1973a, *Nature Phys. Sci.* **243**, 90.
Pringle, J. E.: 1973b, *Nature*, in press.
Pringle, J. E. and Rees, M. J.: 1972, *Astron. Astrophys.* **21**, 1.
Schwartzman, V. F.: 1971, *Soviet Astron.* **15**, 377.
Shakura, N. I. and Sunyaev, R. A.: 1973, *Astron. Astrophys.* **24**, 337.
Shapiro, S. L.: 1973a, *Astrophys. J.* **180**, 531.
Shapiro, S. L.: 1973b, *Astrophys. J.* **185**, 69.
Strittmatter, P. A.: 1974, *Astron. Astrophys.*, in press.
Strittmatter, P. A., Scott, J., Whelan, J., Wickramasinghe, D. T., and Woolf, H. J.: 1973, *Astron. Astrophys.* **25**, 275.
Sunyaev, R. A.: 1973, *Soviet Astron.* **16**, 941.
Trimble, V. L., Rose, W. K., and Weber, J.: 1973, *Monthly Notices Roy. Astron. Soc.* **162**, in press.
Zel'dovich, Y. B. and Guseynov, O. K.: 1965, *Astrophys. J.* **144**, 840.

THE NEXT DECADE IN STELLAR ATMOSPHERES THEORY

P. DELACHE

Observatoire de Nice, Nice, France

It is a pity for all of us that Dimitri Mihalas was unable to come to Sydney, and I wish, in the first place, to thank him for suggesting this discussion, devoted to a prospective study of our research field. There falls to me the task of trying to sketch what might be the axis of progress in stellar atmosphere theory. This subject is pre-eminently one which belongs to D. Mihalas; I am afraid that we are going to miss him now more than ever. Furthermore, I feel myself in a bad position for talking on such a large subject, in front of specialists, while I know so little on it. However, I have accepted the task, at the request of the President of Commission 36, with the goal of giving some physical considerations that may initiate reflections and discussions.

I am not going, in principle, to give a set of topics whose study should be desirable, or possible, arranged according to the internal logic of a specialist of stellar atmospheres, neither shall I look for bibliographical signs allowing the prediction of active zones or flares in our next cycle of activity!

My physical reflections will be based first on a small historical background. This will lead me to consider in the beginning the classical aspect, somewhat restricted, of the theory of stellar atmospheres. We shall find there some well known needs for classical, 'first order', extensions of our competences. In a second part, I shall try to encircle what could be the further extensions of our field, through the two questions:

(i) If it is possible to produce a broad definition of the objects and phenomena that we are to study, what are the competences that we should have, or develop, in order to succeed?

(ii) If we define ourselves by our knowledge, in a wide sense, what are the new problems that we expect to be faced with?

1. Restricted Stellar Atmosphere Theory

The first steps in the understanding of radiation coming from the stars have been possible only after the appearance of radiation physics, that is to say, atomic spectroscopy for the atomic point of view, and thermodynamics of the radiation field and statistical physics for the description of interaction. The synthesis of all that for astrophysical purposes is classically known as radiative transfer theory. It is *uncoupled* from any fluid dynamics, but sometimes it is used in conjunction with the hydrostatic law, and also with some kinematical parameters (the so-called 'velocity field') in order to produce model atmospheres.

Let me make some remarks on radiative transfer theory and on its prospects. I consider that radiative transfer has had bad luck in large portions of its history, and that this fate is still hanging on it. After the two fundamental concepts of source function and optical depth had been worked out, it appeared immediately a very interesting, one-dimensional mathematical problem. First drama, since the physics disappears, swamped by formal flows for years. The LTE vs non-LTE problems could have been posed a long time ago, if the mathematician who is slumbering in every theoretician did not wake up so often! Second drama: the advent of departures from LTE comes just before the advent of computers. The fact that there is no new physics, but simply the recognition that collisions may not dominate the radiative processes and have to be calculated in a self-consistent way, this fact is screened by the quasi-hermetic presentation of the literature. It happened then that, in order to be convinced, most people needed to see good, fully treated examples. But this means going through heavy numerical calculations with a high degree of sophistication. We observe then a tendency to trust FORTRAN more than 'Physics of the Solar Chromosphere', with the result that departure from LTE will be synonymous with machinery open to a few specialists, and too complicated to be operated by oneself. If this difficulty is to remain in the near future, we must be careful to avoid the three following traps:

(i) continue to improve the quality of observations and data reductions, and give them to modelists who do not ask for them.

(ii) take a set of ready-made models and try to fit pieces of data without questioning the legitimacy of the procedure.

(iii) in some cases be satisfied with the internal consistency of LTE calculations without performing the actual non-LTE analysis. Please excuse me if I find it necessary to develop this point which is a question of methodology that arises in other frames, especially in the case of velocity fields in the outer regions of an atmosphere.

You start with a physical description of your medium which includes all the basic phenomena that you want to take into account. Let me take an illustrative simple example in which you assume that in a plane parallel, or spherical geometry the following relations hold:

- statistical equilibrium,
- transfer equation – energy equation,
- momentum equation,
- mass conservation equation.

As you find the problem to be too hard, you make some simplifying hypothesis, e.g. collisions are always dominant (LTE), or velocity=0 (static atmosphere). *But those hypotheses cannot be considered as approximations*, since they suppress at the same time one unknown and one equation (this is readily seen on the static hypothesis example in which the mass conservation equation reduces to $0=0$). Then one arrives at a well posed problem, as far as the initial one was well posed. It is then a cheap satisfaction to verify the internal consistency of one's results. Let me be even more specific on the expansion example: the fact that you are not free to choose $v=0$ is illustrated by the well known result in stellar wind theories that as soon as the static constraint is relaxed, you find the value of v through the so-called criticality condition *which is not an extra parameter with which you may play*. Some interesting problems may arise if we cease to think in terms of non-expanding atmospheres. Let me propose one by the way at the end of this digression.

Suppose that we know that for some reason there is a temperature rise in the outer part of an atmosphere, as is the case, for example, in the Sun. Suppose also that the variation of the temperature is such that it sweeps the region of thermal instability of the hydrogen plasma. In a static regime, one finds immediately the necessary existence of a plateau, followed by a steep rise, as was suggested a long time ago in the solar case. The temperature gradient will then be limited by such things as thermal conduction, and you arrive then at a model of your transition region. But sometimes you go further, and try to study the stability of structures (essentially horizontal), with the hope of finding the source of chromospheric inhomogeneities. My point is the following: if there are instabilities which can develop horizontally and/or if there exists a tenuous region, controlled by conduction, but still thermally unstable, is it not a sign that the local conditions impose a motion, and should we not, in the first place, study the equilibrium situation? We have the greatest chances to find a steady expanding state, in which the instability develops itself smoothly along the outward motion. Only then should we study the behaviour of horizontal perturbations. Put in another way: the existence of an outward temperature rise gives us the necessary ingredients for a thermodynamic machine without violating the second principle. There is no proof that the machine will start off by itself: there is no proof either that it will produce the simplest type of motion, namely steady radial expansion; but maybe we should start sophisticated theories on better grounds, which do not assert a static atmosphere from the beginning, even if it looks satisfactory, and selfconsistent.

Going back to my topic on radiative transfer, let me finish the diagnostic by mentioning the two major extra ingredients that are not at all satisfactory in my opinion: what is called turbulence (either micro-, or macro-), and inhomogeneities. I call them extra ingredients, because they find their physical origin outside the scope of classical stellar atmosphere theory; their description is purely phenomenological, and their role is in most cases reduced to a convenient adjustment of 'free' parameters, with which you can play, sometimes as a virtuoso. Let me be clear: I do not object at all to the use of an extra parameter in order to represent observations, that cannot be interpreted otherwise; I do not object either to the building of model atmosphere that includes formally a source of line broadening of unknown origin. What does not satisfy me, is the fact that our physical understanding of these extra pieces seems to end where it should begin: when they are given a christian name!

Replacing the microturbulence velocity by another local parameter such as the velocity gradient is not better, and in a sense even worse, since microturbulence can be easily incorporated in the procedures of reduction of the spectra, which is not the case for a velocity gradient. Furthermore, the question of the number of free parameters is somewhat misleading: it is true that the physical constraints such as mass conservation prevent the use of arbitrary variations in velocity gradients; but it is only due to our lack of understanding the physics that we think that we can play at leisure with variations of the microturbulence.

One sees in those examples that what I have called classical stellar atmosphere theory must be extended, and should include some hydrodynamics, which is already done by some people, but there are very few studies in which hydrodynamics and radiative transfer are really *coupled*. This subject will be considered in the second part of my talk. For the present time I shall try to bring up some partial conclusions on what can make progress in this restricted field.

(i) The classical calculation of one-dimensional non-LTE models that need big computers and good analysts has still many things to tell us, particularly on the abundances, but not only there. Let me quote a correspondence by D. Mihalas:

"I think that (this kind of calculations) should be pursued vigorously for a wide variety of atoms and ions in as wide a range of stellar temperatures and gravities as possible. ... it will pay enormous rewards both in re-evaluation of the whole 'abundance' question and in delimiting the regions where future work is urgently needed. I am certain that several platoons of graduate students would find this area to be a fruitful source of thesis problems. ... Milkey and Johnson have done some interesting work on the O I lines in 6500° giants and supergiants. They find that much (though apparently not all) of the luminosity effect arises from departures from LTE. ... the LTE abundance required to fit the NLTE computed equivalent width was up by a factor of 1000, or ... the equivalent effect would require 4 km s^{-1} microturbulence (all spurious) to simulate."

From the modelist's point of view, there is no basic conceptual difficulty in introducing the following improvements of the physics: transfer in the presence of a magnetic field and polarisation of the transfered radiation, departure from LTE in molecules, partial coherency of the diffusion. It is however difficult to progress in these directions for various reasons: the handling of the whole set of Stokes parameters is very heavy, and will be reserved to a very few people. Still, the polarization observations are improving, together with the recognition of the importance of magnetic stars, and of large scale magnetic structures on rotating stars. Concerning partial coherency, and molecular structure, we must turn to our usual companions working on atomic physics. It should not be difficult to convince them of the interest of those topics. If the chromosphere problem is really a basic one – as most of us are convinced – we must recognize that we do not understand in enough detail what we see in the H and K lines. Among other open questions, is the one of the scattering process. The discovery of the importance of molecules, essentially CN and CO, in the solar atmosphere as probes of the minimum temperature region and the advent of stellar ultra violet and infrared observations with good spectral resolution are signs of the predictable importance of molecular lines in the future. We are going to need detailed configurations and good radiative and collisional parameters. So we see that the future of classical radiative transfer is wide, worth working on, but *becoming more and more difficult*.

(ii) Some words can be said on the phenomenological approach of the velocity fields problems in connection with radiative transfer. First, if we are dealing with fields that are described in a purely kinematical and deterministic way, there are no conceptual difficulties, since we simply deal with a given variation of the absorption coefficient. However, original methods can be worked out for special cases, in order for example, to see what is the gross effect of a velocity gradient on the line shapes. The use of computers might not be necessary to give us new physical 'feelings' on this type of situation. On the contrary, if the velocity field is basically stochastic, it is most probable that the microturbulence velocity is not the right parameter. Radiative transfer in stochastic media is basically *a problem of physics of turbulence*, before being a problem of radiative transfer. It is likely that the concept of correlation length is a better tool than microturbulence velocity alone, as it can be a link between stochastic hydrodynamics and transfer. Sets of curves of growth in terms of correlations lengths have already been produced in LTE. As expected, they mimic the microturbulent result in the small correlation length limit, and the macroturbulent result in the large length limit. But there exist an infinity of intermediate states that are physically sound which shows that this approach can be of great interest. Since I have just mentioned the curve of growth, which has proved to be such a high value tool in LTE analysis, let me ask a simple question: does there exist an extension, even approximate, or empirically computed, of curves of growth theories to non-LTE diagnostic? (Athay and Skumanich have treated the case of coherent scattering where the source function can be computed exactly).

(iii) Another extension of radiative transfer theory has started to show up, and should develop rapidly: the treatment of geometries far from the plane parallel case. First in a spherical atmosphere, or shell, and second in wholly inhomogeneous structures. This latter problem is essential in the solar chromosphere, and most probably in stellar chromospheres too. To what extent are we making fundamentally serious errors in the diagnostic procedure using homogeneous models? The answer will be eventually given by a careful comparison of homogeneous models giving some kind of 'mean' information with the ultimate inhomogeneous model of the same object. At the present time, the Sun is the only star in which this study can be done, but obviously this remark shows why it is still important to work also on a 'mean solar chromosphere' even if the solar physicists do not share this opinion! Quoting again D. Mihalas: "In fact we may validate (or invalidate) almost all of stellar spectroscopy by knowing the answer to such questions."

It is in place to mention here the extension of radiative transfer in the time dependent case. It can be shown that a quasi-static description of evolving structures can be erroneous in some cases, for example in chromospheric spicules, or in large prestellar contracting clouds. The available literature on neutron transfer can help to understand this kind of problems.

(iv) Less predictable, but highly desirable, is the advent of new useful physical concepts, which will simplify, if not the exact calculations, at least the understanding and permit approximate approach of non-LTE transfer. I refer to the various probabilistic treatments that appeared recently, and also to such things as R. N. Thomas's 'temperature control bracket'. If we could have some new tools that are *at the same time physically meaningful, and simple to operate,* so that a 'do-it-yourself' kit for non-LTE radiative transfer can be put on the market, much progress could be made in what I call the extended theory of stellar atmosphere, in which radiative transfer is an ingredient among others. Of course, we would be satisfied by a kit which does not give exact results, but only a sufficient degree of approximation.

2. Extended Stellar Atmosphere Theory

In the first part of this talk, I limited myself to the classical, restricted stellar atmosphere theory. We were faced with the necessity of including some velocity field theory (or theories). I do not share completely the opinion of the President of Commision 36 about the fundamental need of mass flows, and I shall come back to this point in a few moments, but I salute his unifying work which has permitted clarification and sorting of an impressive bibliography along directions that are very strongly defended. His Report might be arguable, but its content is of a very convenient use, and I am sure all of us are grateful to him for this piece of work.

But why should we limit ourselves to the inclusion of hydrodynamics textbooks in our bedside books? I would like to ask a more general question on the personal status of a specialist in Stellar Atmosphere Theory, who is faced at the same time with astronomical observations and facts, and with a whole batch of physical knowledge. Since the radiation that we receive from the stars, and from interstellar matter gives us information not only on the parameters of classical thermodynamics (temperature, density, chemical composition), but also on data whose physical origin lies elsewhere (gravities, velocity fields, magnetic fields), it is impossible not to wish to incorporate as much of this external physics as possible. We arrive then at the two possible approaches that I outlined in my introduction.

(1) If it is possible to produce a wide definition of the objects and phenomena that we are to study, what are the fields of General Physics that we will be using?

It is a very ambitious project to look for a physical unity, for a continuity, among all the objects, all the zones in which our extended ability should apply. This attempt has been recently made by R. N. Thomas, in his IAU Report, and elsewhere.

In brief, let me summarize what are the leading ideas, as I understand them: we must recognize the following facts: we do observe mass motions, mass exchanges, either directly in the outer layers where for example organized velocities can be seen through line shifts, or less directly in inner layers where, as we have seen previously, we are forced to include kinematical descriptions of the velocity field in modeling the atmospheres of stars. Furthermore in many cases, the theories which could explain the observations of mass flows are lacking. So, to put it simply, we are forced to consider the mass motions as well as the photon motions; i.e. as well as radiative transport. Now the question is: is the analogy meaningful, and are those two types of flux of the same basic importance, in other words:

(i) is this unification well founded, or is it not conveying with it some dangerous simplifications?

(ii) is it useful for the progress of our knowledge?

On the second point, my impression is that, yes, such a tentative approach is very useful, since it makes us think in new terms and in a critical way about physical situations we are familiar with. As a result, we may deepen our physical understanding of interrelations between various parts of a stellar atmosphere, and perhaps discover new theoretical features (as the chromosphere-corona transition zone was theoretically discovered, and the solar wind theoretically understood). As we say in French: 'de la discussion jaillit la lumière'. I am convinced that there will be discussions, but I shall not attempt to predict where the flashes will appear!!

On the first question, namely whether the idea that the mass flow concept plays a role analogous to radiative energy flow is well-founded, I have some comments which, hopefully, may help to define our status vs the General Physics.

Let me state first that I do not think that mass flow is of such a fundamental character as radiative energy flow, and try to explain why: we have a star which is a hot body surrounded by a cool vacuum. This body has at least its internal energy content to release, but it also has generally some energy source (nuclear reactions, contraction in a quasi static description, etc., ...). There is no screen between the star and the vacuum that can prevent radiative energy exchanges, so an irreversible flow sets up, increasing the entropy of the whole system. On the contrary, mass is bounded by the gravitational field, and is not flowing simply because there is a decreasing density outwards. One might argue that, even from a solid surface, evaporation takes place. But this is essentially a microscopic effect, due to the fact that in the microscopic velocity distribution function, there is a certain number of particles running faster than the escape velocity. It simply shows that the gravitational screen is not perfect, but it is in no way comparable to the first order effect of radiative energy flow. (I call it first order because it can be handled with the classical thermodynamics tools.)

Now comes a question: is it possible that a mass flow, even if it is not a direct consequence of the simplified boundary conditions that I have described, always sets up? I think that the answer is in general unknown: as I mentioned in the last paragraph, we have, all the way through our irreversible system, the necessary condition for a thermodynamic machine to work, namely contacts (through radiation) between two sources of different temperatures. Depending upon the type of machine, it may work or not. For example, if you think of convection, then you compare the temperature gradient to the so-called adiabatic gradient, and you work out the details. There is no proof that in general we are going to generate mass motions, and furthermore that these mass motions will induce a flow. This can be seen in the solar example where the convection zone generates non thermal energy, which travels through the photosphere and dissipates somewhere in the chromosphere and corona. It is then this hot base of the corona which, connected to the vacuum interstellar spaces, is at the origin of the solar wind. We see that the Sun's mass flow is the result of two thermodynamic machines, working in two different ways. Another type of argument could be put in favor of mass flows: the outgoing radiation transports with itself a certain amount of momentum; most probably, the non-radiative energy generated in convection zones is also accompanied by some momentum. When those flowing energies interact with matter, they transfer part of their momentum. This is called radiation pressure in the case of electromagnetic or acoustic radiation. One should not draw the conclusion that this transfer of momentum will cause a flow of matter. Again, it is counterbalanced by the gravitational attraction; the complete answer comes only from the study of the whole steady state situation.

To summarize, let me sketch again how we differ in understanding how the stellar machinery is working:

First, we start with an adiabatic star in LTE, in which the temperature and density drop to zero at the surface. The LTE assumption works as a screen for escaping radiation, and the gravity is the screen against outflow. Apparently at the surface our screens are not perfect exactly in the same manner: within a mean free path, for the photons as well as for the atoms. But the radiative screen is wide open, at the surface (half of the photons escape) whereas the gravitational leak is small.

So our star radiates, and slightly evaporates. Of course there will be a point, far away, where most of the particles will be evaporating, as in a planetary atmosphere. Then, things can stay like that, or the modifications induced by the radiation field in the temperature structure may create motions with some more or less high value of internal kinetic energy storage in hydrodynamics modes. As a consequence, the amount of energy eventually transported and dissipated in the atmosphere may lead to important expansion bringing the point of sensible mass motions towards the photospheric layers; or the basic hydrodynamic modes may affect directly the photosphere.

One sees that I have tried to distinguish effects from causes. If it happens that an effect can react upon the cause, then indeed we shall need a self-consistent treatment, and we may not have the right to disentangle what creates what. We arrive then at a picture of a Thomas-type. I simply do not think that it is the rule; and I would have prefered the use of the words 'hydrodynamic forms of energy-storage and flow' to the somewhat misleading term 'mass flux' or 'mass flow'.

I apologize for having spent some time on this picture; but as I have said before, this is a kind of reflection which help to understand better the underlying ideas that are concealed in one's mind.

Several conclusions can now be drawn:
- Mass flow, as a general expansion of the atmosphere, or mass motions, that is matter moving along some kind of pattern, may exist as consequences of the irreversible radiative energy flow. That they always exist remains to be proved (or observed).
- The existence of mass exchange between two different places in the atmosphere is not a simple question, and its solution may necessitate the solution of the steady state of the whole atmosphere.
- It remains that we should be careful to consider, especially in the outer parts of an atmosphere, the possibility of 'material links', that can convey with them kinematical energy, enthalpy flux, abundances peculiarities, patterns, etc., ...

It will be a good thing, if we cease to think in terms of 'zones' more or less arbitrarily defined, connected through 'interfaces' that only the radiation and some acoustical or MHD waves can cross. Moreover, not only can the interfaces be crossed by matter, but they are not necessarily defining boundary conditions, as if one was going always from the causes to the effects in the way out from the center of the star.

I have not given a general answer to my initial question, but I arrive at the conclusion that by trying a unifying description of Stellar Atmosphere Theory, we recognize the necessity of coupling the restricted theory with hydrodynamics, MHD, etc.,..., we recognize the interest of looking at problems with new eyes, we recognize the importance of thermodynamical arguments, and that thermodynamics of irreversible processes should be included in our tools.

But, at the same time, I have personally the impression that I risk becoming dogmatic. If every problem that we are going to encounter in the future were to fall in a large frame or in another that we define today, I am pessimistic; because it nearly means that we will become technicians operating a large factory, with various skills enabling one or the other to treat such and such a topic, leading to a construction, complicated, but with no failure in it, with a final point in perspective.

The second question, even if it is less ambitious at first sight, is fortunately more optimistic. May I add also that I have found it to be more convenient to classify some points of future work?

(2) Let us try to define our physical competences, and imagine what will be the progresses that we are going to make in using them.

One can say that we are specialized in physics of low density matter: our gases are perfect, first order development of statistical physics is satisfactory in most cases, our fields (gravitational as well as electromagnetic) are mild, and we have nearly always direct observations, in one wavelength or in another, of the object that we study. We apply radiative transfer techniques, classical hydrodynamics, and MHD, and we are prepared to incorporate turbulence, and non-equilibrium thermodynamics.

We can first look into progresses expected from extrapolation of our present work. We already met numerous topics, but I think that we have missed some of them, and first let me come back to the abundance problem, disconnected from the modeling process. We do not want good values of abundances per se, or as a product that we deliver to our colleagues working on evolution or nucleosynthesis. We may find interesting problems inside our atmospheres:

- I do not think that the iron solar abundance is a closed question; different values from the photosphere up to the corona are still inside the errors bars, with still a tendancy towards higher abundance in the corona.
- Variation of abundances inside a single star either real, or due to large scale inhomogeneities, seems to be observed.
- The possibility of sorting the elements inside an atmosphere as studied by G. Michaud is very exciting.

. Secular variations of abundances due to selective evaporation, or accretion might be within the scope of future observational evidence.

All those problems will probably be related to transport processes. We have means to attack them, but one element is missing: a comprehension of turbulent transport and mixing. This is a very difficult problem, and most of us are not prepared to consider it. It is likely that meteorologists and oceanographers have the same kind of worries.

Concerning interstellar matter, at least in regions close to stars, the hydrodynamical structure will be a prominent problem. The question of driving mechanisms of winds, essentially by radiation is typical of the desirable coupling between our knowledge in transfer and in hydrodynamics.

However, even without dynamical problems, many questions can be treated in greater detail than is actually done. A circumstellar envelope modifies the diagnostic of the radiation emitted by the central star; the combination of IR, visible, and UV observations of these composite objects will certainly lead to a better comprehension of them. Let us not forget also the treasures of informations that exist in spectra of eclipsing binaries. In the same way, in the case of close binaries, one of them may, particularly by its X-ray emission, affect strongly the atmosphere of the other one; we have here an access to something which is nearly an experimental device that Nature is operating for us, in modifying a medium by an external agent.

If the medium surrounding the star is so close, and so dense that it can have a backward effect on the stellar atmosphere, as mentioned by J.-C. Pecker, we are exactly in the kind of situation where concepts developed by R. N. Thomas and his co-workers may prove useful.

Interactions of a longer range might show up, whose comprehension will be difficult: for example correlations between the maser emission $OH-OH_2$ and the color index $V-I$ of nearby M stars.

Other interesting objects are protostars. In Larson's models one sees situations in which a shock wave stands for a photosphere. Maybe modelists can start to work on such things right now.

Going now to our hydrodynamical skills, I have already mentioned the winds problem. But there is also the whole question of hydrodynamic modes (including convection) of an atmosphere. The granulation and supergranulation of the Sun are well known observationally in space as well as in time. Can we theoretically predict their existence on other stars, and what will be the observational checks, including possible observations of some stellar extension of the solar 5-min oscillations? We cannot avoid considering the inhomogeneities of stellar chromospheres, even if they are not within our observational reach.

But hydrodynamics not only gives us patterns; it also has to deal with the energy problem. On the triptych: generation-transfer-dissipation, only the second point is encountered at the present time, and perhaps slightly the third one. The generation of sound waves lies mostly on Lighthill's theory, but here again, we are faced with the physics of turbulence! The dissipation of waves is still the subject of many discussions, for example in the single case of shocks, not to speak of other types of waves, including MHD waves! Again, we have much to learn from the Sun: we know roughly the amount of energy which is deposited in the low corona, but it is not yet possible to draw from the observations what is the dissipation law with height. So the question of the nature of the energy, and, *a fortiori*, of the dissipating mechanism is a matter of personal theoretical opinion.

Last, but not least, let me mention again turbulence, or rather its hydrodynamical effects: turbulent structures convey with them some gross effects such as pressure, and energy. If there are motions such as expansion, the decay of eddies will be accompanied by energy dissipation, and pressure effects (remember that the 'microturbulent velocity' can be of the order of the thermal velocity). Let me quote here a new result obtained recently in Nice by U. Frisch and coworkers just to show the complexity in which we are to fall: it can be shown theoretically that the decay of eddies in two dimensions is radically different from the 3-dimensional classical Kolmogorov behaviour. In a stellar atmosphere, the gravitational field can induce structures that rely more on a 2-dimensional analysis than on a 3-dimensional one; who knows? But remember some observational distinctions between 'horizontal microturbulence', and 'vertical microturbulence' in the Sun.

As theoreticians, we should also put efforts on the new methods of handling observational data. Speckle interferometry has recently grown up, and it poses many interesting analysis problems. The deconvolution of the effect of rotation can certainly be approached in modern terms. Some tools such as maximum entropy methods can help us to define what we are looking for in the observations. Let me take again a solar example, namely the photospheric velocity field. I do not think that we really need a wholly detailed description of what happens in a calm region with high temporal and spatial and spectral resolution. What is important for the physics are spatiotemporal correlations, spectral density of energy, in one word, statistical informations. Good methods of reduction, and also ingeneous observational devices can give this kind of information without going through intermediate steps of deterministic description or recordings.

To finish, I shall mention briefly other domains of Physics in which we shall have to look.

. Plasma physics of course, but nearly all what I have said on hydrodynamics encompass plasma physics too. So I let you make the generalization. Still, non-thermal processes, such as in flare-stars, or the emission of neutron stars, are important peculiarities of the Sky that deserve separate studies. But they are in the childhood stage, and a prediction of the kind of work that will appear is outside my possibilities.
. Very low density organic chemistry is a necessity. Hopefully once the reactions constants will be known, the diagnostic of interstellar matter through emission of organic material should resemble strongly our practice of atoms and ions.
. We have a very interesting and hard problem to ask to solid state physicists: interstellar matter contains dust particles. Presumably, those particles contain a number of atoms too small for applying extrapolation of classical knowledge. One knows that

the study of thin films is a very peculiar part of solid state physics. 1000 Å is a common dimension in both cases. We badly need to know much about the behaviour of small solid particles in their interactions with themselves (Van der Waals forces, electrostatic attraction, polarisability), with neutral and charged particles, and with the radiation field. There is also the interesting question of the anomalous behaviour of transport coefficients of dust particles in gases, which depends on statistical physics.

. And finally there is the eternal wish to use the powerful tools of thermodynamics, and especially the new ones of non-equilibrium thermodynamics. All the objects that we study are the seats of basically irreversible processes. Why are we unable to produce non-trivial results while powerful means are at our disposal? Those means are now applied to biology; I am not a specialist in biology, but I have the impression that it is not less complicated than stellar atmospheres!

3. Conclusion

I have tried to separate the various problems that one expects to see being solved in the future. They all rely on today experience – some on yesterday's too. Some are more precise than others, some are more difficult than others. There are certainly many gaps that you will fill in the next hour of discussion. But it is a hard game to tell the future, and the facts come very often to deny your predictions (see L. Kuhi's talk.) I am not an astrologer; and if I had been you surely would not have let me make this talk.

Acknowledgements

I am indebted to F. Praderie, H. Frisch, L. Kuhi and R. N. Thomas for valuable discussions during the preparation of this paper.

THE FUTURE OF OBSERVATIONAL STELLAR ATMOSPHERES

L. V. KUHI*

Institut d'Astrophysique, Paris, France

1. Introduction

The title of this talk is, of course, a bit absurd since no one in his right mind can possibly predict what will happen in the future over even one year let alone ten. However nothing is impossible for Dick Thomas and having accepted his invitation to present this talk I have to pretend to have a crystal ball and put up a good front. I think that the best way to proceed would be to look at the past few years in observational stellar atmospheres and point out some of the highlights, especially those advances which seem to hold great promise for our understanding of stellar atmospheres in the future. The observer's work falls neatly into two categories: the collection of data and their subsequent interpretation. The former involves the development and use of instrumentation (usually some type of spectrograph with a photon detection system); the latter is so closely linked to theory that it would be impossible to discuss it without recourse to theoretical assumptions and results. Therefore I would like to first mention a few instrumental developments which I consider to be of great importance and then discuss briefly some of the problems to which they might be applied in the next several years.

2. Instrumentation

There has been a great deal of activity over the past several years in extending the wavelength range of the observed spectrum as well as increasing the sensitivity of the instrumentation. The extension in wavelength range has come about because of two parallel trends: (1) the development of suitable detectors (e.g. the far infrared) and (2) the increased use of balloons, rockets and satellites to get above the Earth's atmosphere to observe the ultraviolet, (both near and far) as well as X-rays. Many remarkable and unexpected discoveries have been made because of this extension in wavelength and we can cite the P Cyg type profiles in the ultraviolet resonance lines of early type supergiants as a good example. In addition the longer baseline in wavelength provides a much better check on the accuracy of model atmospheres especially with regard to opacity sources and line-blanketing as witnessed by the emergent flux distribution. There is no doubt that this spectral interval will continue to pay large dividends in the future and we will return to some of the possibilities later in the talk.

The more efficient use of the radiation after it has been collected has also received considerable attention because of the extravagant waste of photons and/or observing time usually encountered with normal high dispersion spectroscopy and because of the

---

* On leave from the University of California, Berkeley.

natural desire to observe even fainter objects at moderate dispersions. Fabry-Pérot and Michelson interferometers have found increasing use in the study of emission line spectra and interstellar absorption spectra in the optical and of absorption and emission spectra in the infrared. Their big advantage is the lack of an entrance slit with its resultant huge light loss at high dispersion (i.e. a few Å mm^{-1}) but very little application has yet been made to ordinary stellar spectroscopy in the optical. This should be a useful area to exploit in the future for obtaining highly accurate absorption line profiles for comparison with the results from model atmospheres. Another method of obtaining accurate line profiles has been that of photoelectric scanning at the focus of a suitable spectrograph with either a single or multi-slit arrangement to take advantage of the inherently higher accuracy of photoelectric devices compared to photographic plates. However because of their sequential operation (most such scanners can sample only a few spectral regions at a time) they are very inefficient in spite of their higher quantum efficiency. The extreme example of a multislit device is Oke's primefocus scanner which employs 16 photomultipliers behind 16 adjustable slits and enables the rapid determination of spectral energy distributions of very faint stars. However it is not suitable for the determination of accurate line profiles. Image tubes have also been widely used but their gains in the blue are not outstanding; however the gain over photographic plate gets progressively better with increasing wavelength although their photometric properties leave a great deal to be desired.

The major advances have come from the development of instrumentation involving special types of devices such as image dissector photomultipliers, vidicons, silicon diodes, etc. Here the gains over conventional techniques are enormous and the potential for stellar atmospheres (as well as other areas, of course) should be equally enormous for both high and low dispersion spectroscopy. There are a large variety of different systems in various stages of development and use at the present time and a brief description of three of these would give us some idea of how potentially useful they really are. The first of these is the SEC vidicon system developed by Lowrance and coworkers at Princeton which has been used successfully by Morton at a resolution of 0.75 Å over a small spectral range: λ4270 to λ4495 to study a 16.6 mag. quasar, PHL 957. The system has a 25 × 25 mm storage target with an inherent resolution of 25 μ and the ability to integrate for many hours. The quasar required 6 h of integration to get a signal-to-noise ratio of 20 and a background of only 1% of the maximum dynamic range. The transfer function is non-linear so that calibration at a range of intensity levels is required. However it can operate at very low light levels, (the equivalent of ~5 photoelectrons per image element) and consequently doesn't have the low threshold problems of photographic plates. Another vidicon system is that of McCord and Westphal at Caltech which employs a silicon-diode-array as the target. The device has a very high quantum efficiency (85% at λ5000, 30% at λ9000 and 6% at 1.1 μ) with individual elements spaced 15 μ apart in a 256 × 256 element array. However the minimum detectable signal is rather large (about 1000 photons) and something equivalent to reciprocity failure is present. The dark current of a cooled system is ~5% of the maximum signal. However a 1% photometric accuracy after suitable calibration is claimed as is the ability to integrate for several hours. More recently a silicon intensifier tube has been added to overcome the rather large threshold problem, i.e. the photoelectrons emitted by the photocathode of the image tube are accelerated directly onto the silicon-diode-array target. It has been used successfully in narrowband 2 dimensional scanning of Mars and at the 100 in. coudé where the gains over interferometric techniques are claimed to be as large as a factor of 5. The gain over conventional photographic plates and scanners is of course very high. Yet another system is the image-dissector scanner (IDS) developed by Wampler and Robinson at Lick Observatory. This uses the phosphor of a three stage image-tube as a short-term storage element which is scanned by an image-dissector photomultiplier tube with a resolution element of 37×250 μ. Two thousand such elements are scanned every ~ 4 ms with sky and star spectra being scanned alternately. The output goes directly into an on-line computer for immediate processing as well as onto tape for later data treatment. It is presently being used with a Cassegrain spectrograph on the 120-in. with a resolution of 2–3 Å. down to 22nd mag. McNall et al. have coupled such an IDS system to an échelle spectrograph (2 Å mm^{-1}) and claim speed gains of ~ 100 over a conventional slit and a single photomultiplier at a resolution of 0.1 Å. The system is now essentially an area scanner with specific scanning along the spectra of each order.

The advantages and potential advantages (potential, since only the Wampler system is in regular operation) of these systems for stellar atmospheres are quite clear: they should allow high resolution, high accuracy line profiles to be measured in much fainter stars than at present. They should allow lower resolution spectra to be obtained of much fainter stars, e.g. in globular clusters, Magellanic Clouds, etc. for general abundance studies, etc. Finally they convert small telescopes (i.e. ~ 1 m diameter) into the equivalent of large telescopes with conventional coudé systems so that many more observatories can get involved in spectroscopic research. What they could or should do is another question which we will touch upon later but clearly the large increase in sensitivity opens up a whole new world of problems.

Other instrumental developments which to me seem of great future importance are the intensity interferometer of Brown and coworkers and the speckle interferometer of Labeyrie and coworkers which permit the measurement of stellar diameters. The Brown instrument has been in operation for several years now and has produced the largest number (~ 20) of stellar diameters. The technique relies on the fact that the statistical fluctuations of photon intensity also contain a wave interaction term resulting from beats between adjacent frequencies. The maximum resolution of the present 200 m baseline is $\sim 5 \times 10^{-4}{''}$ but the small collecting area now limits the measurements to stars brighter than $V = 2.2$ mag. However a proposed scheme would increase the collecting area as well as making other improvements to extend the magnitude limit to $\sim V \simeq 6.5$ mag. The Labeyrie system makes use of the speckle pattern of a stellar image which is an interference effect caused by random phase and amplitude perturbations impressed upon the incident wavefront by atmospheric turbulence and telescope aberrations. An image tube camera is used to record stellar images with very short exposures (1 to 100 ms). Two-dimensional Fourier transforms of each image are co-added and compared to those of a standard star to obtain the two-dimensional equivalent of the visibility curve obtained by Michelson and Pease. The data reduction and handling appears to be a real problem at present but a practical resolution limit of 0″.01 for $V=9$ mag. is claimed for the 200-in. Ten stars have been measured so far and confirm the measurements of Michelson and Pease using the Michelson interferometer. The technique also allows the determination of stellar oblateness (none detected yet), limb darkening, and separation of close binaries as well as surface detail (non-radially symmetric). The most exciting result has been the measurements of α Orionis as a function of λ and of o Ceti as a function of λ and phase. Both stars are larger in the blue which suggests that continuum scattering is important for cool stars. Mira also has a change in size as a function of phase as shown in Table I.

TABLE I

(a)	α Ori.	$\lambda=$ 4200		dia $=$ 0″.069	
		4880		0.067	
		5700		0.055	
		7190		0.052	
		10400		$\leqslant 0.050$	
(b)	o Cet	$\phi=0.09$		$\phi=0.38$	
		$V=4$		$V=7$	
		$\lambda=$ 4500	dia $=$ 0″.070	$\lambda=$ 6700	dia $=$ 0″.062
		5150	0.057	7000	0.058
		7500	0.051	7500	0.055
		10400	$\leqslant 0.050$	7500	0.055

Clearly the field is just beginning but already a marked improvement in sensitivity seems possible, namely the use in two-dimensions of one of the television devices discussed above to replace the present photographic system. It is claimed that spatial structure will be accessible for objects as faint as 20th mag. with the 200 in. (down to its diffraction limit of $\approx 0″.01$ to $0″.03$) which has perhaps even more importance to nonstellar astronomy than to the narrow field of stellar atmospheres. However objects with circumstellar material certainly fall into this domain and such high resolution capabilities could in principle tell us a great deal about the structure of that material.

3. Observations

In the area of actual observational problems for the future it is of course extremely difficult to say where our interests, calculations, instrumentation and finances will take us in the next ten years. Among these we may cite the following:

(a) The general acceptance of non-LTE representations for the stellar atmosphere along with the inclusion of previously neglected transitions involving two electrons. Such calculations have been (Mihalas and coworkers) very successful in explaining away several so-called selective excitation effects in early type stars. In general the departures from LTE are largest for stars of highest temperature and luminosity, and their effects on abundances and microturbulent velocities can be quite large.

(b) The extension of chromospheric studies to stars other than the sun by Wilson and coworkers with the discovery that the Ca II K chromospheric emission is both a function of the age of the star (from the total flux) and of its visual luminosity (from the width of the line). Thus the presence of chromospheres has been confirmed for most F, G and K stars but virtually no one has included the consequent effects in their model atmosphere calculations.

(c) Evidence for the presence of velocity fields and mass-outflow has been steadily accumulating and now with the striking ultraviolet results (i.e. the P Cygni type profiles in the resonance lines of early type supergiants) mass-outflow is a well established phenomenon common to all luminous stars as well as Miras and pre-main sequence objects. Velocity gradients in the photospheres of supergiants have also been deduced for micro- and macroturbulence.

(d) The discovery of infrared excesses in stars covering a wide range of spectral types which has provided direct evidence for circumstellar material in the form of ionized gas and/or dust particles. The presence of dust has usually been deduced from the infrared energy distribution which often resembles that of a black body of very low temperature ($\lesssim 1000$ K) even though the photospheric temperature is much higher. Free-free emission can usually account for most of the other observations although a large group of objects exists for which the distinction is not so clear. Further evidence for circumstellar material has been the detection of polarization and its variation as a function of wavelength in certain groups of objects with infrared excesses. More recently radio emission has also been detected from some of the more extreme infrared excess objects.

(e) The detection of ultrashort variability in a variety of stars. The periods range from 33 ms for the Crab pulsar to 10's of seconds for many old novae, dwarf novae and similar objects.

(f) The ultraviolet (space) variability of α_2 CVn. Here the amplitude of variation (periodic) is modest ($\lesssim 0.10$ mag.) in the visual, decreases to zero in the near uv and increases again to a few tenths of a magnitude in the far uv but 180° out of phase. This seems explicable in terms of blanketing effects but certainly provides another important handle on atmospheric structure.

(g) The revision of f-values and their resultant change on the solar iron abundance.

(h) Identification of X-ray sources with stellar objects.

(i) The increase in high resolution spectroscopy in the infrared with its resultant determinations of crucial isotope ratios.

We can continue to list such items ad infinitum but it may be more worth while to look at a few of them in more detail before making any comments about the future of observational stellar atmospheres. The role of non-LTE calculations is clearly a very important one for the determination of abundances and other atmospheric parameters. Just how important can be seen from the recent work by Mihalas and collaborators for early type stars. We might first recall some of the problems resulting from conventional LTE analyses of O and B stars:

(1) a helium abundance of $Y=0.15$ to 0.2 and a microturbulent velocity of $\xi_t \simeq 15$–20 km s^{-1} is required with $\log g = 4.5$ (from hydrogen) to get agreement with the observed equivalent widths.

(2) Magnesium abundances are up to ten times the solar value with the most discrepant equivalent widths being 4–5 times too small even with very large ξ_t.

(3) The predicted hydrogen line strengths decrease with temperature, the He I lines disappear and the line profiles do not have deep enough cores.

The non-LTE calculations eliminate all of these difficulties and point out some very interesting general conclusions that observers should keep in mind. These are the following for O and B stars:

(a) the non-LTE abundance is decreased with respect to the LTE value.

(b) the observed behavior of equivalent widths with effective temperature is more closely reproduced.

(c) the microturbulence parameter is reduced, often even to zero, and is likely to be further reduced by including UV line blanketing.

(d) the departures from LTE are largest for low gravity and high effective temperature.

(e) the departure from LTE can be large for weak lines as well as strong ones contrary to the often made statement that non-LTE effects are not important for weak lines.

The later type stars have not been so thoroughly investigated because the required calculations are enormously complicated. However, to me, the trend is clear: curve-of-growth analyses based on LTE models cannot and should not be considered as realistic until they have been shown to be so on the basis of non-LTE calculations. Until that time we must hold suspect the abundances derived from such models.

Another aspect of recent non-LTE work has been the inclusion of all possible transitions into the equations of statistical equilibrium. Chief among the previously neglected processes are those involving two electrons, i.e. autoionization, dielectric recombination, two electron transitions, etc. One example should suffice to illustrate the importance of including such transitions. In the Of stars there has long been the unsolved problem of the origin of the N III 4634, 40, 41 lines which appear in emission whereas the lines from the next cascade down are in absorption. The Bowen fluorescence mechanism has often been proposed as the solution but the intermediate O III lines are not observed in emission. It also requires a large He II flux so that the He II 4686 line strength maybe expected to correlate with the N III lines but this is not observed. Mihalas, Hummer and Conti have shown that the dominant depopulation of the lower level $(3p)$ is via a two-electron transition to $2p^2$ which is so efficient that $3p \to 3s$ remains in absorption. Also the population of $3d$ is dominated by dielectronic recombination. The net result is that $3d \to 3p$ produces the three emission lines observed. However the present calculations with a plane-parallel atmosphere do not predict the observed total fluxes but do predict the qualitative behavior of the emission strength. Clearly an extended atmosphere is needed to increase the emission flux. Similar calculations also seem to explain another 'selectively' excited line C III 5696. The lesson here is very clear: all processes must be included in the calculation of level populations before worrying about selective excitation mechanisms or other fancy processes.

Now what is the role of the observer in all this? Sadly enough the basic data required for checking the predictions and the models are usually lacking. Continuous energy distributions in the optical are usually available and in the near future the ultraviolet region will be well covered as well. However line profiles determined to a precision of 1% are really the final comparison although accurately determined equivalent widths are also very useful even though they throw away all information about the profile. The non-LTE calculations also make specific predictions about expected emission line strengths for the other transitions involved in the so-called 'selective' processes and yet no accurate (i.e. to 1%) total flux measurements exist. This is all of course with regard to the hotter stars but the same basic lack of data also exists for the cooler stars where the theoretical work is much less developed. The measurements would not involve any fancy non-existent equipment but no one seems to be doing anything in this area. I strongly suggest that the observers get busy.

Another area of future interest is the problem of velocity fields in stellar atmospheres, i.e. from microturbulence and macroturbulence to large scale outflow. The microturbulent velocity is one of the most mysterious parameters that has arisen in the analyses of equivalent widths (and line profiles) for stellar abundances. It was introduced many years ago by Struve and Elste to account for the additional broadening that was present in the observed absorption lines, i.e. the thermal Doppler broadening was not large enough to produce the observed line widths. No physical explanation of this parameter was given then nor has any satisfactory explanation been given since. It seems that everyone derives a value for it but nobody knows what it is. At worst it seems to be nothing more than the 'Cook's variable constant' of undergraduate physics labs to make the theoretical and observational results agree. At best it could be some kind of small-scale turbulence related to the convective motions in stars like the Sun. So the outstanding question is, 'What is microturbulence?'

Since most people are familiar with the history of microturbulence I need not repeat it here other than to recall to you that it is usually derived from curve-of-growth analyses under the condition of making the dependence of element abundance on equivalent width as small as possible. This dependence is in the sense that the stronger the line the greater the inferred abundance. In extreme cases the abundance can vary by a factor of 10 from weak to strong lines and deducing a model that will predict entirely consistent abundances is almost impossible. The microturbulent parameter ξ_t is smallest in dwarfs (~ 5 km s^{-1}), larger in giants (~ 10 km s^{-1}) and largest in supergiants ($\gtrsim 20$ km s^{-1}). Its effects on the derived abundances are appreciable: changing ξ_t by a factor of 2 can change the abundance by a factor of $\lesssim 5$. This large sensitivity comes about because ξ_t determines the position of the flat part of the curve-of-growth. Even in the Sun there seems to be a systematic increase in abundance from center to limb and from weak to strong lines by a factor of two. Some light can be shed on this mysterious parameter by the recent work of Smith on six sharp-lined A stars. He finds that ξ_t peaks around 4 km s^{-1}. in mid to late A stars even after all corrections are applied. But even more importantly he suggests that the much higher earlier estimates could arise from the following:

(a) the revision of the f-value scale produces a decrease of ~ 1 km s^{-1} in ξ_t.

(b) errors in broadening (e.g. in the damping parameter **a**) can give rise to spurious ξ_t: a factor of 7.5 in **a** produces a 1.25 km s^{-1} increase in ξ_t.

(c) a 1 kgauss magnetic field mimics an increase of $\sim$ km s^{-1} in ξ_t.

(d) the equivalent width depends on the dispersion of the spectrograph, i.e. it is $\sim 30\%$ larger at 8 Å mm^{-1} than at 2.7 Å mm^{-1} and a factor of ~ 2 larger at 16 Å mm^{-1} than at 2.7 Å mm^{-1}. A factor of 2 increase in equivalent width is equivalent to an increase of 3 km s^{-1} in ξ_t. Finally he points out that ξ_t is also reduced by the proper inclusion of line blanketing effects in the model atmosphere, something which is not very often done. Similar results can no doubt be derived for cooler stars as well.

We should also note a very interesting trend in the values of ξ_t that appears with the increasing complexity of the model atmosphere. As mentioned above one of the consequences of the non-LTE calculations for early type stars is a reduction of the value of ξ_t. For hydrogen and helium ξ_t goes to zero; for Mg some residual (~ 4 km s^{-1}) microturbulence still seems necessary although a slight increase in abundance could get rid of it. The Ca II K line in B stars still requires a few km s^{-1}. Again the complete inclusion of ultraviolet line-blanketing would act to further reduce ξ_t. The same comments can be made about other spectral types but the calculations are not nearly so complete. However the trend is clear: the more complete (and hence the more accurate) the model the smaller the value of ξ_t. Such a result seems only to emphasize the *ad hoc* nature of microturbulence and makes an innocent bystander wonder about its physical significance. It seems very likely that it will either disappear or decrease drastically in model atmospheres in the future.

There may also be another explanation for the large values of ξ_t derived for stars more luminous than dwarfs. The extended atmospheres of supergiants provide us with the interesting possibility that microturbulence may be connected to a general velocity field, i.e. a gentle expansion of the atmosphere. Numerous investigators (e.g. Rosendhal and Wegner, Aydin, Groth, Williams, Lamers, Warren and Peat, Wolf) have arrived at more or less the same conclusions regarding the atmospheres of A and B supergiants:

(a) the atmospheres seem to be in expansion, and a radial velocity gradient is present, e.g. from 5 to 60 km s^{-1}.

(b) the microturbulence correlates with the radial velocity and a gradient is necessary to fit the observations, i.e. ξ_t is small at large optical depth (5–10 km s^{-1}) and large (20–30 km s^{-1}) at small optical depth.

(c) these gradients in v_r and ξ_t are variable with time.

(d) ξ_t increases with luminosity and decreases with lateness of spectral type.

We also note that Karp has compared curves-of-growth computed with and without velocity fields and finds that a curve with $\xi_t = 0$ and a small velocity gradient ($\Delta v_r = 10$ im s^{-1}) is identical to one with no velocity gradient and $\xi_t = 5$ km s^{-1}, until the damping part is reached where a lower value of **a** is mimiced. Absorption profiles produced in the presence of a velocity gradient would be asymmetric but a small amount of microturbulence (~ 30 km s^{-1} as observed) would probably suffice to smear out the asymmetry so that it would not be noticed. The resultant effect could be a misinterpretation in terms of microturbulence. The present data however are so meager that a full-fledged observational attack on this problem is long overdue. Griffin's suggestion of using terrestrial absorption lines as wavelength standards should be explored further and put into use to investigate the velocity fields in normal stars and giants as well as in supergiants. The intrinsically higher accuracy of Griffin's technique should permit measurements of radial velocities to an accuracy at least 10 times greater than at present. Consequently there is no excuse for not making the observations necessary to elucidate the true nature of microturbulence. A fruitful area to also explore is the behavior of ξ_t in those stars for which we already know something about the velocity field e.g. pulsating stars such as the Cepheids. Preliminary work here indicates that the picture is still very murky. Van Paradijs has obtained both v_r and ξ_t as a function of phase for 9 cepheids and finds that ξ_t is largest when the star is most rapidly contracting (e.g. amplitude 2–6 km s^{-1}) but that it is constant for the opposite half of the cycle (~ 3–5 km s^{-1}) contrary to what one might expect if indeed the velocity gradient were solely responsible for the microturbulent parameter. This complication of course serves only to make the problem more interesting and points out very clearly that the next phase of stellar atmosphere research (both observational and theoretical) must be the inclusion and investigation of hydrodynamic effects, i.e. the assumption of hydrostatic equilibrium must be dropped.

The observational prospects here are enormous since the subject of expanding and consequently extended atmospheres has barely been touched. This naturally leads us to the third and final subject that I want to say something about, namely circumstellar matter and by this I mean anything that is in the immediate vicinity of the star and hence very likely intimately connected to it in one way or another, e.g. via mass loss or dynamical collapse or whatever. The observations cover the entire wavelength range accessible to observers, i.e. from X-rays to radio waves. X-ray radiation has been linked to several close binaries with possible mass exchange which results in in X-ray production when the material falls onto a companion neutron star or other very condensed object. In the far UV the P Cyg profiles of resonance lines in OB supergiants suggests high-velocity ($\gtrsim 1000$ km s^{-1}) outflow whereas no evidence for it is observed optically. At optical wavelengths in other stars similar P Cyg profiles and other emission lines have long been interpreted in terms of outflow and circumstellar material, often of great extent. Also displaced absorption lines of Ca II and other elements in late type giants and supergiants led to the idea of low velocity outflow for these stars. The infrared observations have been a great revelation in the detection of infrared excesses from a large variety of stars and have raised many interesting problems. The standard explanations for the excesses involve either thermal reradiation from circumstellar dust grains or free-free emission from ionized hydrogen. Dust around cool stars such as M type supergiants seems quite acceptable but dust around hot stars (such as the Ae and Be stars associated with nebulosity) present the question of not only the survival of the dust but its formation or origin. Here an extreme example might be the Wolf-Rayet stars of type WC9 which have black body peaks corresponding to ~ 1000 K whereas the star itself has an effective temperature in the range 20000 to 30000 K. The emission line widths indicate velocities of expansion of the order of ~ 500 km s^{-1} in the region close to the star (i.e. $\lesssim 30$ stellar radii) but no information exists as to the velocity fields in the region of the dust. Is the dust formed from the ejected material or is it just left over from earlier phases of star formation? How is this possible in the presence of a hot star? Some clues as to the nature of the particles can be provided by high resolution infrared spectroscopy but so far only the diffuse 8–10 μ feature of silicates has been found. Polarization measurements also can serve to distinguish between free-free and thermal re-emission as it seems to have done for the classical Be stars and the M stars. The situation for other stars is very confused at the present time and more measurements are sorely needed.

Microwave and radio measurements have been successfully made for a number of sources. Radio frequency radiation has been detected on the 10–100 mfu level in a number of early type stars with very large infrared excesses and strong emission-line optical spectra. These additional frequences provide a close check as to the nature of the infrared emission (e.g. in MWC 349 it is impossible to get as much infrared radiation as is observed from free-free emission that would be consistent with the radio measurements; dust must be present). As sensitivities increase to lower and lower levels the number of detections should increase dramatically so that similar calculations can be made for other stars as well. OH maser sources have also been mapped out in a $\sim 2 \times 2''$ area around stars like NML Cyg (a late type supergiant) by means of very high resolution interferometry. Several such sources are present and may be connected to an earlier explosion of the star and/or an ejection of material from its surface (Herbig). What the role of such masers may be with regard to the stars with which they are associated is however very unclear to say the least.

We have listed a few of the observations made in each wavelength range that lead to the conclusion that circumstellar material must be a very common phenomenon among a large range of stars. We can generalize a bit and lump them all together in a hypothetical picture of a star. First, surrounding the interior we have the photosphere, chromosphere and corona for stars like the sun. Second, for the objects mentioned in this section we probably still have something akin to the chromosphere and corona but we would call it an extended atmosphere or envelope. It is here that the evidence for mass outflow and large scale velocity fields must be produced. Still further there must exist the region of dust formation and/or destruction i.e. the region responsible for the thermal re-emission. Finally one might run into the OH masers, compact H II regions, etc. but where these are located is not at all clear. In addition many of the objects mentioned are very young and theoretical calculations indicate that we shodul be observing infalling material rather than outflow. Perhaps the speckle interferometer in its TV coupled mode can actually map out the velocity fields of the circumstellar material by producing highly resolved spatial maps in different parts of an emission line, i.e. at different velocities. At present we really do not know anything about the velocity fields involved in the vicinity of the star; is the matter really coming in or going out? As far as the stellar atmosphere problem is concerned I now see it expanded in two directions:

(1) the inclusion of velocity fields, and

(2) the interaction of radiation not only with the photosphere but with the entire gamut of circumstellar material including the grains themselves.

The observations needed to understand these two areas are just beginning to be made in a systematic fashion after the initial spectacular discoveries. However the crucial questions have yet to be asked and the direction of observational attack is very unclear. Nevertheless it is in these two areas that I think our most creative efforts will pay off with the largest dividends not only in understanding the various interactions involved but also in understanding the basic process of star formation itself.

4. Conclusions

In concluding this talk I realize that what I've really presented here is more of a status report of where observational stellar atmospheres are today and not a prediction of the future. The future observers' task lies in several areas:

(a) A general mopping-up type of operation such as the extension of NLTE models to cooler stars and the accompanying refinement of data that is needed to do the job. This alone will keep many astronomers busy for many years.

(b) A full-fledged effort in subjects that have only recently opened up because of new observations usually made with new instrumentation. The problem of circumstellar matter including radiative transfer, velocity fields, mass-loss, grain formation and destruction, and other peculiar phenomena such as OH masers fit into this category. Imaginative new techniques and ideas will be needed to help solve the questions already raised by the very incomplete data presently at hand not to mention the ones we don't even know about yet.

(c) A continuing effort in instrumental development designed to improve sensitivity, accuracy, data handling ability, etc. Accompanying this there must also be a matching sophistication of data reduction techniques in order to avoid being inundated by infinite amounts of data that cannot be adequately analyzed. In addition there must continue to be the development of new and radical ideas which of course are impossible to predict in advance (the speckle interferometer is a good example) but which often can revolutionize an entire branch of astronomy. We should always keep our minds open and receptive to such ideas so that they have a chance of being properly developed. What we should not do is to smother them in conventional objections simply because they are different or because they are not readily understood by most of us. This of course applies to theoretical stellar atmospheres as well as observational ones and perhaps should be a guide for all future work.

The three areas of course go along hand in hand since what is to-day's exciting new discovery is tomorrow's mop-up routine type of work. All three are needed to maintain progress but a strong interaction among the three is very vital and absolutely necessary. Otherwise the subject of stellar atmospheres will stagnate exactly in the manner in which some alarmists have claimed is already the case. This is not true now and shows no sign of becoming true in the future. There is a great deal of exciting new work to be done.

THE NEXT DECADE OF THEORETICAL SOLAR PHYSICS

M. KUPERUS

Astronomical Institute, University of Utrecht, The Netherlands

1. Introduction

Exactly a decade ago an international conference on solar physics took place in Utrecht on the occasion of the late Prof. Minnaert's 70th birthday. Since then, much has happened. Many new instrumental devices were developed or refined. Completely new observational techniques were applied in the XUV and radio wavelengths and in particle detection.

This all resulted in an explosion of new data. Moreover new problems appeared of course. It also resulted in an explosion of papers on solar physics. Actually many more contributions appear than reasonably can be absorbed by the periodical *Solar Physics*, of which the first volume appeared in 1967. At that time the editors were not at all sure that they would receive enough papers to maintain such a journal. But were there major problems solved recently?

At first I would be inclined to think that, notwithstanding the enormous efforts spent in solar physics, little has been achieved and no spectacular breakthroughs have been found. This impression, probably shared by many more colleagues, becomes stronger if one tries to find a line of progress in 'classical' problems of solar physics. We still do not have a satisfactory theory of the Sun's *differential rotation* and the Sun's outer *convection zone*. One may argue that this is due to our bad knowledge of the *interior structure* which is not accessible to direct observation. *Solar activity* and the formation of *sunspots* is now believed to be a direct consequence of differential rotation and cyclonic convective motions and therefore suffers from the same uncertainties.

But also in the outer observable layers we have still a long way to go. We do believe that the *photospheric motions* and the very high temperature of the corona are causally related, but only occasionally a piece of evidence is found that fits the complicated puzzle of the *heating of the corona*.

The greatest mystery on the Sun is the solar flare. Though known for decades it seems that we have not come further than a few controversial theories embedded in thousands of observations of very different and sometimes contradictory character. We only poorly understand how *high energy particles* may be generated and an understanding of the various plasma processes responsible for the sporadic *radio radiation* is just beginning.

This may all sound discouraging but it rather shows how complicated solar physics is at this moment. In Minnaert's talk on solar physics in 1963 he reviewed a period that terminated an era of great discoveries in solar physics: Magnetic fields could be measured on the sun and it had been found that the corona is a very hot plasma.

It was in that period that it became clear that the solar corona was expanding with supersonic velocities: the solar wind. The oscillatory motions were discovered which are in fact the non-thermal energy source which produces the corona and the solar wind. Nowadays we have to do a much harder job to proceed. The pioneers have roughly explored the Sun and an army of settlers is trying to make a living of it.

What is it then that we gained in the last ten years? I would like to make this clear by the following comparison. In 1950 one could describe the Sun as follows:

'The Sun is a thermally radiating gaseous sphere'.

In 1970 I would describe the sun in a more complicated manner:

"The Sun is a rotating magnetic plasma with a complex transient XUV, optical, radio and cosmic ray spectrum"

Around 1950 one was able to construct a spherical model of the solar atmosphere once the source of the continuous opacity was understood. However the abundances could already be determined from the relatively 'simple' optical spectrum. And thus a self consistent picture of the solar atmosphere could be developed. The atmospheric model thus determined was basic in the sense that one could now try to calculate line profiles and use these as diagnostic tools. However, here the first problems already appeared since the Sun is not spherically symmetric on a small scale and certainly in the outer layers the plasma properties are more pronounced than the gaseous properties.

Hence this is present day's situation. A theoretical solar physicist has to study plasmaphysics, magnetohydrodynamics, gasdynamics, non-LTE physics, high-energy physics and general relativity. A non-evolved 1950 solar physicist would feel as ignorant as a schoolboy. This situation can be found in almost all modern branches of astrophysics but it is exceedingly clear in solar physics.

This should elucidate that progress in solar physics should be seen in the context of a mature sophisticated science. There are astronomers for whom the Sun is nothing else than an ordinary main-sequence G2 star. I would classify those people among the non-evolved 1950 astronomers, who do not realize how uninteresting a star or a galaxy is if you are too far away to get the proper observational information.

The next decade in theoretical solar physics will probably be characterized by the understanding that the Sun is the nearest cosmical laboratory in which a large class of astrophysical problems can be studied in detail. A great number of physical processes that are held responsible for phenomena in distant objects occur in the Sun. A study of these processes is absolutely necessary for the understanding of the universe and is thus of general astrophysical importance.

What are these problems? I want to make a personal selection in the hope that others will fill in the gap I leave.

For example I will deliberately omit spectroscopy and line formation. Firstly because I am not well aware of the latest developments in this field but secondly because I do not think it belongs to what I consider theoretical solar physics should be. Spectroscopy should be considered as a general astrophysical technique. It is part of a measuring technique. In fact it is a sophisticated thermometer, barometer, kinometer, voltmeter or magnetometer. This important branch of solar physics belongs to applied-solar physics. Spectroscopy supplies the crucial data after reduction of the observations.

In theoretical (solar) physics the idea should be placed central and should be based on the fundamental laws of physics. Single observations may be of little value as long as there is no idea about the principle physical processes that may take place. At this point I may quote Einstein (*Mein Weltbild*):

"Solange die Prinzipe, die der Deduktion als Basis dienen können, nicht gefunden sind, nützt dem Theoretiker die einzelne Erfahrungstatsache zunächst nichts, ja, er vermag dann nicht einmal mit einzelnen empirisch ermittelten allgemeineren Gesetzmässigkeiten etwas anzufangen."*

I will briefly review the present situation and the expected developments in the subsequent topics which I consider most important for theoretical solar physics.

- Internal structure
  1. neutrino problem
  2. oblateness
  3. rotation and convection.

- Structure of the photosphere and the chromosphere
  1. photospheric structure, granulation, sunspots
  2. chromospheric structure, spicules
  3. waves in the atmosphere.

- Corona
  1. transition region between chromosphere and corona
  2. heating of the corona
  3. solar wind
  4. coronal structure
  5. coronal instabilities and non linear plasma phenomena.

- Solar magnetic fields
  1. solar dynamo
  2. small scale magnetic fields
  3. solar flares and prominences.

2. Internal Structure

2.1. Neutrino problem

One of the most mysterious problems in solar physics is the so called 'Case of the Missing Neutrinos'. Neutrinos are supposedly generated in the solar interior as resulting products of the proton cycle. The proton cycle has several branches. One of them has two major neutrino producing reactions.

---

* As long as the principles, which can serve as a basis for the deduction, have not been found, single empirical facts are at first of no use to a theoretician, in that case he even does not know what to do with single empirically obtained general relations.

The neutrinos produced by the first reaction $p + \bar{e} + p \to v$ are called the 'PEP' neutrinos. The neutrinos produced by the second reaction $B_5^8(e^+ v)\, Be_4^8$ are called the Boron neutrinos. These neutrinos have different energies. The experiment by Davis using the detection reaction $^{37}_{17}Cl\, (v, e^-)\, ^{37}_{18}A$, is able to detect the Boron neutrinos.

It was found that the solar neutrino flux is a factor 10 smaller than predicted by the theory. It is a great step forward that the solar neutrinos have been detected now with a beautiful experiment, but it appears to be one of the mysterious problems in astronomy that there are neutrinos missing.

Several suggestions have been made to solve this enigma.

(a) B^8 is not formed at all in the Sun, but up to now one does not know of a destructive reaction.

(b) The solar interior is not stationary burning but shows a periodic or quasi-periodic kind of activity. Due to the long transfer time for the radiation, the observed solar radiative flux could then be related to some activity in the past, while the neutrino flux reflects the present internal structure.

(c) The neutrinos have already partly decayed before they reach the Earth.

(d) Models with a fast differentially rotating core, show a decrease in neutrino production. However to reconcile a model with the neutrino observations, the oblateness should be larger than observed.

Many more suggestions will be made in the near future. It is a fascinating problem that may give us a direct understanding and check of the nuclear processes in stellar interiors.

2.1. SOLAR OBLATENESS

A second problem that has drawn a lot of attention is the magnitude of the solar oblateness. It started with the experiment by Dicke and Goldenberg in 1966 which demonstrates that the oblateness $\Delta r/r = 5 \times 10^{-5}$ instead of the expected value of 10^{-5}, which would be in agreement with the rotation period of 27 days. An oblateness of 5×10^{-5} means that there is a difference of 35 km between the equatorial and the polar radius. The important thing is that this oblateness would explain the perihelion motion of Mercury of $4''$/century within the framework of the scalar-tensor theory of Brans and Dicke.

The experiment looked very clear and convincing since the measurements seem to show a seasonal variation, which is supposedly due to the variation in the position of the solar rotation axis.

The conclusion must therefore be that the Sun possesses a rapidly rotating inner core.

Several theoretical arguments have been put forward against a rapidly rotating core:

(a) There would be a spin down due to the formation of a viscous Ekman layer. However, we do not yet know sufficiently well what the precise influence of the density gradient is.

(b) The Goldreich, Schubert, Fricke instability would prevent a core from fast rotation. However, if sufficient angular momentum is transported outward through mass loss, the envelope could be braked while the core keeps spinning rapidly.

(c) The observations could be explained by assuming a difference in temperature of 40 K at $\tau_0 < 0.05$. A correlation with faculae has been made. This may explain about 30% of the measured value.

It has recently been shown from a direct measurement of ΔT and a correlation with magnetic fields that a pole-equator temperature difference is not sufficient to explain Dicke's measurements. What is needed is a refinement of measurements of perihelion movements of other planets. Finally a refined analysis of the as yet not published measurements is needed. A repetition of the experiment in some form is certainly useful.

2.3. Solar Rotation and Convection

The rotation of the Sun and especially the differential rotation is an old and still not well understood problem. What is the influence of the convective envelope on the rotation? Are there meridional circulations? Is differential rotation a superficial phenomenon or does it reflect the non-uniform rotation of the Sun's interior layers?

Due to the interaction of convection with the rotation a difference in flux between the pole and the equator can develop inside the convection zone. These differences may become very small in the observable layers due to large-scale meridional circulation. Due to these circulations a differential rotation in the upper part of the convection zone may be generated. The upwelling occurs at the poles and the sinking at the equatorial latitudes. This would however suggest that the polar regions are convectively more unstable than the equatorial regions which is just opposite to the stabilizing influence of the rotation at the poles.

A suggestion that has been made and that will be interesting to work out is that deep inside the convection zone the circulation occurs from a slightly hotter equator towards a slightly cooler pole. A counter cell would then account for the right direction of circulation.

The meridional circulation of course transports angular momentum which could be balanced in a steady state by viscous transport in the direction antiparallel to a gradient in angular velocity.

It is clear from the above discussion that the missing link in any theory of the differential rotation is the absence of a good theory of turbulent convection, or even a good approximation to it.

A different approach is that differential rotation is maintained not by meridional circulation but through momentum transport in the convective eddies.

Calculations, however, show that the meridional currents are in that case directed towards the poles.

A major problem to be solved is compressible non-boussinesq convection and an answer should be given to the question: How deep is the convection zone and what is the effect of coriolis forces on the heat transport and thus in turn on the circulation and on the differential rotation.

3. Structure of the Photosphere and the Chromosphere

3.1. Photospheric Structure, Granulation, Sunspots

The study of the photosphere and chromosphere is the oldest in solar physics. Yet there are a few important unsolved problems.

Is the solar granulation a pure convective mode or does it have an oscillatory character? More high resolution measurements seem needed at this time. The fine structure of the photosphere can give us valuable information about the state of motion and the modes of energy transfer in the subphotospheric layers. Moreover they are an indispensable piece of information for any theory of the structure and heating of the overlying atmospheric layers.

It seems that there are at this moment no outstanding theoretical problems in the physics of the quiet photosphere. However it would be very helpful for the structure of the overlying layers to know the wavenumber spectrum and the frequency spectrum of the photospheric turbulence and the granulation. This would yield the output of mechanical energy that is supposed to maintain the corona.

Again small scale photospheric magnetic fields such as the recently observed subgranular filigree structure are certainly of importance, not in the first place to clarify the photospheric properties, but more to serve as a set of lower boundary conditions, which are necessary to understand the upper chromosphere and the corona (see Section 5.2).

Inhomogeneous models of the quiet photosphere and the faculae should only be constructed if one knows the type of inhomogeneity. It is not very useful to refine solar atmospheric models which depend on too many parameters, which are at this moment unknown. Only models that are physically consistent should be used.

A major problem in theoretical photospheric and subphotospheric physics is the nature of *sunspots*. Our knowledge about the origin of these magnetic markers is very meagre. The formation of sunspots can not be considered apart from the dynamo mechanism of solar magnetic fields, which will be discussed presently (Sections 5.1 and 5.2) Several empirical models have been constructed which are all not fully consistent. The radiation field in the low chromospheric layers of the sunspot is very complex due to lateral deviations in the temperature. It is important to study sunspot problems after the observations have been reduced with a proper theory of line formation applicable to these circumstances. Many peculiarities that are frequently reported may then appear of no value.

Oscillatory motions inside sunspots have been reported. They may be very useful as additional diagnostic aids where the definition of physical properties becomes ambiguous. Oscillations and Evershed flow are probably linked with the energy transport in umbrae and penumbrae.

There is at present no satisfactory theory for the formation and decay of sunspots. What is needed is a complete magnetohydrodynamic analysis of the convection and circulation in the outer convective layers when reasonably strong magnetic fields are present. The exact place where a pore or small magnetic spot occurs in the supergranulation pattern could yield valuable information. As for the photosphere in general, also here any static homogeneous model should be regarded with doubt from a theoretical point of view.

The decay of sunspots occurs too rapidly if one considers Ohmic diffusion as the major process. There are several ways out of this problem. Firstly, the current system is present in thin shells which then diffuse much faster than the whole spot due to the large gradients at the edge. Secondly, the electrical conductivity in the turbulent solar plasma is effectively much smaller than the usual Coulomb conductivity. Observations of the associated velocity fields just around the sunspot may clarify this. In particular the outflow of magnetic flux as is recently observed, is more an indication for a gradual destruction of the sunspot by removal of flux then for the Ohmic dissipation.

The problem of sunspot formation should be studied together with the more general problem of the birth, growth and decay of solar active regions and solar faculae. We will return to this when discussing solar magnetic fields (Sections 5.1 and 5.2).

3.2. Chromospheric Structure, Spicules

Recently chromospheric structure is paid a great amount of attention since high resolution Hα filters and XUV spectroheliographs became available. It may well happen that within a few years not many theoreticians will be interested any more in the current problems of chromospheric structure and chromospheric dynamics, because high resolution extraterrestrial observations in a wide range of wavelengths (Skylab, Spacelab etc.) will probably solve these problems.

A great number of ideas has been developed in the last years and a lot of thinking is going on. It is now the observer who should decide which of the various ideas he considers consistent with the observations.

But it is not yet so far, we are still puzzled by the very nature of the *chromospheric spicules*, which seem to be a physical connection between the chromosphere and the corona, not only positionally but especially as a mode of energy transfer.

At first it seems logical to consider the spicules, driven by the momentum of the chromospheric waves as a kind of visualisation of the shock waves which are required to heat the corona.

But spicules may as well be produced by an energy inbalance in the upper chromospheric layers. In that case they are thermally generated instead of momentum generated. Even radiation pressure due to absorption in strong spectral lines such as Lα is considered as a possible cause for chromospheric instability.

It will be necessary again to decide from the observations which of these ideas can be applied. Magnetic field measurements in the chromosphere or at least good magnetically correlated data would be very welcome.

3.3. Waves in the Atmosphere

A quantity of great uncertainty is the flux of non-thermal energy in the photosphere, the chromosphere and the corona.

The only definite evidence for a flux of mechanical energy is the 5-min oscillatory motion observed in the photosphere and the chromosphere. However the nature of these oscillations is still obscure. It is not yet clear whether they are acoustic or gravity waves or of the evanescent type. In the case the waves are generated by the convective boundary or the granulation elements they are probably gravity waves or magnetically modified gravity waves in regions with enhanced magnetic field.

Waves which are generated subphotospherically in the turbulent convective layers are acoustic waves. These waves may be trapped in the photosphere-chromosphere temperature trough to cause standing oscillations. For waves which are generated in the turbulent convective layers a very strong directional dependence occurs. Much more energy is transmitted in the direction of the mean convective velocity than perpendicular to it. This would result in a very inhomogeneous distribution of wave sources over the solar surface.

Another problem which is not yet solved is the effect of the radiative decay of low frequency waves. The radiative diffusion time is so short in the low photospheric layers that it is hard to maintain 5-min oscillations so low down. This could be used as an argument that these oscillations are actually generated in the higher chromospheric layers where the radiative relaxation time is much longer.

It is possible that the waves are acoustic waves which can easily pass the temperature minimum but are modulated by low frequency oscillations. The carrier waves are then not influenced so much by radiative relaxation.

In medium strong magnetic fields the Alfvén waves must play an important role in the net non-thermal energy transport. When generated in the upper chromosphere they are mainly undamped up to large distances in the corona but in strong magnetic fields they can propagate from the lower chromosphere into the corona. Mode coupling between gravity waves and Alfvén waves enable Alfvén waves to be generated and to propagate upward without frequency limitations.

In a medium with a constant temperature and a constant magnetic field strength the amplitude of Alfvén waves depends on the density as $v \sim \varrho^{-1/4}$ while the amplitude of acoustic waves depends on the density as $v \sim \varrho^{-1/2}$. Therefore non-linear steepening is more pronounced for the acoustic waves. They will become shock waves already low in the atmosphere. In this sense there are two direct lines of coupling.

First: from convection either via gravity waves or directly into Alfvén waves.
Second: from turbulence via acoustic waves into shock waves.
Many interrelations are possible, most of them are not yet well studied.

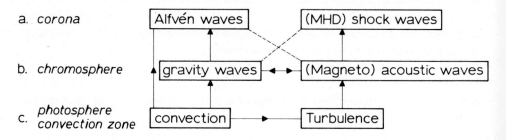

Finally a careful search should be made on the interesting possibility for the thermal generation of waves. Especially in the upper chromospheric layers an energy inbalance may result in motions which predominantly travel upwards and may form shock waves. They may well be the source of the spicules as has been suggested before.

But also hydromagnetic waves may be generated in thermally unstable layers traversed by a magnetic field.

It is worthwhile to investigate whether the kappa-mechanism, well known from the theory of Cepheid and RR-Lyrae pulsations may be operative in the solar ionization zones and thus may drive the photospheric and chromospheric oscillations.

4. The Corona

4.1. The Transition Region Between Chromosphere and Corona

It has long been recognized that the transition from the chromosphere to the corona occurs very abruptly with an extremely large temperature gradient ($dT/dh \approx$ $\approx 10^2$–10^3 K cm^{-1}) and that a large part of the heat dumped into the corona is conducted backwards into the chromosphere, due to the very large coefficient of thermal conduction ($\varkappa \sim T^{5/2}$).

I think it is one of the great achievements in solar physics of the last decade that the very sharp rise in temperature in the transition region has been observationally established, thanks to the interpretation of the XUV observations (OSO-satellites). The method which is used to determine the temperature structure of the transition region from the XUV line intensities also yields the abundance of several elements.

Since the transition region is largely conduction dominated and because the coefficient of thermal conduction is strongly reduced perpendicular to the magnetic field, thereby causing a strong channelling of the heat flow along the magnetic field lines, the transition layer will follow closely the inhomogeneous structure in the magnetic field. The temperature distribution in the supergranular boundaries, where the field is concentrated, is therefore different from the temperature distribution at the region in between the magnetic concentrations.

At this moment it is hard to say whether the temperature gradient above the magnetic mottles is smaller or greater than in between. It depends on many other parameters such as the mechanical energy flux in both regions, which in turn determines the conductive flux. The value of the conductive flux which is frequently found in the literature is $F_c = 3$–6×10^5 erg cm^{-2} s^{-1}. If the XUV observations are interpreted taking reasonable inhomogeneous structure into account, this value may be an order of magnitude lower. High resolution XUV observations must solve this problem.

The thermal stability and the dynamical stability of such a thin layer where the temperature rises two orders of magnitude remains an open question. Is the radiative output of the upper chromosphere able to dispose of the downward conductive energy flux? How does the transition layer respond to fluctuations in energy input? Are the spicules required to match the energy balance? Is the transition region the breeding ground of all kinds of thermally driven motions which may well propagate high into the corona. How does the transition region respond to a disturbance from below such as a shock wave. These questions should be investigated taking into consideration the magnetic field structure, which will certainly dominate the dynamics of the transition region.

A problem of particular interest is the mass flux through the transition region. Spicules seem to provide almost hundred times more mass flux than is needed to supply the quiet day solar wind. This means that most of the mass ejected by spicules in some way must return to the chromosphere. Have these downward motions escaped observation or are there almost as many spicules with downward motions as with upward motions.

4.2. Heating of the corona

It is generally accepted that the corona owes its existence to the dissipation of mechanical energy transported by low frequency waves of acoustic nature from the photospheric layers into the corona.

Nevertheless a detailed theory is lacking which is not surprising if one takes into account on how many detailed physical processes such a theory depends. Moreover there are really no pertinent observations that may give conclusive evidence as to one of these many physical processes, so that one has to rely mainly on theoretical predictions. This situation is likely to remain unchanged for some years. Unfortunately the density structure, which is the easiest to determine, gives very little information on the heating mechanism. What is needed is a detailed knowledge of the temperature structure parallel as well as perpendicular to the magnetic field. One ought to know the proton temperature as well as the electron temperature not only in the outer corona but also at the 'base' of the corona.

In the upper chromosphere and the lower corona the dissipation is very likely provided by ordinary viscosity and thermal conduction in non-linear acoustic or magnetoacoustic (shock) waves. Alfvén waves probably do not contribute to the heating of the chromosphere because they may travel almost undamped into the corona.

Already in the lower corona but certainly in the upper corona the plasma properties are such that even for the low frequency waves the plasma may be considered as collisionless. In that case proton Landau damping of ion acoustic waves (which is the equivalent of an acoustic wave in a neutral gas) is the main energy source for the heating of the corona. Electron heating by direct Landau damping may be efficient in the upper layers of the corona. Such waves are not likely to be the direct remnants of waves, which would have been generated in the photosphere or below. They are probably of local origin generated as a consequence of the interaction of other wave modes or of the instability of plasma streams along the magnetic field.

Besides a study of wave damping, which is wave-particle interaction, in collisionless plasma's, the theory of the heating of the corona can only proceed if a profound study is made of the various possibilities of mode coupling in (collisionless) plasma's.

An interesting possibility to obtain some information on wave propagation in the or ona is from the study of fine structure in high resolution radiospectra. There are a number of non-thermal long-lived phenomena associated with an active region like type I storms, D_m type III storms, long lived fast particle streams, all not flare related, whose understanding would greatly benefit from a study of the collisionless processes during the propagation of magnetohydrodynamic waves in the corona.

4.3. THE SOLAR WIND

There has been a rapid progress in the general knowledge of the solar wind and the structure of the interplanetary medium, mainly from the observational point of view, essentially because of the developments in space research. However, there remain many theoretical problems to be solved since the energy sources are just as badly known as for the heating of the corona.

Collective interactions are fundamental in the interplay between systematic flow, magnetic field, rotation and different particle species, with different distribution functions in various directions with respect to the magnetic field. Evidence for the existence of plasma turbulence in the solar wind is found in the radio scattering data, radio observations of long lived sources and fast particle stream detection.

The study of the excitation, propagation and decay of different wave modes and their influence on the flow-velocity and the particle distribution function is of prime importance for the understanding of the solar wind. It might however be unrewarding to spend too much effort in trying to find the complete distribution function and the electromagnetic fluctuation spectrum in space and time. More physical insight would be gained from the study of the effect on the solar wind from evolutionary and sudden changes of boundary conditions thereby using simplified models. Such studies are necessary to understand e.g. the interplanetary sectors and the evolution and interaction of streamers.

It should be noted here that these problems are not only of interest for solar physics. Hot dilute rotating magnetoplasma occurs near other stars, in H^+ regions, supernovae, pulsars, galactic nuclei and quasars. The solar corona offers a first hand opportunity to test the physical theories applicable to such a plasma, using particle, X-ray and radio data.

4.4. CORONAL STRUCTURE

The lower corona, strongly influenced by inhomogeneous photospheric magnetic fields, exhibits a wide variety of fine structure, which is partially reflected in the inhomogeneous and variable radio radiation. Considerable work has been done on the structure of coronal streamers and I feel that the study of the physics of neutral sheets is of prime importance. The detailed understanding of the plasma physical processes in a neutral sheet is still lacking. Anomalous conductivity, wave-wave, wave-particle processes, wave and particle propagation inside the streamer, the influence of parallel flow on the stability can be handled by the physics known today. As said before, the temporal evolution of inhomogeneous structures is still in a first stage, but especially the birth and decay of a streamer, with and without helmet structure should be investigated.

As was to be expected, the average coronal structure is related to large photospheric regions with weak average magnetic field, and much less to small regions with large field strengths. The exact connection however, the development in time depending on the photospheric boundary conditions, is still quite unknown.

Besides streamers we encounter coronal holes, plasma nodules, filaments, coronal condensations. Interesting tests for the theory are provided if these features are disturbed by flare driven shocks or particle injections.

4.5. Coronal Instabilities and Non-linear Phenomena

Current theories of type III bursts indicate that the growth of plasma waves by a beam-plasma instability has a much smaller time scale than the duration of the beam, and the actual observed profile is more determined by the non-linear processes that stabilize a particle beam pervading the plasma, than by the linear growth and decay of the beam-plasma instability.

It thus appears necessary to calculate non linear coupling coefficients in order to describe the stabilized situation.

A good qualitative study of a particle beam inside or outside a coronal streamer and a comparison with the observations seems possible now. This may well clarify the problem of the occurrence of harmonic-fundamental pairs and fine structure and also to what extent type III bursts can be used as coronal probes revealing the coronal structure.

Above active regions, up to large heights in the corona, a high level of low frequency plasma turbulence is to be expected, which can generate fast particles which in turn can produce turbulence with higher frequency. The effects will be quite different for regions with open radial field lines than for those with closed field lines. In the latter case the energy transport can be blocked at a certain height leading to an explosive situation, while in the case of open field lines the heat flow is unimpeded and a coronal hole may arise.

Although type III bursts seem to offer a well-defined problem, much less is known about stationary type IV bursts with their detailed spectral fine structure. An interesting starting point is to study the relaxation of a hot plasma cloud trapped in a closed magnetic field structure. This is also of importance for understanding the magnetosphere and its interaction with the solar wind.

5. Solar Magnetic Fields

5.1. The Solar Dynamo

The last decade of theoretical solar physics has brought a great understanding of the nature of the 22-yr cycle of solar activity. It is generally accepted that the solar dynamo results from two types of motions: differential rotation and cyclonic convection or/and turbulence.

The basis of any dynamo theory is as follows: At solar minimum the field is essentially poloidal. Differential rotation generates a strong toroidal field which becomes visible (at the surface) first at medium latitudes and then at lower latitudes. The polarity laws can be explained in this way. The toroidal field emerges in active regions of bipolar nature. During the emergence of the toroidal field either by convective motions or through kink instability the field received a poloidal component of opposite polarity as the initial poloidal field during minimum activity. Supergranular motions then spread the fields over the solar surface and a new poloidal field of reversed sign is formed. Then the same processes are repeated with reversed signs to complete the magnetic 22-yr cycle.

It is known that dynamo's must be non-axisymmetric. Cyclonic convection caused by the Coriolis force and helical turbulence (turbulence for which $\langle \mathbf{u} \cdot \mathrm{curl}\,\mathbf{u} \rangle \neq 0$) have this property and are most likely to play an essential role in the formation of a poloidal field out of toroidal fields.

The effect of helical turbulence in a magnetic field is to generate a non curlfree electric field in the direction of the magnetic field. This mechanism, which is called the α-effect, generates a magnetic field perpendicular to the initial magnetic field. Turbulence is very powerful in the upper layers of the convection zone. Hence the α-effect can only be efficient in a shallow layer.

Models have been presented using the differential rotation in the deeper layers to build up toroidal fields and the α-effect to build up the poloidal fields in the upper layers. It is found that the distance between those layers regulates the period of the cycle. These dynamo's are called α-ω dynamo's contrary to the $\alpha\alpha$ dynamo's which only use the α-effect to obtain a cycle. Thus in $\alpha\alpha$ dynamo's the toroidal fields are produced through the helical turbulence from the poloidal fields. Satisfactory butterfly diagrams have been constructed.

However it is still not clear whether the solar dynamo is a mechanism that operates on superficial magnetic fields or whether deep seated fields are involved.

Non linear calculations confirm the observed fact that the period decreases slightly with increasing field amplitude. Moreover they show the asymmetry between rising and declining branch of a cycle.

5.2. SMALL SCALE MAGNETIC FIELDS

Once the toroidal magnetic fields have emerged through the convection zone they are spread over the whole solar surface in a random way. It has become clear that the magnetic field is not diffusively moving over the solar surface but it remains concentrated in magnetic ropes. The convective motions drag the field to the edges of the cells where they are concentrated especially in the corners of the supergranulation. This leads to fields of the order of 100 G.

It is a challenging question how fields can be amplified to values of the order of 3000 G as observed in sunspots. No satisfactory theory of sunspot formation exists at this moment but it seems that the modification of the energy transfer in rather strong fields must play a significant role on the further amplification of these fields. As soon as convection is reasonably suppressed, the atmospheric structure of the overlying layers is changed in such a way that lateral pressure equilibrium can only be achieved by the compression of the magnetic field. The Evershed motions are probably a direct consequence of this process.

The subphotospheric toroidal field strings naturally emerge as bipolar magnetic regions. Some of the field tubes are already heavily twisted when they appear at the surface. It is clear that these fields contain magnetic energy that can be released in the upper atmosphere. Other field tubes may appear less violently. When they merge into the corona the solar wind will drag the weaker fields outward thus forming streamers in which the field direction changes its sign. The neutral line separating the photospheric polarities thus marks the basis of a neutral sheet. The stronger fields can resist the solar wind drag and consequently they remain present as closed loops clearly visible in he lower parts of helmet streamers.

The way in which the small scale fields emerge and the interaction of the solar wind with the coronal fields eventually determines the structure and shape of the solar corona (Section 4.4).

Although it looks as if the corona is a collection of streamers it should be stressed here that in the absence of any kind of solar activity the quiet solar corona exists which is clearly demonstrated by the 1954 eclips photograph.

5.3. Solar flares and prominence

Solar flares are the most complex phenomena on the Sun. Any theory of solar flares ought to explain a great multitude of observations of enhanced radiation in the whole spectral region which takes place in a very short time. The impulsive phase lasts about 2 min, the eruptive phase 15 min and after about one hour the flare eruption is over. On a very short time scale there are type III bursts lasting about 1 s and X-ray and microwave bursts with a duration of 0.5–2 min.

The total amount of energy released is about 2×10^{32} erg which is supposedly extracted from a volume of 10^{28} cm^3. Only a strong magnetic field of the order of 100 G would be sufficient if all its energy could be converted in the electromagnetic and corpuscular radiation.

There are two basic ideas which in some variation always appear in flare theories. Firstly, the X type neutral point or neutral line theory and secondly the current-discharge theory. In the first category of theories the magnetic energy is liberated at the neutral line while in the second type of theory the magnetic energy of a current circuit is liberated at an interruption point. The last theory has the advantage that very large electric fields are created which could accelerate particles up to high energies, but it seems on the other hand very unlikely that coronal currents can be built up to such large values that an interruption occurs. Long before this could happen, the currents become kink unstable and a new configuration appears in which actually a twisted 'neutral line' is present along which the flare eruption can take place. This means that the X-type neutral line theory in a modified geometrical configuration is probably the essential core of the flare model.

But there are many questions to be solved. Which instabilities can occur? How is the plasma heated and where and when precisely are the particles accelerated? However, the basic problem in flare theory remains the storage of energy in suitable magnetic field configurations and the sudden release of this energy after some time.

Prominences are divided in two classes: the quiescent and the active prominences. An element common in all recent prominence theories is a flux tube which is supported against gravity either by its own tension or by ambient magnetic fields.

It can be shown that these flux tubes can originate in coronal neutral sheets during the cooling of coronal matter. Neutral sheets where fields of opposite polarity meet are the places where quiescent prominences naturally occur. There is strong evidence that quiescent prominences are a transition phase during the reconnection of the antiparallel fields.

Since over large parts of the solar surface the field is of one polarity an emerging bipolar magnetic region of much smaller extension than the surrounding unipolar region always somewhere creates a region of oppositely directed magnetic fields. They will thus occur at the border of magnetic regions, preferentially when new born magnetic fields interact with old existing fields.

The evolution and annihilation of magnetic fields which is observable in the form of flares and prominences should have a great priority in the next decade of theoretical solar physics.

6. Conclusion

Solar physics is a living science and one of the most sophisticated of all astronomical sciences. Many physical processes that take place in other cosmical objects but are difficult to observe can be studied in detail on the Sun.

Theoretical solar physics should lead to:

(a) an understanding of observed solar phenomena.

(b) the possibility of extrapolating this knowledge to other stars.

(c) as far as unstable plasma processes are concerned it should be investigated whether results obtained in solar physics can be scaled such as to make them applicable to galactic and extragalactic circumstances.

The subsequent list contains what I would call the major physical problems in solar physics:

I. Rotation and circulation
   1. differential rotation
   2. compressible non-Boussinesq turbulent convection

II. Non-thermal energy transfer
   1. generation and propagation of atmospheric waves
   2. heating of the corona
   3. solar wind

III. Generation and evolution of magnetic fields
   1. solar dynamo

2. evolution of magnetic fields in the atmosphere e.g. sunspots
   3. magnetohydrodynamics of neutral sheets, prominences

IV. Plasma instabilities
   1. the mechanism of solar flares
   2. plasma turbulence in the solar corona
   3. acceleration of particles

I am well aware of the very personal view that is expressed in the above table. But it is still too much extended if it comes to priorities.

If I had to make a selection of what I would think are the important topics in the next decade I would with some hesitation give the highest priority to the following subjects:
 – differential rotation
 – heating of the corona
 – birth and decay of sunspots
 – magnetohydrodynamics of neutral sheets
 – mechanism of solar flares
 – acceleration of particles.

I hope this survey on solar physics may help to concentrate our work around the above mentioned topics.

Acknowledgements

The author thanks Drs Van Bueren, Raadu, Rosenberg, Rutten and Zwaan for many useful information and critical discussions. He is especially grateful for the many interesting discussions he had with Dr Rosenberg on this subject and for his kindness to present this paper at the IAU, General Assembly in Sydney.

THE NEXT DECADE IN OBSERVATIONAL SOLAR RESEARCH

J. M. BECKERS

Sacramento Peak Observatory, AFCRL, Sunspot, N.M. 88349, U.S.A.

1. Introduction

The last two decades have been exceptionally fruitful for astrophysics. Because of very rapid developments, especially in the area of observational astrophysics, our view of the Universe has been substantially completed, challenged and altered. Solar Physics has very much been part of this progress. Developments in space and ground-based technology have virtually completely opened up the entire spectrum of solar electromagnetic and corpuscular radiations. Electromagnetic radiation of the Sun can now be studied from gamma ray wavelengths shorter than 10^{-4} Å to radio wavelengths as long as 10^{14} Å, a range of approximately 60 octaves. Continuing improvements of spatial resolution of the Sun at all these wavelengths and concomitant development of our theoretical understanding of both the radiative and magnetohydrodynamic processes on the Sun have made solar physics an exciting field to those interested in the Sun per se, as well as to those who view the Sun as a star whose study aids us in the understanding of astrophysics in a broader sense.

A committee of The National Academy of Sciences in the United States under the chairmanship of J. L. Greenstein conducted between 1969 and 1971 a study of the needs of astronomy in the next decade. I took part in this study and it is probably because of this that I was asked to give this review on the next decade in observational solar physics. This review is therefore partly based on the results of the discussions in that committee and in its solar panel. Most of this review will reflect, however, my own opinions.

The first part of the Greenstein Committee's Solar Panel study was devoted to the goals of solar research and to the relevance of solar physics to other areas of human endeavor. Specifically, it outlined the significance of solar physics for general astronomy, for pure physics, for the understanding of the solar system and for human activities in the Earth's environment. It is outside the scope of this review to discuss these items in detail. It was felt, however, that it was necessary to put solar research in a broader perspective before discussing the solar physics program. Solar physics is unique among the astrophysical sciences in that in addition to being an area of pure scientific endeavor, it is of practical interest to the inhabitants of this planet. Short term solar variations associated, for example with solar flares, are the dominant factor in determining the weather in the upper atmosphere and nearby space. Longer term variations of the Sun are now considered to be related to weather changes and are a very likely cause for the ice ages.

The remainder of the Greenstein Committee Solar Panel study was devoted to the discussion of a solar physics program for the nineteen seventies. This discussion was so-called *problem oriented*. It is a general trend in discussing future programs in solar physics to outline the main problems and to base on this the requirements for future instrumentation. This way of approaching the future of a science is certainly partly valid. In looking back, however, one notices that many of the discoveries and advances in solar physics, and in astrophysics in general, were the results of technological progress and of exploration. The discovery of quasars and pulsars was not the result of a problem-oriented approach but of the desire to explore and to develop novel and new techniques. So was the discovery of many of the phenomena on the Sun, which form the basis of much of solar physics, the results of the *technological* and *explorative approach*, although one can rightly point at some solar phenomena, like the solar wind, whose discovery was related to a problem orientation.

In the Greenstein study we defined four broad major problem areas in solar physics. These were:
  (a) Flare Instabilities and Particle Acceleration,
  (b) Energy Generating Processes in the Solar Interior,
  (c) Energy and Mass Transport in the Sun, and
  (d) The Large Scale Circulation in the Sun.
All these problem areas are, at least in part, related to the topic of this review which concentrates on the Sun viewed as a star with an emphasis on the observable aspects of the quiet Sun. In this review, however, I will not strictly follow such a problem-oriented approach. Instead I will concentrate on what I will call the needs of observational solar research like, for example, the need for improved spatial resolution. I will then argue for the support of these needs on the basis of problems which are to be solved.

2. Present Status of Observational Solar Research

Before discussing the future objectives of observational solar physics I would like to survey the present status of the field. This will help to outline the limitations of solar observations and it will indicate what the needs are for future improvements at least as far as the purely technological aspects are concerned.

The quality of solar observations can pretty much be described by their spatial, spectral and temporal resolution of solar phenomena. To achieve the highest possible resolution in one of these three dimensions one often has to sacrifice resolution in one or both of the other two. For example, the highest possible spatial resolution observation of photosphere and sunspot fine structures are only possible by very short exposures so that the high spectral resolution becomes impossible. In the ultraviolet region of the spectrum, flux limitations impose a similar limitation.

Figures 1 and 2 summarize the spatial and spectral resolution which can be (and has been) achieved with today's instrumentation. Figure 1 illustrates the spatial resolution on the Sun in arc seconds for the full 18 decade range of the observed electromagnetic spectrum. The dashed lines give the limiting resolution of telescopes of the given aperture. The best resolution on the sun is presently achieved in the optical, and to some extent, in the cm region of the spectrum. In the optical region a spatial resolution

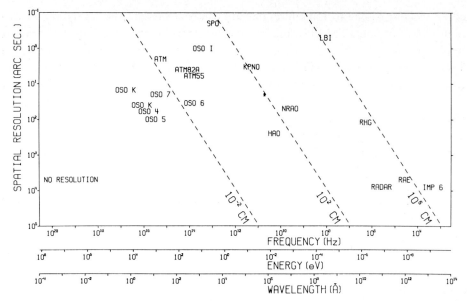

Fig. 1. Best spatial resolution achieved at different wavelengths in the solar spectrum. Dashed lines represent the limiting resolution of telescopes with the indicated aperture. Explanation of symbols: OSO = Orbiting Solar Observatory: ATM = Apollo Telescope Mount; SPO = Sacramento Peak Observatory; KPNO = Kitt Peak National Observatory; HAO = High Altitude Observatory; NRAO = National Radio Astronomy Observatory; RAE = Radio Astronomy Explorer; LBI = Kilometer Baseline Interferometer; RHG = Culgoora Radioheliograph. OSO-I and J points represent the predicted performance. OSO-K points refer to a feasibility study for a solar-flare satellite.

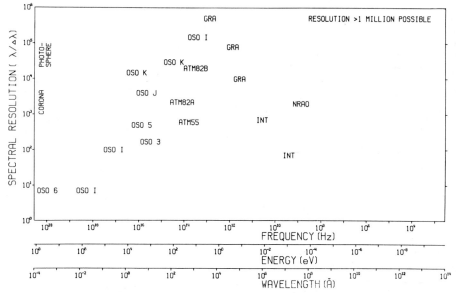

Fig. 2. Best spectral resolution achieved at different wavelengths in the solar spectrum. Explanation of symbols as in Figure 1, except for GRA = Grating Spectrograph; INT = Michelson Interferometer (Fourier spectrometer).

of 0.2″ corresponding to 150 km on the Sun can be achieved. The cm radio observations of 0.3″ are obtained by kilometer baseline interferometry (Kundu, 1973). Their resolution is high only in one spatial dimension. The same is true for occasional eclipse observations, not shown in this slide, which attained high spatial resolution in the ultraviolet, infrared, and mm region of the solar limb spectrum (see e.g. Gabriel *et al.*, 1971; Noyes *et al.*, 1968; Coates *et al.*, 1958) by means of the occultation by the Moon. Only in the optical region and cm region of the solar spectrum is the resolution good enough to see solar details like solar granules and spicules. In the soft X-ray and EUV region we can resolve the chromospheric network and its coarse elements. Single radio dishes and the Culgoora radioheliograph fall just short of resolving the supergranulation and network. In the hard X-ray, and low frequency radio spectrum the resolution becomes so low that only integrated solar radiation can be studied. The same is true for the radar reflections from the Sun. I should point out again that the highest spatial resolution generally corresponds to low spectral resolution. The 0.2″ optical resolution, for example, can only be achieved for very short exposure times as is the case for integrated light observations. Very high spectral resolution solar spectra achieve at best 0.5″ spatial resolution.

Figure 2 summarizes the status of spectral resolution that can be obtained. Spectral resolution sufficient to resolve photospheric, chromospheric and coronal line profiles can be achieved over most of the electromagnetic spectrum. Only in the far infrared,

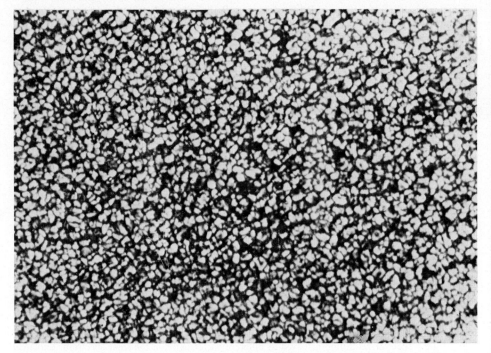

Fig. 3. Observation of the solar photosphere with the Pulkovo balloonborne telescope. (Courtesy: Pulkovo Observatory.)

millimeter and X-ray spectrum is the spectral resolution insufficient to study the shape of solar spectral lines.

I did not prepare a figure to demonstrate the temporal resolution as a function of wavelength. Most solar variations seem to occur on time scales longer than 1 s so that temporal resolution is in principle no limitation. Flux limitations set an effective limit, however, for some experiments if an acceptable signal-to-noise ratio is to be obtained. The only exceptions to this are coronal variations associated with solar activity. In both the hard X-ray fluxes and in the radio emission fast unresolved coronal bursts are observed.

In the last part of this quick review of the present status of observational solar research I want to present a sample of some of the most excellent and up-to-date observations of the solar atmosphere. I will do this successively for photospheric, chromospheric, and coronal observations.

The photosphere is best observed in the optical and near infrared region of the spectrum. Only in these spectral regions is the absorption in the solar atmosphere small enough to make photospheric observations feasible. The best observations so far have been obtained from balloons and from carefully designed ground-based telescopes.

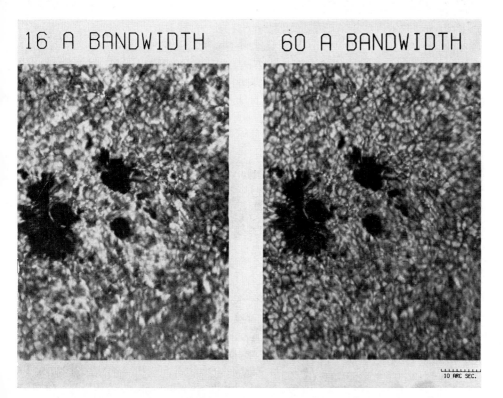

Fig. 4. Observations of the upper solar photosphere near the calcium K line. (Courtesy: Sacramento Peak Observatory, AFCRL.)

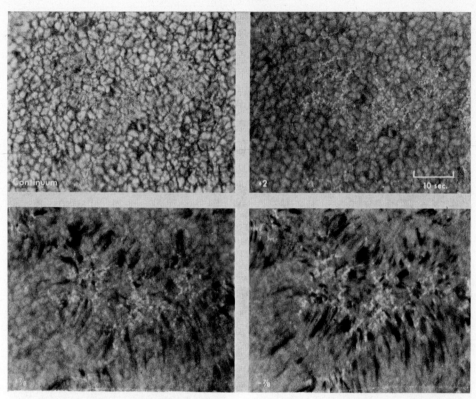

Fig. 5. Observations of solar 'filigree'. (Courtesy: Sacramento Peak Observatory, AFCRL.)

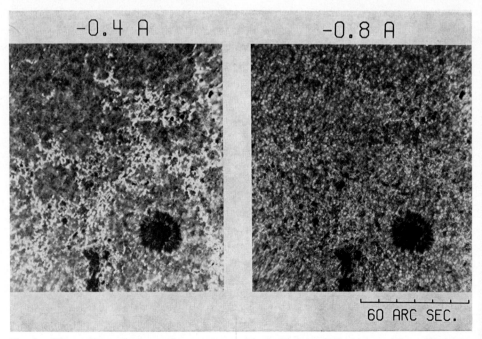

Fig. 6. Observations of subgranular structures with the Universal Birefringent Filter. (Courtesy: Sacramento Peak Observatory, AFCRL.)

Figure 3 was taken with the balloon-borne telescope of the Pulkovo Observatory in the U.S.S.R. This telescope is equipped with a variety of auxiliary devices including a spectrograph with slit jaw monitor and an integrated light camera. This slide was made with that camera. At first sight, it appears similar in quality to those made with the Princeton stratoscope balloon project in the early 1960's. However, when I inspected the films in Leningrad, I noticed the presence of subgranular structures located mainly in the dark intergranular lanes. Because of the excellent performance of the Sacramento Peak Vacuum Telescope, it has been possible to confirm the presence of these structures. Figure 4 shows the best image of the photosphere obtained in integrated light with the vacuum telescope of Dr Mehltretter. It clearly shows the subgranular structures whose size has been measured to be 0.2" or less. Similar structures can be seen in Figure 5 taken by Dr Dunn in the wings of the Hα line, and in many other lines studied with the Universal Birefringent Filter (Beckers, 1972a, b) at the Observatory. Their group appearance has been called by Dr Dunn 'filigree'. Figure 6 shows an image obtained through this fully tunable Lyot filter in the wing of the Magnesium b_1 line. Similar Zeeman observations taken with the Universal Filter in the core of the magnetically very sensitive b_2 line show that the location of the filigree-subgranular structure pattern tends to coincide with regions of magnetic field enhancements. This has led to the picture of these subgranular structures as the result of the bundling of the magnetic field lines at the granule boundaries by the outward flows in the granules. This bundling occurs presumably in very much the same way as the formation of the magnetic network formed by the supergranulation. A spectroscopic investigation by Drs Simon and Zirker (1973) has as yet failed to support this association of the subgranular structure with the magnetic enhancements. Spectra require, however, longer exposure times and therefore have a lower spatial resolution which may explain this failure. The best spectra have a spatial resolution of about 0.5". Figure 7 shows an example of a spectrum in the chromospheric calcium H-line. The strong enhancements in the K_2 reversals (A and B) are the intersection of the calcium chromospheric network. Very well visible are also the bright emission grains in the violet K_2 peak and the dark absorption features which are superimposed on the bright network intersections. These dark features correspond to the dark elongated fine mottles visible in high resolution Hα filtergrams as the one shown in Figure 8. These dark fine mottles, thought by many to be identical to the spicules observed at the solar limb, the Calcium K_{2V} grains and other details visible on chromospheric images are all components of an incredibly complex chromosphere.

The chromosphere is a manifestation of the spectacularly rapid transition zone from the relatively cool photosphere to the hot corona and solar wind. Associated with the complexity of the chromosphere is probably a similar spatial complexity in this transition zone and in order fully to understand the origin of the solar corona, it will be necessary to resolve these spatial complexities. Our best bet for determining the physical conditions of the photosphere-chromosphere-corona transition zone lies in the observations in the ultraviolet and radio region of the spectrum. Line and continuum intensities there give very direct information on the temperature and density

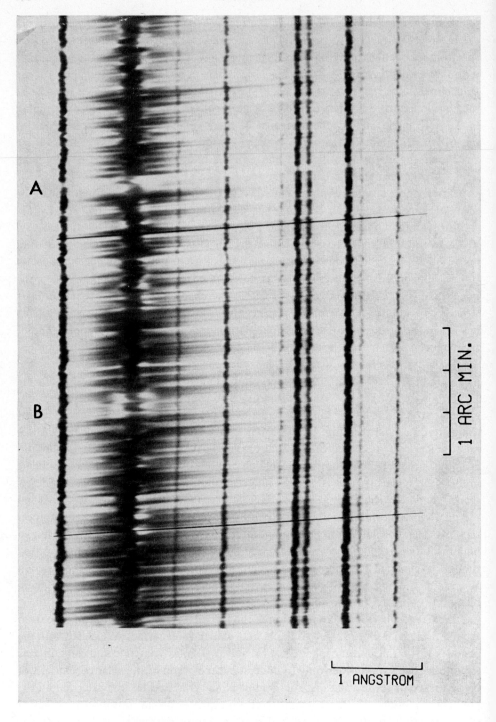

Fig. 7. Spectrum of the center of the solar disk in the Ca^+ H line. Blue is to the right. (Courtesy: Sacramento Peak Observatory, AFCRL.)

Fig. 8. Filtergram in the wing of the Hα line. (Courtesy: Sacramento Peak Observatory, AFCRL.)

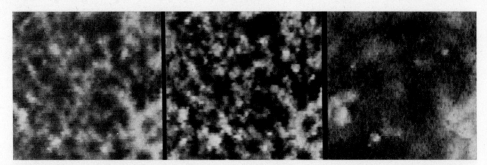

Fig. 9. Observations of the Lα, C III and Mg X chromospheric coronal network with the Harvard ATM experiment. (Courtesy: Harvard College Observatory.)

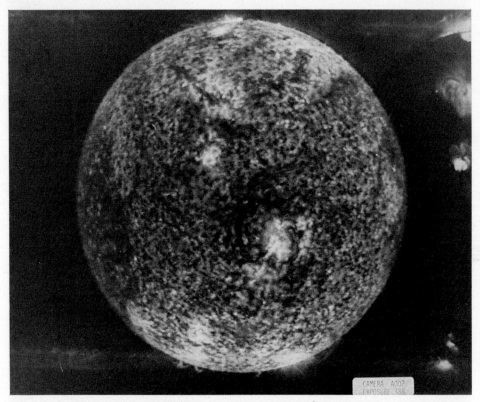

Fig. 10. Observation of the chromospheric network in He II 304 Å with the NRL-ATM experiment. (Courtesy: Naval Research Laboratory.)

structure of the transition region. Recent efforts in UV and radio solar astronomy have, therefore, been in the direction of improving spatial resolution on the Sun. Especially many of the recent ATM-experiments and the experiments on the OSO-I spacecraft to be launched next year are directed toward the study of the fine structure of the transition zone. The ATM experiments have a resolution of about 5″, sufficient to resolve the chromospheric network, and OSO-I will have a resolution of about 1″ which will be sufficient to resolve spatial details like, for example, chromospheric spicules.

Figure 9 shows examples of quiet chromosphere/corona observations with the Harvard ATM telescope. These spectroheliograms were taken simultaneously in the Lα line, a C III line and a Mg X line. The Lα and C III lines clearly show the chromospheric network. When going to lines of ions of higher stages of ionization the network gradually fades out, its elements becoming both weaker and perhaps more diffuse until hardly any network remains in the coronal lines like the 625 Å Mg X.

Figure 10 was taken with the NRL telescope on-board ATM. This telescope gives solar images through a slitless spectrograph which are then photographed. Since it does not use a resolution limiting device like the Harvard 5″ × 5″ slit it is in principle capable of better than 5″ resolution. Again the network elements are clearly visible in this He II 304 Å spectroheliogram.

The ATM experiments are providing solar research with a vast amount of information which will undoubtedly lead to a giant leap in our understanding of the Sun. During the course of this General Assembly and during the symposia that will follow it, we will probably see many data from ATM. We will have to wait for OSO-I to obtain at the same time 1″ spatial resolution and the spectral resolution necessary to study line profiles. From rocket experiments we have, however, already seen a sample of what to expect. Figure 11 shows a spectrum of the Sun in the Lα line and in two lines of the O I triplet at 1304 Å, obtained by Dr Bruner. The spatial resolution of these spectra is about 20″, the spectral resolution is sufficient to resolve the lines very well and to separate them from the central reversals caused by the geocorona.

Kilometer baseline interferometry and image synthesis as applied to solar radio astronomy opens the possibility of giving images in the 1″ resolution class. Dr Kundu is pursuing this route. Figure 12 is an example of an observation at 3.5 cm wavelength. In order to synthesize a fully resolved image it is, however, necessary to record data over a long time (hours) so that the temporal resolution presents a real problem, when observing the short-lived arc second features. Another potentially very exciting development in mm and cm radioastronomy is the study of the solar spectrum with high spectral resolution. Since solar spectral lines were predicted a number of years ago, there have been some, so far unsuccessful, efforts to study the spectrum in this wavelength region. If the nα transitions (Dupree, 1969) or any other atomic transition should be found in the radio spectrum one would have a potentially very powerful way of determining the chromospheric and coronal magnetic fields because of the very large Zeeman splitting for spectral lines in the radio region.

Some experiments on-board Skylab study the coronal structures, one by looking at the white light corona outside the solar limb, two by looking at the corona X-rays on the disk and just outside the limb. Such an X-ray image taken by AS&E with ATM is shown in Figure 13. Spatial resolution is again in the 5″ class. Both types of synoptic coronal observations will lead to vast improvement in our understanding especially of the coronal spatial and magnetic structure. An interesting possibility for studying

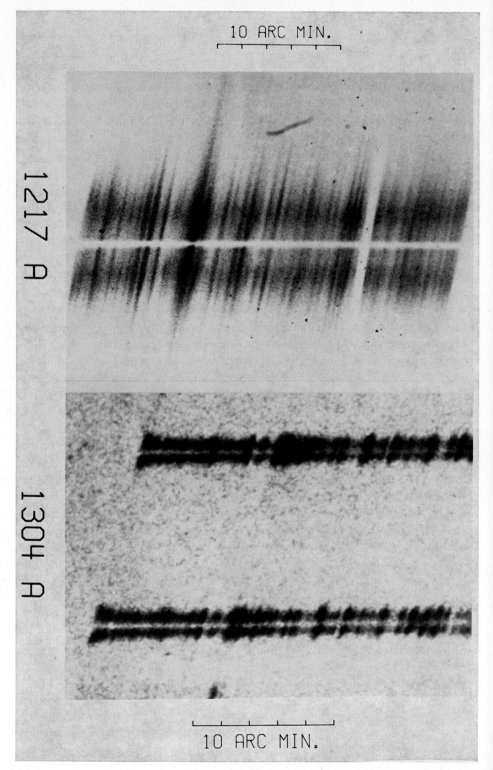

Fig. 11. Hydrogen Lα line and two O I lines at 1304 Å on the solar disk. (Courtesy: LASP, University of Colorado.)

THE NEXT DECADE IN OBSERVATIONAL SOLAR RESEARCH 161

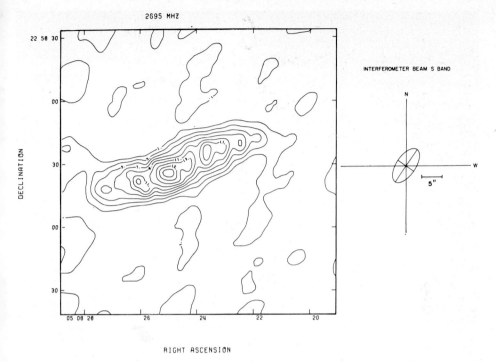

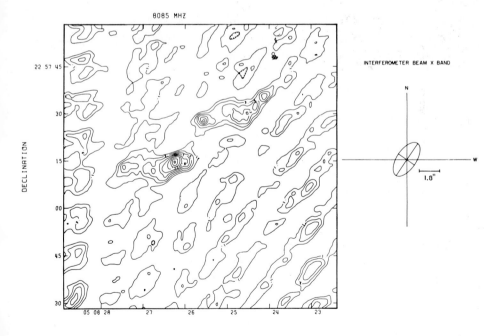

Fig. 12. Map of the solar 3.5 cm radiation as obtained with a long baseline interferometer. (Courtesy: University of Maryland.)

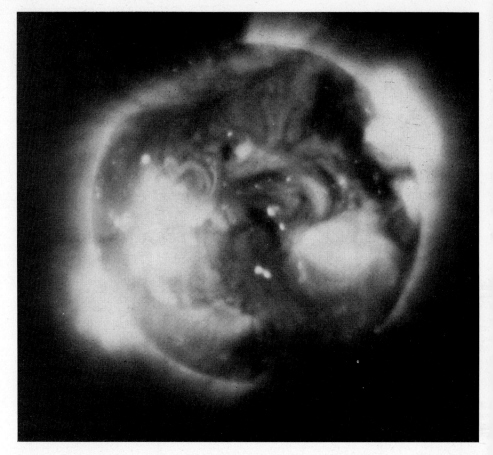

Fig. 13. X-ray image of the Sun taken from Skylab. (Courtesy: American Science and Engineering.)

the coronal temperature structure presented itself, however, during the 1970 solar eclipse when the solar Lα corona was discovered by a Culham-Harvard experiment. Figure 14 shows one of the observations. Dr Gabriel (1971) showed that the Lα emission can be explained by resonance scattering by the very few neutral hydrogen atoms left in the very hot corona. Since these atoms have very large thermal motions, the line width of the coronal Lα line should be a very accurate indicator of the coronal temperature.

With this I terminate what necessarily had to be a rather superficial summary of the present status of solar observational research. Undoubtedly, I am guilty of the omission of some significant efforts in this area on the one hand and maybe of overstressing some of my own interests on the other hand, In summary, I do believe that the main overall stress in observational solar work has been on improving spatial resolution of solar observations with a secondary significance attached to the improvement of spectral resolution.

Fig. 14. Lα image of the solar corona. (Courtesy: Culham Laboratories.)

3. Future Objectives of Observational Solar Research

One obvious way to prophesy on the future developments of observational solar research is to extrapolate from the past. In doing this one would predict a continued improvement, especially in the spatial but also in the spectral, resolution of solar observations. Solar physicists doing the observations tend to follow this line since it leads into one of the major unexplored areas of solar research – that of the subgranular, microstructure of the Sun – in which major discoveries are likely.

Solar physicists who concentrate on the theoretical aspect of the sun are frequently frightened by the complexities resulting from high resolution observations. Theoretical solar physics has not reached a sufficient level of sophistication to deal with these complexities. The need for higher resolution is questioned by both theorists, the sponsoring agencies, and the non-solar astrophysicists who are competing for instrumentation funds. The solar observationalist is thus forced to justify his endeavors beyond the desire to explore the unknown. He is asked to describe the problems which he wants to solve and to show that these problems are relevant in some broader frame of values. Based on this, he then justifies his efforts to obtain novel and new instruments. In his request to me to give this lecture, Dr Athay, a theorist par excellence, indicated his desire to have it given in a problem-oriented framework. I will try to do this by discussing what I believe to be 'the five needs of observational solar astronomers'. These are:

(a) The Need for Improved Spatial Resolution,
(b) The Need for Improved Spectral Resolution,
(c) The Need for Improved Temporal Resolution,
(d) The Need for an Improved Interpretation of Measurements, and
(e) The Need for Novel Observations and Ideas.

3.1. The need for improved spatial resolution

Whenever it has become possible to study the Sun with improved spatial resolution we have seen smaller and smaller details. Most recent examples of this are the observations of the subgranular structures (discussed already) and the results of the solar speckle interferometry by Harvey *et al.* (1973) both of which resolved solar features near 0.2″. There is no reason not to expect much smaller size structures of the solar atmosphere. On the contrary, flare theories, for example, predict that current sheets associated with the flare instability have a scale on the order of 1 km and less. We can also infer from spectral line width observations that small scale structures must exist, since the power in the spatial velocity spectrum for the highest resolution observations falls well short of the total power for all spatial frequencies as derived from the line broadening (see e.g. Beckers *et al.*, 1969).

Very small structures, therefore, undoubtedly exist on the Sun. It may, however, be impossible to observe them since the radiation used to observe the Sun represents an integration over a long effective path along the line of sight. The contribution function half-width is typically about 2 scale heights corresponding to 150 km or 0.2″ in the photosphere. When looking at a 3 dimensionally and isotropically random structure like turbulence it would, therefore, become difficult to resolve anything below this size. This is, however, not necessarily true for all small-scale structures like thread type structures in flares and prominences. Fine structures seen in the corona, in fact, already give an example of observed phenomena which are much smaller than the contribution function halfwidth.

Given the fact that small structures are present and observable, why should an effort be made to study them? I see three broad areas of interest which would require higher resolution observations. The first is the need to know the *three-dimensional model* of the solar atmosphere. Only through the knowledge of the local variations in the atmospheric physical conditions like temperature and density will we be able to properly assess the validity for both stellar and solar physics of the inferences drawn on the basis of simple one-dimensional atmospheric models.

The second reason for improving our spatial resolution lies in the need to fill in major gaps in our understanding of the *energy transport* in the outer layers of the Sun and stars. Radiation carries the bulk of the energy away from the Sun. The motion in the photosphere, resulting in the relatively small kinetic energy flow, is, however, responsible for the existence of the solar chromosphere, corona, and solar wind and associated phenomena like angular momentum loss. What are the characteristics of the motions in the photosphere? We have a fair idea as to the properties of the supergranulation although there is still some disagreement as to the nature of these flow patterns. It has been customary in the last decade to think of the photospheric granulation as a direct display of the convection elements of the solar hydrogen convection zone. Recently, however, the suggestion that the granulation is the result of waves has been getting renewed attention. It seems to me that new observations of exploding granules and of rapid horizontal granular motions lend support to the notion of the granulation as the results of waves which in turn are excited by the subphotospheric turbulence or convection. High resolution observations are needed to decide between these two granule models. The nature of the subgranular structures is entirely a matter of speculation.

Which size elements are responsible for the heating of the transition region and the corona? As far as I am aware, all coronal heating models are one-dimensional. The effect on coronal heating of the horizontal scale size of the waves that excite the shock waves which in turn heat the corona by dissipation is not clear to me. It could well be that the interaction and collision of small-scale shocks would enhance the effectiveness of this heating mechanism, in which case the granular and subgranular elements, which contain much of the kinetic energy flow, are dominant. This is an area of speculation on which I invite comments. There are, however, also observational indications that the heating occurs at a small scale. Both Liu (1973) and Cram (1973) have suggested that the small grains visible in the violet wing of the H and K lines are the results of shocks.

Knowledge of the temperature and density structure of the transition zone is of great value when evaluating the corona heating processes. It permits us to study the rate of energy dissipation and of energy transport by conduction and perhaps kinetic processes. Observations of the emission measures of lines of widely varying ionization and excitation potential in the ultraviolet have given us an excellent tool for the study of this temperature and density structure. So far, most studies have been one-dimensional, although it is clear from satellite and rocket observations that there are very large horizontal fluctuations in the emission measure. Hence there are probably variations in the energy transport down to the scale of spicules and calcium grains and perhaps smaller structures. The emission measure technique of probing the transition zone is directly applicable to any size structure so that high resolution observations in the ultraviolet will give direct information on the small scale structure of the transition zone.

The third reason for improving the spatial resolution comes from the need for understanding the *mass transport* in the Sun. The loss of mass in solar and stellar winds is directly related to the presence of non-radiative energy transport. Mass and energy transport are therefore related. Solar spicules carry two orders of magnitude more matter into the corona than is lost by the solar wind. They are, therefore, very significant in the mass balance. We know nothing about the mass transfer by other photospheric, chromospheric and coronal structures. Nor is the relation between mass and energy input in the corona clear. Do they occur in the same region or do they occur spatially quite independently?

In order to increase our understanding of the structure of the solar atmosphere and of the energy and mass transfer in solar and stellar atmospheres, it is necessary to improve our spatial resolution. Large aperture optical telescopes are needed for photospheric studies. The introduction of ground-based vacuum telescopes has been very successful and it is likely that proper site selection will make it possible fully to exploit vacuum telescopes with apertures larger than 1 meter. Good time sequences will probably require more expensive balloon and satellite-borne telescopes. The study of the transition zone requires high resolution satellite-borne ultraviolet telescopes. ATM and OSO-I will produce 1″ class observations. The Greenstein Report recommended higher resolution OSO-type spacecraft, especially to study the transition zone and solar flares. It seems also that a continued effort for better resolution in solar radio astronomy might supply very significant additional information on the solar chromosphere.

3.2. The need for improved spectral resolution

In contrast to the spatial resolution it is possible to give definite limits to the desired spectral resolution. Thermal broadening limits the required spectral resolution to $\approx 10^5$ in the photosphere and $\approx 2 \times 10^3$ in the corona unless extremely precise profiles are needed in which case an order of magnitude larger resolution is desirable. OSO-I will give spectral resolutions approaching these values down to about 1000 Å. From the OSO-I investigation one hopes to obtain additional information on velocity patterns in the transition zone which in turn will lead to information regarding mass transport and wave dissipation. In the optical region we have now all the spectral resolution which one could possibly desire. The areas of future developments seem to be (a) continued improvements of the spectral resolution especially in the far ultraviolet and X-ray spectrum, especially when combined with high spatial resolution, and (b) the analysis of the microwave and mm radio spectrum, especially as concerned with the search of solar absorption and emission lines there.

3.3. The need for improved temporal resolution

Smaller and smaller structures have shorter and shorter lifetimes. The temporal resolution of solar telescopes is generally, however, sufficient to resolve the temporal variations occurring on the Sun. Flux limitations sometimes present a problem which requires larger telescope apertures for a solution. Frequently image variations caused by non-solar origin like seeing and spacecraft motions also require high time resolution.

3.4. THE NEED FOR IMPROVED INTERPRETATION OF THE MEASUREMENTS

Solar observations consist in the first instance of intensity measurements as a function of position on the Sun, time, wavelength and polarization mode. To infer from these the physical conditions on the Sun is often difficult even in such commonly considered simple cases as Doppler shift and Zeeman splitting observations. One real need for the observational solar physicist is a better insight into the meaning of his measurements so that he may avoid the various traps associated with simple-minded interpretation. Let me illustrate this need in the case of Doppler shift and Zeeman splitting measurements.

All measurements are out of necessity a spatial average over an area of the Sun. All indications are that any area which can presently be resolved must contain velocity (or magnetic field) inhomogeneities so that the measurement gives a velocity (or magnetic field) which represents a spatial average, weighted by factors which include local variations in line strength and continuum intensity. One may not, therefore, conclude from the existence of an average blue shift that this corresponds to an average mass flux in the direction of the observer. The most spectacular example of this trap is the blue shift of the Sun itself which is believed to be caused by the upward motions of the bright granules dominating, as far as the average spectral intensities are concerned, the downward motion of the intergranular regions. In weak lines this blue shift amounts to 0.5 km s^{-1}. If interpreted as an expansion of the Sun this would mean that the sun would expand out to 1 AU in about 10 years. Since the blue shift observations were made over 17 years ago we can be sure that this interpretation is wrong. Other examples of trouble caused by spatial averaging are the inconsistencies in the magnetic fluxes observed in different lines (Harvey *et al.*, 1969) and the measurements of the Evershed Effect. The only real solution to the problem of spatial averaging is to avoid the averaging by improving the spatial resolution. This is another good reason for improving the spatial resolution. Since it is unlikely that one will ever resolve the Sun fully we will also have to continue the past approach which makes sophisticated guesses as to the spatial structure based on the observations in different lines.

Temporal averaging presents, in principle, a similar trap. If velocity variations with time are so fast that we cannot resolve them, one can have the same trouble as with spatial averaging. One example of this effect is perhaps the Evershed Effect which Maltby *et al.* (1967) explained as a temporally and spatially unresolved wave phenomenon.

There are other difficulties in the interpretation of data even when the temporal and spatial resolution is sufficient. First it should be remembered that downward motions, observed in the Hα lines at the supergranule boundaries, only refer to the motions of the neutral hydrogen atoms responsible for the Hα line and not to an overall mass flux as has often been suggested. The downward Hα motions may in fact be compensated by upward motions in gases which are invisible in the Hα line. Then there is averaging over height which causes real problems in case of a variation of the Doppler velocity along the line of sight. Frequently contribution function arguments are being used to estimate the effective optical depth to which a Doppler shift measurement in a particular line refers. It can be shown (Parnell *et al.*, 1969) that the effective optical depth is a function of the velocity variation along the line of sight and that it very well can fall outside the main part of the contribution function. In addition, there are cases in which the apparent measured velocity does not correspond to an actual velocity at any point along the line of sight. Athay (1970) and myself (Beckers, 1968) showed that this may frequently be the case in the Hα line where even the direction of an apparent motion or magnetic field can be opposite to the real motion or magnetic field.

Apart from improving the spatial and temporal resolution, the best progress in this area can be made in line profile calculations of two- and three-dimensional solar model atmospheres in which various velocity and magnetic field distributions are assumed. Hopefully, one may gain some insight as to the meaning of the measurements by comparing the profiles and shifts of carefully selected lines with these calculations.

3.5. The need for novel observations and ideas

The final need for observational solar research lies in the need for new types of observations and for new ideas. So far I have just extrapolated from past experiences into the future. The most exciting progress is often made, however, with entirely new experiments and ideas. Examples are the solar neutrino experiment and the attempts to detect solar oblateness. Less spectacular examples of new experiments are perhaps the various developments of Stokes polarimeters and an attempt to study the intensity and profile of the coronal Lα line at the 1974 solar eclipse. New ideas and experiments can be the result of a close interaction with other fields of scientific endeavor including theoretical astrophysics, fluid dynamics, optics, etc.

4. Conclusion

I have refrained in this review from discussing details of potential future solar instruments. Instead, I indicated what, in my opinion, the general direction should be in which solar observational research should go. There are already a number of efforts underway which are consistent with my suggestions. It takes a lead time on the order of a decade to obtain most instruments of the magnitude suggested in this review. It is in a way, therefore, a futile effort to discuss the aims for the next decade since most of the plans for the next decade are already firm.

Most new instrumental efforts are expensive. As Greenstein repeatedly stated in his study 'Astronomy is Big Science'. In recommending a program for science, the big items are commonly stressed since these require the main promotion effort. Much can, however, be done with more modest means – especially in connection with the last two needs of observational solar astronomy, the need for novel observations and ideas, and the need for an improved understanding of the observations.

References

Athay, R. G.: 1970, *Solar Phys.* **12**, 175.
Beckers, J. M.: 1968, *Solar Phys.* **3**, 367.
Beckers, J. M.: 1973a, *Bull. Am. Astron. Soc.* **5**, 269.
Beckers, J. M.: 1973b, *J. Opt. Soc. Am.* **63**, 484.
Beckers, J. M. and Parnell, R. L.: 1969, *Solar Phys.* **9**, 39.
Coates, R. J., Gibson, J. E., and Hagen, J. P.: 1958, *Astrophys. J.* **128**, 406.
Cram, L.: 1973, private communication.
Dupree, A.: 1969, *Astrophys. J.* **152**, L125.
Gabriel, A.: 1971, *Solar Phys.* **21**, 392.
Gabriel, A. H. *et al.*: 1971, *Astrophys. J.* **169**, 595.
Harvey, J. W. and Breckenridge, J. B.: 1973, *Astrophys. J.* **182**, L 137.
Harvey, J. W. and Livingston, W.: 1969, *Solar Phys.* **10**, 283.
Kundu, M.: 1973, in R. G. Athay (ed.), 'Chromospheric Fine Structure', *IAU Symp.* **56**, in press.
Liu, Su Yang: 1973, *Astrophys. J.*, in press.
Maltby, P. and Erickson, G.: 1967, *Solar Phys.* **2**, 249.
Noyes, R. W., Beckers, J. M., and Low, F. J.: 1968, *Solar Phys*, **3**, 36.
Parnell, R. L. and Beckers, J. M.: 1969, *Solar Phys.* **9**, 35.
Simon, G. W. and Zirker, J. B.: 1973, *Solar Phys.*, in press.

ASYMMETRY IN SOLAR SPECTRAL LINES*

C. MAGNAN and J. C. PECKER

Institut d'Astrophysique de Paris du C.N.R.S.

and

Collège de France, Paris, France

Abstract. After reviewing observations of the spectral solar features originated either in the chromospheric layers or in the photospheric layers, from the point of view of the observations, and after having shown the strikingly discrepant set of interpretations that can be found currently in literature, a numerical experiment is performed in a case not too different from the solar case. It is shown that the use of the line bisector to determine, from the asymmetry of a single line, the trend of the velocity field might be considerably misleading, a fact which explains partly the results published in literature.

Clearly, asymmetries, in emission or in absorption lines, in a stellar or in a solar spectrum, can be due (if one excludes blends, or asymmetries of the instrumental profile, unsuitably corrected for) to *velocity* fields of some kind. On the other hand, the symmetry of a line does not exclude velocity fields, either 'macrovelocity fields', which can be such as to produce symmetric lines; or 'microvelocity fields', such that the integration along the line-of-sight, at any wavelength in the line gives place to a symmetrically broadened feature.

Therefore, the diagnostic of asymmetries might be insufficient to derive velocity fields; moreover, as we shall see, it will be quite difficult to make it unambiguous.

Both statements are leading to the conclusion that, in addition to asymmetry, other observable features will have to be observed: broadening, center-to-limb variation of broadening, intensification such as the one displayed by the height of the plateau of the curve of growth, and the like. Above all, very high spatial resolving power spectrograms should be able to allow us to disconnect the determination of the usual 'macro-velocity-fields', and to obtain them separately.

We shall briefly examine the two main types of observations – chromospheric and photospheric – of line asymmetries. We shall then look into the diagnostic problem, as it appears through the literature, and how we can see it now.

1. Chromospheric Features

1.1. The H and K lines of Ca II and Mg II

We shall refer to Linsky and Avrett's paper (1970) as one of the most complete bibliographical studies of the very numerous observations of H and K lines of Ca II in the solar spectrum.

The H and K lines, first observed (and named) in 1814 by Fraunhofer, are well known as the most conspicuous lines of the observable spectrum of almost all stars

---

* This paper was presented as an invited review paper at the General Assembly of the IAU in Sydney (August 1973), at a meeting of Commission 12.

and galaxies. Figure 1 reminds the reader of some well known characteristics and notations referring to these lines.

The most remarkable feature of these two lines is their doubly reversed profile, well known since Hale and Deslandres, in the late eighties. The behaviour of this double reversal above spots, faculae, or at the limb of the Sun, has been studied in great details. We shall not mention these questions any more and shall send the reader back to Linsky and Avrett, as well as to the original literature.

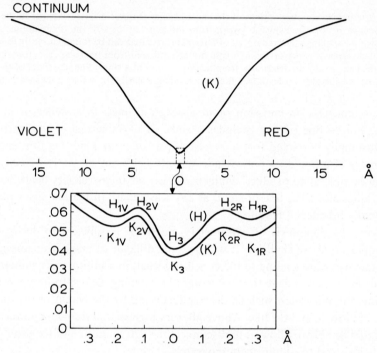

Fig. 1. *Usual notations relative to K-line. Top:* Small scale profile. The dotted rectangle is enlarged in the bottom part of the figure. *Bottom:* Large scale profile of both H and K (central parts of the profile).

The asymmetry of the doubled reversed peaks has been first observed by Jewell (1896), and studied in more details by Saint-John (1910). As a rule, it seems clear that the violet emissions K_{2V} is more intense than that of the red emission K_{2R}. *On the average*, the K_{2V} is displaced (to the blue, i.e. towards the observer, – if interpreted in such a simple-minded way) by 1.97 km s^{-1}, the K_{2R} component being displaced to the red by 1.14 km s^{-1}, the displacements being measured with respect to the center K_3 of the line. One interpretation is that the matter responsible for K_2 is rising, whenever the matter responsible for K_3 is falling. Of course, we shall come back on the analysis of this easy and early diagnostic.

Amongst the most significative studies made in the recent years, after decades of research, of such a dissymmetry, is the study by Pasachoff (1969, 1970). A rather good spatial resolving power allowed him to show that the above-described profile is only an *average* profile (this, we knew), but especially that the local profiles are extremely different from average, the standard deviation being considerable. According to the study by Pasachoff (a study which has been superseded in some way to the authors quoted at the end of this section), it seems that: (a) the 'normal' profile has only one peak, on the violet side; (b) the double peak feature occurs only in about 10% of the cases; (c) often, there are no emission peaks at all.

These characteristics can be derived from spectra such as the ones represented on Figure 2.

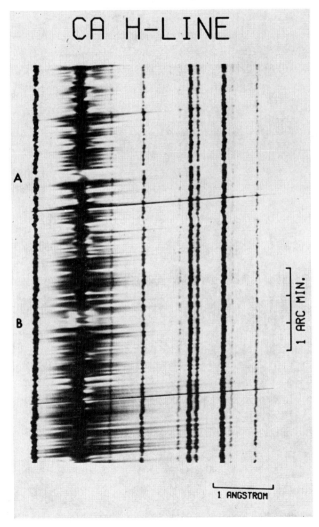

Fig. 2. *The H line (a section of it)* (from J. Beckers). Note the various aspects of the double peak asymmetries when going across the solar surface.

The asymmetry, with a given spectral, and spatial resolving power, at a point of the disk, is a function of time. This has been demonstrated clearly by Jensen and Orrall (1963) with a limited resolving power on the disk, and later confirmed by Pasachoff (1969), who found much larger variations. Spatial fluctuations of the two emission peaks K_{2V} and K_{2R} are badly correlated, – as could be expected from the above description by Pasachoff. The position of these peaks fluctuate; the rms value of these fluctuations is of the order of 0.04 for K_2, of 0.02 for K_3. The asymmetry of the center of the K line thus affects the spectral location of the peaks, and their relative intensity.

It should be noted that the K line asymmetry is diminishing towards the limb. However, off the limb, where the line appears only in emission, there are little indications on how the asymmetry behaves. It would be of course quite interesting to know this better.

High resolution (both spatial and temporal) observations are now the field of intensive work; those by Bappu and Siravaman (1971), Wilson and Evans (1971), and Wilson *et al.* (1972) are worth mentioning. In some cases the evolutionary behaviour of individual features is reported, and this offers clearly the possibility of testing in a more precise way the various theories of formation of the lines.

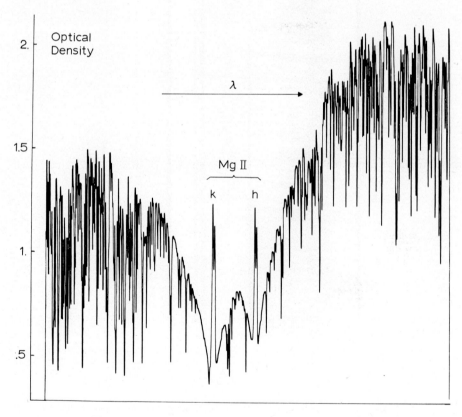

Fig. 3a. *Small scale profile of H and K lines* (after Lemaire).

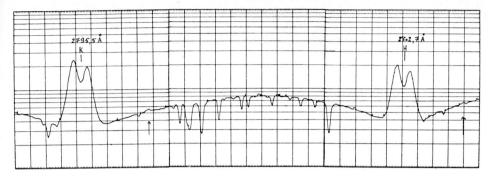

Fig. 3b. *Large scale profile of H and K lines.* Note the dissymmetries of these 'averaged-on-surface' profiles.

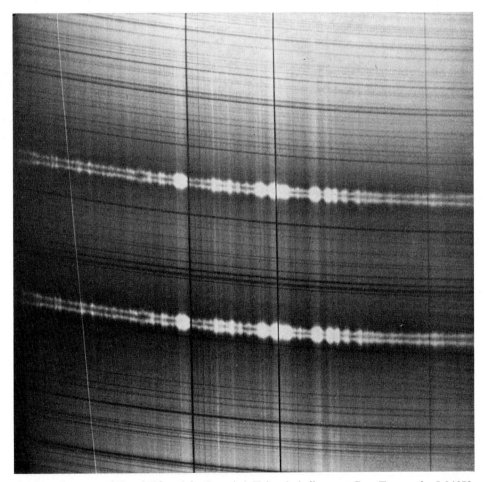

Fig. 4a. *Spectrum of H and K lines* (after Lemaire). Taken in balloon, at Gap, France, the 5.6.1972. Resolving power: spectral: 25 m Å; spatial: 2–3″. Exposure time: 17 s. Note the limb at the right side, and the increase separation between the two emission peaks.

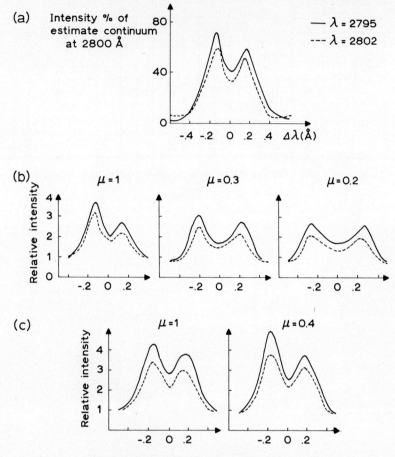

Fig. 4b. *Profiles of H and K lines* (after Lemaire). (a) Average of the central third of the disk on August 21, 1961 (Purcell *et al.*, 1963); (b) (c) (Lemaire, 1969) with a spatial resolution of 10″; (b) corresponds to quiet regions, (c) to faculae.

1.2. The H and K lines of Mg II

For these lines, formed higher in the chromosphere than the H and K lines of Ca II, the emission peaks are much more conspicuous (Figure 3). Their behaviour on the disk, notably from center to limb, with a relatively good resolution on the disk, is reproduced on Figures 4a, b. Both Figures 3 and 4 are taken from Lemaire (1969a, b, 1971).

Again, we cannot possibly attempt to describe fully all the observations, balloon-borne, or rocket-borne, dealing with the Mg II h and k lines. Lemaire (1969a, b, 1971) notes, as do Bates *et al.* (1969) the marked asymmetry of the line. Here, as in the case of Ca II, and certainly for similar reasons (but note that the Mg II lines are formed higher in the solar atmosphere), the violet peak is more intense than the red one. The asymmetry is decreasing towards the limb; in the meanwhile, the separation $K_{2V} K_{2R}$ is increasing from 0.28 Å (at the center) to 0.40 Å (near the limb).

2. Chromospheric Features: the Interpretations

The discussion by Linsky and Avrett (1970) is so pertinent that we do not try to elaborate it much further. But their discussion is almost limited to one-dimensional models, and as we feel, unfortunately, that no satisfactory picture has been so far given to the observations abstracted in the above section, we will report here the analysis concerned with multi-components models. The actual suggestions follow those by Cram (1972). We shall certainly not consider them as satisfactory (they are even contradictory to each other) but, at least, we hope to reach partial (negative and positive) conclusions which will demand for complementary tests, or complementary analysis, both observational and theoretical.

(1) We have mentioned the idea that the region where K_2 is originated is moving upwards, the region where K_3 is originated being moving downwards. This interpretation, if we follow the suggested behaviour of homogeneous models, such as Dumont (1967), leads us to admit that, around $h=300$ km, matter is moving up,

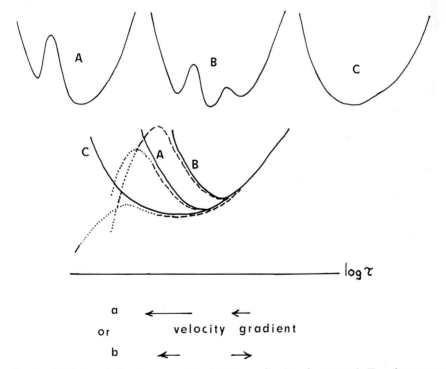

Fig. 5. *Possible interpretation for asymmetrical K-line profiles* (two layers type). *Top:* three types of profiles: A, B, C. *Center:* models corresponding to A, B, C; dotted line: source function. *Bottom:* possible distribution, along the τ-scale, of the gradient of velocity.

whenever it is moving down at the higher level $h = 600$ km. As no local permanent increase of density, or no averaged increase of density, is indeed permissible, we have to exclude the combination of the 'homogeneous model' and the classical double motion interpretation of the asymmetrical emission.

(2) But can we exclude a 'two-columns inhomogeneous model'? It is less obvious that the idea cannot work either.... Indeed one can consider this suggestion in many ways. Either we can assume (as strongly suggested by Pasachoff's experimental evidence) that single emission peaks are the rule. Then, at a given point, it implies a gradient of velocity, in most of the cases, with an increase outwards, *and* a higher rise of temperature (in order to annihilate the classical decrease of the source function in classical models, built for some homogeneous chromosphere). In some of the cases, rise of temperature would occur higher, in others lower. This is shown on the Figure 5, highly schematically.

As interesting as this model may qualitatively be, we could not buy it, at various points of the disk unless: (a) the analysis of each observed profile gives weight to the

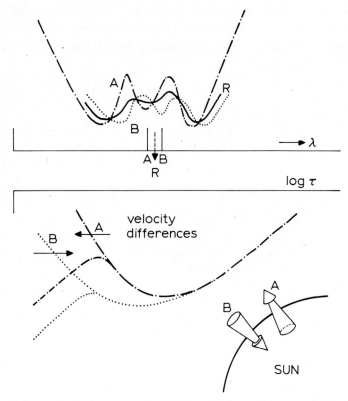

Fig. 6. *Possible interpretation for asymmetrical K-line profile* (two columns type). *Top:* Combination of profile A (rising) with profile B (descending). *Bottom:* Models on points A and B (the lower curve, in each case, represents the source-function). Obviously, this gives a profile of which the asymmetry is contrary to what is observed; an inversion of A and B and of violet and red gives a result similar to observation.

assumed behaviour for the source-function, in a quantitative manner, as a function of the optical depth in the line center; (b) some tentative explanation is proposed on how is a point such as A or C heated more than point B. An evaluation, according to Pasachoff's statistical suggestions, on how much energy is needed to heat the 'usual' chromosphere described by situation A, is obviously a need at this stage. Let us note (Figure 5) that such a model implies, in any case, a continuous mass loss, at chromospheric levels. Let us note also that, in this model, the contradiction mentioned above (Section 2(2)) has also to be eliminated in a region such as B.

(3) Another way to look at the averaged profile is to admit indeed that there are some columns – i.e., essentially, that chromosphere differs from point to point. This point of view is not essentially, so it seems, different from that described in the preceding paragraph. However, in Section 2(2), we have suggested a gradient of velocity at any of the points A, B, C, suggesting that the effects observed, essentially, are local, and that there are asymmetries everywhere, locally. Now, we assume that the asymmetry comes from the averaging of different spectra, each corresponding to a point of the disk where the profile is perfectly symmetrical. The observed asymmetry of the profile comes here only from combinations of displaced symmetric profiles. This model is described on Figure 6, highly schematical also.

Of course, this behaviour seems contrary to Pasachoff's observations. But the latter are still questioned, as to their statistical significance: actually they should be, by some appropriate averaging process, equivalent to the 'normal' profile of Figure 1 – which they do not seem to be, according to their characteristic statistical features, quoted above in Section 1. The k Mg II lines observed earlier described, do not seem either – at the first glance – to agree with Pasachoff's description; but there, the lack of resolving power on the disk may be responsible.

Apart from this point, the model in question is quite plausible, in terms of the averaged profiles; and in terms of the center-to limb observed decrease of the asymmetry. But again, we have been only qualitative, and one should show, in a forthcoming study, that the source-functions in situations A or B are reasonably well in agreement with plausible models. Moreover, the 'infalling model', in B, should be understood, compared to the 'outflowing model', in A, in the sense that one should explain why the first one is heated in a different way, and how.

At this stage, our preferences do not go to any of the two suggested models, the 'locally asymmetrical' one, or the 'locally symmetrical' one, unless better computations are performed, and more significantly statistical study of the first structure of H, K, h, k lines achieved.

3. Photospheric Features: the Observations

Measurements of line asymmetry dealing with photospheric layers have been performed with an increased accuracy, from the earlier studies by Voigt (1956), till the recent works by Roddier (1965), to quote the most accurate one, in the authors mind. The Table I reminds the reader of the observational studies in question, and of the lines measured by the various authors.

TABLE I
Asymmetry in photospheric lines profiles

Author	Date	Designation	Line	E.P. (eV)	Remark
Voigt	1956	V1	Ni I 7789	1.9	center-to-limb
	1959	V2	Ni I 7798	3.5	
		V3	O I 7772	9.1	
		V4	O I 7774	9.1	
		V5	O I 7775	9.1	
Delbouille *et al.*	1960	D1	O I 7772	9.1	
		D2	O I 7774	9.1	
		D3	O I 7775	9.1	
Higgs	1960	H1	Fe I 6297.8	2.21	center-to-limb
	1962	H2	Fe I 6301.5	3.64	
		H3	Fe I 6302.5	3.67	
Müller	1961	M1	O I 7772	9.1	
		M2	O I 7774	9.1	
		M3	O I 7775	9.1	
Brault	1962	BR1	Na(D) I 5896	0.00	center-to-limb
Olson	1962	O1	Fe I 6173	2.21	
		O2	Ca I 6166	2.51	
		O3	Fe I 5935	3.94	
		O4	Fe I 7090	4.21	
		O5	Fe I 5929.7	4.53	
		O6	Fe I 5930.2	4.63	
		O7	Fe I 5927.8	4.63	
		O8	O I 7772	9.1	
		O9	O I 7774	9.1	
	1966	O10	C I 10700	7.4	5 lines
		O11	O I 7770	9.1	3 lines
		O12	Fe I 5930	4.63	
Roddier	1965	R1	Sr I 4607	0.00	center-to-limb
De Jager and Neven	1967	J1	C I 10754	7.46	
Boyer	1969	B1	Ti I 4535	0.83	
		B2	Ti I 4563	0.02	
		B3	Ti I 5210	0.05	
		B4	Ti I 5239	0.84	

In most of these studies, the type of asymmetry observed at the center of the disk is of a similar shape, essentially described by the 'bisector' line of the spectral feature. On Figure 7, we have superposed several of these bisector lines, and one sees clearly their *C*-shape, which seems to be function of the excitation potential of the line.

However, we must be very cautious in using these various observations; the instrumental errors are of many kinds. They have been well, and rather extensively, discussed by Boyer (1969), and *a priori* eliminated by the very astute experimental devices used by Roddier (1965) in his own measurements of the line profiles. By using a direct atomic beam (in which the dispersion of velocities is less than would be given by a temperature of 3 K), Roddier excites the Sr I resonance line by the solar light. The solar absorption line is swept by the beam, using Zeeman displacement. The wavelength measured is obviously the solar wavelength with respect to the laboratory wavelength –

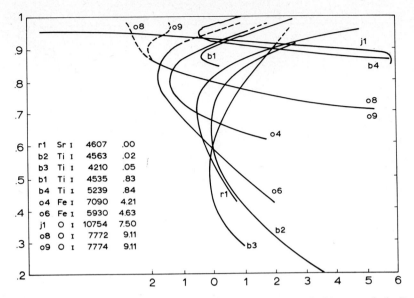

Fig. 7. *Dissymmetric observed profile bisectors* (composite picture, complied from results by Roddier, Boyer, Olson, de Jager and Neven, and Olson). Note the systematic behaviour with excitation potential (in electron-volts in the last column on the left of the figure, together with identification of the various curves).

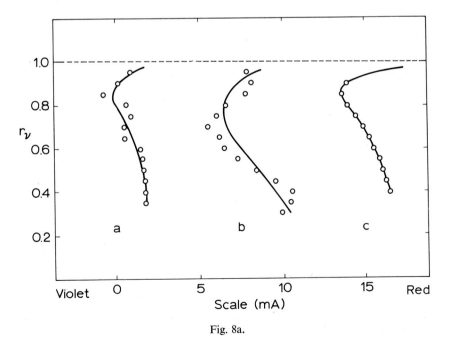

Fig. 8a.

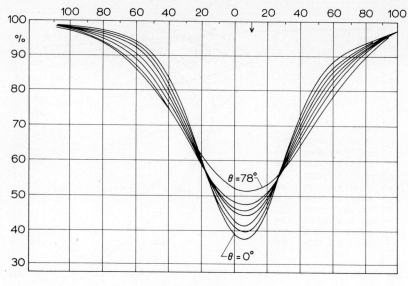

Fig. 8b.

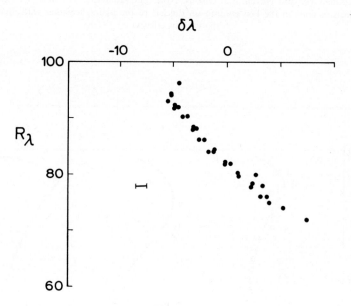

Fig. 8c.

Figs. 8a–c. *Dissymmetric observed profiles (bisector of profile)*. Some original data on some Fe I lines (a: Higgs), on the Si I resonance line (b: Roddier) and the infrared O I lines (c: Olson).

a difference not so well determined in the classical conventional spectrographical studies (not with the same accuracy, at least...). We should, we believe, give the greatest weight to Roddier's measurements.

An additional important data, not always measured by the quoted authors is the center-to-limb behaviour of asymmetry. When available, the data have been plotted against μ, as $A(\mu) = (h_V - h_R)/(h_V + h_R)$, h being the half-width at half-line depth. It is probably necessary in this discussion to disregard the earlier measurements by Voigt (1956) so discrepant they were from the results by Delbouille *et al.* (1960) obtained later. We shall confine the relevant observations to Higgs' (1960, 1962) and Roddier's (1965), which seem more reliable so far as center-to-limb variation is concerned (Figures 8 and 9) – but they find quite a different result each from the other, quite systematic also. Our personal tendency, knowing the exceptional quality of Roddier's instrumentation, is to believe the results of this author. But, obviously, measurements are difficult, and we should worry here very much about the data.

We should also note (and this remark may still increase our ill feelings) that in some cases, noted apparently only by Higgs, the asymmetry is changing, at certain points if the disk, or at certain moments, in such a way as to even completely change the sense of the asymmetry itself. This comment may be linked with some of our conclusions in Section 5 hereafter.

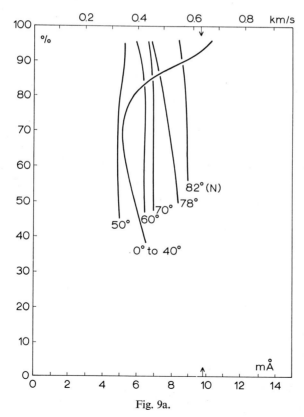

Fig. 9a.

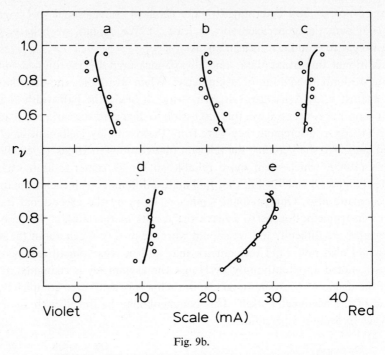

Fig. 9b.

Fig. 9a–b. *Center-to-limb variation of asymmetries.* (a) After Roddier (1965). The bisector of Sr I resonance line, from center to limb. (b) After Higgs (1962). Note the difference with Roddier's results.

4. Photospheric Measurements: The Interpretations

Naturally, the authors themselves have, in some cases, either alone or in cooperation, tried to achieve a satisfactory understanding of their data. However, one is struck by the severe factual indetermination, or uniformity, that is stemming from these papers. One cannot avoid being struck, also, by the lack of concern that is displayed by most of the authors about this non-uniformity, as if they had considered, in most of the cases, that they were the only ones to bring a sensible solution to a delicate problem.

We shall certainly not bring ourselves here any additional solution, and we shall limit our efforts to go through the literature and to gather more or less logically the various arguments, often accepted as proofs.

The first work of some relevance, in our opinion, is probably H. K. Böhm's theoretical paper of 1954. In this paper, he has shown first that three-temperature models (or three column models) although very unsophisticated (they were in LTE, RE, etc.) were not badly suited to interpret the abnormal limb red-shift earlier observed by Allen (1937) and by Adam (1959), and by others.

But Böhm did not study all the possibilities of such models, as he was not aware of the observed asymmetries of lines. Hence, we should consider that the more detailed computations concerning Böhm's model, or at least models similar to Böhm's, and relevant here, papers concerning not only indeed the amount of asymmetry, but also its behaviour, are those by Voigt (already quoted) and Schröter, 1957. The papers by Voigt and Schröter have been written almost simultaneously, and independently from each other, as clearly indicated in footnote 2, p. 172, of Schröter's paper.

On the other hand, it has been shown later by Jorand (1962) that several component models, as good they may be for some other purpose, could indeed correctly describe the shifts of line near the Sun's center, but that the extra redshift measured near the limb, in addition to the gravitational Einstein redshift, could not be accounted for (if real) by any such a model.

At least, Schröter (1957) was successful in predicting, or confirming, the behaviour of the asymmetry of spectral lines, and of some specific ones (Figure 10). Clearly, the behaviour he predicted for the disk's center was quite similar to that observed (see above, Figure 7). This theoretical work of Schröter was in fact the starting point of a whole series of experimental work, already quoted in Section 3. The Figure 11, due to Schröter (1955) shows finally how strongly different models can give indeed resulting spectral lines the symmetry of which is almost (but not completely) identical

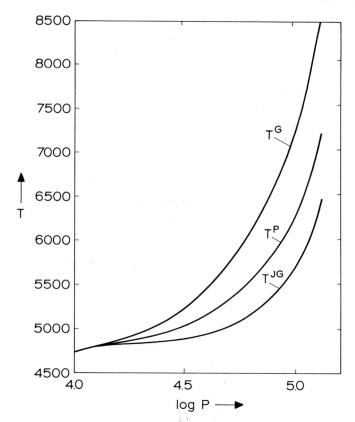

Fig. 10. *A two-column model.* After Schröter (1956). The letters G and JG correspond respectively to granular and intergranular region. T_P designates the one-column equivalent model. In abscissa, logarithm of pressure.

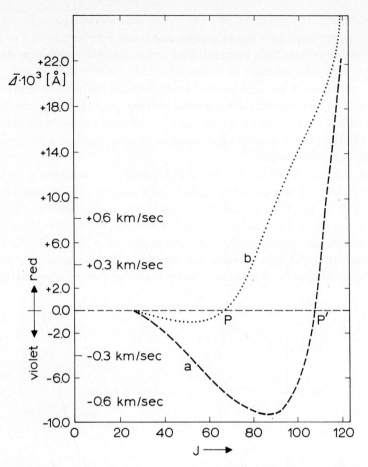

Fig. 11. *Asymmetrical profiles computed with two columns models.* After Schröter (1956). The figure reproduces the bisector line. In abscissa, the zero corresponds to the continuum; J is the line intensity ($J = 122$ corresponds to continuum).

to the observed one. But still the fitting is not as good as it should be, showing at the same time the impossibility of a correct diagnostic, and the underdetermination of an 'almost correct' diagnostic... A point, which, applied to many other problems, is undoubtedly rich of deep significance....

Moreover, when Olson (1962, 1966) tried to perform similar work, the ΔT (difference between the temperatures of the rising and falling columns) had to be very different from line to line, a fact certainly far from being satisfactory (Figure 12). Olson, who was aware of the fine structure of lines, as revealed from high spatial resolving power spectrograms such as Schwarzschild's (1961), as deduced in any case from the high spatial pictures from Stratoscope I, studied by Bahng and Schwarzschild (1961), or by Edmonds (1964), did not find a good agreement between tthe temperature-velocity distribution that is needed to explain the dissymetries and the velocity

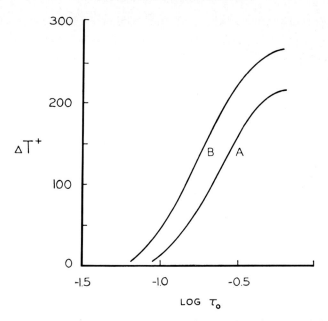

Fig. 12. *The temperature difference between the two columns.* After Olson (1962). A: for O I lines; B: for Fe I lines.

distribution which is inferred from direct fine resolution photographs. After elimination of telluric effects, or of differential Zeeman asymmetrical effects, Olson sadly concludes, at the end of his first paper on this question: 'the discrepancies are apparently still to be resolved'. By reading the subsequent literature, we do feel this conclusion as being still completely valid!

Let us note, incidentally, that another effect studied by Olson (1966) is to be eliminated from the analysis: the very interesting suggestion by Noyes and Leighton (1963) of a mechanism of acoustic waves propagation does not seem to lead to the production of any real asymmetry of the profiles.

We must obviously, at this stage, notice that the difficulties of the fitting were obvious even without reference to center-to-limb observations; that all computations were strictly in LTE; that they introduced velocity fields as ad hoc parameters in an often non-physically consistent way. The whole problem of velocity fields of macro- and micro-scale has of course to be discussed at length in a broader context. But insomuch as profile asymmetries are considered, the undetermination in inferred velocity fields is showing in still a less pleasant manner than in the interpretation of the general symmetrized shape of the profile for reasons we hope to clarify in Section 5, hereafter. Let us, not leaving out, from the analysis, models that are trying to describe center-to-limb variations, now refer to the results of Boyer (1969, 1970).

This author comes back to the three columns model (in LTE), and is using the best 'up to date' models, at the time, i.e. the URP model (Heintze *et al.*, 1964) for the temperature distribution. But he lets free, as parameters, the respective area A_j of the three components (a parameter which indeed should not be free in such a way, because of the absolute need to correctly represent the continuum as well as the line intensities), and the microturbulence law, $\xi(\tau_c)$, considered as isotropic, and identical (but why?) in the three columns. The 'macroturbulent' velocities (V_1, V_2, V_2) are fixed in part by the area A_j and the law of mass conservation; assuming $V_2 = 0$, the only parameter $W = \frac{1}{2}(|V_1| + |V_3|)$ (different from $V = V_1 + V_2 + V_3$) is thus defining the convection. Boyer admits $W = 0$ for $\tau < \tau_l$; $W = $ const for $\tau > \tau_l$, i.e. in the 'convective layer'. All computations have been done in LTE.

The numerical experimentation performed by Boyer is interesting in that, in spite of an *a priori* very crude physical description, a good enough fitting is done with some adequate choice of the parameters. However, this is true only if one fits a *single* line. The second line does not fit any more – as already essentially noted by Olson (1962). Again, we hope that Section 5 will allow to understand why it is so. Boyer's model is looking very much like Olson's (Figure 13) .

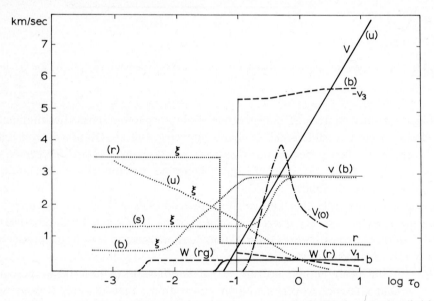

Fig. 13. *The variation of macro and micro-fields of velocities with depth.* This figure is compiled from some models obtained by different authors: r:Roddier, rg Roddier-Gonczi; o:Olson; b:Boyer; s:Schmalberger; u:Utrecht reference atmosphere. In dotted line, the 'micro-turbulence'. V_1, V_2 represents the velocity in moving columns (Boyer). V the average $(V_1 + V_2 = 2V)$ velocity. w the velocity in rising column ($w = V_1$).

Roddier (1965) and Gonczi and Roddier (1970) have, seemingly, looked into the problem with an attitude less obviously 'numerological'. In addition, they take into account the center-to-limb observations. They first noted the V-shape (in contradiction with the U-shape, generally predicted by theory) typical of the intervention of large velocity differences. Then they remark, as we did earlier, hereabove, that two simple types of models can account for observations: either large gradients of velocity;

or multicolumn models. But they note also that the line being generally violet shifted (with respect to the line center, properly located, account being taken of the Einstein shift), a neat convection is necessary, the non-zero upward observed velocity being an average value weighted by inhomogeneities. Then the center-to-limb studies impose anisotropic velocity fields: the convective cells indeed are not isotropic –, and this is showing clearly through the reduction near the limb of the violet shift. The widening of the profiles, contrarily, is due to an apparent increase of the line-of-sight component of microturbulence (either anisotropy in one of the n components – or decrease, inwards, of microvelocity fields).

For the various reasons listed above, the authors do not think that an easy conclusion can be reached quickly. Before going into the numerical experimentation, which obviously is needed, they can only suggest that according to published work, two kinds of models seem possible:

(a) *a two-column model* (or multi-column model). Essentially one column would move upwards, being slightly microturbulent; the other one would be moving downwards, the microvelocity field being there somewhat anisotropic.

(b) *a two-layers* (or with a strong velocity gradient) model: Essentially, the outer layer, highly microturbulent, would have very little, if any, average motion; the inner one has an average upward motion and is weakly microturbulent; no anisotropy is necessary; but evidently, in both regions, an inhomogeneous model has to be considered.

Model (b) was presented first by Roddier in 1965. In 1970, from about the same data, Gonczi and Roddier are favouring the model (a).

However, we may note several simplifications they made in their assumption before any experiment with the free parameters of the problem: (i) they assume that ΔT between the columns is not an essential parameter, and they give it the value zero; (ii) the area occupied by the two columns on the solar surface are equal and independent of height.

Gonczi and Roddier note first that, at the center of the disk, the two models are essentially equivalent; hence they do not accept easily the too limited conclusions of Olson and Boyer. But obviously, from center to limb, the model (a) is by far better. Hence their final choice. We must note that, as others already mentioned, they could not fit another line, (they have attempted to do it for C I 10691) with the same model....

Turning into details of the numerology of Gonczi and Roddier, they finally present the model of Figure 14, which indeed fits remarkably the observations made from center to limb by Roddier. There exists a small indication that $V \cos \theta$ (the apparent violet shift) is indeed to be supplemented, to fit the observations, by an additional violet shift term of the order of $-150 \, (\sin \theta)^2 \mathrm{~m~s}^{-1}$, the origin of which is far from clear: the authors think that supergranulation horizontal cells cannot account for it.

A source of error noted by Gonczi and Roddier is the existence of NLTE source-function: the abundance of Sr is found at the limb twice as low as at the center of the disk; but they comment that departures from LTE have no influence on the asymmetries and shifts.

We shall not try here to criticize heavily this model. We have, so we believe, no

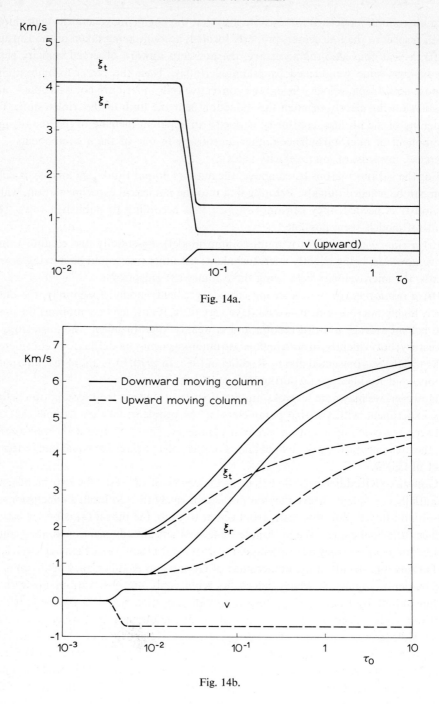

Fig. 14a–b. *The Gonczi-Roddier discussion.* (a) A possible model (of two-layers type) with anisotropic micro-turbulence. (b) A probable model (of two-column type) with anisotropic micro-turbulence.

better one at hand, and many worse ones.... Indeed the convective velocity values, lower than usual, are fitting better the rms measurements of fine structure spectra. Also this model agrees with the observed fact that, in the wings, the contrast between dark and bright regions is larger on the red side than it is on the blue; this difference of contrast is reversed in the line center, and in the line core; this has been well observed and is compatible with the model of Gonczi and Roddier. Another important fact, observed, and of which the model gives a fair account, is also the fact that the profiles are widened in the dark intergranular regions.

To our knowledge (which is limited) no new attempts have been done in order to get new observations of asymmetries, and to deduce from them new models for the velocity fields of the solar photosphere; new velocity models certainly have been produced, but from other considerations. Obviously, many observations of high quality, and recently obtained, could be analyzed for it; but we did not find any relevant information from the best possible bibliographical report at the date, i.e. the 'draft' report of commission 12, prepared for the XVth General Assembly of the IAU (Sydney 1973)!

We shall now take the problem from an entirely different point of view, that of the general methodology of diagnostic in 'artificial' asymmetric line profiles.

5. Theoretical Line Asymmetries

5.1. INTRODUCTION

In view of the confusing state of the various interpretations of observed line asymmetries, it is clearly impossible to propound a new 'marvellous' and all-purpose methodology. The simplest diagnosis procedure, the so-called 'bi-sector' method, has been discussed by Kulander and Jefferies (1966) and, in the same spirit, we intend to set the problems by analyzing, from a theoretical point of view, the influences of some physical parameters upon the emergent profiles. The peculiar case of the solar H and K (or h and k)lines has been reviewed by Linsky and Avrett (1970) and here above, in such a way that we do not feel it necessary to consider these lines in detail, – but in a more general context, we hope to help everyone to ask himself the 'right' questions. In fact, these questions are always simple – though the answer is not! – and must not be hidden in the intricacy of the calculations. Fortunately the observations of line asymmetries lie in the heart of the problem of diagnosis, because they force us to introduce velocity fields very explicitely.

There, we cannot content ourselves by calling upon the help of that obscure parameter ξ: it is more natural and more consistent to search for velocity fields that account simultaneously for the width of the line and for its asymmetry. In this respect, it is to be expected that the study of asymmetries will help us in return to clarify the questions related to the formation of symmetric lines, because these symmetric lines are likely to be some average of many asymmetric ones.

At this point of the discussion, we cannot yet consider the mixing of two or more different emerging profiles: if we are trying to understand what is going on, then we must evidently limit ourselves first to the *intensity* emergent from a single line-of-sight. In other terms, the intensity is what concerns us, not the *flux*. This is, incidently, appropriate to solar problems at high spatial resolution, but also to some special cases. Let us consider the result of an integration along a simple straight line, and in order to introduce the notations, let us recall what this 'integration' consists of.

Consider first a local frame of reference moving with the fluid at some given point. The local frequency v_L of a photon in this frame will be characterized by the distance

$$\Delta v_L = v_L - v_0 \tag{1}$$

from the absolute rest frequency v_0 of the transition being considered. The thermal Doppler width $v_0 w/c$ corresponding to the thermal velocity $w = (2kT/m)^{1/2}$ will be denoted by Δ. If the distribution of the atoms in the upper state of the transition is a maxwellian distribution at the temperature T, the probability that a photon emitted in the line will be emitted at about the frequency Δv_L, in the range $d(\Delta v_L)$, is:

$$\frac{1}{\Delta} \Phi\left(\frac{\Delta v_L}{\Delta}\right) d(\Delta v_L), \tag{2}$$

where the profile $\Phi(x)$ is usually the Voigt profile, normalized to unity over $x = (\Delta v_L)/\Delta$ in the range $(-\infty, +\infty)$. This profile depends only on the parameter $a = \Gamma/4\pi\Delta$, equal to the ratio of the atomic to the thermal broadening width. In what follows, a is always taken as constant, and for the sake of simplicity, we have dropped out from our notation the dependence of the profile upon this parameter a. In the numerical examples reported here, we have taken the well known approximate form

$$\Phi(x) = \frac{1}{\sqrt{\pi}} e^{-x^2} + \frac{a}{\pi x^2}, \tag{3}$$

where the second term on the right-hand side is to be added only for $x^2 > 1$. We have chosen $(a/\pi) = 10^{-3}$, so that the second term dominates for $x > 3$.

The absorbing properties of the medium at the same point are characterized by the same frequency dependence given by the expression (2). We take as the depth coordinate the continuum optical depth τ_c between the surface and the point in consideration. This quantity τ_c can safely be assumed to be independent on the velocity field. The probability that a photon Δv_L is absorbed in the line along the path $d\tau_c$ is

$$d\tau_L = \frac{\eta}{\Delta} \Phi\left(\frac{\Delta v_L}{\Delta}\right) d\tau_c, \tag{4}$$

where (neglecting the induced emissions)

$$\eta = \frac{1}{\varkappa_c} \frac{h v_0}{4\pi} N_1 B_{12} \tag{5}$$

is the ratio of the mean line opacity to the continuum opacity $\varkappa_c$. The parameter depends only upon atomic coefficients and upon the population N_1 of the lower state but not upon the distribution function of the velocity of the atoms in that state. We shall take this parameter η as constant.

5.2. Adding a velocity field

Up to this point, the things are symmetric with respect to Δv_L. We assume now that each layer situated at depth τ_c has a velocity $v = v(\tau_c)$ with respect to an observer lying outside the medium and counted as positive when directed towards this observer. Along its path, the photon emitted towards the surface at point τ_c with local frequency v_L is capable of being absorbed at point $\tau'_c (\tau'_c < \tau_c)$ where it is seen in the local frame moving with the absorbing material at the local frequency:

$$v'_L = v_L + v_0 (v - v')/c, \tag{6}$$

where $(v - v')$ is simply the velocity of the slab τ_c with respect to the slab τ'_c. Ultimately, this photon will be seen by the observer at the frequency:

$$v = v_L + v_0 v/c. \tag{7}$$

From the preceding formulae, we write the line optical depth up to the surface for the photon emitted at point τ_c with the local frequency Δv_L as:

$$\tau_L = \int_0^{\tau_c} \frac{\eta'}{\Delta'} \, \Phi \left(\frac{\Delta v_L}{\Delta'} + \frac{v - v'}{w'} \right) d\tau'_c. \tag{8}$$

It is clear from this expression (8) that the photons emitted *symmetrically*, at local frequencies $\Delta v_L = +\delta$ and $\Delta v_L = -\delta$ suffer now a *non-symmetrical* history during the subsequent way to the surface. The reason is that the shift in the argument of Φ is never symmetric. A situation which is often considered consists in taking Δ' constant in the medium. Then for an expansion (i.e. $v' > v$ and $v > 0$), the argument of Φ is always algebraically decreasing so that its absolute value first increases on the violet side $+\delta$ and decreases on the red side $-\delta$. In that case, the optical depth is larger for the photon $+\delta$ than for the photon $-\delta$. But we want to insist on the fact that, even in this quite simple situation, we cannot infer immediately the sign of the asymmetry of the emergent profile. In fact, it must be realized that the local frequencies $+\delta$ and $-\delta$ are received by the observer at frequencies $+\delta + v_0 \, v/c$ and $-\delta + v_0 \, v/c$, i.e. symmetrically with respect to a quantity $v_0 v/c$ which depends on the velocity of the slab which has emitted the photons. So, by adding the contributions of many slabs of many local frequencies we lose inexorably the possibility to refer the various frequencies to some fixed frequency. The important consequence is that, in general: *speaking of the red side or the violet side of the emergent profile has no immediate theoretical signification*. In other terms, the 'line center' may be chosen at the point where the emergent intensity is minimum (i.e. 'observationally' chosen), but we have no way of saying, from a general and theoretical point of view, at which frequency this minimum will occur, except by treating completely the specific problems we may have to solve.

5.3. Introducing an Important Parameter

We now want to emphasize very strongly another physical fact expressed by the formulae (6) and (8). It is quite possible (and, generally, it is to be expected), that at 'some distance' from the emitting point, the value of the argument of Φ will depend primarily on the distribution of the velocities along the path, this distribution being or not of a random character. But this is absolutely wrong 'in the vicinity' of the emitting slab, where $v' \cong v$, so that the absorbing properties near the emitting point depends primarily upon the value of the local frequency Δv_L which has random characteristics (see the formula (2)) independent of the velocity fields. In order to make the words 'at some distance', or 'in the vicinity', more precise, it is then necessary, even in a first crude analysis, to introduce some *characteristic length for the variation of the velocity* (or equivalently some *characteristic value of the velocity gradient*). In nearly all cases, this characteristic length is assumed to be 'small', but this is always a priori assumed: we think that this parameter must be introduced *explicitly* and that a convenient diagnosis must then be applied in order to determine its value. Let us illustrate these considerations with the aid of a model very close from that considered by Frisch (1969) and associates (Auvergne *et al.*, 1973). The medium is represented, for convenience by succession of different slabs of equal continuum optical depth Λ – (in Frisch's model, these lengths have a distribution which follows Poisson's law; in Auvergne *et al.* it is suggested that it can depend upon depth in the atmosphere), each slab i having a velocity v_i; the distribution of the v_i's is random so that the probability of finding a velocity in the range $(v_i, v_i + dv_i)$ is:

$$\frac{1}{\sqrt{\pi}} \frac{1}{v_0} e^{-(v_i/v_0)^2} dv_i. \tag{9}$$

The thermal velocity w is assumed to be constant, so that the line optical depth from the deeper end of the slab n up to the surface from a photon emitted at local frequency Δv_L in this slab n is simply:

$$\tau_L = \frac{\eta \Lambda}{\Delta} \sum_{i=n}^{0} \Phi\left(\frac{\Delta v_L}{\Delta} + \frac{v_n - v_i}{w}\right). \tag{10}$$

So, at a given value of Δ_L, the quantity $(v_n - v_i)$ is effectively randomly distributed, but not for the first value, which is of course always zero. We expect then that the contribution of this first slab is the largest and thus cannot be dropped out without great care.

The quantity which is usually considered is the line optical depth of all subsequent slabs from $(n-1)$ to zero. When n is large, this line optical depth takes the value adopted in so-called 'microturbulent' situations:

$$\tau_L = \frac{n\eta \Lambda}{\Delta_1} \Phi\left(\frac{\Delta v_L + v_n v_0/c}{\Delta_1}\right), \tag{11}$$

where Δ_1 is a new width which takes now into account the value of the characteristic velocity v_0 by the relation:

$$\Delta_1^2 = \Delta^2 + (v_0 v_0/c)^2. \tag{12}$$

But the important point is that the residual optical depth of the emitting slab is independent of v_0 and is always equal to:

$$\frac{\eta \Lambda}{\Delta} \Phi \left(\frac{\Delta v_L}{\Delta}\right). \tag{13}$$

To give an example of application of these simple formulae, let us consider an LTE situation with a source-function $B = B(\tau_c)$, taking the values $B_0, B_1, \ldots$ in the different slabs. Then the slab n is emitting at the local frequency Δv_L the total intensity:

$$I_n = B_n \left\{ 1 - \exp\left[-\Lambda \left[1 + \frac{\eta}{\Delta} \Phi \left(\frac{\Delta v_L}{\Delta}\right) \right] \right] \right\}, \tag{14}$$

which has then to be attenuated by the factor:

$$\exp\left[-n\Lambda \left[1 + \frac{\eta}{\Delta_1} \Phi \left(\frac{\Delta v}{\Delta_1}\right) \right] \right], \tag{15}$$

where Δv denotes the frequency $\Delta v_L + v_0 v_n/c$ seen by the observer. The emergent intensity is then expressed as:

$$I(\Delta v) = \sum_{n=0}^{\infty} B_n \left\{ 1 - \exp\left[-\Lambda \left[1 + \frac{\eta}{\Delta} \Phi \left(\frac{\Delta v}{\Delta} - \frac{v_n}{w}\right) \right] \right] \right\} \times$$

$$\times \exp\left[-n\Lambda \left[1 + \frac{\eta}{\Delta_1} \Phi \left(\frac{\Delta v}{\Delta_1}\right) \right] \right]. \tag{16}$$

The great advantage of a formula of that type is that it contains explicitely two quantities: the value of a characteristic velocity v_0 but also the very important parameter Λ directly correlated to the scale height of the velocity distribution.

It is immediately apparent that the 'width' of the profile depends essentially on Δ_1 (i.e. on v_0), but that the value of the intensity is very sensitive to Λ. We expect to get, upon a 'background' given by the second exponential term of the formula (16), a very wiggly, and of course absolutely unsymmetrical, profile. We suggest that the 'true' situation is very similar to that one and we think that the majority of profiles that are observed are in fact the result of averaging such wiggly profiles. This is clearly demonstrated by spectra taken at higher and higher time resolution, both for the Sun and the stars. In that case, the 'theoretical' averaging must be done very carefully and in particular the kind of average (either over space or over time, or both) must be precisely specified. Formulae like (16) seem to be a sound basis for this task (we shall incidently remark that the papers by Huang (1952) and De Jager and Pecker (1951) were early attempts in these directions). From an observational point of view, it would be very desirable to obtain instantaneous and precisely located profiles: it is to be expected that the diagnosis would be easier if we were able to do ourselves our own averages, at our own will....

The formula (16) has been written for an LTE line, but it is easy to include NLTE effects in a model of that kind, by replacing the B_n's by appropriate S_n, in the line, different from that in the continuum. In fact, when the relative velocities of the different slabs are large (i.e. $v_0 \gg w$), each slab tends to be isolated and to build its own radiation field. In the limiting case, the source-function will depend no more on the characteristic velocity v_0, but only on the optical depth of each slab in this representation of the variation of velocity. This is of course a further argument to demonstrate the importance of that parameter Λ, which has the meaning of a 'correlation' distance. Even if this limiting case is seldom encountered (in the case of the Sun, we rather expect that v_0 and w are of the same order of magnitude), these kinds of effects may be present, especially for strong lines.

5.4. Varying the Thermal Doppler Width

We submit now to each one's thinking the results of simple calculations based on another model which is very schematic but, we hope, instructive, and all things considered, may be not very far from real situations. We have seen in the formula (8) that the argument of the profile $\Phi(x)$ is in fact a function of the velocity, but also of the thermal Doppler width. In order to illustrate the influence of this last parameter in the presence of velocity fields, we take a model in which the thermal width Δ increases towards the surface according to the law

$$\Delta = \Delta_0 \, e^{-(\tau_c/T)}, \tag{17}$$

where T is some scale height which has been chosen equal to 10^{-2}, i.e. just of the order of magnitude of the continuum optical depth before the chromospheric rise of temperature which we try to mimic by the law (17). We further assume that Δ remains constant when it has attained the value $\Delta_0/10$ (Figure 15). We consider now a velocity directed towards the observer and equal at each point to the Doppler velocity so that

$$v = w. \tag{18}$$

This choice is not entirely arbitrary: in fact, a rough relation between the thermal and the non-thermal velocities is to be expected from physical considerations, and moreover it is just when the equality between the two velocities holds that the effects become very significant. In fact, Sobolev's work (1966) demonstrates clearly that, when $v \gg w$, the problems become simple, because the photons do not diffuse in space. But when $v \cong w$, Sobolev's theory breaks down, and the entire problem must be reconsidered.

The formula (8) gives now for the line optical depth:

$$\tau_L(\Delta v) = \int_0^{\tau_c} \frac{\eta}{\Delta'} \, \Phi\left(\frac{\Delta v}{\Delta'} - 1\right) d\tau_c', \tag{19}$$

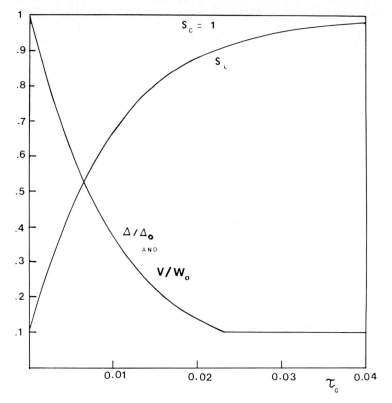

Fig. 15. *A Test Model* for source-function in continuum, and line, and for the variation of Doppler width. (Notations as in the text.)

where $\Delta v = v - v_0$, with the formula (7) being taken into account. From the formula (17), in the region where the thermal width is varying, we have

$$d\tau_c = T \, d\xi/\xi, \tag{20}$$

where $\xi = \Delta_0/\Delta$ is the dimensionless parameter giving the ratio of the surface thermal width to the actual one. It is then easy to express the total optical depth as

$$\tau(\xi, \Delta v) = \tau_c + \frac{\eta T}{\Delta_0} \int_1^\xi \Phi\left(\xi' \frac{\Delta v}{\Delta_0} - 1\right) d\xi'. \tag{21}$$

In the region where $\Delta = (\Delta_0/10)$ is constant, the optical depth is of course computed according to the formula

$$d\tau_L(\Delta v) = \frac{10\,\eta}{\Delta_0} \Phi\left(10\frac{\Delta v}{\Delta_0} - 1\right) d\tau_c \tag{22}$$

Concerning the source function, we have taken a NLTE case, with a continuum source function $S_c = B = 1$, and a line source function S_L decreasing towards the surface according to the law (Figure 15)

$$S_L = B[1 - 0.9\, e^{-(\tau_c/T)}] \tag{23}$$

so that the scale height of variation of S_L is the same as the scale height of variation of the thermal width. This fact also is likely to occur in the case of the Sun, even if the exact law is not exponential and even if the two scale heights are not exactly the same. The total source function at frequency Δv is then

$$S_v = \frac{B\, d\tau_c + S_L\, d\tau_L(\Delta v)}{d\tau_c + d\tau_L(\Delta v)}. \tag{24}$$

The strength of the line is characterized by the quantity η, or better:

$$T_0 = \eta T/\Delta_0 \tag{25}$$

which is $\sqrt{\pi}$ times the optical depth at the center of the line in a static situation over a distance equal to the scale height T and for a constant thermal width Δ_0. We have considered some values of T_0 ranging from 0.5 up to 10. The value of this parameter labels the emergent profiles shown in Figure 16. The more usual parameter $\eta_0 = \eta/\Delta_0$ takes then values from 50 up to 1000: the latter corresponds to fairly strong lines.

The Figure 16 shows the emergent intensity plotted vs the dimensionless parameter $\Delta v/\Delta_0$. The profiles are shifted as a whole towards the violet, as expected, indicating an overall approaching velocity. From an analytical point of view this is simply a

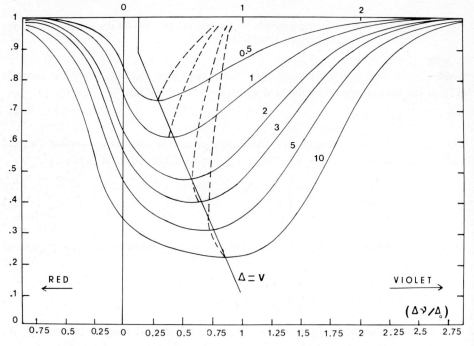

Fig. 16. *Asymmetrical profiles* resulting the test from model of Figure 15. (Notations: see the text.)

consequence of the (-1) term in the argument of the profile (see formula (21)): this negative term must be compensated by positive values of Δv in order to produce a larger optical depth.

The second most evident feature in Figure 16 is the general trend for the bisector lines to be directed towards the violet when going from the line minimum up to the wings. So if we think that the wings represent deeper layers, we infer that the velocity *increases* in the direction of the increasing optical depth: this is just the contrary of what has been assumed. In other words, the bisector procedure fails entirely by giving the wrong sign for the velocity gradient. Again from the examination of the formula (21), it is to be expected that τ varies more rapidly with Δv for negative values of Δv (the argument is more and more negative) than for positive values of Δv. Thus the red part of the emergent profile is much steeper than the violet one and the bisector ought to have a positive slope.

The third noticeable feature is that, when increasing the strength of the line (see for example the curve $T_0 = 10$) there is a tendency to recover the true velocity gradient in the central part of the line but not in the wings. This leads immediately to the C-shape of the bisector, a fact so often quoted by the observers (see Figure 7). This point is of importance and we want to discuss it in more details. For purpose of comparison, we have plotted in the same Figure 16 the 'true' velocity (v/w_0) when a Barbier-Eddington-like relation is assumed, i.e. by simply saying that an observed intensity I is the value of S_L at a certain depth τ_c where the velocity has the value $v(\tau_c)$. By taking into account the formulae (20), (17), and (18), the relation between intensity and the 'true' velocity is then:

$$I = 1 - 0.9\,(v/w_0). \tag{26}$$

We see that the agreement between the inferred velocity and the one that we call the 'true' one is very good at the line profile minimum. This implies that in this part of the line the integration runs over a small part of the medium, so that the velocity gradients have little importance, and we observe simply a certain depth at the rest frequency of the line in the *local* frame, but at a frequency shifted for the observer by the velocity of the given depth. This argument may be extended in the core of a strong line: there again, the velocity gradient is not seen but a non-zero local frequency (Δv_L) such as to produce an optical depth of the order of unity is simply shifted by the velocity of the corresponding depth. This is of course the justification of the bisector procedure, which implies thus both the validity of the Barbier-Eddington approximation and the fact that the integration runs over such a narrow region that the effects of the velocity gradient on the argument of the profile Φ are eliminated. In the wings these two conditions tend to break down. First the line source-function reaches the Planck value (the continuum source-function) so that the Barbier-Eddington relation is no more valid (it is essentially a 'one-layer' situation, with $\tau_v = \tau^*$, and, on the contrary, we recover rather an opposite situation, of which the most appropriate approximation is by a single layer at a given geometrical – and not optical – depth $(z = z^*)$. Second, the region of integration becomes so large that now the variation of the argument of the profile Φ is also very large and we tend to 'see' the region which has the largest absorption coefficient: in the model under study, this region is just the surface, since the thermal width has its largest value Δ_0 thus giving smaller values of the argument $(\Delta v/\Delta)$ of the profile Φ. To sum up, we are seeing the *same* part of the atmosphere, both at the line profile minimum and in the wings: it is this situation which leads immediately to the C-shape of the bisector. These facts have already been noticed by Kulander and Jefferies (1966) and by Roddier (1964) in other cases.

We must then conclude that the observation of a C-shape for the bisector is just a strong indication of the failure of the bisector procedure, especially in the line wings, and we emphasize that this C-shape is more likely to be attributed to transfer effects along a single line of sight than to a complex mixing of two or more profiles emerging from different points of the solar surface. In other terms, we tend to be sceptical in front of the 'many-columns' models.

5.5. Conclusions

The first overall conclusion of this section is that all calculations show now very clearly that, in order to reproduce a given observed profile, the relation $\tau(v, z)$, i.e. optical depth vs frequency and depth, is, generally, of much greater importance than the relation $S(z)$, i.e. source function, vs depth. In other words it is often unnecessary to compute precisely the source-function if the first relation is unknown. Of course, one must be fully aware of the possible effects of velocity gradients upon the value of the source function, but we are now prepared to evaluate these effects by means of various methods such as those reported by Rybicki (1970). Actually, the work by Magnan (1968, 1970) shows that we can handle complex geometrical situations and physical conditions such as implying various frequency redistribution functions. But, as far as the observed profiles are concerned, and on the iterative path from one relation (z, v) to the other $S(z)$, we are inclined to think that the most promising studies must now concentrate on the first one. This was often remarked by Athay (1970, 1972).

The second firm conclusion is the necessity of thinking always in terms of 'velocity gradients' and not in terms of 'velocities' alone: all this section was based on this point of view and in fact the introduction of the characteristic length parameter Λ is one way to do this. Related to this point is the importance of varying Doppler widths cause large variations of the dimensionless parameter $(\Delta v_L/\Delta)$ entering the profile Φ. We have demonstrated that this kind of effect must be taken into account even for 'photospheric' lines.

The third remark is that one cannot argue upon the simplicity of the models that we have discussed there to turn towards more sophisticated models including many parameters, but ignoring the fundamental facts that we have considered. In particular, we have restricted our discussion to the profile emergent from a single line-of-sight, but it must be realized that a subsequent summation upon many lines-of-sight includes implicitly the complexity of the situation along a single one! This seems evident, but it is also evident that the large majority of studies do not take properly into account the transfer problem along a single line-of-sight. It is thus an urgent need for the observers to specify what kind of average is truly observed, and this only requirement demands of course larger and larger resolution, both in space and in time. A high spatial resolution is possible only in the solar case; for the stars the situation is more difficult: there the geometry of the model becomes very crucial and determines for a large part the shape of the profile. The difficulties in establishing a diagnosis in this case have been discussed by Simonneau (1973).

6. General Conclusions

We shall conclude, obviously, by tempering the note of warning, which dominates this paper, by a message of long range hope.

(1) First of all, let us note that asymmetries in lines, often well determined from observations, are proving the existence of some velocity fields, either varying at a large depth scale, or else varying from point to point of the fraction of the solar surface under study (both cases commonly described as 'macro-velocity fields'). But, if we can assert that such fields exist, their diagnosis is terribly ambiguous. It is possible and even sometimes easy, to reproduce well some asymmetric profile by modeling the velocity fields. But we are forced to admit that the various solutions do not, even qualitatively resemble one another. The worse ambiguity is precisely the one above suggested: between a law $v(z)$ and a law $v(x, y)$, it is practically impossible to decide on the only basis of the existing observations, this being true both in the chromospheric and photospheric layers.

(2) Of course, this being said, we know that, *a priori*, a 'parametric' diagnosis is not what is needed. In the present state of knowledge of the physics of stellar atmospheres, taken in its broader sense (as suggested by Pecker *et al.*, 1973), it is relatively easy to solve the transfer equations, in NLTE, in complex geometrical or physical situations – *assuming* a given velocity field, even a complicated one. This is the basic reason why one does use a good physics of the source-function, and one derives from the observations the velocities, properly parametrized.

But the problem symmetric to that one, i.e. solving the hydrodynamical (or even MHD) equations in the stellar atmosphere, is in its very infancy! However, we might conceive that progress to come in this field will be done, hence removing a large part of the ambiguity. Then the diagnosis of measurements, using physical theory in a more and more self-consistent and complete way, will, ultimately use observations essentially as tests, or, possibly, to determine physical parameters of atoms and ions, the Sun becoming then nothing more than a big laboratory furnace, of known physical properties. This is still far from being the case, but we may at least hope that the earliest progress to come, not yet refined enough to make obsolete the use of diagnosis for the determination of velocity fields, will at least help to remove the ambiguities and undeterminations.

In shorter terms, now the theory gives $S(z)$, provided we know $v(x, y, z)$ and $\tau(z, v)$. The theory to come should allow us to determine $v(x, y, z)$ in a self-consistent way, not deriving it, rather badly, from observations, but more soundly from the equations of hydrodynamics and physics.

(3) The progress does not have to come only from improved theory. As we have said earlier, in Section 5 notably, better observations should complement the theoretical approach. A good guide to a better theory would indeed be given by very high resolving power (time, space, wave-length) observations of the solar lines, on the disk or off the limb. The problems evoked about the fine structure of K line are typical of what questions could be asked from the spectrographs. We know well that, fortunately, every observer is aware of this need. The efforts now under way are certainly going in that direction, and will soon provide us with new material to diagnose, with new guides to less ambiguous physical theories.

Therefore, our final conclusion will be very simple. It is a definite insistence on the necessarily parallel development on the three types of approaches already mentioned: the hydrodynamical (even MHD) approach; the NLTE transfer solutions, in presence of supposedly known velocity fields; and last but not least, the improvements of the resolution of the observations. Some kind of iterative process, injecting in the iterative loop the three types of data, is necessary in order to reach the same unified description of the astrophysical reality. A close coordination of work is necessary to improve the efficiency of the iterations. This implies from the part of the transfer people a clear conscience of the limitation of their theories, due to unadequate treatment of HD or MHD equations, from the part of the HD-MHD people a clear conscience of the coupling of radiation field with dynamical problems; from both a clear expression of what observations can provide, or must provide, them with what they need, and from the part of the observers, a coherent wish to observe quantities which add really new information according the requests of theory, instead of a bunch of unnecessary data.

References

Adam, M. G.: 1959, *Monthly Notices Roy. Astron. Soc.* **119**, 460.
Allen, C. W.: 1937, *Astrophys. J.* **85**, 156.
Athay, R. G.: 1970, *Solar Phys.* **12**, 175.
Athay, R. G.: 1972, *Radiation Transport in Spectral Lines*, D. Reidel Publ. Co., Dordrecht.
Auvergne, M., Frisch, H., Frisch, U., Froeschlé, Ch., and Ponquet, A.: 1973, *Astron. Astrophys.* **29**, 93.
Bahng, J. and Schwarzschild, M.: 1971, *Astrophys. J.* **134**, 337.
Bappu, M. K. V. and Sivaraman, K. R.: 1971, *Solar Phys.* **17**, 316.
Bates, B., Bradley, D. J., McKeith, C. D., McKeith, N. E., Burton, W. M., Paxton, H. J. B., Shenton, D. B., and Wilson, R.: 1969, *Nature* **224**, 161.
Böhm, K. H.: 1954, *Z. Astrophys.* **34**, 182 and **35**, 179.
Boyer, R.: 1969, *Astron. Astrophys.* **2**, 37.
Boyer, R.: 1970, *Astron. Astrophys.* **8**, 134.
Brault, J. W.: 1962, Thesis (unpublished).
Cram, L. E.: 1972, *Solar Phys.* **22**, 375.
De Jager, C. and Neven, L.: 1967, *Solar Phys.* **1**, 27.
De Jager, C. and Pecker, J. C.: 1951, *Compt. Rend. Acad. Sci. Paris* **232**, 1645.
Delbouille, L., De Jager, C., and Neven, L.: 1960, *Ann. Astrophys.* **23**, 949.
Dumont, S.: 1967, *Ann. Astrophys.* **30**, 861.
Edmonds, F. N.: 1964, *Astrophys. J.* **139**, 1358.
Frisch, U.: 1969, Ecole d'été de Montpellier, unpublished.

Gonczi, P. and Roddier, F.: 1971, *Astron. Astrophys.* **11**, 28.
Heintze, J. R. W., Hubenet, H., and De Jager, C.: 1964, *Bull. Astron. Inst. Neth.* **17**, 443.
Higgs, L. A.: 1960, *Monthly Notices Roy. Astron. Soc.* **121**, 421.
Higgs, L. A.: 1962, *Monthly Notices Roy. Astron. Soc.* **124**, 51.
Huang, S. S.: 1952, *Astrophys. J.* **115**, 529.
Jensen, E. and Orrall, F. Q.: 1963, *Astrophys. J.* **138**, 252.
Jewell, L. E.: 1896, *Astrophys. J.* **3**, 89.
Jorand, M.: 1962, *Ann. Astrophys.* **25**, 57.
Kulander, J. L. and Jefferies, J. T.: 1966, *Astrophys. J.* **146**, 194.
Lemaire, P.: 1969a, *Astrophys. Letters* **3**, 43.
Lemaire, P.: 1969b, in L. Houziaux and H. E. Butler (eds.), 'Ultraviolet Stellar Spectra and Related Ground-Based Observations', *IAU Symp.* **36**, 250.
Lemaire, P.: 1971, Thèse, Paris.
Linsky, J. L. and Avrett, E. H.: 1970, *Publ. Astron. Soc. Pacific* **82**, 169.
Magnan, C.: 1968, *Astrophys. Letters* **2**, 213.
Magnan, C.: 1970, *J. Quant. Spectrosc. Radiat. Transfer* **10**, 1.
Müller, E. A.: 1961, *Astron. J.* **66**, 458.
Noyes, R. W. and Leighton, R. B.: 1963, *Astrophys. J.* **138**, 631.
Olson, E. C.: 1962, *Astrophys. J.* **136**, 946.
Olson, E. C.: 1966, *Astrophys. J.* **143**, 904.
Pasachoff, J.: 1969, Dissertation, Harvard Univ.
Pasachoff, J.: 1970, *Solar Phys.* **12**, 202.
Pecker, J. C.: 1970, Colloque de Nice, *Dynamique et transfert en astrophysique*.
Pecker, J. C., Praderie, and F. Thomas, R. W.: 1973, *Astron. Astrophys.* **29**, 289.
Purcell, J. D., Garrett, D. L., and Tousey, R.: 1963, *Space Research* **III**, 781.
Roddier, F.: 1964, Thèse, Paris.
Roddier, F.: 1964, *Ann. Astrophys.* **28**, 463.
Rybicki, G. B.: 1970, in *Spectrum Formation in Stars with Steady State Extended Atmospheres*, NBS spec. Publ. 332.
Saint-John: 1910, *Astrophys. J.* **32**, 36.
Schmalberger, D. C.: 1963, *Astrophys. J.* **138**, 693.
Schröter, E. H.: 1957, *Z. Astrophys.* **41**, 141.
Schwarzschild, M.: 1961, *Astrophys. J.* **134**, 1.
Simonneau, E.: 1973, *Astron. Astrophys.* **29**, 357.
Sobolev, V. V.: 1960, *Moving Envelopes of Stars*, Harvard Univ. Press, Cambridge, Mass.
Voigt, H. H.: 1956, *Z. Astrophys.* **40**, 157.
Voigt, H. H.: 1959, *Z. Astrophys.* **47**, 144.
Wilson, P. R. and Evans, C. D.: 1971, *Solar Phys.* **18**, 29.
Wilson, P. R., Rees, D. E., Beckers, J. M., and Brown, D. R.: 1972, *Solar Phys.* **25**, 86.

JOINT DISCUSSIONS

I. PRECESSION, PLANETARY EPHEMERIDES AND TIME SCALES

(Edited by J. Kovalevsky)

Organizing Committee

J. Kovalevsky (Chairman), R. L. Duncombe, W. Fricke, B. Morando, G. A. Wilkins.
Secretary for discussions: B. Morando

INTRODUCTORY REMARKS

J. KOVALEVSKY
Bureau des Longitudes, Paris, France

May I open with a couple of remarks this joint discussion No. 1 on Precession, Planetary Ephemerides and Time scales. This discussion has a rather long story that I think it is good to remind.

As you all know, it is in 1964, that the IAU has adopted a new system of astronomical constants, that were introduced in most ephemerides like the apparent places in 1968 and fully introduced in all the Ephemerides in 1972. But already in 1964, it was clear that the work was left unfinished. What actually the IAU did in 1964 was to replace those constants that were too widely away from the values made known by the current observations. This was the case of the constant of aberration, the astronomical unit as expressed in kilometers, the masses of the Moon and the Earth, lunar parallax, geocentric constants of gravitation and the ellipticity of the Earth's figure. Some of these were afterwards adopted also by the IUGG.

Now, already in 1964, it was clear that the system of planetary masses will have to change some time but it was felt that the years to come should bring so much new information about the masses, essentially, but not only, through the tracking of deep space planetary probes.

Also, in 1964, as well actually as in 1950 when the case was taken up for the first time, it was well known that the constant of precession was in error by about 1″ per century. Minor but not negligible inaccuracies existed also in the constant of nutation and in the obliquity of the ecliptic. Finally it was also apparent that the ephemerides of various bodies, in particular the Sun and the Moon, were not referred to the same position of equinox. However, the difficulties, inherent in the application of such changes were considered too large and no change in these constants was proposed.

Since then, several new factors arose. The most important is certainly the inadequacy of the ephemeris time to cope with the high precision lunar laser and planetary radar observations as well as the precise tracking of space probes.

Atomic time is now used as a clock for ET... Should it be like that or should we adopt a new time-scale for the ephemerides in the solar system? Another important point is the construction of the FK5. The date when the fundamental catalogue is changed is an appropriate date for changes in other constants defining the system of reference.

The catalogue FK5 should be completed around 1980. The question arises then: should we also change the constant of precession?

The current planetary ephemerides are also very insufficient and should (and could actually) be greatly improved using better masses, but also better theories.

The deadline of 1980 was given by Commission 4 three years ago and three working groups were formed. These W.G. has started to debug the complex problem and the joint discussion is essentially aimed at a critical discussion of their preliminary conclusions.

This is why there is only one contributed paper, but three invited papers, that I invite you to discuss throughly in the light of the goal we wish to achieve in the nearest future: to establish a really comprehensive and coherent system of astronomical constants and units that is also consistent with the actual precision of theories and observations.

PRO AND CONTRA CHANGES IN THE CONVENTIONAL VALUES OF PRECESSION

W. FRICKE

Astronomisches Rechen-Institut, Heidelberg, F.R.G.

Abstract. From an evaluation of the arguments pro and contra changes in the precessional constants it is concluded that corrections should be made simultaneously with changes in the values of planetary masses at the time when the new fundamental reference system, the FK5, is introduced into the ephemerides. In this case the location and motion of the equinox would be corrected at the same time. An IAU decision on changes in the constants should be made in 1976. Since the completion of the FK5 may be expected around 1980, the introduction into the ephemerides appears practicable in 1984.

1. Introduction

Whenever an improved fundamental reference frame shall be compiled and introduced into the ephemerides, a decision has to be made on the values of precession. Since 1960 the positions and proper motions of the fundamental stars given in the 'Fourth Fundamental Catalogue' (abbreviated FK4) represent the conventional reference system. This system requires a revision, and there is general agreement that work at the improvement must start as soon as possible on the basis of the observations that have become available since the completion of the FK4 and on observations of stars fainter than magnitude 7.5 which are not included in the FK4, but may be suitable for an extension of the fundamental system to a fainter magnitude limit. The goal of the programme that will result in a new fundamental catalogue, the FK5, is the improvement of the systematic and internal accuracy of the data of the FK4 and the inclusion of stars down to about magnitude 9.0 in the FK5. More details about the plans for the revision and extension of the FK4 and about observations relevant to this task have been reported by Fricke (1974) and by Gliese (1974). From the large observational material that has to be exploited it can hardly be expected that the FK5 will be completed before 1980.

The first international agreement on the use of precessional values was reached by the *Conférence internationale des étoiles fondamentales de 1896* in Paris (Procès-Verbaux, Gauthier-Villars, Paris, p. 54). This conference adopted Newcomb's (1898) values which are still in use at present. The compilers of the FK4 have applied Newcomb's precession on the basis of a recommendation made by the international colloquium on 'Constantes fondamentales de l'astronomie' held in Paris in 1950 (*Publ. Centre Nat. Rech. Sci.*, Paris 1950, p. 129). After the completion of the FK4 no urgency has been seen for changes in precession. Consequently, in 1963, the *IAU Symp.* 21 on 'The System of Astronomical Constants' held in Paris recommended that no change be made at that time, see Kovalevsky (1965), p. 322. It was, however, recognized that a reconsideration would be desirable at a time when the 'IAU System of Astronomical Constants' proposed by the Paris Symposium in 1963 and adopted by the IAU in 1964 (*Trans. IAU* **XIIB**, 593, 1966) should require amendments. The IAU Colloquium No. 9 on 'The IAU System of Astronomical Constants' held at Heidelberg in 1970 has considered this question and resolved "that any changes in the precessional constants and in the system of planetary masses be introduced into the national and international almanacs together, at a time that is closely linked with the introduction of the next fundamental star catalogue" (*Celes. Mech.* **4**, 147, 1971). This resolution was endorsed by IAU Com. 4 in Brighton in 1970 (*Trans. IAU* **XIIB**, 81, 1971), and a Working Group was set up to report, in time for consideration in 1973, on the consequences of changes in the precessional constants and on the procedure for the introduction of new values at a later date. The term 'precessional constants' is used here in a wide sense to refer to a set of numerical expressions that define the mean positions of the equinox, equator and ecliptic at any epoch with respect to their positions at some standard epoch. In this connection the possible removal of the E-terms of aberration from the mean places of stars and the possible adoption of a new series for the nutation has to be considered.

The Working Group consisted of W. Fricke (chairman), T. Lederle (secretary), S. Aoki, S. V. M. Clube, J. Kovalevsky, J. H. Lieske, A. A. Nemiro, C. A. Lundquist, P. K. Seidelmann, S. Vasilevskis, and R. O. Vicente. The results of an exchange of views between the members of the Working Group are included in this report. Concerning the date for an IAU resolution on precessional constants, the Group has agreed that the decision should be made simultaneously with those on planetary masses, and that no final decision should be made at the IAU General Assembly in Sydney. From the view of the compilers of the FK5, the IAU should therefore decide on precession at one of its next General Assemblies, preferably in 1976. Hence, there is still an opportunity for considerations pro and contra changes.

2. Strength of Evidence for New Precessional Values

The following effects and quantities have to be discussed: (a) planetary precession, (b) lunisolar precession and a non-precessional motion of the equinox, (c) the obliquity and its secular changes, and (d) nutation.

2.1. PLANETARY PRECESSION

Two methods of determination are available. These are (1) the dynamical method based on planetary masses, and (2) the method of stellar kinematics which uses proper motions in a fundamental system and yields a simultaneous solution for lunisolar and planetary precession. The first method was applied by Newcomb, and it is even now the more powerful one. The second method is applicable, if the fundamental proper motions in right ascension are neither affected by any spurious motion of the equinox arising from errors in equinox determinations, nor by any component of real motions of stars parallel to the equator. At present, there is strong evidence for a fictitious motion of the equinox and still uncertainty about the 'real motions' of stars. Under these circumstances it is fortunate that the dynamical method yields results of an accuracy which may be satisfactory for a long time to come. Laubscher (1972) has determined the correction to Newcomb's value of the first order term in planetary precession on the basis of the planetary masses given by Klepczynski *et al.* (1971), and he has estimated its uncertainty by taking into account the ranges of the values for various planetary masses given by Kovalevsky (1971). Laubscher's result is

$$\Delta\lambda = -0''.03 \pm 0''.01 \text{ per century},$$

which means, (1) that the improvements made in the determination of planetary masses since the time of Newcomb have been of small effect on planetary precession, and (2) that future improvements will hardly change the correction significantly. I understand that an investigation made by Dr J. D. Mulholland is in agreement with this result.

2.2. LUNISOLAR PRECESSION AND A FICTITIOUS MOTION OF THE EQUINOX

Appreciable corrections to Newcomb's lunisolar precession have been derived by many authors since about 1910. Although it is not intended to review here the numerous determinations, it should be mentioned that Boss (1910) was the first to derive values on the basis of a proper motion system that was superior to Newcomb's system. Boss had no knowledge of the effects of galactic rotation at that time, but his results were later on found to be in the right direction. He also confirmed what Newcomb already had suspected, namely, that the equinox determined from observations of the Sun and planets prior to 1870 must be considerably in error.

Later on, when improved proper motion systems had become available – the GC, FK3, and the N30 –, and the effects of galactic rotation had been introduced into the solutions as additional unknowns, corrections Δn to Newcomb's value of precession in declination were derived in the range

$$+0''.30 \lesssim \Delta n \lesssim +0''.60 \text{ per century}.$$

A list of such determinations has been given by Böhme and Fricke (1965). Among these are some values which were not derived for the purpose of contributing to a better knowledge of astronomical constants but resulted as a by-product of investigations on stellar motions. Some other determinations in the list are affected by traceable errors. Two investigations deserve mentioning in which the authors made attempts to estimate the most probable values on the basis of older determinations. These are the investigations by Gordon (1952) and by Morgan and Oort (1951). Gordon applied a fairly primitive averaging procedure, while Morgan and Oort used a much more refined one, but made a traceable error in the reduction of results from one reference system to the other. More details on the comparison of precessional corrections derived on the basis of different systems were given by Fricke (1967a, b). Gordon's result of averaging is

$$\Delta p_1 = +1''.09 \pm 0''.03 \text{ per century, (corresponding to } \Delta n = +0''.44)$$
$$\Delta\lambda + \Delta e = +1''.24 \pm 0''.03 \text{ per century}.$$

(Δp_1, correction to lunisolar precession; $\Delta\lambda$, correction to planetary precession; Δe, correction to all proper motions in right ascension due to a fictitious motion of the equinox.)

The completion of the FK4 in 1963 suggested new investigations with the aim to find the precessional corrections in the FK4 system, and, by the application of the same stars, in the older systems, the GC, FK3, and N30. In order to avoid errors due to large proper motions, stars had to be selected in FK4 and FK4 Sup with distances larger than about 100 pc. The result was unexpected and surprising. On the basis of FK4 proper motions Fricke (1967b) found

$$\Delta n = +0''.44 \pm 0''.06 \text{ per century,}$$
$$\Delta\lambda + \Delta e = +1''.20 \pm 0''.15 \text{ per century.}$$

The errors are standard deviations. The correction Δn turned out to have the same value, within $0''.01$ per century, in the systems FK3, N30, and FK4, while $\Delta\lambda + \Delta e$ is slightly but not significantly different in the three systems.

In an effort to ascertain the reason for the deviation of Newcomb's results from those derived on the basis of FK3, N30, and FK4, Fricke (1971a) has made a rediscussion of Newcomb's determination. For a summary of the results reference is made to Fricke (1971b). It has been found that the value Δn given above is almost entirely the consequence of deficiencies of Newcomb's proper motion system in declination, and that the value $\Delta\lambda + \Delta e$ originates from deficiencies in Newcomb's determination of the equinox. The widespread suspicion that Newcomb's results were mainly affected by the neglect of galactic rotation or by an erroneous analysis of the proper motions has not been confirmed. On the contrary, Newcomb's analysis can be considered as an example of great mastery, and Newcomb cannot be blamed for the imperfections in the proper motion system available to him. The equality of the values of lunisolar precession in the systems FK3, N30, and FK4 demonstrates that in the 20th century the fundamental declination system has become an almost stable one. This has been confirmed by observations made within the past 20 years, which will hardly change the FK4 system in declination significantly. Some serious deficiencies in the FK4 which have recently become apparent lie in the right ascension system in the southern sky; but their nature is such that they cannot affect determinations of precession. These facts may be considered as strong arguments in favour of changes in the precessional constants.

The explanation of the large fictitious motion of the equinox

$$\Delta e = +1''.23 \text{ per century,}$$

where $\Delta\lambda = -0''.03$ is taken into account, must come from the observations of the Sun and other bodies of the planetary system. Their analysis is being carried out at Heidelberg. This is one of the most urgent tasks within the work at the improvement of the fundamental right ascension system. There are clear indications that Δe is the consequence of systematic errors of observations. One of the more recent investigations confirming these indications has been carried out by Martin and van Flandern

(1970) and van Flandern (1971) who have determined the motion of the FK4 equinox on the basis of lunar occultation observations. They found $\Delta e = +1\overset{"}{.}36 \pm 0\overset{"}{.}06$ per century which is in good agreement with the results from proper motions. The observations of the Sun, of planets and of lunar occultations may also help to decide whether part of the equinox motion arises from some systematic motions of the stars as suggested by Clube (1972) and by Vasilevskis and McNamara (1973). Undoubtedly this possibility cannot be entirely excluded on the basis of fundamental proper motions alone.

There are other methods which, in principle, allow the determination of precessional constants. None of these has so far given satisfactory results or any result that could cast doubt on the values derived from fundamental proper motions. At the Jet Propulsion Laboratory it is planned to apply the dynamical method based on planetary observations. Dr Lieske has communicated that, at JPL, the optical planetary data from the U.S. Naval Observatory have been reduced to the FK4 for the purpose of exploring the possibilities of solving for precessional corrections. Since the radar data are insensitive to the correction to general precession, it is hoped that the precessional correction and Δe can be separated.

Another method consists in the comparison of proper motions measured with respect to galaxies with those in the fundamental system. This method which has been applied by Fatchikhin (1970) and by Vasilevskis and Klemola (1971) has given some encouraging results. Apart from still unexplained discrepancies between the right ascension systems measured at the Lick Observatory and at Pulkovo, both determinations have yielded corrections Δn in the direction of the value on the FK4 (Fatchikhin's value is, in fact, identical with the FK4 value). Reference is made to the brief review of these results by Fricke (1972). Final conclusions from proper motions with respect to galaxies will have to wait for the completion of the programmes at both observatories and for measurements in the southern sky, and thus cannot be expected soon.

For the future, there remains the possibility of determining precession from absolute radio astrometric measurements of compact extragalactic radio sources by means of radio interferometers. A study of the potentialities of the method has been made by Walter (1974), but so far no plans for observing programmes have become known.

In summarizing this review of the strength of evidence for new precessional values it must be concluded that, if a decision for changes in precession is made in 1976, a value for lunisolar precession would have to be adopted from the determinations based on proper motions in the systems FK3, N30, and FK4, and the value for planetary precession would have to be adopted from a determination based on planetary masses. The elimination of a non-precessional motion of the equinox has to result from the discussion of observations of the Sun, Moon, and planets, and it must be done in the FK5.

Arguments against a change in the precessional values are, first, the hypothetical nature of the method of determination by means of proper motions, and, second, that erroneous values of precession do not result in erroneous ephemerides, if the ephemerides are computed rigorously. Letters from Dr Clube and Dr Vasilevskis have reminded me that these arguments should not be forgotten. In fact, I have used these arguments myself in 1964, (see Fricke, 1966), and given full consideration to these questions since that time. The first argument is correct, since precession cannot be determined without any hypothesis. In using fundamental proper motions the question as to the origin of the inherent rotation has to be answered. The rotation observed in the proper motions is the cumulative effect of various physically comprehensible and some spurious rotations. The hypothesis is made that the rotation arises from incorrect precession, galactic rotation and a non-precessional motion of the equinox caused by observational errors. From my own investigations I would not exclude that additional rotational effects may be present, but these effects are of the second order except when unsuitable observational material is applied. Concerning the second argument I have convinced myself that in the past incorrect precessional values have done harm in the computation of ephemerides and particularly in the reduction of observations by observers. The fatality of incorrect constants is the unforeseen confusion they produce and the great effort required for tracing their effects in the various applications.

2.3. OBLIQUITY AND ITS SECULAR CHANGE

For determining the value of the obliquity a rediscussion of the results obtained from the observations of the sun and the planets is necessary. The secular change of the obliquity has to follow, in conformity with the procedure applied by Newcomb, from planetary precession. Should the observations indicate a significant excess secular change, as it was taken for granted by Aoki (1967), then the origin of the effect would have to be found. As a preliminary result of own investigations I can report that the individual values of $\Delta\varepsilon$ from observations of the Sun and planets show indeed a decrease from about 1780 to about 1900. The values of $\Delta\varepsilon$ derived from observations after 1900, however, do not deviate sensibly from zero and do not show a significant secular change. There are similarities between the problems existing in the determinations of the equinox and the obliquity. In both, the older observations give rise to difficulties which cannot be easily solved. It may be mentioned that absolute observations carried out before about 1890 show not only a large scatter but also clear indications of neglected instrumental errors, so that the right ascension system of the FK4 had to be based on observations from about 1900 onwards exclusively, and that observations before 1900 entered the proper motion system of the FK4 with very small weight. It is to be expected that the FK5 will not bring a revival of the observations of the 19th century. This means that we will have to determine the equinox and the obliquity from observations made during the 20th century, and there is every indication that no significant variation of the obliquity in excess of the theoretical one is present.

2.4. NUTATION

Determinations of the principal nutation in obliquity, the constant of nutation, have been reviewed by Vicente (1971). Observations after 1900 have led to a value which is smaller than Newcomb's ($N=9\rlap{.}{''}21$) by about $0\rlap{.}{''}01$ with a mean error of a few units in the third decimal place. The smallness of the correction, the incomplete analysis of available observations, and the continuation of theoretical studies suggest that, at present, no change in the conventional value should be considered. I have noticed that Prof. Vicente supports this opinion, and no other suggestion has become known to me. The series developments of the nutation which have been used in the past in a consistent way have turned out to be of great advantage.

3. Changes in the Procedures for Computing Precise Places

When changes in the ephemerides are made, various suggestions for changes in the procedures for computing precise places may be taken into account. It is not intended to discuss here all such changes which may be considered. But, without going into details, two proposals made by Atkinson (1951, 1973) shall be mentioned here (cited from Dr Atkinson's letters).

(1) It would be helpful in all observational work, if all Apparent Right Ascensions were computed for the mean equinox of date, and not for the true equinox. At present, whether they are tabulated in the Ephemeris (or in the APFS) or are computed in the standard way by an observer, they must all have the Equation of Equinoxes subtracted from them, before they can be compared with transit-times read on a modern clock.

(2) It would benefit all observers of precise places, if the nutation were computed for the Earth's Pole of Figure and not for its Instantaneous Pole of Rotation.

4. Form of Changes in the Precessional Constants

New precessional constants can be introduced in a number of ways, and a particular method has to be adopted. One of the possibilities is to change explicitly the values of the general precession in longitude per tropical century and of the obliquity of the ecliptic for 1900.0 – these are the quantities included in the current IAU System of Astronomical Constants –, and the corresponding secular variations. Dr Seidelmann has drawn my attention to the desirability of a new definition of general precession that should include the relativistic effect and that would be clearly stated at the time when a decision on the constants is made. Work on the form of changes and definitions is ungratifying before a decision on the desirability of changes of the constants has been made, in particular, since the amount of work involved is considerable.

The consequences of changes have to be made known to those who prepare ephemerides and to all users, and formulae can easily be given for the reduction from the old to a new system.

Since the completion of the new fundamental reference system, the FK5, can hardly be expected before 1980, a realistic date for the simultaneous introduction of all changes into the ephemerides appears to be 1984.

References

Aoki, S.: 1967, *Publ. Astron. Soc. Japan* **19**, 585.
Atkinson, R. d'E.: 1951, *Monthly Notices Roy. Astron. Soc.* **111**, 619.
Atkinson, R. d'E.: 1973, *Astron. J.* **78**, 147.
Böhme, S. and Fricke, W.: 1965, in J. Kovalevsky (ed.), 'The System of Astronomical Constants', *IAU Symp.* **21** ≡ *Bull Astron.* **25**, 269.
Boss, L.: 1910, *Astron. J.* **26**, 111.
Clube, S. V. M.: 1972, *Monthly Notices Roy. Astron. Soc.* **159**, 289.
Fatchikhin, N. V.: 1970, *Astron. Zh.* **47**, 619 ≡ *Soviet Astron.* **14**, 495.
Fricke, W.: 1966, *Trans. IAU* **XIIB**, p. 604.
Fricke, W.: 1967a, *Astron. J.* **72**, 642.
Fricke, W.: 1967b, *Astron. J.* **72**, 1368.
Fricke, W.: 1971a, *Astron. Astrophys.* **13**, 298.
Fricke, W.: 1971b, *Celes. Mech.* **4**, 150.
Fricke, W.: 1972, *Ann. Rev. Astron. Astrophys.* **10**, 101.
Fricke, W.: 1974, in W. Gliese, C. A. Murray, and R. H. Tucker (eds.), 'New Problems in Astrometry', *IAU Symp.* **61**, 23.
Gliese, W.: 1974, in W. Gliese, C. A. Murray, and R. H. Tucker (eds.), 'New Problems in Astrometry', *IAU Symp.* **61**, 31.
Gordon, J. E.: 1952, *Izv. Glav. Astron. Obs. Pulkovo* **19**, No. 148, 72.
Klepczynski, W. J., Seidelmann, P. K., and Duncombe, R. L.: 1971, *Celes. Mech.* **4**, 253.
Kovalevsky, J. (ed.): 1965,. 'The System of Astronomical Constants', *IAU Symp.* **21**, 322.
Kovalevsky, J.: 1971, *Celes. Mech.* **4**, 213.
Laubscher, R. E.: 1972, *Astron. Astrophys.* **20**, 407.
Morgan, H. R. and Oort, J. H.: 1951, *Bull. Astron. Inst. Neth.* **11**, 379.
Newcomb, S.: 1898, *Astron. Papers Washington*, **8**, Part 1.
Van Flandern, T. C.: 1971, *Celes. Mech.* **4**, 182.
Vasilevskis, S. and Klemola, A. R.: 1971, *Astron. J.* **76**, 508.
Vasilevskis, S. and McNamara, B. J.: 1973, *Astron. J.* **78**, 639.
Vicente, R. O.: 1971, *Celes. Mech.* **4**, 186.
Walter, H. G.: 1974, in W. Gliese, C. A. Murray, and R. H. Tucker (eds.), 'New Problems in Astrometry', *IAU Symp.* **61**, 131.

DISCUSSION

Mulholland: The argument against change that an incorrect value of the precession need not affect the accuracy of ephemerides was true in 1964, but is no longer. We now have observations in the third coordinate, distance, for three planets and the Moon. In addition, independent means exist for determination of masses. The effect of these new tools is that an error in precession cannot so readily be absorbed into erroneous mean motions.

Murray: I am not in favour of a change in the adopted constant of precession. Any value derived from analyses of proper motions depends on hypotheses about the stellar velocity pattern, and can never be truly fundamental.

A so-called fundamental system, such as for example FK5, must be regarded as a conventional system. It does not matter if it has a small net rotation relation to the inertial system.

Vicente: About the needs of a certain IAU commission to have a change of the system of astronomical constants only in 20 years' time, I should like to emphasize that we have to consider the problem of changing the constants not only in relation to the needs of one of our commissions but the needs in the whole field of astronomy.

Eichhorn: Two years ago, I would have welcomed a change of the constant of luni-solar precession to the best available value. This is not true anymore, and there are two reasons for this.

(1) During the last two years, the advent of radio-astrometry, and especially its potentialities, allows one to expect that these measurements will make an important contribution toward a more accurate value of the constant of precession. It may be indicated to wait for 20 years and incorporate these data.

(2) We have been reminded that stellar kinematics is more complicated than the Oort model, and

that its description requires more parameters than this model provides. Inspite of all precautions, this fact may influence the best possible values of n and k which can be obtained from the material available even at this time.

Lastly, it is well known that a catalogue system is different from an inertial system. If the FK5 were too good an approximation to an inertial system, this fact might be overlooked by some investigators who should not overlook it.

For these reasons, it seems that the next 5 or 10 years might not be the most auspicious time for a change of the constant of luni-solar precession.

Fricke: The argument that more accurate determinations of luni-solar precession may become available in the near future can be used at any time. With our present knowledge, the accuracy of the value can be increased by one order of magnitude. Second order effects may be detected by radio astrometric methods or on the basis of improved stellar kinematics. The argument that the adoption of corrections to Newcomb's precession would lead to an FK5, which is a too good approximation to an inertial system is one which I would use in favour of changes.

Eichhorn: The contribution of radio-astronomers in the next 20 years will be spectacular as more and more definite material is being accumulated, so we can wait as the difference between the FK4 system and an inertial system is known.

Kovalevsky: All the consequences of having a wrong value of the constant of precession on a very precise dynamical theory have not been completely studied and I believe that it is safer to have a newer and better value for this constant.

Tucker: Dynamical people should do what they like when necessary.

Vasilevskis: Do not forget the need of stellar astronomy and I concur with the remark by Dr Murray.

Vicente: When I mentioned that we cannot consider only the objectives of stellar astronomy, I did not mean they are not important objectives, but the working group has to take into account the needs and objectives of all IAU commissions and, therefore, I consider the date 1980 as a convenient one for a change of the system of astronomical constants.

THE CALCULATION OF THE NUTATIONS

R. O. VICENTE
University of Lisbon, Portugal

It is well known that the knowledge of precession and nutation is essential for the computation of astronomical coordinates and the comparison of values obtained at different dates. It is therefore important to compute the nutations from the best available observations.

Unfortunately, there are not many long series of reliable observations that can be used for the calculation of the several nutations. Nowadays, we need more accurate values and, therefore, it is fundamental to have observations reduced in an homogeneous way. For this purpose, Commission 19 (Rotation of the Earth) set up a 'Working Group on Pole Coordinates', during the last IAU meeting in 1970 (Vicente, 1972), with the objective of reducing the 70 years of variation of latitude observations done by the International Latitude Service (called the International Polar Motion Service at the present time) that constitute a remarkable set of astronomical data. It is expected to obtain more reliable values for the coordinates of the pole and be able to calculate the nutations.

The Working Group on Pole Coordinates is transferring to punched cards all the observations registered in the original observation books and that involves nearly 2 million cards. This work has been hampered by financial difficulties, but it should be supported by the international astronomical community in order to obtain the best results from so many years of observations, done by international cooperation.

The theoretical researches done in the last decades have shown that the values of the nutations depend on the structure of the Earth (Jeffreys and Vicente, 1957). Lately, the researches done in seismology have resulted in a better knowledge about the structure of the Earth, leading to the setting up of many Earth models due to the availability of computers. This fact has led to the situation where one cannot propose better theoretical values for the nutations because they depend on the model adopted for the structure of the Earth.

In order to avoid such difficulties, the International Union of Geodesy and Geophysics has set up, during the 1971 General Assembly (Vicente, 1973), a committee formed by members of the International Association of Geodesy (IAG) and the International Association of Seismology and Physics of the Earth Interior (IASPEI), denominated 'Standard Earth Model Committee', with the purpose of recommending an Earth model that could be adopted as a standard in any studies that depend on the knowledge of the Earth's structure. This reference model will be important not only for astronomical purposes but also in geodesy and geophysics.

The Standard Earth Model Committee has set up a number of sub-committees composed by specialists concerned with different layers of the Earth. The reports of these sub-committees will be published in order to be discussed by all scientists interested in these subjects.

References

Jeffreys, H. and Vicente, R. O.: 1957, *Monthly Notices Roy Astron. Soc.* **117**, 142.
Vicente, R. O.: 1972, *Revista Fac. Ciencias Lisboa*, **A14**, 5.
Vicente, R. O.: 1973, *Bull. Geodes.* No. 107, 105.

DISCUSSION

Melchior: The flattening of the core in a reference Earth Model is very important as the difference between the two axes is of the order of 10 km and this fact plays a great role in the theory of nutations.

PLANETARY EPHEMERIDES

R. L. DUNCOMBE, P. K. SEIDELMANN, and P. M. JANICZEK

U.S. Naval Observatory, Washington, D.C., U.S.A.

At the present time the planetary ephemerides in the Astronomical Ephemeris and in the American Ephemeris and Nautical Almanac (both hereinafter referred to as the AE), the Astronomical Ephemeris of the U.S.S.R. and most other national almanacs have the following basis: For Mercury, Venus, Earth, and Mars the general theories of Simon Newcomb (1898a), the ephemeris of Mars including the empirical corrections determined by Ross (1917); for the five outer planets, the numerical integration of Eckert *et al.* (1951); the Connaissance de Temps publishes ephemerides of Mercury, Venus, Earth, and Mars based on the theories of Leverrier (1858, 1859, 1861a, b); for Jupiter, Saturn, Uranus, and Neptune the ephemerides are based on Leverrier's (1876a, b, 1877a, b) expressions as modified by Gaillot (1904, 1910, 1913). The ephemeris of Pluto is based on the numerical integration of Eckert *et al.* In all of the above publications the ephemeris of the Moon is now based on the Improved Lunar Ephemeris which is derived from the theory of Brown (1919). Newcomb's theories and the numerical integration of the orbits of the five outer planets all rest primarily on the system of astronomical constants and planetary masses adopted at the Paris conferences of 1896 and 1911 (*Monthly Notices Roy. Astron. Soc.*, 1912).

With the passage of time these basic theories and their underlying constants have been found to be defective in numerous respects. Analyses of extended series of precise meridian circle observations, as well as the advent of radar and laser observing techniques and the use of space probes, have permitted the determination of more precise values of many of the basic astronomical constants. The availability of these new data precipitated the Paris Symposium of 1963 on the System of Astronomical Constants. The results of this Conference were formally adopted by the 13th General Assembly in 1964 as the IAU System of Astronomical Constants, and commencing in 1968 the effects of the new system were incorporated into the national ephemerides (Supplement to AE 1968). Several constants remained unchanged in the new system, however, and it was recommended that they be considered later. Precession and the masses of the principal planets are among them. As a consequence of discussions at IAU Colloquium No. 9, Heidelberg 1970, and in accordance with the recommendations of Commission 4 of the IAU, the 14th General Assembly set up working groups to study the subjects of Precessional Constants, Units and Time Scales, and Planetary Ephemerides. Commission 4 has requested that the Working Group on Planetary Ephemerides consider the system of planetary masses to be used in a new set of fundamental ephemerides as well as other factors that it considers to be relevant to the adoption of a new set of ephemerides, including choice of orbital elements or starting values, coordinate systems, form of equations and precision in computation.

The membership comprises R. Duncombe, convener; P. Janiczek, secretary;

J. Kovalevsky, V. Abalakin, D. O'Handley, C. Oesterwinter, B. Morando, A. Sinclair, J. Schubart, W. Klepczynski, and S. Herrick. Through correspondence there has been an exchange of opinions among members of the Working Group and, in addition, opinions and recommendations have been solicited from the following consultants: L. Carpenter, G. M. Clemence, A. Deprit, L. E. Doggett, D. Dunham, A. D. Fiala, J. Griffith, P. Herget, H. Hertz, R. Laubscher, J. H. Lieske, B. Marsden, J. D. Mulholland, D. Pascu, P. K. Seidelmann, I. I. Shapiro, T. C. Van Flandern, W. Zielenbach.

Discussions and correspondence concerning the form of the fundamental theories seem inevitably to divide the participants into two groups, one defending the methods of analytic and general perturbations, the other favouring special perturbations by numerical means. It would be possible to objectively present a detailed list of the advantages and disadvantages of the two classic methods; but to do so would invite the continuation of a debate which has had, and is likely to have, no resolution that would satisfy every need and purpose. Among the national ephemeris offices charged with the responsibility of fundamental ephemerides, however, there is one practical requirement that is unavoidable. The interval of time between the adoption of new fundamental ephemerides and their publication is several years. For example, in order to introduce new ephemerides in the 1980 AE, the fundamental theories must be established and available by 1976. Moreover, the astronomical constants which enter the ephemerides should be available when the calculations begin. This means that the time for decisions to be made is almost at hand. As a consequence of the scheduling requirements, a choice may be made to utilize the method of numerical integration, to be followed later by consistent analytic theories which would serve over a more extended period.

At this time it seems appropriate to comment on the frequently stated opinion that ephemerides will be subject to frequent revisions dictated by highly accurate observations, such as radar and laser ranging. This is currently an unrealistic assessment. In order to exploit the accuracies of ranging techniques to their fullest potential, they must be conducted within the framework of continuing and systematic observational programs. Only after extended series of observations have accumulated will it become possible to determine the various separate effects of a complex planetary system which influence observations made from within the system. It is noted that there are presently no systematic programs of planetary radar observations; and, in fact, there are very few such observations being made at all for fundamental purposes.

For the accuracy required by the published ephemerides, a newly calculated system of ephemerides may be satisfactory for a period of time. A more accurate set of ephemerides, available only in machine readable form, may change more frequently, but this would not be influenced by nor affect the restraints of publication deadlines.

In addition to the method of calculating the ephemerides and the requirement of a series of observations, the underlying constants must be specified. The system of planetary masses used in the presently published ephemerides is basically that derived and introduced by Newcomb (1898a, b). The availability of extended series of meridian circle observations of the planets since the date of his analysis and the advent of radar, laser, and spacecraft data have made possible several refinements in our knowledge of the masses of the principal planets. Thus far the new observing techniques have been confined to the Moon, Mercury, Venus, and Mars; and therefore the most dramatic refinement of our knowledge would be expected with reference to these particular bodies.

A few of those consulted by the Working Group favor the adoption of a system of planetary masses arising from a single global investigation, while at the other extreme an opinion was given that the system of masses should be left flexible and unspecified. A system of planetary masses stemming from a single global investigation may in a sense be self-consistent, but it will normally not reflect all the types of data which provide information on the masses. Also, the resulting mass values are not necessarily of uniform accuracy, nor is the possibility eliminated for one value to be systematically in error. Generally speaking, the mass of a principal planet may be determined from the perturbations induced in the motion of a neighboring planet or a nearby spacecraft, or from the orbits of its natural or artificial satellites. It has been the established practice in the past to derive a system of planetary masses by the judicious combination of all the observational determinations. The majority opinion of the working group respondents favors the consideration of such a compiled system of planetary masses. Since the advent of new data techniques for the determination of planetary masses, such as spacecraft flybys and artificial satellites, two independent comprehensive discussions of all observationally determined values have been made (Kovalevsky, 1971; Duncombe et al., 1973).

The first discussion reflects the experienced scientific judgment of the investigator in rating the various observational determinations to arrive at suggested values of the masses. In the second discussion an effort was made to apply statistical techniques to the combination of all observational determinations in order to minimize the effect of systematic errors known to be present in the various observational types. Due to the paucity of observational determinations in a few cases, the statistical approach was not entirely successful; and therefore to some extent the subjective judgments of the compilers are involved in both systems.

Table I presents the currently adopted IAU values, the statistically based mean values, Kovalevsky's suggested values, and the authors' recommendations for values to be adopted in preparing new ephemerides. Also, comments are given to explain the selections. It is hoped that this list of reciprocal masses will form the basis for further discussion.

Before new ephemerides can be determined, the other Working Groups must make their recommendations. To meet the desired deadline for the new ephemerides of 1980, it is suggested that a colloquium be held in 1974 to tentatively adopt new constants, planetary masses, the gravitational model, the fundamental equator and equinox, and other guidelines for calculating new ephemerides. Cognizant organizations could calculate ephemerides, fit them to observations and intercompare the results. The

TABLE I

Planetary reciprocal masses

Planet	Present value	DKS	Kovalevsky proposed	Recommended	Comments
Mercury	6000000	5972000 ± 45000	6000000	6000000	systematic variations in recent determinations bracket old value
Venus	408000	408520 ± 9	408522	408522	spacecraft determinations
Earth/Moon	329390				
Mars	328912	328900.12 ± 0.20	328912	328900.1	consistent spacecraft data
Jupiter	3093500	3098709 ± 9	3098714	3098709	spacecraft data
Saturn	1047.355	1047.357 ± 0.005	1047.355	1047.355	still satisfies most recent evidence
Uranus	3501.6	3498.1 ± 0.4	3498.5	3498.5	recent planetary determinations
Neptune	22869	22759 ± 87	22869	22869	systematic differences preclude determining a new value
Pluto	19314	19332 ± 27	19314	19314	old value sufficiently close to recent determinations
	360000	3000000 ± 500000	2000000	3000000	most recent determinations

system of masses, constants and ephemerides, including changes to tentatively adopted constants, could be adopted at the 1976 IAU. This would permit adoption of new ephemerides in time for publication in the 1980 editions of the AE.

References

Brown, E. W.: 1919, *Tables of the Motion of the Moon*, New Haven.
Duncombe, R. L., Klepczynski, W. J., and Seidelmann, P. K.: 1973, *Fundamentals of Cosmic Physics*, Gordon and Breach, New York.
Eckert, W. J., Brouwer, D., and Clemence, G. M.: 1951, *Astron. Papers Am. Ephem.* **XII**.
Gaillot, A.: 1904, *Mem. Obs. Paris* **24**, 172.
Gaillot, A.: 1910, *Ann. Obs. Paris* **28**, A1.
Gaillot, A.: 1913, *Ann. Obs. Paris* **31**, 105.
Kovalevsky, J.: 1971, *Celes. Mech.* **4**, 213.
Leverrier, U. J. J.: 1858, *Ann. Obs. Paris* **4**, 1.
Leverrier, U. J. J.: 1859, *Ann. Obs. Paris* **5**, 1.
Leverrier, U. J. J.: 1861a, *Ann. Obs. Paris* **6**, 1.
Leverrier, U. J. J.: 1861b, *Ann. Obs. Paris* **6**, 185.
Leverrier, U. J. J.: 1876a, *Ann. Obs. Paris* **12**, 1.
Leverrier, U. J. J.: 1876b, *Ann. Obs. Paris* **12**, A1.
Leverrier, U. J. J.: 1877a, *Ann. Obs. Paris* **14**, A1.
Leverrier, U. J. J.: 1877b, *Ann. Obs. Paris* **14**, 1.
Newcomb, S.: 1898a, *Astron. Papers Am. Ephem.* **VI**.
Newcomb, S.: 1898b, *Astron. Papers Am. Ephem.* **VII**.
Ross, F. E.: 1912, *Monthly Notices Roy. Astron. Soc.* **72**, 342.
Ross, F. E.: 1917, *Astron. Papers Am. Ephem.* **IX**, Part II.

DISCUSSION

Kovalevsky: As far as the mass of Pluto is concerned, I refer to I. I. Shapiro who states that almost any value can be derived from the observations available at present.

My second remark has to do with the mass of the Earth-Moon system. If a value of 1/328 900.1 is adopted and the 1964 value of 1/328 912 is changed, then, there will be other consequences on the already adopted IUGG system for the Earth, and this may be unwise.

Mulholland: It seems inconsistent to be considering seriously an improvement of one order of magnitude in precession and then decline to accept a possible two orders of magnitude improvement of the Earth mass.

Vicente: I should like to support the remark of our chairman about the value he proposed for the mass of the Earth, in spite that it is different from Duncombe's recommended value.

I should like to emphasize that our Union has to acknowledge the values of certain constants proposed by other unions. In this case, the International Union of Geodesy and Geophysics has adopted a set of conventional constants defining the Geodetic Reference System 1967: equatorial radius for the Earth a_e, geocentric gravitational constant of the Earth GM, dynamical form factor of the Earth J_2.

We have to accept the values proposed in the same way that other unions will have to accept the new system of astronomical constants we intend to set up in 1980.

ASTRONOMICAL UNITS, CONSTANTS AND TIME-SCALES

G. A. WILKINS
H. M. Nautical Almanac Office, Herstmonceux, England

As I have already reported (*Trans. IAU* **XVA**, 9, 1973), the Working Group on Units and Time-scales was not able to reach any firm conclusions on whether the concepts of the astronomical unit and ephemeris time should be retained or replaced by the SI unit of length (metre) and atomic time. This disagreement was not unexpected as the membership was deliberately chosen to cover a wide spectrum of opinion. The following comments and suggestions represent a middle-of-the-road approach that is intended to meet the requirements of most astronomers and others who use the astronomical data. It has been suggested to me that this approach is not appropriate for use in those precise applications where relativistic concepts are necessary, and that therefore we should abandon completely the use of astronomical units of mass, length and time. There are, however, many applications where Newtonian concepts are perfectly adequate, and astronomical units are widely in use because of their general suitability and convenience. We should therefore continue to define a system of astronomical units, but should state quite clearly their relationships to SI units. An individual may use either system according to the circumstances, and standard conversion factors will be available for use by those who prefer the other system.

The system of astronomical units may be defined and related to SI units in the following way: the currently adopted values of the relevant constants have been used to illustrate the form of the system, but more accurate values should be substituted if the new system is adopted.

Definition of the Astronomical Units

- The astronomical unit of mass = mass of Sun, S.
- The astronomical unit of time = 1 day, D, of 86400 SI seconds. In what follows the terms day and second will be used without qualification.
- The astronomical unit of length is that length (A) for which the Gaussian constant of gravitation, k, takes the value of 0.017202098950 when the units of measurement are the astronomical units of mass, length and time. The dimensions of k^2 are those of the (Newtonian) constant of gravitation, G, i.e. $L^3 M^{-1} T^{-2}$

Primary Constants in SI Units

- Speed of light $\quad c = 299\,792\,500$ m s^{-1}
- Constant of gravitation $\quad G = 6.670 \times 10^{-11}$ m^3 kg^{-1} s^{-2}
- Light-time for unit distance $\quad \tau = 499.012$ s

Derived Constants in SI Units

- Astronomical unit of length $\quad A = c\tau = 149\,600$ Mm

- Helio-gravitational constant $\quad GS = k^2\,A^3/D^2 = 132.718$ Mm^3 s^{-2}
- Astronomical unit of mass $\quad\quad S\ \ = 1.990 \times 10^{30}$ kg

For some purposes it is convenient to use the light-second as a unit of length, and this unit is related to the SI and astronomical units as follows:

   1 light-second $\quad\quad\quad\quad\quad\quad = 299\,792\,500$ m
   1 astronomical unit of length $= 499.012$ light-seconds.

The numerical values are identical to those of constants c and τ above. It may also be useful to state the values of c (173.142) and τ (0.005 775 6) in astronomical units in order to show clearly that the astronomical units form a self-contained system.

   It is now generally recognized that the difference between the international atomic time-scale (TAI) and the ephemeris time-scale may be regarded as essentially constant over the period for which both scales are known (i.e. from 1956 to the present time). The adoption of the astronomical unit of time as defined above, namely the day of 86 400 SI seconds, for the unit of ephemeris time would not introduce any practical change in the use of ephemeris time. It would, on the contrary, remove the uncertainty that arises from the present necessity of attempting to relate current intervals of time to the tropical year of 1900.0. Similarly, it would be convenient for future work if the difference between the scale values for a particular instant were to be zero, so that for most practical astronomical applications the two scales could be regarded as equivalent. The arbitrary constants of new theories of motion would be adjusted so as to give the best fit between observations recorded against the scale of TAI, and the ephemerides based on the theories. In due time further comparisons would be expected to reveal systematic departures between new observations and the ephemerides. Such departures could arise merely from the deficiencies of the initial adjustments, but they could also be due to deficiencies in the theories, such as inadequate allowance for the effects of tidal friction. It is also conceivable that the constant of gravitation could be varying with respect to atomic time. If this were found to be the case the equations of motion would be modified and any additional parameters determined from the comparison with observation. The differences between the scale of TAI and the implied scales of the ephemerides could easily be taken into account by developing tables of the differences between TAI and the ET-scale of each ephemeris, as is now the practice for UT and the various lunar ephemerides.

   It is clear that we cannot hope to determine any scale of ephemeris time that will not be subject to later adjustment as new observational data are obtained and compared with ephemerides based on the theory for the dynamical system concerned. We can, however, visualize an ideal scale that is defined by adopting a particular definition of the unit of time-interval and a particular scale-value for some identifiable instant. The most precise way in which this can now be done is by reference to the scale of international atomic time. This idealized scale could be called the reference scale of ephemeris time and denoted by, say, the acronym TER. It could be carefully defined so that it would be appropriate for use in conjunction with relativistic theories. It could be regarded as being of unlimited extent, whereas it is meaningless to refer to TAI before about 1956 since it is a scale that is defined operationally in terms of the weighted mean of the system of national atomic time-scales.

If we define TER so that the difference between it and TAI was zero at a particular instant (e.g. 1958 Jan. $1\overset{d}{.}0$ TAI), the difference between it and the current scales of ET will be about 32 seconds. For the Moon this corresponds to an adjustment of about $1\overset{s}{.}3$ in right ascension of the current ephemeris, but for the Sun and planets the effects are much smaller and can easily be treated differentially.

My conclusion is that it would be worthwhile to retain the general concept of ephemeris time, but to determine the constants of new theories in such a way that the time-scales of the ephemerides would correspond closely to TER and hence to TAI. We would need to prepare tabulations of the best estimates of the difference between UT and TER and, where appropriate, of the differences between TER and the time-scales implied in the published ephemerides. The publication of such tabulations should be the responsibility of some internationally-recognized organisation, such as BIH, although individual astronomers or groups could submit or publish their estimates as has already been done by Morrison (1973).

It also appears to be desirable that the following changes should also be made in the interests of simplicity. Firstly, that Greenwich mean sidereal time should be defined so that it is unique; this implies that the corrections for polar motion should be applied in the reduction of observed sidereal time to GMST, rather than as a correction in universal time. Secondly, that universal time (UT or TU) should be defined so that it is unique; it should be identified with the scale now denoted by UT1 (TU1) and the scales now denoted by UT0 and UT2 should no longer be used.

Reference

Morrison, L. V.: 1973, *Nature* **241**, 519.

DISCUSSION

Mulholland: I am in complete agreement with the proposals relating to the astronomical system of units. I have some cautionary remarks, however, about possible hazards associated with the proposals relative to time scales. First, it is proposed to introduce a new time scale, based on the SI second, as the fundamental argument of the ephemerides. This seems a particularly good idea, since it will only be a recognition of what is already now done in critical applications. It seems likely, however, that the present construct now called 'ephemeris time' will continue to be useful for studying common features of ephemeris inadequacies. Thus, if the new scale is to be called 'ephemeris time', a new name will have to be found for the old concept.

This seems an unnecessary invitation to confusion. I suggest that a new name be applied instead to the new scale, perhaps dynamical time. Second, if the name 'ephemeris time' is applied to the new scale, it seems very desirable that the epoch of the new scale not differ significantly from the presently-used epoch of ET. The potential hazards of introducing a discontinuity in epoch between ET and TER are strikingly illustrated by the experience following the 1925 epoch change of 12 h in astronomical time. As late as 40 years after, a major paper appeared in which the pre-1925 residuals were very bad, because the observation times had been mishandled. While it is true that a 32^s discontinuity will not be as serious a hazard as 12 h, there seems no good reason to introduce this risk.

Markowitz: I think that ephemeris time should be dropped from the Almanacs. Only UT and atomic time should be used for all purposes.

Williams: I would like to point out that defining the difference between the new ephemeris time (TER) and TAI to be an integral number of seconds at a particular date will cause their mean difference to be non-integral. This is because there are periodic terms in the transformation from one to the other. It seems much better to me to make their mean values differ by an integral number of seconds.

Winkler: If we change the definition of ephemeris time, it will be related to SI units, and it is better to have a single ideal time parameter.

We should be concerned about what should be the time-keeping mechanism.

If we base our scale on atomic time, this will be evidently solved. One should take advantage of the fact that the atomic time scale is new, to absorb the 32-s difference in the epochs. The 180 ms can be easily absorbed in the incertainties in the ET.

Herrick: May I remind that 'ephemeris time' has been invented to suppress the errors in the tables of the Sun.

Fliegel: I have two remarks with respect to what Dr Winkler has just said:

(1) There is already some confusion over the 32^s... offset between TAI and the existing ET. When AI was the standard of atomic time in the USA, the ET-AI offset was universally adopted as $32^s.15$.

When the change was made to TAI, the U.S. Naval Observatory approximated ET-TAI = $32^s.180$ exactly.

However, at JPL, we changed to ET-TAI = 32.18438..., not because we thought that ET was so well known, but because it was necessary to prevent any discontinuity in ET in computer operations.

Rather than adopt an offset of 32 s exactly when new definitions are adopted, thus producing a discontinuity of about 184 ms, we would rather see an offset of zero, simply because a small discontinuity is more likely to cause mistakes that will be overlooked by future workers than a large one.

(2) If the gravitational time scale proves to be different from the atomic scale, is it proposed to use a new parameter t in the dynamical equations for planetary motions?

Winkler: Yes.

Becker: In the paper circulated among us, Dr Winkler described ephemeris time to be coordinate time and TAI to be 'proper time'. This cannot be accepted: many different coordinate time scales are feasable. TAI is designed as a coordinate time with respect to the Earth's surface. I should like to mention that at the IAU General Assembly 1967 in Prague I proposed to define a new ephemeris time on the basis of astronomical time by making to it relativistic corrections in such a way that a coordinate time results, having a reference point outside the solar system, where the gravitational potential of the solar system is zero, and with no movement relative to the solar system. Apparently Dr Winkler's ideas are on the line of my proposal of 1967.

At the conclusion of the meeting, a motion was proposed and accepted that, at this point, the Working Groups set up by Commission 4 should continue their work.

II. STELLAR INFRARED SPECTROSCOPY

(Edited by Y. Fujita)

Organizing Committee

Y. Fujita (Chairman), P. Connes, H. R. Johnson, H. Spinrad, N. J. Woolf

INTRODUCTION

Y. FUJITA

Dept. of Astronomy, University of Tokyo, Japan

Stellar infrared spectroscopic survey has advanced in two directions in recent years: in the number of different celestial sources investigated, and in the range of the wavelength region used. It is remarkable that, in the latter case, the scope of the survey is now extending from the photographic infrared region to the far infrared.

When I was asked by the Executive Committee of the IAU to organize the Joint Discussion on 'Stellar Infrared Spectroscopy' in the IAU General Assembly in Sydney, I thought that it would be rather wise and reasonable to restrict it in both directions in order to make our talks more effective and concentrated.

This is the reason why our subjects today are concerned with the study of the near-infrared region of stars only. Even so, I do hope this meeting may be fruitful with stimulating talks followed by very profitable discussions.

Finally I should like to express my special thanks to each member of the Organizing Committee who has given me many useful suggestions as regards the speakers and the subjects discussed.

HIGH RESOLUTION INTERFEROMETRY OF COOL STARS

D. L. LAMBERT

McDonald Observatory
and
Dept. of Astronomy, The University of Texas at Austin, Austin, Tex., U.S.A.

Abstract. This paper reviews results obtained in a program of high resolution infrared spectroscopy of K and M giants. The reported results include (i) a preliminary analysis of the C, N and O abundances for α Ori, (ii) determinations of the $^{12}C/^{13}C$ and $^{16}O/^{18}O$ ratios for α Ori ($^{12}C/^{13}C \sim 6$ and $^{16}O/^{18}O \sim 500$), (iii) the derivation of a lower limit, $^{16}O/^{18}O \gtrsim 300$, for the $^{16}O/^{18}O$ ratio in α Boo, (iv) a discussion of observations of the SiO first-overtone vibration-rotation bands in M stars, and (v) a search for the fundamental quadrupole vibration-rotation lines of H_2.

1. Introduction

This paper describes a selection of the results obtained in a program of infrared high resolution spectroscopy of cool stars. This program is a collaborative venture between University of Texas astronomers and Dr Reinhard Beer of the Jet Propulsion Laboratory in Pasadena, California.

The high resolution spectra are obtained with a Connes-type Fourier transform spectrometer located at the coudé focus of the McDonald Observatory's 107-in. reflector. This instrument was constructed at the Jet Propulsion Laboratory and a full description is available (Beer *et al.*, 1971). The highest resolution attainable is 0.03 cm^{-1}. Presently available stellar spectra correspond to a resolution of 0.1 cm^{-1} for the brightest objects. The useful spectral range is either 1.4 to 2.3 μ or 3.0 to 5.0 μ.

The goal of the observing program is to obtain spectra at a resolution of about 0.1 cm^{-1} for bright infrared stars; the working list includes giants and supergiants of spectral types K and M, long period variables and carbon stars. Since the observing time required to achieve a certain signal-to-noise ratio is inversely proportional to the square of the resolution, a compromise must be reached between resolution and signal-to-noise ratio. The desirable resolution depends on the star and the problems under investigation. The following example illustrates this problem. The absorption lines of M-type supergiants (e.g. α Ori, α Sco) are very broad. A direct interpretation is that the atmospheres are very turbulent with velocities of the order of 10 km s^{-1}. If the velocity distribution is Gaussian with a full halfwidth at half maximum (FWHM) of 10 km s^{-1}, a weak absorption line has a profile with a FWHM of 0.2 cm^{-1} at 6000 cm^{-1} (1.7 μ) or 0.1 cm^{-1} at 3000 cm^{-1} (3.3 μ). In this example, a resolution of 0.1 cm^{-1} in the 1.4 to 2.3 μ atmospheric windows will yield a spectrum which is not seriously degraded by the instrumental profile. Furthermore, the instrumental profile is well defined so that a correction for instrumental smearing may be applied. Line widths and turbulent motions are substantially smaller for the typical K giant star and a resolution of better than 0.1 cm^{-1} appears necessary in order to fully resolve the line spectrum.

2. The Nature of Infrared Stellar Spectra

Infrared stellar spectra contain classes of transitions which are sparsely represented in the previously analyzed visible spectra. An outline of the general nature of the infrared spectra is sketched here.

Low resolution infrared spectra of cool stars are dominated by the R branch heads of the CO vibration-rotation transitions. Although other molecules may not provide so spectacular a contribution, vibration-rotation transitions of molecules composed of the astrophysically abundant elements provide a rich and vital contribution to the spectra. The fundamental series of bands for the lighter molecules are located in or near the 3 to 5 μ atmospheric windows. Overtones for the abundant molecules can be detected at shorter wavelengths; for example, the 1.4 to 2.3 μ interval includes the first and second overtone bands of CO and the first-overtone bands of the OH radical. Heavier molecules have their fundamental transitions at longer wavelengths; SiO has first overtone bands near 4 μ. Vibration-rotation bands have not been detected in the visible region of stellar spectra; high order overtones would be necessary but these are intrinsically very weak. All of the presently observed transitions involve the ground electronic state of a molecule.

In the visible region, molecules contribute to the line spectrum through electronic transitions. In general, the separations of the lowest electronic states in molecules (also, in atoms) are such that transitions between them are located in the visible or ultraviolet. However, a few electronic transitions do contribute to the infrared stellar spectra. Examples are the CN red system and the Phillips and Ballik-Ramsay systems of C_2.

The contribution from atoms to the stellar infrared spectra is substantial. As the wavelength of observation increases, the average excitation potential of the contributing lines must increase. This result is a reflection of the fact that infrared transitions requiring energy level spacings of a few thousand wavenumbers must occur relatively near the ionization potential of the atom or ion. A typical infrared line of Si I has a lower state excitation potential of about 6 eV. This rather high value might prompt an expectation that the lines should be weaker in cool stars than in the Sun. In general, this expectation is not fulfilled because the increased scale heights for the luminous cool stars can compensate for the reduced excitation at the lower temperatures: weak lines in the solar spectrum appear as strong lines in the spectrum of an M type supergiant. The complex rare-earths can provide infrared transitions of low excitation potential.

A comment on the continuum radiation is not out of place among these general remarks on the spectrum. A typical cool star is now recognized to have a circumstellar shell which is detectable through the infrared excess across the 8 to 13 μ atmospheric window. In extreme cases, this shell may contribute to the spectrum at shorter wavelengths and produce a dilution of the photospheric lines.

3. Astrophysics and Stellar Infrared Spectroscopy

Detailed analyses of infrared spectra of cool stars will provide answers to many astrophysical problems. Two general problem areas are
- the physics of the stellar atmospheres
- cool stars and problems of stellar evolution.

The atmospheres of evolved cool stars represent a fierce challenge to the astrophysicist. The following selection should suffice to demonstrate the wealth of problems.

– The low gas density in giant and supergiant atmospheres provides near optimum conditions for the appearance of departures from local thermodynamic equilibrium. These departures must be considered in abundance analyses and in the treatment of atomic and molecular line blanketing in model atmosphere construction.

– The atmospheres appear to be very turbulent. Supergiants like α Ori and α Sco have very broad absorption lines which may correspond to supersonic turbulent velocities. The velocity fields have to be defined and their influence on the atmosphere evaluated; the turbulence will provide a pressure term in the photosphere and, also, provide a source of mechanical energy for chromospheric heating.

– A substantial chromosphere has been inferred from observations of strong Ca II H and K emission. The chromospheric structure is unknown. The possibility of a corona has not been definitively excluded. These particular questions may be answered from searches for the ultraviolet emission lines from the chromospheric-coronal gas. The infrared studies should provide the photospheric boundary conditions including estimates for the mechanical energy reservoir.

– Circumstellar shells are detected by the infrared excess and the narrow violet-displaced absorption components which accompany resonance lines of atoms and ions. Mass-loss rates have been estimated from analyses of the expansion rates for the shell. The physics of the shell is incomplete; for example, the origin of the expansion is presently uncertain.

– The outer atmosphere may be a site for dust grain formation. The influence of grains on the atmosphere has to be explored fully.

The stellar evolutionary steps which transfer a main sequence star to the red giant region are moderately well understood. A detailed mapping of evolution during the red giant stage is beginning and feedback from observational programs is likely to prove very productive.

In many applications of spectroscopic abundance analyses, the assumption – often implicit – is made that the surface composition is representative of the initial composition of the star. This is an invalid assumption for the cool evolved stars. Their initial surface layers may be mixed with processed material from the interior. Convective mixing can occur as the star evolves up the red giant branch and a more violent mixing is predicted to occur at the helium flash stage. No doubt, other possibilities may become apparent as theoretical efforts are expanded in this area. Direct observational checks are needed. The key elements and their isotopes are H, He, Li, C, N and O. The abundant elements, C, N, O and H provide molecules with vibration-rotation transitions in the infrared. The frequency of a vibration-rotation transition is dependent on the masses of the vibrating -rotating nuclei and an isotopic substitution can give a readily detectable frequency shift. The important $^{12}C/^{13}C$ ratio is an example of a quantity which can be extracted from the infrared spectra and which is a measure of the mixing experienced by the star.

4. An Abundance Analysis for α Ori (Betelgeuse)

Calculations based upon single slab models (Spinrad *et al.*, 1970; Beer *et al.*, 1972) had suggested that the present composition represented strong evidence for abundance changes wrought by mixing: i.e. C was underabundant by a large factor and N was overabundant. These results have been reexamined using a model atmosphere. The full analysis will be published elsewhere.

Model atmospheres for α Ori were described by Faÿ and Johnson (1973). The abundance calculations are based upon a model kindly supplied by Professor Hollis Johnson. Calculations of the absorption line profiles are based on the customary set of assumptions: local thermodynamic equilibrium, plane-parallel homogeneous layers and hydrostatic equilibrium. *All* of these assumptions deserve a critical examination when applied to an M supergiant atmosphere.

The initial goal was to derive abundances for the light elements C, N and O. The spectral features of relevance are
- CO: the fundamental, first and second overtone vibration-rotation bands
- OH: The fundamental and first-overtone vibration-rotation bands
- CN: the red $(A\,^2\Pi - X\,^2\Sigma)$ system
- SiO: the first-overtone $(4\,\mu)$ vibration-rotation bands.

These transitions have been identified on the infrared spectra. In addition, the 2–0 band of the CN red system near 8000 Å has been observed with a photoelectric coudé scanner. This scanner is also being used to observe permitted and forbidden lines of C and O in the visible and near infrared.

The analysis procedure is approximately as follows:
- the CO line intensities provide an estimate for the C abundance (C is associated into CO throughout the greater part of the atmosphere).
- the OH provides the O (in effect, O−C) abundance.
- the CN gives the N abundance when the C is known.

The available infrared spectra correspond to a resolution of 0.1 cm^{-1}. The interval 5500–6600 cm^{-1} is available at a signal-to-noise ratio of about 60 to 1. The 3 to 5 μ region spectrum has a rather poor signal-to-noise ratio of 15 to 1. A spectrum of the 3 to 4 μ region (Beer *et al.*, 1972) is available at a resolution of 0.29 cm^{-1} and a signal-to-noise ratio of 150 to 1.

The CO fundamental and first-overtone lines are predominantly highly saturated. The abundance analysis is based on the relatively unsaturated second-overtone lines near 1.6 μ. The oscillator strengths were computed from matrix elements provided by Young and Eachus (1966) which are based on room temperature measurements of

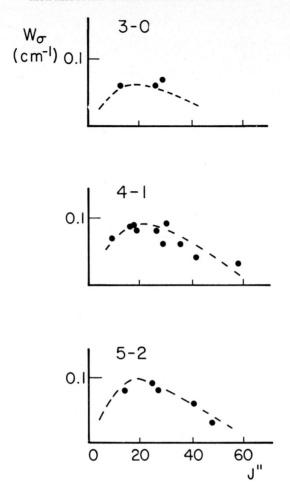

Fig. 1. Equivalent widths for P branch members of the $^{12}C^{16}O$ second-overtone bands 3–0, 4–1 and 5–2 in the spectrum of α Ori. The predictions (dashed lines), which are fitted to the observations (filled circles), are described in the text.

the absorption coefficients. The equivalent width for single unblended lines was measured and model atmosphere predictions were fitted to these observations (Figure 1). The R branch band heads were synthesized. Observed and predicted band heads for four bands are illustrated in Figures 2 to 5. The predicted profiles correspond to a reduction in the carbon abundance by a factor of two relative to the solar value. Oxygen was assumed to have a solar abundance.

The OH fundamental vibration-rotation bands in α Ori were discovered by Beer *et al.* (1972). First-overtone lines near 1.6 μ are prominent. The four $P(11)$ 2–0 lines are shown in Figure 6. The fundamental lines and the strongest first-overtone lines are subject to saturation effects. The absolute oscillator strengths for the 1–0 and 2–1 bands were measured by d'Incan *et al.* (1971). Since relative transition probabilities

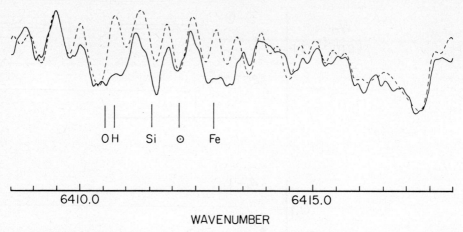

Fig. 2. The R branch head of the 3–0 $^{12}C^{16}O$ vibration-rotation band in the spectrum of α Ori. The synthesized spectrum (dashed line) is fitted to the observed spectrum (solid line). The former includes the CO and CN lines but not the OH and atomic lines identified below the spectrum.

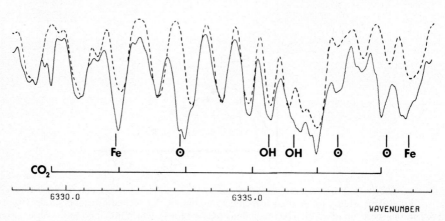

Fig. 3. The R branch head of the 4–1 $^{12}C^{16}O$ vibration-rotation band in the spectrum of α Ori (see caption to Figure 2). Telluric CO_2 lines are identified: their narrow profiles are to be contrasted with the broad stellar line profiles.

for fundamental and first-overtone bands were measured by Murphy (1971), absolute oscillator strengths for the first-overtone bands may be obtained. The model atmosphere predictions suggest that a factor of two increase in the oxygen abundance is required to explain the observed line intensities. A careful error analysis will have to be made before the oxygen overabundance is clearly established. In particular, the OH lines are gravity and temperature sensitive; modest decreases in surface gravity or effective temperature would remove the required overabundance.

The CN analysis is presently based on the observations of the 2–0 band near 8000 Å. The photoelectric scans were obtained with the coudé scanner at the 107-in. reflector (Tull, 1972). The resolution is about 0.1 Å and a single scan covers approxi-

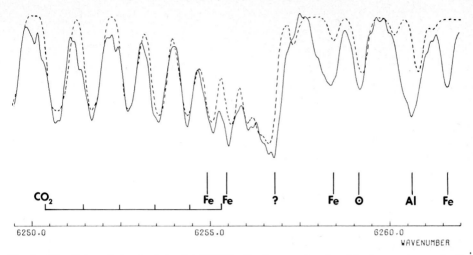

Fig. 4. The R branch head of the 5-2 $^{12}C^{16}O$ vibration-rotation band in the spectrum of α Ori (see caption to Figure 2).

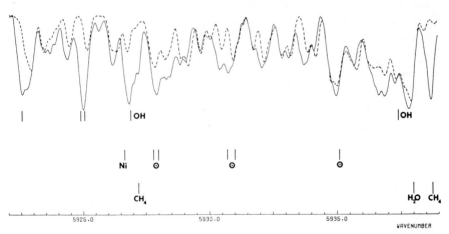

Fig. 5. The R branch head of the 9-6 $^{12}C^{16}O$ vibration-rotation band in the spectrum of α Ori (see caption to Figure 2). Telluric H_2O and CH_4 lines are identified.

mately 30 Å. A difficulty arises in the choice of dissociation energy and oscillator strength. Several recent independent experiments (see a summary by Arnold and Nicholls, 1973) have led to a preferred dissociation energy $D_o = 7.85$ eV. However, this value with the measured oscillator strengths leads to predicted solar CN line intensities which are inconsistent with the observed intensities for *both* the violet ($B\ ^2\Sigma - X\ ^2\Sigma$) and the red ($A\ ^2\Pi - X\ ^2\Sigma$) systems (Lambert and Mallia, 1974). This discrepancy is approximately a factor of two. The origin of this discrepancy is unclear. This preliminary analysis for α Ori is based on the dissociation energy $D_0 = 7.85$ eV and the oscillator strength for the 0–0 band which is required to fit the observed solar equivalent widths ($f_{00} = 1.0 \times 10^{-3}$). Relative band oscillator strengths ($f_{V'V''}/f_{00}$)

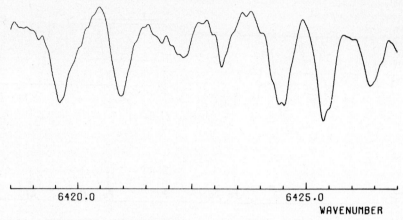

Fig. 6. First-overtone vibration-rotation lines of the hydroxyl (OH) radical in the spectrum of the α Ori: the $P(11)$ 2–0 lines are at 6419.6, 6420.9, 6424.4 and 6425.4 cm^{-1} in this spectrum.

were taken from Arnold and Nicholls (1972). An alternative procedure would be to adopt $D_0 = 7.5$ eV and to use the measured oscillator strength ($f_{00} = 3.4 \times 10^{-3}$). This would lead to a slightly higher nitrogen abundance. With the higher dissociation energy, the model atmosphere calculations can be fitted to the observations with an approximately solar nitrogen abundance.

The results of the preliminary abundance analysis show that the C, N and O abundances have their solar values to within a factor of two. This contradicts results of earlier analyses which showed considerable departures from solar values. These earlier analyses were based on a simple slab model for the atmosphere. The CO first-overtone and the OH fundamental lines were analyzed and both are severely saturated.

The preliminary analysis described here will be extended to include permitted and forbidden lines of O I and C I. The effects of changes in the model atmosphere parameters will also be studied.

5. The ^{12}C/^{13}C Abundance Ratio

As material processed in the interior is brought to the surface, the ^{12}C/^{13}C ratio will change. An increase in ^{13}C is anticipated. The terrestrial ratio is ^{12}C/^{13}C = 89. If the ratio for a stellar atmosphere can be shown to be significantly different, the mixing hypothesis would be confirmed and quantitative tests can be attempted. If certain conditions are satisfied, an isotopic abundance ratio can be measured with substantially greater accuracy than an element abundance. The ideal situation involves the comparison of ^{12}CX and ^{13}CX lines of similar intensity and excitation potential. A program of ^{12}C/^{13}C measurements is in progress based on the infrared spectra and near infrared photoelectric scans. Initial results for α Ori are described here.

The bandheads of both ^{12}C^{16}O and ^{13}C^{16}O molecules are prominent in even low resolution spectra of the 2.3 μ first overtone bands. The problem is not in the identifi-

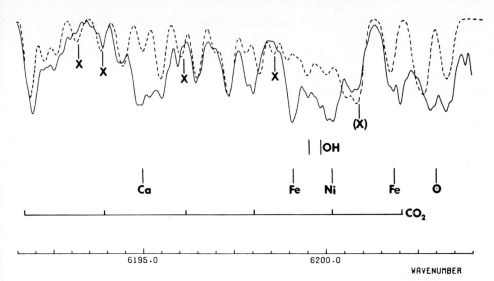

Fig. 7. The R branch head of the 4–1 $^{13}C^{16}O$ vibration-rotation band in the spectrum of α Ori. The synthesized spectrum (dashed line) includes the $^{12}C^{16}O$, $^{13}C^{16}O$, ^{12}CN and ^{13}CN lines but not the OH and atomic lines identified below the spectrum. This synthesis assumes $^{12}C/^{13}C = 5$ and the $^{13}C^{16}O$ transitions are the dominant contributions to the regions labelled 'X'.

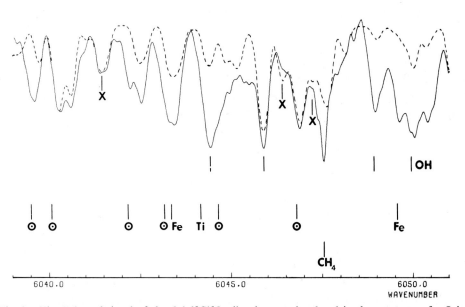

Fig. 8. The R branch head of the 6–3 $^{13}C^{16}O$ vibration-rotation band in the spectrum of α Ori. The synthesized spectrum (dashed line) includes only the $^{12}C^{16}O$, $^{13}C^{16}O$, ^{12}CN and ^{13}CN lines. This synthesis is for $^{12}C/^{13}C = 6$ with $^{13}C^{16}O$ transitions dominant in regions labelled 'X'.

cation but in the interpretation of these saturated lines. Weak lines are to be preferred for abundance estimates.

This discussion looks at the possibility of extracting the $^{12}C/^{13}C$ ratio from the weak second-overtone bands near 1.6 μ. The $^{12}C^{16}O$ bandheads are moderately saturated but weak unsaturated lines can be identified (see Figure 1). The $^{13}C^{16}O$ bandheads are visible but analysis is far from straight-forward on account of blending by atomic and molecular lines. At lower resolution, a $^{12}C/^{13}C$ analysis from this region would be very difficult. The 4–1 and 6–3 $^{13}C^{16}O$ bandheads are illustrated in Figures 7 and 8. The predicted spectrum includes CO and CN lines but not the various atomic and OH lines which are identified on the figures. Telluric CO_2 and CH_4 lines are indicated.

The 4–1 bandhead is heavily blended. The region beyond the bandhead shows almost as much structure as the CO bandhead itself! The predicted spectrum corresponds to a ratio $^{12}C/^{13}C=5$ and the regions marked by '$\mathbf{X}$' were given greatest weight in determining this ratio. In fact, a ratio $^{12}C/^{13}C \sim 6$ would provide a better fit. The region used in the derivation of the $^{12}C/^{13}C$ ratio extended beyond the portion illustrated in Figure 7.

The 6–3 bandhead is similarly confused by blends. The predicted spectrum with $^{12}C/^{13}C=6$ provides a reasonable match to the observed spectrum. Other bands have been examined and confirm the results for the 4–1 and 6–3 bands. A study of possible sources of error is in progress; for example, the microturbulent velocity might be increased from the assumed 7 km s^{-1} to improve the fit to the observed spectrum. The effect on the $^{12}C/^{13}C$ ratio will be small because the $^{13}C^{16}O$ lines are being compared with $^{12}C/^{16}O$ lines of very similar equivalent width. The result of the analysis is $^{12}C/^{13}C \sim 6 \pm 2$ where the uncertainty is estimated from the deterioration of the fit to the observed spectrum.

The $^{12}C/^{13}C$ ratio may also be derived from the first-overtone band near 2.3 μ. special attention must be given to a correct treatment of saturation. One approach is to search for weak lines in these bands; members of the P and R branches with very low or very high rotational quantum number should be weak. Comparison of lines of equal intensity can be attempted. The non-linearity resulting from saturation was illustrated by Lambert and Dearborn (1972) in a discussion of the bands in α Boo. Analysis of these bands is underway and the initial sample of stars includes α Boo, α Her and α Sco.

Another approach to the $^{12}C/^{13}C$ ratio is to consider other molecules. The 2–0 band of the CN red system near 8000 Å is being observed with the Tull coudé scanner. Initial results of this program are described elsewhere (Day et al., 1973). A typical scan for the CN rich K giant α Ser is shown in Figure 9. The ^{13}CN lines are readily identifiable. A ratio $^{12}C/^{13}C=12\pm2$ is derived. Analysis of similar scans for α Boo gave $^{12}C/^{13}C=7.2\pm1.5$.

Scans of CN in α Ori were obtained. Figure 10 covers the same interval as Figure 9 and ^{13}CN lines in α Ori are clearly present in considerable strength. The extremely broad lines introduce great difficulties in locating the continuum from scans as short as the one illustrated in Figure 9. To alleviate this difficulty, scans at a resolution of

0.1 Å and covering 30 Å have been obtained. Analysis gives $^{12}C/^{13}C \sim 5$ to confirm the result from the CO second-overtone bands. A paper describing the CO and CN analyses is being completed for publication (Lambert et al., 1974).

The CN line intensities in α Ori show an odd behavior relative to the model atmosphere predictions. In the observed spectrum the P, Q and R ^{12}CN lines of similar rotational quantum number have very similar intensities with Q branch lines appearing only slightly stronger than adjacent P and R lines. The model atmosphere predictions show the Q lines to be markedly stronger than the P and R lines. This is the anticipated behavior based upon the relative oscillator strengths. The correct explanation is unknown. A possibility is that these molecular lines are formed by scattering. Then, the line source function, S_l, involves an integration over the radiation field which reduces to the mean continuum intensity, $S_l = J_\lambda^c$ in the limit where the atmosphere is optically thin at the center of the CN lines. The present calculations assume the lines are formed by pure absorption or $S_l = B_\lambda(T)$. Calculations show that at 8000 Å, the mean intensity J_λ^c exceeds the Planck function $B_\lambda(T)$ and the gradient of J_λ^c through the atmosphere is smaller than that for $B_\lambda(T)$. Since several vibrational bands can excite the upper state of the 2–0 band, the appropriate mean intensity is a

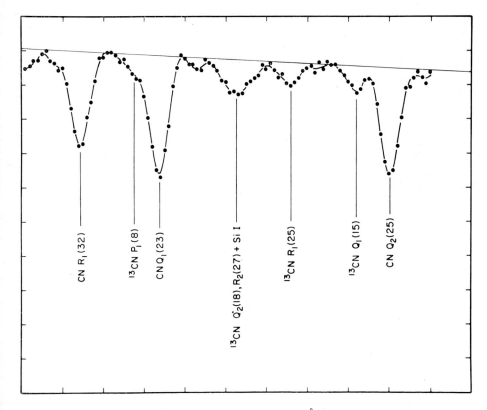

Fig. 9. A photoelectric scan of the spectrum of α Ser near 7975 Å. Four ^{13}CN features are prominent. The line ^{13}CN P_1 (8) is blended with a ^{12}CN line.

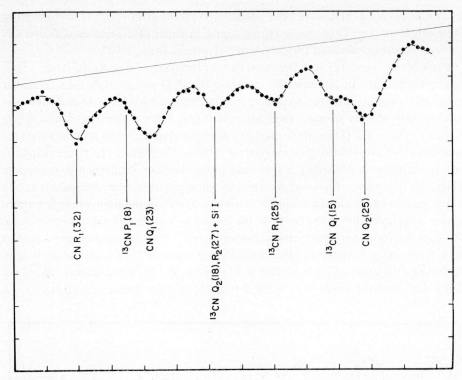

Fig. 10. A photoelectric scan of the spectrum of α Ori for the interval near 7975 Å. A comparison with Figure 9 shows that the ^{13}CN lines in the α Ser spectrum are readily identifiable in this scan of α Ori.

summation over mean intensities at several wavelengths. The difference between the mean J_λ^c and $B_\lambda(T)$ persists when this summation is carried out. As J_λ^c exceeds $B_\lambda(T)$, a line formed by scattering is weaker than the pure absorption equivalent and the shallower gradient of J_λ^c results in smaller intensity differences between lines. While this idea provides a qualitative explanation, detailed calculations show that the CN lines in the present model atmosphere are formed in deep layers such that the mean intensity closely approaches the local Planck function. A general discussion of the appropriate source function for molecular lines in stellar atmospheres is given by Hinkle and Lambert (1974). The scattering hypothesis and alternative explanations for the CN intensity anomaly are described by Lambert et al. (1974).

6. The ^{16}O/^{18}O and ^{16}O/^{17}O Abundance Ratios

The spectrum of α Ori in the region of the CO fundamental lines has been examined for evidence of the oxygen isotopes ^{17}O and ^{18}O. The terrestrial ratios are ^{16}O/^{17}O = = 2700 and ^{16}O/^{18}O = 490. The column density of CO is sufficiently large that many ^{12}C^{16}O and ^{13}C^{16}O are highly saturated and detection of ^{12}C^{18}O and even ^{13}C^{18}O lines can be anticipated (Lambert, 1973). Adverse consequences of the considerable saturation are readily apparent: the spectrum is extremely rich in $^{12}C^{16}O$ and $^{13}C^{16}O$ lines and the extraction of an abundance ratio from a line intensity ratio must be performed with caution. Regions free from $^{12}C^{16}O$ and $^{13}C^{16}O$ lines are rare. In addition, the rotational structure is developed in the stellar spectrum beyond the point to which the available molecular constants can be used to predict accurate wavenumbers. This introduces the possibility that lines attributed to rare isotopes are merely unrecognized $^{12}C^{16}O$ or $^{13}C^{16}O$ lines.

The 5 μ spectrum of α Ori was searched for $^{12}C^{17}O$ and $^{12}C^{18}O$ lines. $^{12}C^{18}O$ was positively identified through about 10 well resolved lines. The rich spectrum demands an analysis by spectrum synthesis techniques. A segment of the spectrum is illustrated in Figure 11. The well resolved $^{12}C^{18}O$ lines at 1977.5 cm^{-1} is the $P(20)$ 1-0 transition. The predicted spectrum (broken line) has been multiplied by the transmission of the Earth's atmosphere. The predictions were made for $^{16}O/^{18}O = 490$ and ∞ (no ^{18}O present). The stellar abundance ratio $^{16}O/^{18}O$ is considered equal to the terrestrial

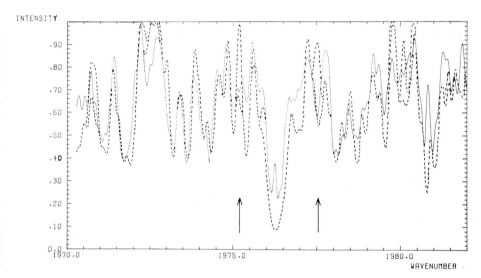

Fig. 11. The spectrum of α Ori between 1970 and 1982 cm^{-1}. The synthesized spectra (dashed lines) include all CO lines; the lower and upper traces corresponds to ratios $^{16}O/^{18}O = 480$ and ∞ respectively. The upper trace is coincident with the lower trace except in regions where $^{12}C^{18}O$ or $^{13}C^{18}O$ provides a dominant contribution and two such prominent regions are identified by the arrows.

ratio within an uncertainty of a factor of 3. The $^{12}C^{17}O$ molecule cannot be identified. Analysis shows that the observed spectrum is not inconsistent with a terrestrial $^{16}O/^{17}O$ ratio; the strongest $^{12}C^{17}O$ lines are severely blended.

A careful comparison of predicted and observed spectra reveals some interesting discrepancies. A few unidentified stellar lines are present. Some CO lines appear weaker than predicted. This is possibly attributable to departures from local thermodynamic equilibrium. A study of collision rates (Thompson, 1973; Hinkle and Lambert, 1974) shows that rotational states for a vibrational level should be kept in equilibrium while vibrational levels may show departures. Tentative evidence from the systematics of the discrepant CO lines supports this conclusion.

Interpretation of fundamental CO lines is difficult. According to the present model atmosphere, the $^{12}C^{18}O$ lines are not too seriously affected by saturation. Saturation is very severe for slightly stronger lines as the following numerical result illustrates: a typical $^{12}C^{16}O$ fundamental line may have a predicted equivalent width $W_\sigma=0.3$ cm^{-1} and the identical line from an isotopic species reduced in abundance by a factor of 500 has $W_\sigma=0.15$ cm^{-1}, i.e. reduction by only one-half. The boundary temperature for the atmosphere effectively determines the value of W_σ for which saturation begins to be severe; a high boundary temperature can produce weak lines which are strongly saturated. The risk of deducing incorrect isotopic abundance ratios may be minimized by comparing lines of similar intensity and weak lines should be used whenever possible. A possible approach would be to compare $^{12}C^{18}O$ fundamental lines with the weak $^{12}C^{16}O$ second-overtone lines.

Another approach is to search for isotopic lines among the first-overtone CO lines. This was attempted for Arcturus. A high ratio $^{16}O/^{18}O \sim 40$ was indicated by Geballe *et al.* (1972) in an analysis of a high resolution scan for a small interval of the

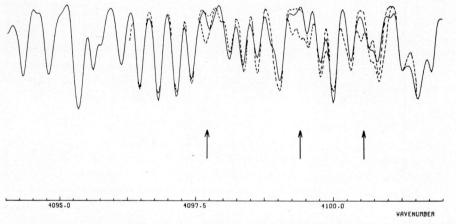

Fig. 12. The spectrum of α Boo between 4093 and 4104 cm^{-1}. The synthesized spectra (dashed lines) include only CO lines; the lower and upper traces correspond to high ($^{16}O/^{18}O = 40$) and normal ($^{16}O/^{18}O = 480$) abundances of ^{18}O. The arrows identify regions in which the observed $^{12}C^{18}O$ contribution is incompatible with the predicted spectrum for $^{16}O/^{18}O = 40$.

CO fundamental region. Model atmosphere calculations showed that $^{12}C^{18}O$ lines should be readily visible in the first-overtone region for $^{16}O/^{18}O \sim 40$. A thorough analysis of the observed spectrum reveals no acceptable identifications for $^{12}C^{18}O$ or $^{12}C^{17}O$ lines. The lack of $^{12}C^{18}O$ lines is well shown in Figure 2. The predicted spectra (dashed line) correspond to $^{16}O/^{18}O=40$ and $^{16}O/^{18}O=490$ (terrestrial). Several excellent regions are available to provide a firm lower limit to the $^{16}O/^{18}O$ ratio: the preliminary estimate is $^{16}O/^{18}O \gtrsim 300$. A similar result is obtained for the $^{16}O/^{17}O$ ratio.

This spectral region in stars with a larger column density of CO can be expected to provide positive identifications for both ^{17}O and ^{18}O. The M supergiants show considerably stronger CO both lines in rough accord with predictions. A difficulty is that some supergiants have very broad lines and many $^{12}C^{18}O$ lines will be unresolved. The $^{12}C^{17}O$ isotope is more favorably placed in that part of the 2–0 R branch between 4280 and 4295 cm^{-1} is in a region containing a minimum of $^{12}C^{16}O$ lines and no $^{13}C^{16}O$ lines. Figure 13 shows how a part of this region appears in Arcturus. In the spectrum of α Her, relatively strong lines appear at the positions of the $^{12}C^{17}O$ lines: 9 R branch lines between $J=20$ and $J=32$ are well resolved. By happy circumstance, the favorably located part of the 2–0 band contains the intensity maximum in the R

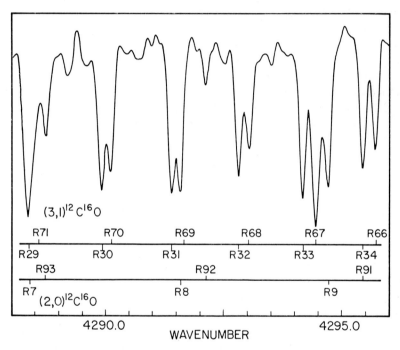

Fig. 13. The spectrum of α Boo between 4288 and 4296 cm^{-1}. $^{12}C^{16}O$ lines are identified. The 2–0 R branch $^{12}C^{17}O$ lines are located across this and neighboring regions. The relatively low density of $^{12}C^{16}O$ lines and the absence of $^{13}C^{16}O$ lines eases the detection of weaker $^{12}C^{17}O$ lines. In this particular region, the $^{12}C^{17}O$ lines with the exception of $R(32)$ at 4291.8 cm^{-1} are blended with stronger $^{12}C^{16}O$ lines.

branch. A few other $^{12}C^{17}O$ lines in other vibrational bands have been identified. The $^{12}C^{18}O$ lines are not so favorably placed but a few weak lines can be identified. A simple analysis suggests $^{17}O/^{18}O > 1$ which is to be compared with the terrestrial ratio $\simeq \frac{1}{5}$. The relative overabundance of ^{17}O can be linked to the CNO-cycle. These identifications confirm the results reported by Maillard (1973) who presents a more extensive abundance analysis.

7. M Stars and SiO Vibration-Rotation Bands

The fundamental vibration-rotation bands for SiO are located at about 8 μ. The first-overtone bands near 4 μ were discovered by Cudaback et al. (1971) in spectra of α Ori at a resolution of 2 cm^{-1}. The individual rotational lines can be identified away from the bandhead in the 0.1 cm^{-1} resolution spectra of α Ori. Indeed, these spectra provided a sufficient number of identified lines with accurate line positions that the molecular constants for the SiO electronic ground state were improved (Beer et al., 1973). The strength of the SiO lines in α Ori is approximately consistent with model atmosphere calculations, abundance expectations and the transition probabilities computed by Hedelund and Lambert (1972).

The SiO column density is predicted to increase for later spectral types. SiO is detected for α Her but the bandheads are absent from spectra of the Mira variables o Cet and R Leo obtained at several phases. This absence is not in accord with expectation.

Several possible explanations are under consideration: abundance adjustments, depletion of SiO by grain formation, atmospheric structure peculiarities (i.e. a flat temperature profile), and dilution of the photospheric 4 μ spectrum by continuum radiation from the circumstellar dust shell.

8. The H_2 Quadrupole Vibration-Rotation Lines

Although H_2 is an abundant molecule in cool stars, this symmetric molecule does not have an electric dipole vibration-rotation spectrum. Weak quadrupole transitions do exist and have been detected in the spectra of the major planets. Spinrad (1966) identified two H_2 2–0 quadrupole lines in spectra of the sequence of M stars from α Ori to o Cet. Later, it was noted that stronger 1–0 lines were not seen in a high resolution spectrum of α Ori (Spinrad and Wing, 1969). Convincing explanations for this discrepancy were unavailable (Spinrad, 1973) and a reexamination was undertaken.

A search for 1–0 lines was made using interferometric spectra of α Ori, α Sio, α Her, R Leo, o Cet and W Hya. Tentative identifications were established for 3 lines in the α Her spectra and upper limits were set for all other stars. These measurements are consistent with the model atmosphere predictions provided by Goon and Auman (1970). These results are at odds with the presence of 2–0 lines in these stars. The 2–0 identifications were examined and alternative identifications were proposed. A full discussion appears in a paper by Lambert et al. (1973).

9. Summary

In this paper, an attempt has been made to display the diverse contributions of high resolution infrared spectroscopy to a fuller understanding of the physics of cool stars.

The important conclusions of a general nature appear to be:

- infrared spectra are rich in information: the illustrations given here can be supplemented by radial velocity studies and detailed analyses of line profiles.

- the infrared spectra should not be considered in isolation: other spectral regions contain essential information, i.e. the [C I] and [O I] lines should be included in an abundance analysis.

- laboratory and theoretical studies on atoms and molecules are needed in order to extend available information on the wavelengths and intensities to provide a complete coverage for the spectrum analyses.

Acknowledgements

The active collaboration of the author's colleagues is deeply appreciated: important contributions have been made by Drs R. Beer, T. G. Barnes, A. L. Brooke and C. Sneden and Messrs. D. Dearborn and A. Bernat.

This paper presents one phase of activity in the joint Jet Propulsion Laboratory-University of Texas Infrared Astronomy Program supported, in part, by National Science Foundation Grant GP-32322X with the University of Texas at Austin, and, in part, by National Aeronautics and Space Administration Contract NAS 7-100 with the Jet Propulsion Laboratory, California Institute of Technology.

References

Arnold, J. O. and Nicholls, R. W.: 1972, *J. Quant. Spectrosc. Radiat. Transfer* **12**, 1435.
Arnold, J. O. and Nicholls, R. W.: 1973, *J. Quant. Spectrosc. Radiat. Transfer* **13**, 115.
Beer, R., Barnes, T. G., and Lambert, D. L.: 1973, *Proc. Conf. on Red Giant Stars*, Indiana Univ. Press, p. 84.
Beer, R., Hutchison, R. B., Norton, R. H., and Lambert, D. L.: 1972, *Astrophys. J.* **172**, 89.
Beer, R., Norton, R. H., and Seaman, C. H.: 1971, *Rev. Sci. Instr.* **42**, 1393.
Cudaback, D. D., Gaustad, J. E., and Knacke, R. F.: 1971, *Astrophys. J. Letters* **166**, L490.
Day, R. W., Lambert, D. L., and Sneden, C.: 1973, *Astrophys. J.* **185**, 213.
D'Incan, J., Effatin, C., and Roux, F.: 1971, *J. Quant. Spectrosc. Radiat. Transfer* **11**, 1215.
Faÿ, T. D. and Johnson, H. R.: 1973, *Astrophys. J.* **181**, 851.
Geballe, T. R., Wollman, E. R., and Rank, D. M.: 1972, *Astrophys. J. Letters* **177**, L27.
Goon, G. and Auman, J. R.: 1970, *Astrophys. J.* **161**, 533.
Hedelund, J. and Lambert, D. L.: 1972, *Astrophys. J. Letters* **11**, 71.
Hinkle, K. H. and Lambert, D. L.: 1973, in preparation.
Lambert, D. L.: 1973, *Proc. Conf. on Red Giant Stars*, Indiana Univ. Press, p. 350.
Lambert, D. L. and Dearborn, D. S.: 1972, *Mem. Soc. Roy. Sci Liège, 6th Ser.* **3**, 147.
Lambert, D. L. and Mallia, E. A.: 1974, *Adv. Astron. Astrophys.*, in preparation.
Lambert, D. L., Brooke, A. L., and Barnes, T. G.: 1973, *Astrophys. J.* **186**, 513.
Lambert, D. L., Dearborn, D. S., and Sneden, C. A.: 1974, in preparation.
Maillard, J. P.: 1973, this volume, p. 269.
Murphy, R. E.: 1971, *J. Chem. Phys.* **54**, 4852.
Spinrad, H.: 1966, *Astrophys. J.* **145**, 195.
Spinrad, H.: 1973, *Proc. Conf. on Red Giant Stars*, Indiana Univ. Press, p. 9.
Spinrad, H. and Wing, R. F.: 1969, *Ann. Rev. Astron. Astrophys.* **7**, 249.
Spinrad, H., Kaplan, L. D., Connes, P., Connes, J., Kunde, V. G., and Maillard, J. P.: 1970 in G. W. Lockwood and H. M. Dyck (eds.), *Proc. Conf. on Late-Type Stars*, KPNO Contribution, No. 54, Tucson, Arizona.

Thompson, R. I.: 1973, *Astrophys. J.* **181**, 1039.
Tull, R. G.: 1972, 'Auxiliary Instrumentation for large Telescopes', in S. Laustsen and A. Reiz (eds.) *Proceedings of ESO/CERN Conference of May* 1972, Geneva.
Young, L. A. and Eachus, W. J.: 1966, *J. Chem. Phys.* **44**, 4195.

DISCUSSION

Underhill: I notice you said in fitting some of your lines you used a microturbulent velocity of the order of 10 km s^{-1}. This seems hardly the thing to do because that would be supersonic at the temperature of your atmosphere. Supersonic microturbulence is a no-no.

Lambert: I don't think it is actually supersonic.

Underhill: What temperatures are you using?

Lambert: I suppose between about 2000 and 3000 deg.

Underhill: Well the velocity of your hydrogen line is far under 10 km s^{-1} at 3000 deg.

Lambert: There is a problem which I am trying to resolve at the present time, which is that the lines are definitely broad – of that there is no doubt – and some velocity field is broadening them; the question is whether you want to call it micro or macro turbulence.

Underhill: It makes a difference in computing your spectrum.

Lambert: When the lines are separate, yes I realize that.

Underhill: It throws suspicion on your methods of computing synthetic spectra.

Lambert: Well, we have to make a detailed comparison of line shapes. I have some very nice profiles of some well isolated atomic lines around one micron to do this sort of thing.

Hyland: Those synthetic spectra look very good and seem to fit beautifully but can you tell me how the model atmosphere calculations compare with the overall continuous energy distribution of the stars.

Lambert: It compares very well. This is described in a paper by Hollis Johnson in the *Astrophysical Journal*.

Hyland: Personally I didn't think they fitted very well.

Lambert: I did. I think the point is that the model atmosphere may not be too crucial because when you are talking about isotopic abundace ratios you are comparing $^{12}C/^{16}O$ lines with $^{13}C/^{16}O$ lines of about the same intensity and in the same wavelength region.

Hyland: That is all right when you are talking about isotopic abundances but when you are comparing the carbon and hydrogen abundance, that does not apply.

Lambert: Yes you are right.

PRESENTATION AND INTERPRETATION OF HIGH RESOLUTION INFRARED SPECTRA OF LATE-TYPE STARS

R. I. THOMPSON

Steward Observatory, University of Arizona, Ariz., U.S.A.

Current interest in stellar evolution is concentrated on the life of a star after it has left the main sequence. Of particular interest are the red giant or supergiant periods during the hydrogen and helium shell burning phases. Convective mixing during these stages can mix nuclear processed material to the surface where it may be viewed by spectroscopic methods. It is imperative that this rare chance to view processed material be exploited fully to increase our knowledge of stellar evolution.

The observation and interpretation of cool star spectra has its own particular set of problems and advantages. A particular difficulty is the formation of molecules at the low temperatures which occur in the atmospheres of late stars. Not only must the particularly complex spectra of molecules be dealt with but the problem of chemical equilibrium in the atmosphere must be solved accurately before quantitative analysis may be performed. The formation of molecules, however, has one advantage in that it very dramatically separates those stars with carbon to oxygen ratios greater than one from those with ratios less than one. It is the very high dissociation energy of 11.1 eV for the CO molecule which performs this separation. If carbon is less abundant than oxygen all of the carbon is tied up in CO and only oxides are formed in the stellar atmosphere which produce typical M star spectra. If, however, carbon is more abundant than oxygen then carbon compounds such as C_2 are formed in place of the oxides and a carbon star spectrum is formed. One of the great advantages of infrared stellar spectra is that it is the only ground based technique for observing CO in stellar atmospheres.

Figure 1 shows a portion of the CO spectrum for four M giants and supergiants. These spectra have a resolution of 0.5 cm^{-1} or $\Delta\lambda/\lambda$ equal to 10^{-4} at a wavelength of two microns. A Michelson interferometer at the *Cassegrain* focus of the Steward Observatory 90″ telescope and the Baja Mexico 60″ telescope was used to obtain these spectra. Rapid scanning techniques with two liquid nitrogen cooled lead sulfide detectors were used to eliminate scintillation noise. Full sky subtraction was obtained by using both inputs and outputs of the interferometer and switching the source input after each scan. All of the spectra shown here were taken by the observing team of Dr Harold L. Johnson of the Universities of Arizona and Mexico, Fred F. Forbes, David L. Steinmetz and the author of the University of Arizona.

All of the CO bands in Figure 1 are due to the first overtone vibration-rotation transitions in which the vibrational quantum number changes by 2 units. Individual lines may be observed in the right most (2, 0) band although the line shapes are not

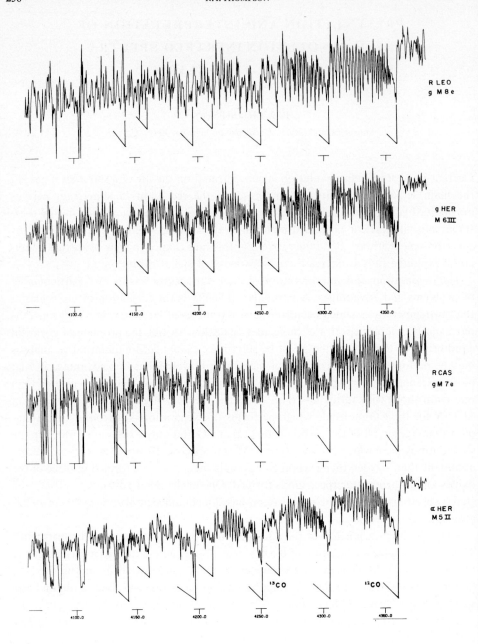

Fig. 1. Spectrum of four M stars with ^{12}CO and ^{13}CO positions marked. The lowermost markings are for ^{12}CO and the uppermost ^{13}CO. The first band head to the right is the (2, O) head for each isotope. The bottom scale is in wave number cm^{-1}.

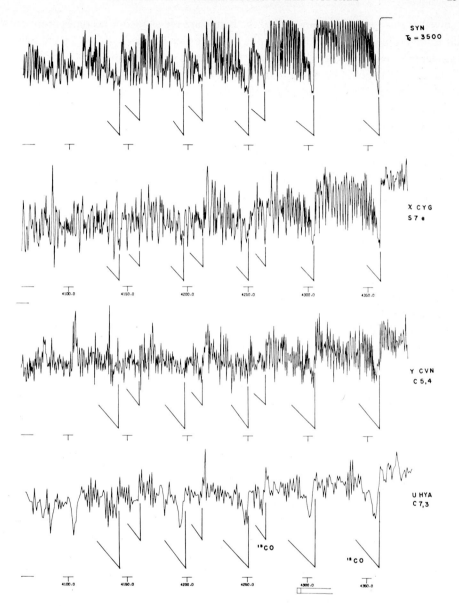

Fig. 2. The same region as Figure 1 for χ Cyg and two carbon stars. The top spectrum is a synthetic spectrum of CO described in the text. The spectrum of U Hya is 1.0 cm^{-1} as opposed to 0.5 cm^{-1} for all others.

resolved at this resolution. Since all of the bands are red shaded it becomes more difficult to determine individual lines for the higher order bands (3, 1), (4, 2), etc. where considerable overlap occurs. Individual band heads of ^{12}CO and ^{13}CO are marked on the figure. It should be noted that the ^{13}CO (2, 0) and (3, 1) band heads appear in all of the spectra. The significance of this is discussed below.

Figure 2 shows the same spectral region for the S star χ Cyg and the two carbon stars Y CVn and U Hya. χ Cyg shows a very strong and clean spectrum of CO whereas the two carbon stars show evidence of blanketing by other opacity sources such as CN and C_2 in this spectral region. Above the spectrum of χ Cyg is presented a synthetic spectrum of CO for comparison purposes. This spectrum was produced by integrating the flux equation

$$F_v = 2 \int_0^\infty S_v(\tau_v) E_2(\tau_v) \, d\tau_v$$

through a computed model atmosphere with $T_e = 3500$ K, $\log g = 0$, solar abundance, turbulent velocity = 4 km s^{-1} and a ^{12}C/^{13}C ratio of 27. The usual continuum opacity sources and H_2O along with CO as a line opacity source were used to form τ_v. Local thermodynamic equilibrium was assumed so that the source function $S_v(\tau)$ became the Planck function at the local temperature. The calculated spectrum was then convolved with the 0.5 cm^{-1} instrumental profile to give the spectrum shown in Figure 2.

As can be seen by a comparison of the synthetic spectrum with that of χ Cyg below, most of the spectral features can be accounted for by CO alone except in the case of the carbon stars. A local minimum in the (2, 0) band can be observed at about $J = 30$ for both the observed and synthetic spectra. This minimum is produced at the point where the low J lines coincide with the high J returning lines of the band. Since the lines are unresolved and highly saturated the coincidence greatly reduces the observed equivalent width.

Of particular interest is the good agreement between the observed and synthetic spectrum for the ^{13}CO (2, 0) and (3. 1) band heads. A ^{12}C/^{13}C ratio of 27 was used in the synthetic spectrum as it is the number predicted by convective mixing during the hydrogen shell burning red giant stage. Equally good agreement, however, can be obtained with ^{12}C/^{13}C ratios in the range from 89 to 5 due to the highly saturated nature of the lines. Figure 3 shows the (high resolution unconvolved) result of (synthetic spectrum) calculations for the ^{13}CO (2, 0) band head with ^{12}C/^{13}C ratios of 89, 27, and 5. When more than one line is visible the top is for a ratio of 89, the middle 27 and the lower 5. It is easily seen that there is very little difference in the spectrum produced by the three ratios. The differences in the convolved spectra are less than the error in current spectra of this region. Previous determinations of low ^{12}C/^{13}C ratios in M stars form the CO first harmonic bands are most likely in error and are by no means definitive measures of the ratio.

An alternative to the first harmonic CO bands for ^{12}C/^{13}C ratio determination are the second overtone bands which begin near 6400 cm^{-1}. These bands have oscillator strengths approximately 10^2 less than the first overtone bands and therefore are much less saturated. Figure 4 shows a region dominated by CO second overtone absorption for the four M stars. The ^{12}CO band heads are clearly visible although the ^{13}CO band heads are either very weak or not present. Preliminary analysis shows that a ^{12}C/^{13}C ratio on the order of 30 can account for the observed spectrum but a definite conclusion must await further work.

The two carbon stars and χ Cyg are shown in Figure 5 for the same spectral region. Again χ Cyg displays strong, clean CO features whereas the carbon stars show the presence of other molecules such as CN and C_2. It is difficult in the presence of strong C_2 and CN absorption to determine the ^{12}C/^{13}C ratio from the CO second overtone bands in this region. The spectra, however, do not seem inconsistent with the low ^{12}C/^{13}C ratios found by other means for many carbon stars.

A puzzling part of the CO spectrum in late stars is the (3, 0) band head at 6417.77 cm^{-1}. In most stars this band head is either missing or greatly reduced from its expected strength. Figure 6 shows this spectral region for the four M stars. De-

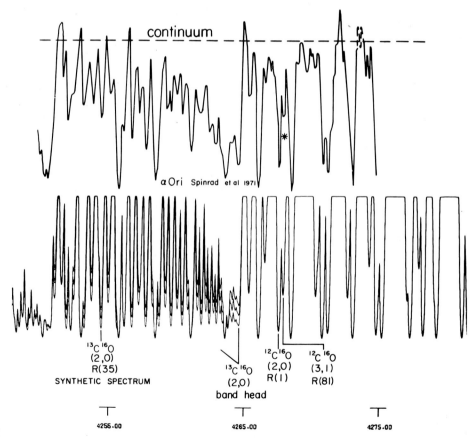

Fig. 3. A synthetic spectrum of the ^{13}CO (2, 0) band head is shown for the ^{12}C/^{13}C ratios of 89, 27 and 5. The top line is for 89, the middle 27 and the lower 5.

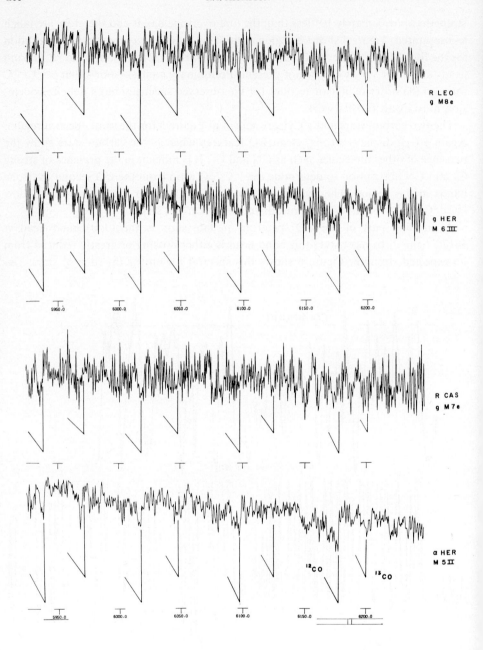

Fig. 4. This spectral region is dominated by ^{12}CO and ^{13}CO second overtone bands. The upper markings are for ^{13}CO and the lower ^{12}CO.

population of the ground state cannot be the cause of the reduced strength because as is shown in Figures 1 and 2 the (2, 0) band is present in all of these stars with its normal strength. A most likely explanation is that an unidentified source of opacity exists in this region which masks the CO absorption. The spectra of the carbon stars and χ Cyg shown in Figure 7 add more intriguing aspects to the puzzle. It is easily seen that the peculiar S star χ Cyg shows very strong (3, 0) ^{12}CO absorption. In Y CVn no (3, 0) absorption is observed but in U Hya a strong feature is observed at the position of the (3, 0) band head. If the same source of opacity exists in both Y CVn and the M stars it would then have to contain neither carbon nor oxygen. This opacity source is clearly not present in χ Cyg and may not be in U Hya. Higher resolution spectra may eventually solve this problem.

High resolution infrared spectroscopy is not limited to problems in molecular spectroscopy. A particularly clean region of the spectrum for M stars is shown in Figure 8 in which several high excitation atomic lines are present. A few of these lines are indicated on the figures. No systematic attempts has been made at this time to analyze

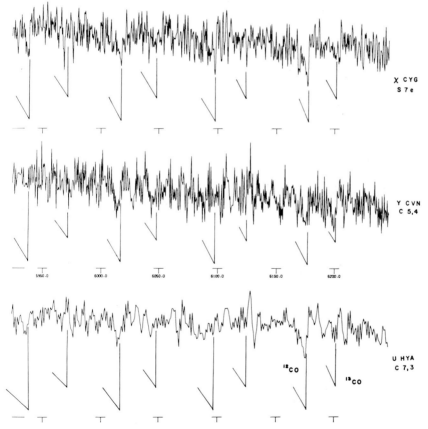

Fig. 5. The same region as Figure 4 for χ Cyg and the two carbon stars. The carbon stars show evidence of C$_2$ and CN absorption.

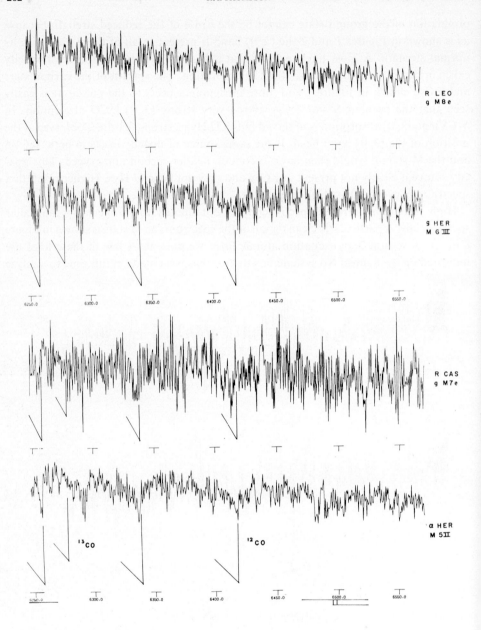

Fig. 6. This region contains the (3, 0) ^{12}CO band head which is greatly reduced in these M stars. It is the right most feature marked.

these lines in terms of abundance or variation with period of the star. The same region is shown in Figure 9 for χ Cyg and the carbon stars. χ Cyg shows the same general spectrum as that of the M stars but Y CVn and U Hya show strong absorption due to CN and C_2. Positions for some CN band heads are indicated but the complicated nature of the CN spectrum tends to wash out band head features. An analysis of U Hya by Duane Carbon, Steven Ridgway and the author based on a 0.5 cm^{-1} spectrum obtained by Ridgway indicates that a good fit can be obtained with CN alone. The more complicated spectrum of Y CVn however suggests that significant C_2 as well as CN absorption is present.

The spectra in Figure 10 confirm that significantly more C_2 is present in Y CVn than in U Hya. A strong absorption feature is present in both stars at the position of the (0, 0) C_2 Ballik-Ramsay band head but the feature is much stronger in Y CVn. The strength of this feature is due both to a very close packing of lines of $^{12}C^{12}C$, $^{12}C^{13}C$

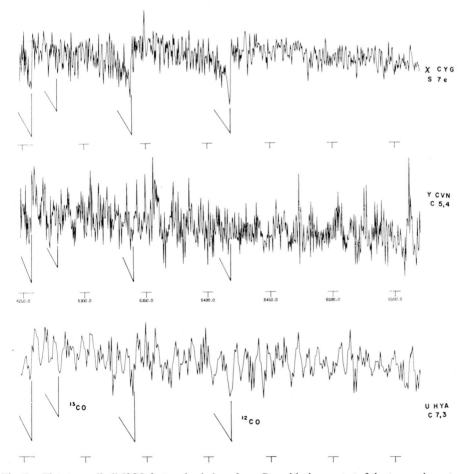

Fig. 7. The strong (3, 0) ^{12}CO feature is obvious for χ Cyg with the spectra of the two carbon stars shown below.

and $^{13}C^{13}C$ as well as a minimum in continuous opacity in this spectral region. This feature has great significance in the theory of carbon star evolution.

At present the two dominant theories of carbon star evolution are: (a) carbon star abundances represent CNO bi-cycle processed material in which case the carbon is depleted from its solar value or (b) carbon stars are the result of the mixing of triple-alpha produced carbon into a hydrogen shell during the Schwarzschild-Harm instability phase of helium shell burning. In case (b), the subsequent mixing of the carbon to the surface would enhance the carbon abundance over the solar abundance. Since the C_2 abundance varies essentially as the square of the C/O value it can serve as a sensitive monitor of the carbon abundance and thus an indicator of the correct theory.

A preliminary analysis of the C_2 bands by spectral synthesis techniques is shown in Figure 11. Along with the sources of continuous opacity mentioned earlier, CO, CN and both the Phillips and Ballik-Ramsay system of C_2 are included as line opacity sour-

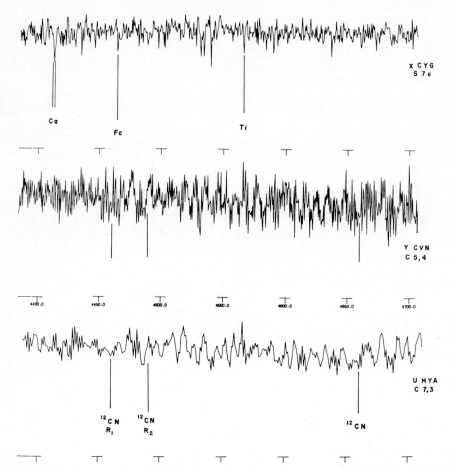

Fig. 8. This region is relatively clean of molecular lines and some atomic features are marked.

ces. All isotopic combinations of ^{12}C and ^{13}C are included for all of the molecules. The top spectrum represents a model with enhanced carbon which is indicative of triple-alpha mixed carbon. The number ratios of H:C:N:O are 1.0: 8.4×10^{-4}: 1.7×10^{-4}: 5.9×10^{-4}. An effective temperature of 3250 K and a log g value of 0 is used in the model. A turbulent velocity of 4 km s^{-1} and a ^{12}C/^{13}C ratio of 4 is used in the synthetic spectrum production. As can be seen in the figure the computed spectrum matches the observed spectrum for Y CVn directly below it quite well. The depressed region to the right of the main band head is due to several satellite bands, notably the R_{21} band.

At the bottom of Figure 11 is a computed synthetic spectrum with a low value of carbon. The number ratios for H:C:N:O in this computation are 1.0: 1.8×10^{-4}: 6.6×10^{-3}: 5.9×10^{-5}. All other parameters are the same as for the high carbon spectrum. This abundance of carbon although low is not a CNO bi-cycle abundance which would be about a factor of five lower. At present the synthetic spectrum program does not contain He Rayleigh and He$^-$ opacity terms which are needed for a

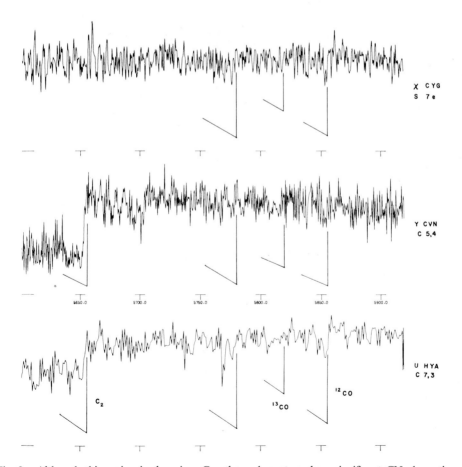

Fig. 9. Although this region is clean in χ Cyg the carbon stars show significant CN absorption.

true low H and high He CNO bi-cycle abundance. The computed low carbon spectrum is similar to that of U Hya in that no significant satellite band absorption is observed. Also if the synthetic spectrum resolution is reduced by a factor of two to match the resolution of U Hya, the main band intensities match well.

As it is the parameter C/O which determines the C_2 abundance the low C spectrum may be mimicked by an enhanced C spectrum in which C is just barely greater than O·

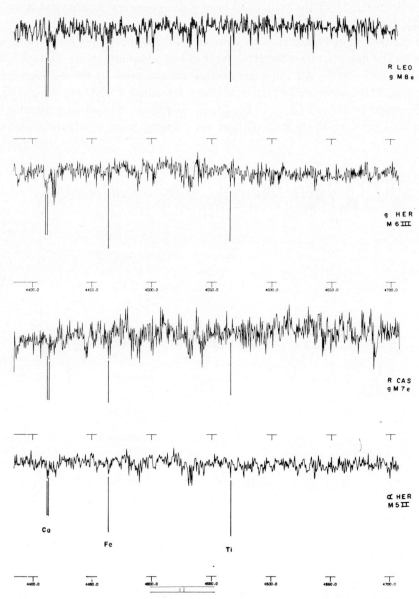

Fig. 10. This region shows the strong Ballik-Ramsay C_2 (0, 0) band in the carbon stars.

The high C spectrum which matches Y CVn, however, can not be reproduced by the CNO bi-cycle and is good evidence that in this star at least, a high carbon abundance is correct.

It should be cautioned that the above results are preliminary in that a full grid of temperatures, compositions and gravities has not been produced. The Ballik-Ramsay

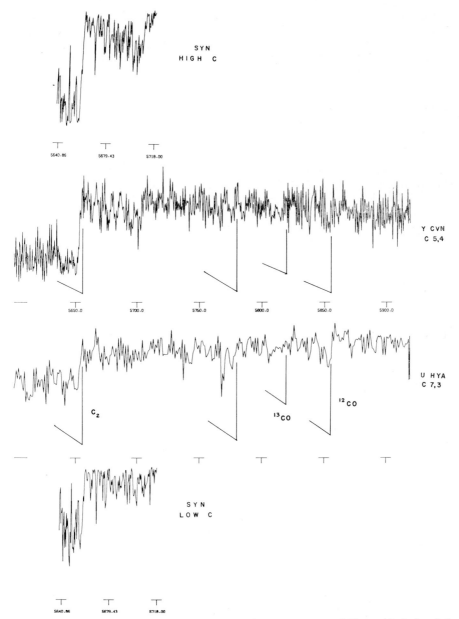

Fig. 11. Synthetic spectra are compared with the carbon star spectra of Figure 10. A description of the synthetic spectra is given in the text.

oscillator strength has been obtained as an unpublished result of Cooper and may be subject to revision. At this time, however, the CNO bi-cycle theory of carbon star formation seems to be in deep trouble.

DISCUSSION

Bell: Could you tell me where you got the stellar temperature for these stars because I would think when you are discussing the C_2 Ballik-Ramsay bands that the strength of the band is extraordinarily sensitive to the temperature as well as the carbon oxygen difference.

Thompson: Well it is sensitive but not as sensitive to the temperature as you might think. We have not had a good determination of the temperature but we are looking to Dr Lockwood whose scanner spectra of carbon stars is just becoming available and we intend to match very carefully his scanner data to determine the temperature. Right now we have used a rather high temperature of 3200 deg. Of course if you drop the temperature you associate more C_2 but you also increase your opacities. We have tried two or three different temperatures. The point is that the differences between the CNO bi-cycle and the Ulrich model is so great; a factor of 50 different in carbon abundance which is almost 50 squared different in C_2 abundance. It is sort of an off/on phenomenon rather than trying to get just the exact amount of carbon. When we do try a CNO bi-cycle and keep the hydrogen up to approximate the opacity that the helium minus woud give us we just see no C_2 at all. It just completely goes away. We feel that though there are certainly temperature effects the abundance effects are so strong that it completely outweighs the temperature effect.

HIGH RESOLUTION SPECTRA OF M AND C STARS BY FOURIER TRANSFORM SPECTROSCOPY

J. P. MAILLARD

Laboratoire Aimé Cotton, C.N.R.S. II, Orsay, France

A Michelson interferometer was put into operation during the year 1972 at the Coudé focus of the Haute-Provence Observatory's 76-in. telescope. We built this instrument at Aimé Cotton Laboratory (Orsay-France) where the method of Fourier Transform Spectroscopy has been largely developed in the direction of very high resolution work. A complete description of this device has been given elsewhere (Guelachvili and Maillard, 1970) and further details on the present study are reported by Maillard (1973). We will only describe the main characteristic elements of the interferometer (Figure 1).

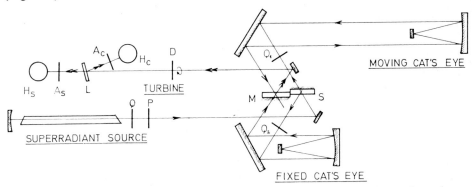

Fig. 1. Optical system of the interferometer showing the light path of the reference beam (xenon superradiant source at 3.5 μ). The path difference is measured by phase modulation system whose optical elements are: P, polarizer; Q_1, Q_2, quarter-wave plates; D, half-wave plate rotating at high speed; A_s, A_c, crossed analyzers; H_s, H_c, detectors.

1. Experimental

Because it was originally devoted to laboratory work, the maximum path-difference is two meters. The beam has a diameter of 80 mm. It is equipped with two cat's eyes which provide two output beams. They are focused on two (dry-ice-acetone) cooled PbS detectors of minimum area. The path difference is interferometrically measured by means of a very sharp and stable line produced by the superradiant effect in xenon at 3.5 μ. This beam is small in diameter and uses the central portion of the beam of the source under study. The error signal is generated by a phase modulation system to be insensitive to the intensity fluctuations of the reference source. The interferogram is recorded by a stepping technique. So, the cycle of the operations is the following: (a) positioning, (b) modulation of the signal, (c) integration, (d) storage on magnetic tape in six BCD characters, (e) displacement. The modulation is performed by a square wave modulation of the path difference. This technique has been demonstrated to be very more effective in action, cancelling the large amount of noise produced by atmosphere turbulence in the interferogram (Connes *et al.*, 1967). During all these operations the path difference is continually servo-controlled, which gives a very reliable instrument.

2. Diagram Magnitude / Resolving Power

This important effort to make available a high resolution instrument was justified because high resolution spectra of astronomical sources in the infrared are urgently

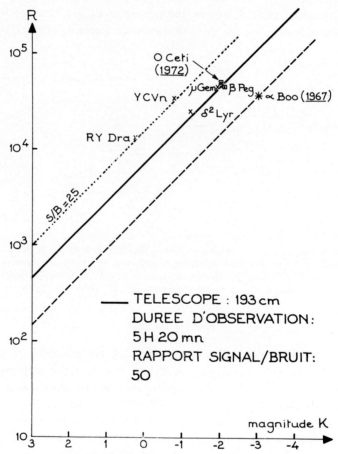

Fig. 2. Resolving power at $2\,\mu$ of spectra obtained with an interferometer as a function of the K magnitude of the stars, for given observational conditions. The resolution is limited by the incident energy. Full line: present interferometer (1972); dotted line: for the same instrument, with a signal-to-noise ratio of 25; dashed line: comparison with a previous instrument (1967) for $S/N=50$. Experimental points from recorded spectra are reported in the diagram.

required, and the spectra must be of sufficient quality to permit accurate and unambiguous analysis.

The more suitable sources for high resolution work in the 1–2.5 μ range are cool stars. So, within two periods of observation, I have used this instrument in a program devoted to the study of several stars of M and C type.

The majority of the previous observations of these stars in this spectral range was at relatively low or middle resolution, from classical slit-spectrometers or from commercial interferometers, the more recent limit being 0.5 cm^{-1} (Mertz, 1972). Apart from that, a previous interferometer (Connes et al., 1967), less sophisticated than this new one, built in the same laboratory, was operating in the years 1966–67. It had produced the first high resolution spectra but of a small number of bright cool stars (Chauville et al., 1970). Compared to this original instrument we have obtained a gain in available resolving power. That can be more explained by considering the diagram of the Figure 2. The K magnitudes (M) are plotted on the x-axis (a similar diagram is possible with other infrared magnitudes), the log of the resolving power (R) at 2 μ on the y-axis. For a given telescope, the same time of observation and the same signal-to-noise ratio, all the points in this diagram are on a straight line, according to the formula*:

$$\log R = -\frac{M}{2.5} + \text{const}.$$

A dashed line is plotted from a result obtained on α Boo in 1967, with the first interferometer, and the following observational conditions-telescope size: 193 cm; time of observation: 5^h20^m; signal-to-noise ratio: 50. With the same conditions, all the stellar points are raised by a factor 3 from our new results. If a signal-to-noise ratio of 25 is accepted, an upper line can be plotted.

This gain is due to the improvement of the stepping system, to the reduction of transmission losses and most importantly to the use of carefully selected detectors.

It must be noted also that this diagram can have an important use before observations. The resolving power achievable on a source of given K magnitude can be predicted and/or the best experimental conditions deduced.

---

* The signal-to-noise ratio in an absorption spectrum obtained by Fourier Transform Spectroscopy is expressed by:

$$\frac{S}{N} = \frac{I\,d\sigma}{p}\sqrt{T}$$

with I, the mean energy of the source in the recorded spectral range (W cm^{-1}); $d\sigma$, the limit of resolution (cm^{-1}); T, the total observing time (s); P, the noise-equivalent-power of the detectors (W Hz$^{-1/2}$).

The resolving power $R = \sigma/d\sigma$ is therefore proportional to the incident flux by:

$$\log R = \log I + \text{const}$$

It can be related to the magnitude through the usual definition, which gives:

$$\log R = -\frac{M}{2.5} + \text{const}.$$

3. Observations

Our observations of cool stars are summarized in Table I, II, III where the stars are listed with their spectral type and the luminosity class in general use, the V and K magnitude taken from the 'Two-Micron Sky Survey' (Neugebauer and Leighton, 1969) or from the photometric measurements of Johnson (1964), the recording time, the date of observation and the limit of resolution reached (by taking into account the apodization used). For all the presented spectra, the signal-to-noise ratio has a figure of 25 to 50 in the 1.6 and 2 μ regions.

3.1. Normal M stars

Table I gives a set of M stars of class II and III, in which spectral type goes from M0 to M6.

From these spectra, all having approximately the same resolution, a coarse analysis can be achieved, to study the evolution of the C/O, C/H, $^{12}C/^{13}C$ ratios and of other parameters such as the turbulent velocity and the temperature, according to the spectral type. Other specific problems and more detailed analysis of each spectrum are also possible as will be shown on α Her.

TABLE I
Observational conditions of normal M stars

Star	Spectral type	Magnitude		Date of observation	Recording time	Limit of resolution (cm^{-1})
		V (0.55 μ)	K (2.2 μ)			
μU Maj	M0 III	3.04	-0.82	29.04.72	3^{h}20^m	0.20
β Peg	M2 II	2.54	-2.16	15.12.72	4^{h}20^m	0.12
μ Gem	M3 III	2.83	-1.89	15.12.72	4^{h}50^m	0.11
δ^2 Lyr	M4 II	4.31	-1.21	06.05.72	3^{h}40^m	0.24
α Her	M5 II	3.05	-3.44	23.04.72 / 04.05.72	2^{h}25^m / 3^{h}05^m	0.12 / 0.10
g Her	M6 III	4.4	-2	03.05.72	2^{h}30^m	0.17

TABLE II
Observational conditions of C stars

Star	Spectral type[a]	Magnitude		Date of observation	Recording time	Limit of resolution (cm^{-1})
		V	K			
UU Aur	C5 SR	5.3	-0.71	26.04.72	3^h	0.17
Y CVn	C5 SR	4.8	-0.77	24.04.72 / 04.05.72	3^{h}30^m / 3^{h}45^m	0.17 / 0.17
RY Dra	C4 SR	6.3	$+0.27$	26.04.72	2^{h}30^m	0.55
U Hya	C7 I	4.8	-0.67	07.05.72	2^{h}30^m	0.55

[a] SR: semi-regular; I: irregular.

TABLE III

Observational conditions of variable M stars

Star	Spectral type	Magnitude V	K	Phase at observation	Date of observation	Recording time	Limit of resolution (cm^{-1})
RX Boo	M8	6.5	−1.85		02.05.72	4^h	0.12
Type Mira							
R Leo	M7e–M9	5.4–10.5	∼ −2	0.23	03.05.72	3^h15^m	0.11
				0.85	19.12.72	6^h	0.11
R LMi	M7e–M9	6.3–13.2	−0.67	0.76	23.12.72	3^h05^m	0.22
U Ori	M6.5e–M9	5.2–12.6	−0.49	0.34	24.12.72	3^h10^m	0.19
R Cas	M7e–M10	5.5–13.0	−1.84	0.37	21.12.72	3^h25^m	0.12
o Cet	M5.5e–M9	2.0–10.1	−2 à −3	0.70	23.12.72	5^h20^m	0.10
χ Cyg	S7.2e–M9	5.4–13.6	−1.5 à −2.5	0.42	07.05.72	2^h	0.19
Irregular variable							
VY CMa	M3–5 I	7.4	−0.72		21.12.72	2^h15^m	0.24

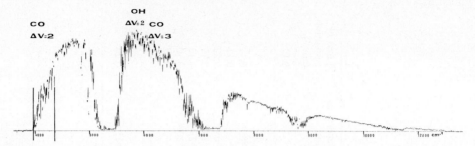

Fig. 3. Low resolution spectrum of α Her, computed from the 4000 first points of the high-resolution interferogram (d$\sigma \simeq 5.5$ cm^{-1}), showing the covered spectral range and the available windows, as do for all of the spectra. The region of this spectrum included between the marks has been extensively studied and is presented at high resolution in Figure 6.

Figure 3 shows the low resolution spectrum of α Her computed from the beginning of the interferogram. The whole spectral range is divided into windows by the strong atmospheric water vapor absorptions, as are all of our spectra. Other important features are due to telluric CO_2 and O_2. The general shape of the continuum is due to the decreasing sensitivity of the PbS detector toward the red part of the spectrum.

3.2. Variable M Stars – Mira Variables

The variale M stars which have been observed are presented in Table III.

One non-Mira star of M8 type, the semi-regular variable RX Boo was observed. In addition to a particular study the data can be analysed according to the preceding program.

But among the later M type stars, the Mira variables comprise an attractive class. So, I have recorded spectra of several stars of this type. Phase, computed from the Lockwood's data (Lockwood and Wing, 1971) have been specified, as well as the mean spectral type at minimum and maximum light. It would be necessary to record spectra at high resolution for different phases in the course of one cycle to account for the changes in stellar atmosphere. Two distinct observations were made during a cycle for R Leo. An analysis of this star is in progress and a preliminary result with regards to the structure of its atmosphere will be presented.

In addition, these spectra can provide important information on the role of H_2O in connection with OH in these stellar atmospheres. The rotational structure of these constituents is well resolved, particularly in the o Cet spectrum where they are abundant and where the highest resolution has been reached with 0.1 cm^{-1}. High resolution is essential for analysis of these very complex spectra.

A very different, but exciting object has also been observed: the irregular variable star VY CMa. One spectrum only has been previously recorded at low resolution (8 cm^{-1}) in the PbS region (Johnson et al., 1968). Our new spectrum has a limit of resolution of 0.24 cm^{-1}. Figure 4 shows the low resolution spectrum of VY CMa (5.5 cm^{-1}) computed from the beginning of the interferogram, which confirms the abnormal brightness at 2 μ of this star, compared to a normal M star.

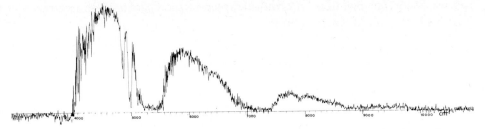

Fig. 4. Similar plot to Figure 3 for VY CMa, showing comparatively the abnormal brightness of the 2 μ window.

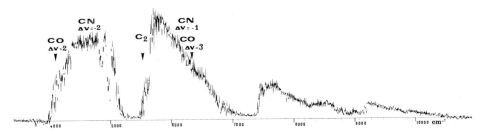

Fig. 5. Low resolution spectrum of Y CVn showing the location of the main stellar features.

3.3. C stars

The first spectroscopic observations in the 2 μ region of some carbon rich stars were begun by Boyce and Sinton (1964) with a birefringent interferometer with a limit of resolution of 20 cm^{-1}. Other low resolution interferometric spectra were recorded afterwards at 8 cm^{-1} (Johnson et al., 1968). Also at low resolution, (around $R=600$), spectra of carbon stars were scanned by using an Ebert-Fastie spectrometer, by McCammon et al. (1967) and more recently by Frogel and Hyland (1972). Only the Y CVn spectrum had been previously recorded in 1966 at about 0.8 cm^{-1} with the first interferometer evocated before (Connes et al., 1968).

Table II lists the observational details associated with four C Stars. The limit of resolution for UU Aur and Y CVn is 0.17 cm^{-1}. It is slightly lower than for some M stars because we cannot find C stars as bright as M stars.

It must be noted that for U Hya a higher resolution can be reached with reference to the other results. This relatively low value is only the fact of the limited telescope time which was available.

Figure 5 gives from a low resolution spectrum the general shape of Y CVn spectrum. As it is now well-known the most important features in this spectral range are due to diatomic molecules involving carbon. CO is present in the first and the second overtone. CN, particularly strong in Y CVn, extends along all the spectrum with $\Delta v=0$, $\Delta v=-1$, $\Delta v=-2$, $\Delta v=-3$ bands in an extreme overlapping. Ballik-Ramsay system of C_2 gives sharp discontinuity at about 5660 cm^{-1}.

High resolution spectra are essential for the analysis of these very complex atmospheres. Model computations and synthetic spectra to compare to these spectra are in progress by F. and M. Querci (1974).

4. Analysis of the First Overtone of CO in α Her

We now present the first results which were obtained from the analysis of a small fraction of the high resolution spectra recorded. α Her provides a typical example for M stars.

The first high resolution spectrum of α Her had been taken with the first interferometer in 1966, with a limit of resolution of 0.17 cm^{-1} in the windows 5500–7000 cm^{-1} and 7300–8700 cm^{-1}. Since this date, no new spectra at similar resolution were recorded. (with all due deference to R. I. Thompson (Thompson *et al.*, 1972) 0.5 cm^{-1} on α Her is not for me a high resolution!) Hence, with our interferometer we have recorded a new spectrum at 0.1 cm^{-1} over the range 4000–10000 cm^{-1} as it is presented in Figure 3. Figure 6 shows a part of the high resolution spectrum in the region of the first-overtone band sequences of CO.

4.1. Method of identification

First, we have made an identification of the CO lines. For this purpose we have used a program computing the CO lines positions and the intensities, derived from Kunde's (1967) program by F. Querci and T. Tsuji. Molecular constants determined by Mantz *et al.* (1970) were introduced. We have plotted a synthetic spectrum with the positions of the ^{12}CO and ^{13}CO lines for the 2–0, 3–1, 4–2, 5–3, 6–4 bands as is shown in Figure 7 to make the identification easier because CO bands presents very complex overlapping. This synthetic spectrum could be directly superimposed on the stellar spectrum. A comparison solar spectrum gave the telluric lines. In this way a great number of ^{12}CO and ^{13}CO lines were identified, and, the equivalent widths were measured, for the purest lines.

4.2. Detection of ^{12}C^{17}O

In spite of the innumerable lines of CO we have been able to detect some other lines which obviously were stellar features but not due to the most abundant isotopes of CO. Some of them were atomic lines and we have also identified in this region the lines of the 1–0 band of HF. But also we have identified a set of lines due to an isotopic species of CO: ^{12}C^{17}O. It is the first time that isotopic oxygen ^{17}O is clearly identified in a stellar atmosphere. To make this identification I have computed the ^{12}C^{17}O lines positions from the preceding program by introducing ^{12}C^{17}O molecular constants formed from ^{12}C^{16}O constants by using expressions which relate the molecular constants of isotopes. The wavenumbers are in very good agreement for 13 lines of the R branch 2–0 band as it is shown in Table IV. The rms value of the error for each line between the observed and the computed positions is 0.015 cm^{-1}. All other predicted lines are either blended with telluric lines or more often with lines of other isotopic CO. The 2–0 band head of ^{12}C^{17}O is confused with the 3–1 band head of ^{12}C^{16}O. Figure 8 shows the ^{12}C^{17}O identified lines.

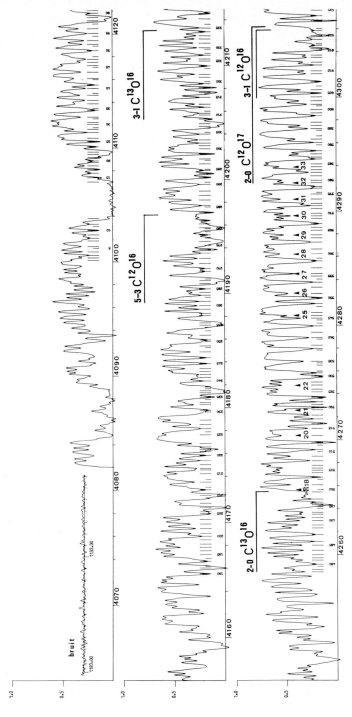

Fig. 6. Part of the high resolution spectrum of α Her in the region of the CO first overtone (cf. Figure 3) with a limit of resolution of 0.10 cm^{-1}. The resolved lines, automatically detected by a program, are marked by a dash below the trace. At the top of the figure is given the level of the noise from a no-signal region.

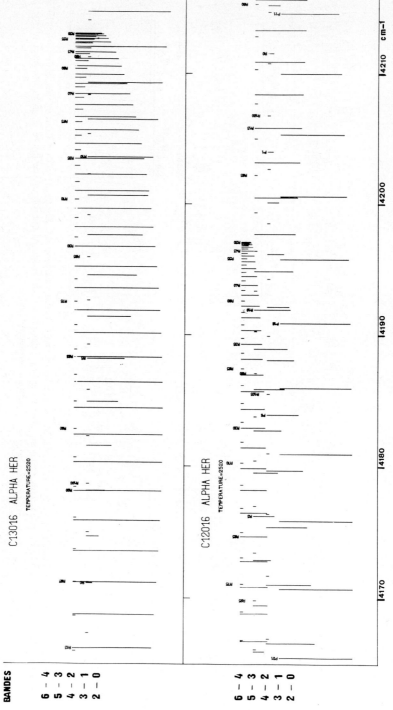

Fig. 7. Synthetic spectrum of the first overtone of $^{12}C^{16}O$ and $^{13}C^{16}O$ with only the line positions. The length of each line is proportional to the intensity of the spectral line computed from an isothermal absorbing slab model at 2520 K.

TABLE IV

Positions of $^{12}C^{17}O$ identified lines in α Her after correction of the shift due to the radial velocity of the star, compared to the computed values and measurements of the equivalent width

J	σ calculated (cm^{-1})	σ observed (after corr. VR) (cm^{-1})	σ$_{obs.}$ − σ$_{cal.}$ (cm^{-1})	Remarks	Equiv. width (W cm^{-1})
3	4221.095	4221.120	+0.025	Blend sans doute avec autre raie	0.091
4	4224.504	4224.498	−0.006		0.055
5	4227.846				
6	4231.119				
7	4234.324				
8	4237.461				
9	4240.529				
10	4243.528				
11	4246.458				
12	4249.319	4250.886		Tête de la bande 4–2 de $C^{12}O^{16}$	
13	4252.111				
14	4254.833				
15	4257.487				
16	4260.070				
17	4262.584				
18	4265.028	4264.993	−0.035	tête de la bande 2–0 de $C^{13}O^{16}$ à gauche	
19	4267.402	4267.518		+ R1(2–0) $C^{12}O^{16}$	
20	4269.706	4269.701	−0.005		0.090
21	4271.939	4271.944	+0.005	raie ☉ s à droite	0.090
22	4274.102	4274.109	+0.007		0.060
23	4276.195	4276.170		+ doublet de Na I	
24	4278.217	4278.243		+ R4(2–0) $C^{12}O^{16}$	
25	4280.168	4280.183	+0.015	raie * m à gauche	0.061
26	4282.048	4282.115		+ raie ☉ m, R5(2–0) $C^{12}O^{16}$ à gauche	
27	4283.857	4283.846	−0.011	+ raie ☉ f	0.095
28	4285.595	4285.586	−0.009		0.082
29	4287.261	4287.246	−0.015		0.078
30	4288.856	4288.915		+ raie ☉ m	
31	4290.379	4290.386	+0.007		0.072
32	4291.830	4291.841	+0.011	R8(2–0) $C^{12}O^{16}$ à g. R92(2–0) $C^{12}O^{16}$ à d.	0.070
33	4293.209	4293.235	+0.026	R68(3–1) $C^{12}O^{16}$ à g.	
34					
50	4305.521			Tête de la bande 3–1 de $C^{12}O^{16}$ + tête de la bande 2–0 de $C^{12}O^{17}$	
51	4305.582	4305.367			
52	4305.569				
53	4305.481				

In the column 'Remarks' are given the lines blending the $^{12}C^{17}O$ lines
☉ = telluric line; * = stellar line.
s = strong, m = medium, f = faint.
à droite (at right) = line bluer than the considered line.
à gauche (at left) = line redder than the considered line.

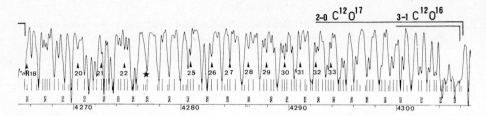

Fig. 8. Part of α Her spectrum showing, among the lines due to $^{12}C^{16}O$, some identified lines of the R 2–0 branch of $^{12}C^{17}O$ (the line marked by an asterisk is a Na doublet).

TABLE V

Possible lines of $^{12}C^{18}O$ with the same indication as in the previous table

J	σ observed (cm^{-1})	σ calculated (cm^{-1})	Δσ σ$_{obs.}$ − σ$_{cal.}$ (cm^{-1})	Remarks	Equiv. value (W cm^{-1})
23	4227.120	4227.086	+ 0.034	$P8(2–0)$ $C^{12}O^{16}$ at right	0.050
24					
25					
26					
27					
28					
29	4238.138	4238.031	+ 0.107	$R8(3–1)$ $C^{12}O^{16}$ at right	
30					
31	4241.182	4241.129	+ 0.053	$R9(3–1) + R34(4–2)$ $C^{12}O^{16}$ at right	
32					
33					
34	4245.163	4245.255	− 0.092	$R62(4–2)$ $C^{12}O^{16}$ at left	

4.3. Detection of $^{12}C^{18}O$

Having $^{12}C^{17}O$ detected I have carried out a search of $^{12}C^{18}O$ by the same technique. But the discovery is less clear because in the $^{12}C^{18}O$ region the mixing of the bands of other isotopic CO species, largely stronger than $^{12}C^{18}O$, is inextricable. Hence, only four lines may be possibly due to this isotopic variant. The measurements are listed in Table V.

4.4. Estimation of the rotational temperature of CO

In order to give a rough estimation of the isotopic ratios I have attempted to determine the rotational temperature of CO. By plotting the variation of the equivalent widths with the J quantum number of the R 2–0 branch of each isotopic species (Figure 9) it was clearly shown that the measurements cannot be described by absorption in a simple slab model at uniform temperature. Weakest lines (around $J = 90$) fall on a curve-of-growth giving $T \simeq 6500$ K. This high temperature is likely attributable to a significant temperature gradient in the atmosphere. The strongest lines can define a mean apparent excitation temperature of the CO layer. By plotting the

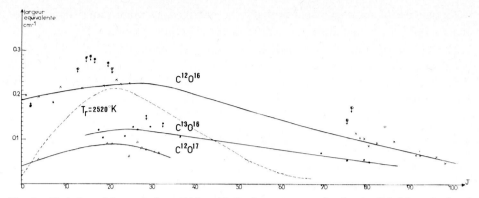

Fig. 9. Variation of the equivalent widths with the J quantum number for the R 2–0 branch of each isotopic species of CO in the α Her spectrum. In comparison, the curve obtained from an isothermal absorbing slab model at 2520K (dashed line).

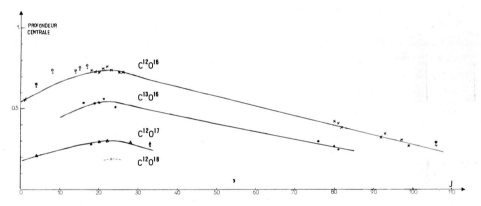

Fig. 10. Similar plot to Figure 9 but with the central line depths; the intensity of the continuum is taken equal to 1. For both sets of curves (Figures 9 and 10) the large range without points is due to the effects of the folding of the R 2–0 branch, where it is impossible to find unblended lines. The points with an arrow indicate an upper limit for the equivalent width or the central depth and correspond to lines which are clearly partially blended.

variation of apparent central depth for the least blended single lines against rotational quantum number (Figure 10) a common maximum appears for J around 24. It corresponds to an isothermal absorbing slab model a temperature: $T_{ex} \simeq 2950$ K.

4.5. Tentative estimation of $^{12}C/^{13}C$, $^{16}O/^{17}O$, $^{16}O/^{18}O$ ratios

From an empirical method detailed by Fujita (1970) and suggested to me by T. Tsuji, based on use of lines of same intensity, we have determined a preliminary estimation of isotopic ratios. With the temperature discussed before we have obtained:

$$^{12}C/^{13}C \simeq 5.3 \pm 1$$
$$^{16}O/^{17}O \simeq 450 \pm 50$$
$$^{16}O/^{18}O \simeq 700 \pm 100$$

An increase in accuracy could probably be obtained by using more sophisticated analysis. In first, temperature gradient must be taken into account.

The most interesting result is, without any doubt, the detection of ^{17}O. This isotope has theoretically an essential role in the nucleosynthesis of the stars. It takes place in the CNO bi-cycle. As a consequence of the extreme temperature sensitivity of $^{16}O/^{17}O$ ratio, a further detailed analysis must be performed.

5. Structure of the CO Atmosphere in R Leo

There are special problems concerning Mira variables related to their changing spectrum. An analysis of R Leo is in progress and first results are presented. The two different spectra recorded at phase 0.23 and 0.85 during the same cycle show some evident difference visible from the low resolution spectra. In the first spectrum (Figure 11a) an unidentified molecular band (likely metallic oxyde) appears clearly in the 8000 cm^{-1} region. In the second spectrum (Figure 11b) at post minimum light, this band has disappeared. It is therefore a strongly phase-dependent band. On the other hand, emission in the Paschen lines of hydrogen are observable with P_β and P_α. The both spectra have in common a large depression between 9200–9350 cm^{-1}, like other M-type Mira variables, due to molecular bands.

The analysis from the high resolution spectrum of the 2–0 band of CO in the first spectrum (phase = 0.23) shows an unexpected phenomenon, never related in molec-

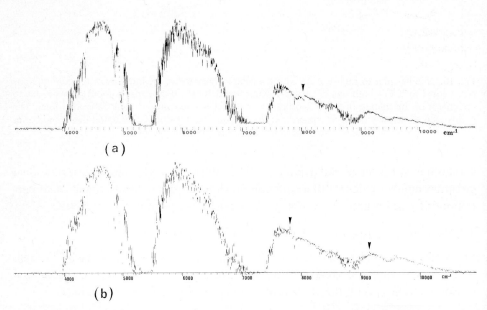

Fig. 11a–b. Low resolution spectra of R Leo at two different phases during the same cycle, showing some differences. At post-maximum light (Figure 11a) an absorption band in the 8000 cm^{-1} region appears. It has disappeared at post-minimum light (Figure 11b) and Paschen lines in emission are observable.

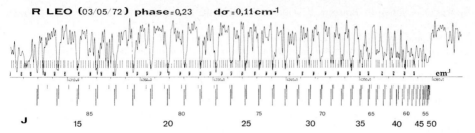

Fig. 12. Part of high resolution spectrum of R Leo in the R 2–0 branch region of $^{12}C^{16}O$, showing the doubling of the low J number lines. Both systems are schematically represented by series of dashes at different heights, below the spectrum. The smallest marks are for the high J number lines.

ular features of Mira variables spectra. As it is shown in Figure 12 CO band presents a line doubling. Below the spectrum are drawn the positions of each component for the R 2–0 branch. But it must be noticed that the lines of high J number, above about 50 are not distinguishable for a CO band while the corresponding components are very strong. This can be explained by the assumption of a stratified CO atmosphere with differentially expanding layers one being a cool layer and the other a hot layer.

We can also measure with accuracy from the well-defined doublets a difference of radial velocity of:

$$12.8 \pm 0.1 \text{ km s}^{-1}.$$

One CO layer seems moving at the same velocity as the star according to the Doppler shift computed by using radial heliocentric velocity of R Leo reported in Wilson's (1953) catalog, while the other layer, with strong high J number lines is in expansion. The expanding layer seems to be hotter than the stable CO layer.

The atomic lines located in the same spectral region are not double.

A more complete analysis will be carried on, with a comparison with the second spectrum.

6. Conclusion

18 stars of M and C type have been observed at maximum resolution possible with the instrument employed. These spectra, which extend from 4000 to about 10000 cm^{-1} represent an enormous bulk of information. They provide the most powerful way for an accurate quantitative analysis of cool stars as has been illustrated here. But to be more effective laboratory work on molecules is urgently required in connection with advances in theoretical treatments of late-type star problems.

References

Boyce, P. B. and Sinton, W. M.: 1964, *Astron. J.* **69**, 534.
Chauville, J., Querci, F., Connes, J., and Connes, P.: 1970, *Astron. Astrophys. Suppl.* **2**, 181.
Connes, J., Connes, P., and Maillard, J.-P.: 1967, Coll. 'Méthodes Nouvelles de Spectroscopie Instrumentale', *J. Phys.* **28**, C2. (English transl. in *New Methods in Instrumental Spectroscopy*, Gordon and Breach, 1972).

Connes, J., Connes, P., Bouigue, R., Querci, F., Chauville J., and Querci M.: 1968, *Ann. Astrophys* **31**, 485.
Frogel, J. and Hyland A.: 1972, *Mem. Soc. Roy. Sci. Liège, 6th Ser.* **3**, 111.
Fujita, Y.: 1970, *Interpretation of Spectra and Atmospheric Structure in Cool Stars*, Univeristy Park Press.
Guelachvili, G. and Maillard J.-P.: 1970, Aspen International Conference on Fourier Spectroscopy. AFCRL Special Report No. 114, p. 151.
Johnson, H. L.: 1964, *Bol. Obs. Tonantzintla Tacubaya* **3**, 305.
Johnson, H. L., Coleman, J., Mitchell, R. J., and Steinmetz, D. L.: 1968, *Comm. Lunar Planetary Lab.* **7**, 83.
Kunde, V. G.: 1967, 'Tables of Theoretical Line Positions and Intensities of the $\Delta v = 1$, $\Delta v = 2$, $\Delta v = 3$ Vibration-Rotation Bands of CO', NASA, X-622-67-248.
Lockwood, G. W. and Wing, R. F.: 1971, *Astrophys. J.* **169**, 63.
Maillard, J.-P.: 1973, Thèse d'Etat, Université de Paris-Sud, No. 1157.
Mantz, A. W., Nichols, F. R., Alpert, B. D., and Rao ,V. N.: 1970, *J. Mol. Spectrosc.* **35**, 325.
McCammon, D., Münch, G., and Neugebauer, G.: 1967, *Astrophys. J.* **147**, 575.
Mertz, L.: 1972, *Mem. Soc. Roy. Sci. Liège, 6th Ser.* **3**, 101.
Neugebauer, G. and Leighton, R.: 1969, 'Two-Micron Sky Survey – Preliminary Catalog', Caltech, Pasadena, Cal., NASA, Washington, D. C.
Querci, F.: 1974, this volume, p. 341.
Thompson, R. I., Johnson, H. L., Forbes, F. F., and Steinmetz, D. L.: 1972, *Publ. Astron. Soc. Pacific* **84**, 775 and 779.
Wilson, R. E.: 1953, *General Catalogue of Stellar Radial Velocities*, Carnegie Institution of Washington, Washington, D.C., Pub. No. 601.

DISCUSSION

Hyland: Regarding OH lines do you actually see OH lines in the 2.8 μ region or is it too cut out by the terrestrial atmosphere. If you do see any OH lines do these show the doubling that the CO lines do.
Maillard: In R Leo they are strong and are not doubled.
Vardya: Going back to Dr Hyland's question I was wondering if one interprets this doubling of the CO line due to the two layers of the model having a different radial velocity. No; you are not finding it in OH now and maybe one has to think of some other interpretation.

SCANS AND NARROW-BAND PHOTOMETRY OF LATE-TYPE STARS IN THE ONE-MICRON REGION

R. F. WING

Perkins Observatory, The Ohio State
and
Ohio Wesleyan Universities, Delaware, Ohio, U.S.A.

Abstract. A summary is given of the molecular bands occurring in the near-infrared spectra of cool stars, especially those having sufficient strength and freedom from contamination to be measurable by narrow-band photometry. In some cases useful indices of both temperature and luminosity can be obtained from such measurements. Several bands remain unidentified, including the 9910 Å band in late M dwarfs and at least nine bands in cool S stars.

Three topics of a spectroscopic nature are discussed. (1) In Mira variables, grossly different spectral types are sometimes obtained from zero-volt and excited TiO bands of the same band system. (2) A few M stars have been found to show bands of both VO and CN at the same time. They may be the coolest known supergiants, although there remains some doubt as to their luminosities. (3) The first results are given from a program of measuring crude C^{12}/C^{13} ratios from narrow-band photometry of sensitive points on the profile of the $\Delta v = +2$ band sequence of CN. The observations require only a few minutes per star, and the method can be applied to G and K giants and supergiants as well as to carbon stars.

1. Introduction

Narrow-band photometry obtained with scanners or interference filters is well suited to the measurement of molecular bands in the spectra of late-type stars. The method has great speed and accuracy, and the low spectral resolution is not a disadvantage in studies which only require indices of the integrated band strength. In the $1-\mu$ region, where the continuum is usually better defined than in other photoelectrically-accessible regions, the method has been applied to a variety of spectroscopic problems as well as to the determination of color temperatures and spectral types.

The present discussion will be limited to molecular bands, since the $1-\mu$ region contains only a few atomic lines strong enough for measurement by narrow-band photometry, and these have been avoided on most of the photometric systems used in this region. An important exception is the scanner system of Spinrad and Taylor (1969), which includes the temperature-sensitive Ca II line at 8662 Å and the luminosity-sensitive Na I doublet near 8190 Å. Consideration will be given to bands occurring throughout the near infrared from roughly 7000 Å to 11 000 Å, since most of the observing programs involving the $1-\mu$ region span this entire interval.

1.1. Temperature classification

Among the ordinary oxygen-rich stars of types G, K, and M, molecular bands sensitive primarily to temperature appear in the $1-\mu$ region only in types K4 and later: the strongest TiO bands are first measurable at K4 and the VO bands appear at about M5. CN bands can be measured in the G and early K stars but they convey little information about the temperature, being more useful as indicators of the luminosity and/or composition. The TiO band strengths, on the other hand, are particularly good temperature indicators since they show relatively little dependence on the luminosity and – rather surprisingly – are also nearly independent of the metal abundance (Wing, 1973b; Glass and Feast, 1973). There are numerous TiO bands in the near infrared, as shown in Figure 1 (below). Any of them can be used equivalently to classify the nonvariable or small-range semi-regular or irregular variable stars; however, this is not the case for Mira variables, for which different bands often indicate different spectral types (see Section 4 below). Maximum sensitivity to spectral type is obtained by using the strongest band, the three-headed (0, 0) band of the γ system near 7100 Å, which is also the only TiO band to appear in K4 stars. The growth of this band with decreasing temperature is so rapid that all decimal subdivisions of the spectral subclasses (e.g. M2.1, M2.2, etc.) can meaningfully be used for observations having the normal photometric accuracy of 1 or 2%.

In the late M stars the TiO bands are very strong and are accompanied by bands of VO and H_2O. The H_2O bands have not been used for classification purposes, since their measurement is affected by the same bands in the Earth's atmosphere, but their behavior seems to be parallel to that of the VO bands near 1.05 μ, which serve as a very sensitive temperature criterion for classes M8–M10. There is evidence from the weak TiO bands near 1 μ (Lockwood, 1973) that the TiO abundance continues to increase beyond M8, although the photometric indices measuring stronger TiO bands at shorter wavelengths become saturated as a result of deterioration of the continuum. For this reason, we recommend using VO alone for photometric classification in the range M8–M10 (Wing and Lockwood, 1973). The VO strength is certainly very sensitive to temperature in this range, as is shown by its behavior in Mira variables, but its dependence upon the luminosity has not been established observationally. Fortunately (for classification purposes), stars with abnormal VO/TiO ratios such as those discussed in Section 5 appear to be relatively rare.

1.2. LUMINOSITY CLASSIFICATION

Of the various molecules with bands in the near infrared, only CN is of much use as a luminosity indicator; the others either show little dependence upon luminosity or appear in only certain types of stars. The positive correlation between CN strength and luminosity has long been known from observations of bands in the violet system. Griffin and Redman (1960), who made photoelectric measurements of the 4215 Å band, confirmed that the main luminosity classes of G and K stars are well separated by their CN strengths, but they cautioned against using CN as a luminosity criterion because many stars were found to have abnormally strong or weak CN strengths for their temperatures and luminosities.

The infrared CN bands show the same behavior as the violet bands, but whereas the violet bands cannot be measured in stars later than about M0 (because they are then much weaker than the strong atomic lines in their vicinity), the infrared bands can be observed to type M4 or even M5. Wing (1967c), using infrared scanner observations, found that the M giants, like the G and K giants, often have anomalous CN strengths, so that the CN bands cannot be trusted to give more than a coarse indication of the luminosity.

A more encouraging result was obtained in the case of M supergiants by White (1971, 1972). He measured the infrared CN bands in all M supergiants that had been classified on the MK system and found that the MK luminosity classes Ia, Iab, and Ib could be distinguished with a high degree of confidence on the basis of CN strength. We really don't understand why the CN bands 'behave themselves' better in supergiants than in giants, but it may reflect the fact that all the supergiants are young Population I stars, while the giants are from a mixture of populations with a wide range in age and mass.

Late-type dwarfs can be recognized in the infrared by the absence or near-absence of CN absorption. Significant CN strengths are sometimes seen in K dwarfs but never, to my knowledge, in M dwarfs. Dwarfs later than about M3 do show a single luminosity-sensitive band at 9910 Å (Wing and Ford, 1969), but it is still unidentified (see Section 2.1 below) and has not been included on the narrow-band photometric systems used to date for general classification purposes. Apart from the 9910 Å band, the M dwarfs show only the TiO bands, since even Wolf 359, which has the strongest TiO bands ever measured in a dwarf (Joy, 1947; Wing, 1973a), is not cool enough to show VO bands or the $1-\mu$ bands of water.

1.3. Carbon and S-type stars

The $1-\mu$ spectra of carbon stars are dominated by the red system of CN to such an extent that it is difficult even to establish the presence of anything else. Relatively weak bands of the C_2 Phillips system have, however, been identified in a few stars of high carbon abundance class (McKellar, 1960a). A few additional absorptions have been noticed occasionally on scanner records (Wing, 1967c) and on image-tube spectrograms (Wing, 1972), but they have not been studied systematically. Several attempts have been made to indentify HCN and C_2H_2, both of which have favorable bands from the ground state near 1.03 μ. The earlier results were inconclusive, but according to Fujita (1973), M. Hirai recently has found strong evidence for C_2H_2 (and weaker evidence for HCN) on high-dispersion spectrograms. Once the identifications of these features are firmly established, narrow-band photometry could be used effectively for a survey of the occurrence and behavior of these interesting molecules.

The S stars are difficult to study by narrow-band photometry because of the great variety in their molecular spectra (Wing, 1972). Every molecule found in the infrared spectra of M stars can also be found in S-star spectra, which in addition show bands of ZrO near 9300 Å (always) and of LaO near 7400 Å and 7900 Å (if sufficiently cool). Those showing the LaO bands are further complicated by several unidentified bands which are discussed in the next section.

2. Unidentified Molecular Bands

Among the giants and supergiants of types G, K, and M, I am not aware of any unidentified spectral features that can be seen at low resolution in the 7000–11000 Å range. This is not to say that all the little wiggles in the spectra have been accounted for, since such detailed studies have only been carried out for a few stars. It would not be surprising if the very late M stars have detectable absorptions which cannot be attributed to TiO, VO, or H_2O; but if such features exist, they have not been noticed to date, and indeed they would be very hard to find owing to the great strength and spectral coverage of the known bands.

In the near-infrared spectra of other types of cool stars, there are ten molecular bands which remain unidentified – one in M dwarfs and nine in S stars – and which are definite enough to have been specifically discussed in the literature. This I think is a good number of unidentified features: not depressingly large, yet large enough to prevent us from becoming bored, overconfident, or unemployed.

2.1. The 9910 Å band in M dwarfs

M dwarfs differ from M giants in the $1-\mu$ region in having greatly enhanced features due to atomic sodium and potassium and an unidentified molecular band centered at 9910 Å (Wing and Ford, 1969). These features are illustrated in Figure 1, where

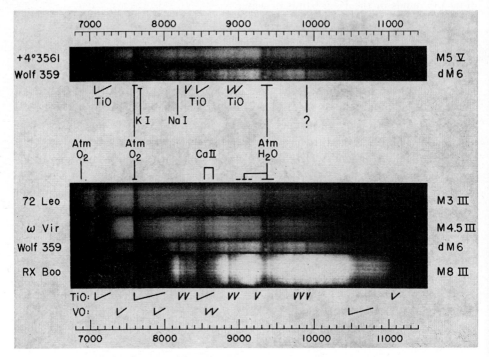

Fig. 1. Spectrograms of M dwarfs and giants in the 7000–11000 Å region taken with a Carnegie S-1 image tube at the 72-in. Perkins telescope in collaboration with W. K. Ford, Jr. Wolf 359, which shows great enhancements of lines of neutral potassium and sodium and an unidentified band at 9910 Å, is compared to Barnard's star = BD +4°3561a (*above*) and to a set of giant stars (*below*). Bands of TiO and VO are marked beneath the spectra.

the spectrogram of Wolf 359 on which the 9910 Å band was discovered is compared to a sequence of M giants (*below*) and to the somewhat warmer dwarf, Barnard's star (*above*). These plates were taken with the Perkins 72-in. telescope of the Ohio State and Ohio Wesleyan Universities and an S-1 Carnegie image tube; the original dispersion was 250 Å mm^{-1}. Note that the 9910 Å band does not appear in giants of any spectral type and is considerably stronger in Wolf 359 ($M_V = 16.7$) than in Barnard's star ($M_V = 13.3$).

Whitford (1972, 1974) has used scanner observations of this band to evaluate the contribution of late M dwarfs to the integrated spectra of external galaxies. It is a very useful band for that particular problem because it is exclusively an M dwarf feature, and it occurs in the spectral region where M dwarfs might be expected to make their greatest contribution to a composite spectrum.

In stars the 9910 Å band shows the same behavior as one in the visible region which Pesch (1972) has identified with CaOH. It seemed natural to suppose that the 9910 Å band might be produced by the same molecule, and Dr Paul Byard and I checked this possibility with a laboratory experiment. By holding a piece of calcium carbide in an oxygen-enriched bunsen flame, we obtained a bright yellow flame, a spectrogram of which showed that most of the visible light was radiated in the Pesch band. An infrared spectrogram of the same flame showed some bands of calcium monoxide but, unfortunately, showed nothing attributable to the hydroxide and nothing at 9910 Å. The band thus remains unidentified. We are planning to study other metallic hydroxides in a similar manner, and we certainly would also encourage any other attempts to identify this important feature.

2.2. Unidentified bands in S stars

The S-type stars – particularly the S-type Mira variables at minimum light – are the richest source of unidentified molecular bands in the $1-\mu$ region. The four Keenan bands near 8500 Å (Keenan, 1950) have defied all attempts at identification for more than two decades, despite their sharp, easily-measurable bandheads and the considerable strength that they occasionally attain. It has, however, been established that they are not all produced by the same molecule (Wyckoff, 1970; Wing ,1972). Five additional bands – not as strong or sharp, but occurring in the same stars – have been found by the writer at longer wavelengths. They, too, are unidentified. Their wavelengths, appearance, and behavior are reported in Wing (1972). One of them is centered at about 9900 Å; it seems unlikely that it is the same as the band appearing in M dwarfs, but the available observations do not rule out this possibility. Between the unidentified bands in cool S stars evidence is sometimes seen of additional structure which should be examined at higher spectral resolution than has been employed to date.

2.3. The importance of the unidentified bands

The reasons for being interested in the bands shown by cool S stars and in the 9910 Å band in M dwarfs are quite different. The 9910 Å band, as far as we know, occurs in *all* stars of sufficiently low temperature and high gas pressure; identifying this band is therefore not likely to teach us anything about chemical abundances or nucleosynthesis. The usefulness of the band lies in the fact that it uniquely identifies cool dwarfs and the contribution they make to integrated spectra. For this purpose we can go ahead and use the band without waiting to learn its identification, which is not even relevant.

In contrast, the bands in S stars are of little practical use as long as they remain unidentified, but their identifications may prove to be extremely interesting. The S stars show great enhancements of features due to the heavy s-process metals and rare earths, abnormal abundance ratios among the light elements C, N, and O, and the presence of the unstable element technetium and possibly even the still more unstable promethium. We need all the observational clues we can get to help us understand the nuclear processes which have occurred in S stars. Since the unidentified bands found in these stars do not appear in normal M stars of any temperature, they probably involve elements that are greatly overabundant in S stars. Thus their eventual identification may well have significance reaching far outside the relatively narrow field of infrared spectroscopy.

3. The Eight-Color Photometry

A large part of the currently-available data on molecular band strengths in the near infrared has been obtained on an eight-color system employing interference filters of approximately 50 Å width (Wing, 1971a). The system measures five quantities: the $I(104)$ magnitude (Wing, 1967a), the continuum color, and the strengths of TiO, VO, and CN. The filters are centred at 7117, 7544, 7809, 8122, 10395, 10544, 10804, and 10975 Å; they measure the strongest bands of all three molecules as well as the best available continuum points in most kinds of cool stars. Since 1969, the writer and several collaborators at the Ohio State University have made more than 4000 observations on this system using various telescopes at the Kitt Peak, Cerro Tololo, and Lowell Observatories.

In a related development, a five-color system measuring $I(104)$ magnitude, color, TiO, and VO has been set up by Lockwood at Kitt Peak. It too has been used extensively, primarily for studies of Mira variables (Lockwood, 1972) and infrared stars.

Here I would like to present a brief summary of the projects involving narrow-band photometry of molecular bands in the near infrared that have been carried out or are now in progress. The observations discussed here have been made on the Ohio State eight-color system unless otherwise stated.

3.1. Bright stars

In order to provide a large body of homogeneous data on the colors, spectral types, and CN strengths of bright stars, I am observing all stars expected to be brighter than $I(104) = +2.5$ at 1.04 μ. There are approximately 1000 such stars, mostly late-type giants, and the observations are 70% complete. Stars showing TiO absorption are being classified according to an index of the strength of the (0, 0) band near 7100 Å, calibrated with giant stars classified on the MK system by Keenan. The same calibration is also being used for stars of other luminosity classes.

3.2. M supergiants

Colors and spectral classifications for 135 M supergiants are being prepared for publication by White and the writer. The strong correlation between CN strength and MK luminosity subclass found by White (1971, 1972) greatly enhances the value of these stars for galactic structure studies, particularly in the case of obscured regions which are best observed in the infrared. White is now studying the variability and intrinsic colors of these stars.

3.3. M dwarfs

Since a new list of standards for the spectral classification of M dwarfs was presented at Córdoba in 1971 (Wing, 1973a), the project has been enlarged, with the collaboration of C. A. Dean, to include essentially all stars within 10 pc of the Sun. The observations are more than 50% complete and will be used to study the shape and intrinsic width of the lower main sequence.

3.4. Carbon stars

Approximately 350 of the brighter carbon stars have been observed by Baumert (1972) and the writer on the eight-color system, some of them repeatedly. Baumert has shown that carbon stars of the various variability classes tend to lie in different regions of a diagram of CN strength vs color temperature; in particular the Mira variables have substantially weaker CN than the semi-regular or irregular variables of the same temperature. Baumert (1974) has also derived mean absolute infrared magnitudes for carbon stars grouped in various ways, using the $I(104)$ magnitudes and a statistical analysis of their proper motions.

3.5. R Coronae Borealis stars

Eight-color photometry of R CrB and RY Sgr has been obtained at both maximum and minimum light. At faint phases both stars show an excess flux in the seventh filter attributed to emission in the He I line at 10830 Å; it will, of course, be important to secure a spectrogram or continuous scan to confirm this identification. On one occasion emission by the CN molecule was observed in R CrB (Wing *et al.*, 1972).

3.6. Red giants in globular clusters

Fourteen red variable stars belonging to the southern globular clusters 47 Tuc, ω Cen, and NGC 362 have been classified by narrow-band photometry (Wing, 1973b, c). The types observed in 47 Tuc are in the range M3.1 to M7.5; stars as late as M5 also occur in ω Cen although several of its variables are of type K. Despite the great difference in the metallicities of these clusters, there is no clear systematic difference in the TiO strengths at a given temperature. Three carbon stars in the field of ω Cen have been studied by Wing and Stock (1973), although one of them has been found not to belong to the cluster (Smith and Wing, 1973).

3.7. Infrared stars

Objects which are bright enough at 2 μ to be included in the IRC catalogue (Neugebauer and Leighton, 1969) and yet faint enough visually to have been excluded from older catalogues nearly always turn out to be late M stars or (much less frequently) carbon stars. Narrow-band photometry is an efficient means of identifying and classifying such stars. Lockwood has used his five-color system extensively for this purpose (Lockwood and McMillan, 1971) and will soon publish additional finding charts and spectral classifications.

The general area of the sky in the direction of the galactic center contains a concentration of IRC objects and the largest proportion of unidentified sources. Some 50 of these have been observed on the eight-color system by Warner and the writer. Only one carbon star has been found, and one object proved to be the combined signal from an M8 star and the heavily-reddened globular cluster Terzan 5 (Wing *et al.*, 1973). All the others observed to date are M giants and supergiants, having types in the range M3–M9 and varying degrees of interstellar reddening. Attention has already been called to the three objects which appear to have the highest luminosities, greatest distances, and closest proximities to the galactic center of the stars in this sample (Wing and Warner, 1972).

3.8. Mira variables

Lockwood's (1972) large body of five-color photometry has been combined with observations on the Lick 27-color scanner system (Wing, 1967b, c) and the Ohio State eight-color system (Baumert and Wing, 1974) to produce a catalogue of the ranges in spectral type and $I(104)$ magnitude of approximately 300 Mira variables. This is now being prepared for the printer by Lockwood and the writer. Light curves in $I(104)$ and concurrent spectral types have been published for 25 variables (Lockwood and Wing, 1971).

Of all the Miras studied, the infrared star IK Tau (NML Tau) is the only one which consistently reaches spectral type M10 at minimum light, if our rather stringent criterion for this type is employed (Wing and Lockwood, 1973). In fact we have considered it necessary to introduce types later than M10.0 to accomodate the extraordinary spectrum of this star.

4. Anomalous TiO Band Strengths in Miras

Classification by narrow-band photometry is generally based upon the strengths of a relatively small number of spectral features, and often only one. For example, the types published to date for K4–M6 stars observed on the eight-color system depend solely upon the strength of the (0, 0) band of TiO near 7100 Å. In such a case the type can be assigned to the same accuracy as the photometric measurement, since there is no conflicting evidence. The question arises, however, whether the same precise type would have been assigned if some other spectral feature had been used as the criterion. The spectroscopic classifier, after all, normally inspects many features before deciding on the type and is frequently aware that the types indicated by the various criteria are not identical.

Lockwood and I have looked into this question by comparing the types assigned to M stars on our respective photometric systems, which use different bands (and different reduction procedures) to determine the TiO strength. Lockwood's (1972) five-color system measures the (2, 3) band of the γ (triplet) system near 7800 Å and the (0, 0) band of an infrared singlet system near 8900 Å. These are also the features that I used on the Lick scanner system (Wing, 1967b; Wing and Spinrad, 1970); neither arises from the ground state of the molecule, since the former band is excited vibrationally and the latter electronically. The eight-color system, on the other hand, uses the (0, 0) band of the γ system, which does arise from the ground state. A program of simultaneous observations on all three systems was carried out, so that the types assigned could be compared for variable stars as well as non-variables. Systematic differences should be negligible since the same stars have been used to calibrate the spectral-type indices on all three systems, the primary standards being those of Keenan (1963).

For the non-variable and small-range variable M stars earlier than M8, the agreement in the types was excellent, the differences being no larger than expected from the uncertainties quoted on the respective systems. For stars later than M8 the agreement was again excellent if the types were based on VO alone; however, as mentioned earlier, we found that our systems do not give a reliable TiO index in the very late types because of absorption in the filters used as continuum points.

The surprise came when comparing results for Mira variables in the range M4–M7, for which the types depend on the TiO strength. The agreement, frankly, was terrible, with discordances up to two subclasses, usually in the sense that Lockwood's types were earlier than mine. Typical differences for Miras were five to ten times as large as those for non-Miras in the same range of spectral type. It is to be noted that the problem is present even at relatively early spectral types, for which the level of the continuum is well established on all three photometric systems. It is necessary to conclude that the different bands of TiO are simply telling different stories about the TiO abundance, and hence the temperature, of the star.

As it happens, the (2, 3) band used on Lockwood's system is also measured on the eight-color system, by the third filter whose primary function is as a continuum point in carbon stars. It has therefore been possible to confirm that the measurements of this band on the different systems are self-consistent, and it is consequently possible to assign types for Miras from the eight-color photometry that are the same as would be given on Lockwood's system. Further, although the discrepancies were first noted by comparing results from two different systems, the effect can be studied by means of the eight-color photometry alone.

The phenomenon is illustrated in Figure 2, where the Mira variable U Cet, observed one month past maximum on 5 November 1969, is compared to 56 Leo, a small-range irregular variable also known as VY Leo. Although the two stars have similar TiO (0, 0) band strengths at filter 1, U Cet has a much weaker (2, 3) band than 56 Leo and a much bluer energy distribution. From comparisons with many other stars we can state that 56 Leo is a normal, unreddened giant, in that its color and both of its TiO bands all indicate the same spectral type, M5.8 III. For U Cet, on the other hand, we obtain M6.0 from the (0, 0) band but M4.5 from the (2, 3) band, while the energy distribution is that of a normal giant of type M3! What, then, should we say was the spectral type of U Cet on 5 November 1969?

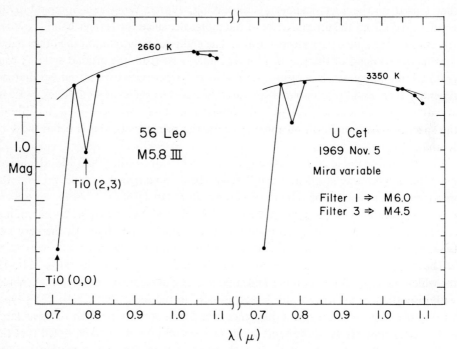

Fig. 2. Eight-color spectra of the normal giant 56 Leo (*left*) and the Mira variable U Cet (*right*). F_λ is plotted on a magnitude scale against the wavelength in microns. Blackbody curves of the indicated temperatures have been fitted to the spectra. Filters 1 and 3 are depressed by TiO; note that the relative strengths of these bands are very different in the two stars.

While it is clear that the absolute strength of at least one of the TiO bands observed in U Cet is affected by some process which renders it unreliable as a spectral-type indicator, it should be equally clear that one would do no better to use their relative strengths, as is done in the spectroscopic classification of Miras (Merrill *et al.*, 1962; Keenan, 1966). In the example shown in Figure 2, the excited band is so much weaker than the zero-volt band that one would have to assign a type later than M8 to U Cet on the basis of the relative band strengths. This is another way of saying that the two bands imply an implausibly low vibrational temperature.

Although the discrepancies shown in Figure 2 are fairly typical, it would be wrong to imply that the (0, 0) band always indicates a later type than the excited band in Miras, or that the bands are always too strong for the color. This is *usually* the sense of the discrepancy, but as we have shown in previous work (Wing, 1967c; Spinrad and Wing, 1969; Lockwood, 1972), each Mira executes large loops in diagrams of band strengths vs color, so that occasionally the bands appear of normal strength or even weak for the color.

In discussing the loop phenomenon, we were led to conclude that the measurements of band strength and of continuum color refer to two different layers of the atmosphere having appreciably different temperatures. Now it seems that we must go a step farther and conclude that contributions to the zero-volt and excited TiO bands are likewise coming from different layers, the cooler one of which is considerably cooler than the photosphere. It does not seem unreasonable to suppose that a cloud of TiO molecules formed at minimum light (the phase of lowest temperature, greatest band strength, and greatest physical extent) is left behind as the star contracts, becoming brighter and warmer in the photosphere. Such a cloud would quickly cool, so that it would absorb mostly from the ground state and very little from excited states, thus altering the relative and absolute band strengths of the underlying photospheric spectrum. It is not clear, however, how long this cloud should be expected to survive as a significant contributor to the (0, 0) band. By following several Miras with eight-color photometry throughout their cycles, it may be possible to determine whether this picture is basically correct.

Another factor which may come into play is a differential radial velocity between two layers contributing to the absorption. Since the lines of the (0, 0) band are no doubt often saturated while those of a weaker excited band may not be, the effect of a velocity difference would be to increase the strength of the (0, 0) band without affecting the excited band. Observations of line doubling in Mira variables (Merrill and Greenstein, 1958; Maehara, 1968; Spinrad and Wing, 1969) show that such effects do occur, and instances of line broadening are presumably much more common than clear cases of line doubling.

An interesting example of discrepant types from different TiO bands in a Mira variable has been noted in the spectroscopic literature by Wyckoff (1970), who observed the 1966 minimum of Z Oph in the near infrared. It is not clear to me, however, whether this is an example of the same effect that we have been discussing here, since she did not observe the bands from the ground state. Since the classification of M stars by photographic spectroscopy can have an accuracy of one-quarter of a subclass (Keenan, 1963), while the discrepancies in the types of Miras indicated by different bands commonly amount to one or two full subclasses, one might wonder why this effect was not already well known. The answer probably lies in the fact that the TiO (0, 0) bands are in general too strong to measure on photographic plates whenever the excited bands are strong enough to use for classification. It might nevertheless be worthwhile to re-examine plate collections of early-type Miras to try to find further examples of this phenomenon.

5. Stars Showing Both VO and CN Bands

In normal giant stars the CN bands are fairly strong throughout the interval G5–M3, passing through an intensity maximum in the middle K types and fading to invisibility after about M5. The VO bands show completely different behavior, appearing only in the late M stars and becoming strongest in the latest types. It is sometimes possible to detect both molecules in stars of types M5.5 or M6.0, but the region of overlap is very small and both molecules are then very weak.

Supergiants of types G, K, and M show stronger CN bands than giants of the same spectral types. The enhancement is most pronounced in the M stars since the bands do not fade with decreasing temperature as fast in the supergiants as in the giants. The coolest supergiants known – and which are certain from cluster or association membership to be post-main-sequence supergiants – have types around M4.5 or M5.0 and quite substantial CN strengths. These stars are not cool enough to show appreciable VO absorption, but their CN strength leads one to suspect that if sufficiently cool supergiants exist, they might show both VO and CN bands strongly. When the type is later than M4, the presence of CN can really only be tested at the (0, 0) band near 11 000 Å, since the bands at shorter wavelengths become swamped by the strong TiO absorption.

5.1. VX SAGITTARII

This unusual star seems to combine some of the properties of supergiants and Mira variables without being a certain member of either category. Its long mean cycle length (732 days) and the semi-regular nature of its variations are characteristic of late-type supergiants, as are some of the spectral details which show that the atmosphere is greatly extended; however, its large amplitude, its late spectral type at minimum, and the irregular decrement of its Balmer emission lines are properties generally associated with Mira variables. Recent discussions of its spectrum have been given by Wallerstein (1971) and by Humphreys and Lockwood (1972). Most of the attention recently given this star was stimulated by the discovery of its very strong radio OH emission lines (Caswell and Robinson, 1970) and far-infrared excess (Humphreys *et al.*, 1972). In these respects VX Sgr resembles two other very remarkable objects, VY CMa and NML Cyg.

When VX Sgr was first observed on the eight-color system on 15 June 1971, I was so struck by the unusual shape of its spectrum in the last four filters that I was afraid something had gone wrong with the equipment. This observation is shown in Figure 3, where the depressions attributed to VO and CN are labeled and are seen to be quite strong. Although I have given reasons above for thinking that these molecules could conceivably both be strong in the same spectrum, no spectrum like this had ever been observed before.

It should be stated clearly that the observation of CN absorption does not necessarily prove that VX Sgr is a supergiant. It is hard to prove anything about an object that is unique. All I can say is that this is how I think the spectrum of an extremely cool supergiant should look, if such stars exist.

The spectrum of VX Sgr can be classified if the VO strength is used for the temperature class (Wing and Lockwood, 1973) and the CN strength for the luminosity class (White, 1971). The result for 15 June 1971 is M8.4 Ia. I prefer to give this type as M8.4 Iap, where the 'p' is a reminder of the peculiar nature of this object and a warning that the usual calibrations in terms of temperature and absolute magnitude may not apply to it.

An interesting series of observations of VX Sgr, also made in 1971, has been reported by Humphreys and Lockwood (1972), who observed the spectral type to progress from M4 in February to M9.5 in October, the latter type being entirely unprecedented for supergiants and very unusual even for Mira variables. One of their spectrograms was taken on the same night as the eight-color observation shown here, and they classified it as M5e I, much earlier than my classification. It appears that this discrepancy is the result of yet another spectral peculiarity of this star that has not previously been pointed out: the VO bands are abnormally strong relative to the TiO bands. All the types earlier than M8 given by Humphreys and Lockwood are based on TiO absorption seen on spectrograms in the blue, while their latest types are from photometry on the five-color system and thus depend primarily on the VO strength. Both molecules were individually observed to strengthen as the star became fainter, but there was probably a discontinuity in the classifications when the criterion changed.

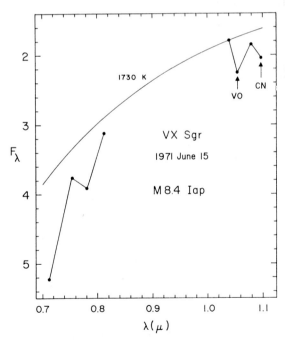

Fig. 3. The eight-color spectrum of VX Sgr on 1971 June 15, showing the simultaneous presence of bands of VO and CN. On this date the TiO type was M5 and the VO type M8.4. The position of the blackbody continuum was estimated in the manner discussed in the text. No corrections for interstellar reddening have been applied to these data.

Although it is the fortunate circumstance of simultaneous observations on 15 June 1971 that gives the best evidence for a systematic difference between the TiO types and VO types of VX Sgr, some supporting evidence can be found in the eight-color observation of Figure 3 alone. It is difficult to measure the TiO strength in the infrared when VO is also present because none of the first four filters is then a good continuum point; the additional presence of CN in this case makes the situation still worse. Nevertheless, we have estimated the position of the continuum (represented in the figure by a blackbody curve) in this region by requiring the depressions of filters 2 and 4 due to VO and CN, respectively, to be of normal strength relative to the depressions at filters 6 and 8 by the same molecules. Then the absorption by TiO at filters 1 and 3, measured relative to this continuum, indicates a spectral type of M5, in agreement with the result of Humphreys and Lockwood (1972) for the same date and confirming that a discrepancy exists between the TiO types and VO types.

5.2. Related objects

We now ask whether any other stars show signs of the two spectral peculiarities – the simultaneous presence of VO and CN, and the abnormally great strength of VO relative to TiO – which can be detected by the eight-color photometry and which set VX Sgr apart from the vast majority of late M stars. It is natural to look first at VY CMa in view of its many similarities to VX Sgr (Wallerstein, 1971), but I have never observed VY CMa when it was cool enough to show the VO bands, and I don't know if it ever is.

W Hya, one of the brightest stars in the infrared (Wing, 1971b), has long been known to have abnormally strong VO bands for its TiO strength (Cameron and Nassau, 1955). Furthermore its variability type, like that of VX Sgr, is difficult to assign: Keenan (1966) considers it a Mira variable for spectroscopic reasons, but it has the small amplitude of a semi-regular variable. It seems to me possible that it is a very cool supergiant, although I have not been able to detect CN in its spectrum. This possibility can be checked directly by a trigonometric parallax measurement; if it is only a giant, it should be one of the nearest M giants to the Sun with an easily measurable parallax.

The star whose infrared molecular spectrum seems most similar to that of VX Sgr is none other than the famous infrared star and intense OH emitter, NML Cyg. Its classification as an M6 star by Wing *et al.* (1967) was based upon the sum of the TiO and VO indices, but as we pointed out, the VO bands are slightly stronger and the TiO bands slightly weaker than in a normal M6 giant. The discrepancy is not as large as shown by VX Sgr, but it is another reason for considering these stars to be closely related. Furthermore, although this was not clear from the raw data used by Wing *et al.* (1967), the (0, 0) band of CN is fairly strong in NML Cyg, and its identification has been confirmed by continuous scans obtained by Spinrad and the writer. Neither VO nor CN is as strong in NML Cyg as in VX Sgr, however. Herbig and Zappala (1970) have classified this star as M6 III since no indicators of high luminosity were seen in the 7000–9000 Å region, but I believe that this simply shows once again the importance of observing the uncontaminated CN (0, 0) band near 11 000 Å whenever the oxide bands are strong. Most authors (e.g. Johnson, 1967), on the basis of its very red color, have considered NML Cyg to be a distant and heavily-reddened supergiant, and this interpretation would seem to be supported both by the presence of CN and by the abnormal relative strengths of VO and TiO. At least we can say that NML Cyg resembles VX Sgr more closely than an ordinary giant.

Figure 4 shows the eight-color spectrum of V1804 Sgr = IRC -30350, which we suggest may be a cooler version of VX Sgr (Wing and Warner, 1972). The CN strength indicated by the last filter is less than in VX Sgr but still enough for a Ia luminosity classification; the VO bands indicate type M9.1. The position of the 'continuum' has been estimated in the same manner as described above for VX Sgr. This star is extremely red, and at the time of this observation its visual magnitude was estimated to be $V \approx 17.5$.

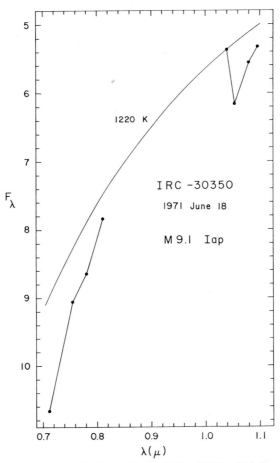

Fig. 4. The eight-color spectrum of the infrared star IRC -30350, which has been identified with the little-studied variable V1804 Sgr. It has stronger VO bands and a redder color than the observation of VX Sgr shown in Figure 3, but it also seems to have substantial absorption by the (0, 0) band of CN in the last filter. The star was barely visible at $V \approx 17.5$ when observed on 1971 June 18 with the 150-cm telescope at Cerro Tololo Inter-American Observatory.

We have seen that measurements of the near-infrared bands of TiO, VO, and CN provide a means of singling out stars that have spectacular characteristics in the far-infrared and radio regions. They also give us a way of knowing that stars like NML Cyg and V1804 Sgr, which are so faint at short wavelengths that their spectras are largely unknown, are related to better-known objects like VX Sgr and VY CMa. At the present time we don't even know if these stars are pre- or post-main-sequence objects. However, the recognition of similarities and differences in their molecular spectra, by indicating their relationship to each other and to other stars, may prove to be an important step towards understanding their evolutionary state.

6. The Measurement of C^{12}/C^{13} Ratios by Narrow-Band Photometry

Until quite recently, the only observational data on carbon isotope ratios outside the solar system had come from carbon stars, whose C_2 bands in the visible region made the detection of C^{13} very easy. Most carbon stars were found to have substantial amounts of C^{13}, with C^{12}/C^{13} ratios in the range 4 to 20 or so, while a few did not show detectable C^{13} features. Since carbon stars have obviously abnormal compositions, it was perhaps natural to regard their low C^{12}/C^{13} ratios as abnormal, and it was generally assumed (for lack of evidence to the contrary) that normal giant stars have C^{12}/C^{13} ratios near the solar/terrestrial value of 90.

The first identifications of features involving C^{13} in non-carbon stars, except for the Sun, were made in 1968 by Connes *et al.* (1968) and by Spinrad (Spinrad and Wing, 1969), both independently analyzing the same high-resolution infrared spectra of α Ori obtained by Connes. They identified numerous lines of $C^{13}O^{16}$ near 2.3 μ and found that the bandheads were strong enough to be visible at much lower resolution. Shortly thereafter, the paper by Johnson *et al.* (1968) showing medium-resolution infrared spectra of various late-type giants and supergiants appeared just as Spinrad and I were completing a review of infrared spectra (Spinrad and Wing, 1969), and we saw that star after star showed the same little dips that had been identified with $C^{13}O^{16}$ in α Ori. The implication that normal evolved giants and supergiants have mixed processed material to their surfaces was not hard for me to accept since I had already drawn this conclusion from the weakness of CN in T Tau stars.

During the past few years, C^{12}/C^{13} ratios have been derived for a few K giants from high-resolution material (Lambert and Dearborn, 1972; Krupp, 1973; Upson, 1973) using bands of CH and CN as well as CO. While detailed studies such as these no doubt provide the most reliable results, they are feasible for only the brightest stars. A somewhat larger number of stars can be reached by means of medium-resolution spectroscopy of the infrared CO bands (Ridgway, 1974). However, to investigate the incidence of various isotope ratios among stars of all luminosities, metallicities, and populations, including peculiar stars and members of clusters, it is necessary to develop a method for obtaining crude isotope ratios for much fainter stars. Dr Irene Marenin-Little and I have found that narrow-band photometry can provide this information in an efficient manner.

6.1. The Photometric System

There are problems associated with using any of the various molecular bands containing carbon for photometric measurements of the isotope ratio. The C_2 bands are not strong enough for this purpose except in the carbon stars, and thus would provide no new information. The violet CN bands and the G band of CH are badly contaminated by atomic lines, while the CO bands lie beyond the range of photoelectric detectors. Of the various bands of the CN red system, those lying longward of the (0, 0) band at 11 000 Å are favorable from the standpoint of the direction of the isotope shift, but unfortunately the strongest $C^{13}N^{14}$ bandheads lie in regions of strong atmospheric water absorption and have been detected only in scans made from the Stratoscope balloon (Wing and Spinrad, 1970).

We are left with the red CN bands lying shortward of the (0, 0) band. Their $C^{13}N^{14}$ components are hard to identify on spectrograms because their shifts are in the wrong direction (i.e. the $C^{13}N^{14}$ bands lie within the $C^{12}N^{14}$ bands), but they must nonetheless affect the absorption profile of the band. The idea is to measure photoelectrically a small number of points on the band profile that are sensitive to the contribution by $C^{13}N^{14}$. We have chosen the (2, 0) band near 7900 Å as being the best combination of band strength and freedom from contamination by stellar atomic lines and atmospheric water lines, while also having an adequately large isotope shift.

In order to choose the optimum bandwidths and central wavelengths for these measurements, Dr Marenin-Little computed synthetic spectra for the $\Delta v = +2$ sequence of CN for various temperatures, CN abundances, and C^{12}/C^{13} ratios. Two of these spectra are illustrated in Figure 5; both were computed for a temperature of 3000 K and for the same amount of $C^{12}N^{14}$ (more than in a K giant, but less than in a typical carbon star), but they differ in the amount of $C^{13}N^{14}$, corresponding to $C^{12}/$

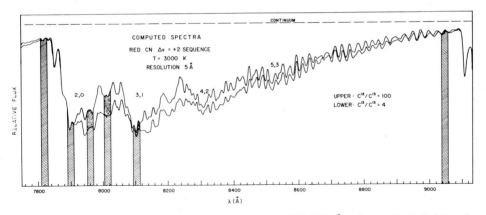

Fig. 5. Synthetic spectra of the CN molecule in the region 7800–9100 Å, calculated by I. R. Marenin-Little and plotted on a linear scale. They show the effect of varying the amount of $C^{13}N^{14}$ while holding $C^{12}N^{14}$ constant. The upper curve has $C^{12}/C^{13} = 100$ (close to the solar value) and the effect of $C^{13}N^{14}$ is negligible; the lower curve has $C^{12}/C^{13} = 4$. The vertical columns represent 20 Å bandpasses selected for scanner measurements of the isotope ratio.

C^{13} ratios of 100 and 4. Details of the calculations will be given elsewhere; suffice it to say here that the positions, strengths and profiles of approximately 5000 lines were computed for each spectrum before a smearing function was applied to degrade the resolution to 5 Å.

Our choices of the regions to be measured are indicated by the six vertical columns in Figure 5, which represent 20 Å bandpasses centred at 7820, 7900, 7960, 8014, 8104, and 9048 Å. The first and sixth points have been placed just shortward of the first heads of the (2, 0) and (1, 0) bands, and they serve to establish the level of the best available approximation to the continuum. The second point measures $C^{12}N^{14}$ and is almost completely independent of the isotope ratio, while the third point is very sensitive to the presence of $C^{13}N^{14}$. Thus the relative depressions at points 2 and 3, taken with respect to a continuum fitted to points 1 and 6, yield an index of the isotope ratio. Points 4 and 5 are intended to provide redundant information, and to serve as a check that the isotope ratios obtained are not a function of the band strength.

Observations on this system can most conveniently be made with a photoelectric scanner having an instrumental resolution of 1 or 2 Å. Because of the importance of the continuum point at 9048 Å, it is necessary to use an S-1 cell, although the other points could be measured with much more sensitive detectors. Reductions follow the procedures used for the eight-color photometry: the data are reduced to an absolute flux system by means of observations of standard stars which have been tied to α Lyr, which in turn is assumed to follow the energy distribution of the model adopted for it by Schild *et al.* (1971). A blackbody curve is passed through the continuum points, and the depressions at the other points measured from this continuum become the raw data from which CN abundances and carbon isotope ratios can be derived.

6.2. Observations and Results

The only observations obtained to date on this system were made on four nights in March and April 1972, with the Harvard College Observatory spectrum scanner mounted on a 90-cm telescope at Kitt Peak National Observatory. Because of imperfect weather and the time needed to establish a new system of standard stars, only two nights were really used for observations of program stars; nevertheless it was possible to obtain carbon isotope ratio indices for 47 stars, most of which were observed twice. Despite the small size of the telescope, stars of the ninth visual magnitude could be observed in about 1 h. The probable errors were about 1% in the magnitudes and somewhat less in the indices.

The observations of 17 stars are plotted in Figure 6. The uppermost star is ϕ Aur, a strong-CN K3 giant; in this case the six absolute fluxes F_λ are plotted against wavelength, and the fitted blackbody continuum is also shown. For the remaining stars – for greater convenience in interpreting the spectral shapes, and to save paper – we have plotted only the depressions at the first five points, i.e. the magnitude differences between the observed fluxes and the blackbody continua.

The four stars in the left half of Figure 6 are carbon stars. From the relative depressions at the second and third points, it is obvious that HD 52432 has a much higher proportion of C^{13} than X Cnc, which in turn has more than TT CVn or HD 137613. This agrees with what was already known from inspection of the C_2 Swan bands: TT CVn and HD 137613 have no detectable $C^{12}C^{13}$ bands, implying a C^{12}/C^{13} ratio greater than 50 (McKellar, 1960b); X Cnc is a rather typical carbon star whose isotope ratio is probably between 5 and 10 (its CN profile is similar to those of several other carbon stars we have observed); and HD 52432 has outstandingly strong $C^{12}C^{13}$ and $C^{13}C^{13}$ features (Yamashita, 1967). Ultimately we should be able to derive our own C^{12}/C^{13} ratios by comparing the photometric observations to synthetic spectrum calculations, but I would prefer not to give numerical values on the basis of the preliminary com-

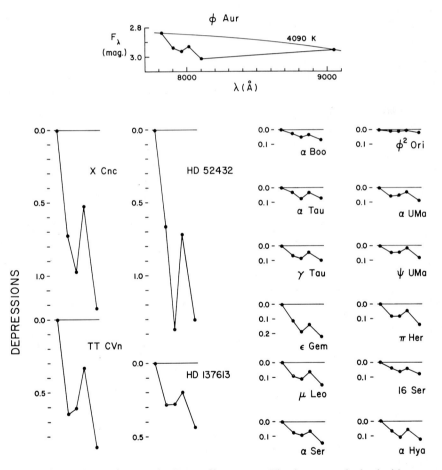

Fig. 6. Results of observations on the CN profile program. The data were obtained with a scanner at Kitt Peak National Observatory on the photometric system defined by the vertical columns in Figure 5. For ϕ Aur (*above*) the fluxes are plotted against wavelength and have been fitted with a blackbody curve. For the four carbon stars (*left*) and 12 G and K stars (*right*), the depressions at the first five wavelengths relative to the blackbody continuum are plotted on the same scale as for ϕ Aur. The relative depressions at the second and third wavelengths are indicative of the carbon isotope ratio.

parisons made to date. It seems safe to conclude, however, that the C^{12}/C^{13} ratio of HD 52432 can hardly be greater than 2, and it may be closer to unity.

Twelve G and K stars are shown in the right half of Figure 6. Again, the relative depressions at the second and third points indicate that their isotope ratios are not all the same. It would appear that α UMa (K0$^-$ IIIa), ψ UMa (K1 III), and π Her (K3 IIab) have very little C^{13}, whereas α Boo (K2 IIIp, weak CN), α Tau (K5 III), ε Gem (G8 Ib), and α Hya (K3 IIIa) are C^{13}-rich. The Hyades G-giant γ Tau and the strong-CN stars μ Leo and α Ser have intermediate C^{12}/C^{13} ratios. There seems to be no correlation between the CN strength and the C^{12}/C^{13} ratio. The barium star 16 Ser is not distinguished from normal giants by either its CN strength or its isotope ratio. The peculiar star ϕ^2 Ori has such weak CN bands that nothing can be said about its isotope ratio.

The synthetic spectrum calculations indicate that, for the observational accuracy achieved here (the probable errors are approximately equal to the sizes of the dots in Figure 6) and for the CN strength of a typical K giant, we should easily be able to distinguish isotope ratios differing by factors of 2, i.e. C^{12}/C^{13} ratios of 2, 4, 8, 16, 32, and >60. Thus we should be able to divide these stars into at least 5 groups – a degree of refinement that we hope will be adequate for a preliminary study of the correlations that may exist between the carbon isotope ratio and other observable parameters.

Acknowledgement

I would like to acknowledge that many of the observations discussed in this paper were made at Cerro Tololo Inter-American Observatory and Kitt Peak National Observatory. Special thanks go to Drs J. H. Baumert, G. W. Lockwood, I. R. Marenin-Little, and N. M. White for permitting me to discuss unpublished results from collaborative efforts. My research has been supported by the U.S. National Science Foundation.

References

Baumert, J. H.: 1972, Ph.D. Dissertation, The Ohio State University.
Baumert, J. H.: 1974, in press.
Baumert, J. H. and Wing, R. F.: 1974, in preparation.
Cameron, D. M. and Nassau, J. J.: 1955, *Astrophys. J.* **122**, 177.
Caswell, J. L. and Robinson, B. J.: 1970, *Astrophys. Letters* **7**, 75.
Connes, P., Connes, J., Bouigue, R., Querci, M., Chauville, J., and Querci, F.: 1968, *Ann. Astrophys.* **31**, 485.
Fujita, Y.: 1973, 'Identification of Spectra in Violet and Infrared Regions of Carbon Stars' (read at 15th IAU General Assembly).
Glass, I. S. and Feast, M. W.: 1973, *Monthly Notices Roy. Astron. Soc.* **163**, 245.
Griffin, R. F. and Redman, R. O.: 1960, *Monthly Notices Roy. Astron. Soc.* **120**, 287.
Herbig, G. H. and Zappala, R. R.: 1970, *Astrophys. J.* **162**, L15.
Humphreys, R. M. and Lockwood, G. W.: 1972, *Astrophys. J.* **172**, L59.
Humphreys, R. M., Strecker, D. W., and Ney, E. P.: 1972, *Astrophys. J.* **172**, 75.
Johnson, H. L.: 1967, *Astrophys. J.* **149**, 345.
Johnson, H. L., Coleman, I., Mitchell, R. I., and Steinmetz, D. L.: 1968, *Comm. Lunar Planet. Lab.* **7**, 83.
Joy, A. H.: 1947, *Astrophys. J.* **105**, 96.
Keenan, P. C.: 1950, *Astron. J.* **55**, 74.

Keenan, P. C.: 1963, *Stars and Stellar Systems* **3**, 78.
Keenan, P. C.: 1966, *Astrophys. J. Suppl.* **13**, 333.
Krupp, B. M.: 1973, *Bull. Am. Astron. Soc.* **5**, 336.
Lambert, D. L. and Dearborn, D. S.: 1972, *Mem. Soc. Roy. Sci. Liège, 6th Ser.* **3**, 147.
Lockwood, G. W.: 1972, *Astrophys. J. Suppl.* **24**, 375.
Lockwood, G. W.: 1973, *Astrophys. J.* **180**, 845.
Lockwood, G. W. and McMillan, R. S.: 1971, in G. W. Lockwood and H. M. Dyck (eds.), *Proc. Conf. on Late-Type Stars*, Kitt Peak National Observatory Contr. 554, p. 171.
Lockwood, G. W. and Wing, R. F.: 1971, *Astrophys. J.* **169**, 63.
Maehara, H.: 1968, *Publ. Astron. Soc. Japan* **20**, 77.
McKellar, A.: 1960a, *J. Roy. Astron. Soc. Canada* **54**, 97.
McKellar, A.: 1960b, *Stars and Stellar Systems* **6**, 569.
Merrill, P. W. and Greenstein, J. L.: 1958, *Publ. Astron. Soc. Pacific* **70**, 98.
Merrill, P. W., Deutsch, A. J., and Keenan, P. C.: 1962, *Astrophys. J.* **136**, 21.
Neugebauer, G. and Leighton, R. B.: 1969, *Two-Micron Sky Survey – A Preliminary Catalog*, NASA SP-3047.
Pesch, P.: 1972, *Astrophys. J.* **174**, L155.
Ridgway, S. T.: 1974, this volume, p. 327.
Schild, R., Peterson, D. M., and Oke, J. B.: 1971, *Astrophys. J.* **166**, 95.
Smith, M. G. and Wing, R. F.: 1973, *Publ. Astron. Soc. Pacific* **85**, 659.
Spinrad, H. and Taylor, B. J.: 1969, *Astrophys. J.* **157**, 1279.
Spinrad, H. and Wing, R. F.: 1969, *Ann. Rev. Astron. Astrophys.* **7**, 249.
Upson, W. L.: 1973, Ph.D. Dissertation, University of Maryland.
Wallerstein, G.: 1971, *Astrophys. Letters* **7**, 199.
White, N. M.: 1971, Ph.D. Dissertation, The Ohio State University.
White, N. M.: 1972, in M. Hack (ed.), *Colloquium on Supergiant Stars*, Trieste, p. 160.
Whitford, A. E.: 1972, *Bull. Am. Astron. Soc.* **4**, 230.
Whitford, A. E.: 1974, in J. R. Shakeshaft (ed.), 'The Formation and Dynamics of Galaxies', *IAU Symp.* **58**, in press.
Wing, R. F.: 1967a, in M. Hack (ed.), *Colloquium on Late-Type Stars* Trieste, p. 205.
Wing, R. F.: 1967b, in M. Hack (ed.), *Colloquium on Late-Type Stars*, Trieste, p. 231.
Wing, R. F.: 1967c, Ph. D. Dissertation, University of California, Berkeley.
Wing, R. F.: 1971a, in G. W. Lockwood and H. M. Dyck (eds.), *Proc. Conf. on Late-Type Stars*, Kitt Peak National Observatory Contr. 554, p. 145.
Wing, R. F.: 1971b, *Publ. Astron. Soc. Pacific* **83**, 301.
Wing, R. F.: 1972, *Mem. Soc. Roy. Sci. Liège, 6th Ser.* **3**, 123.
Wing, R. F.: 1973a, in Ch. Fehrenbach and B. E. Westerlund (eds.), 'Spectral Classification and Multicolour Photometry', *IAU Symp.* **50**, 209.
Wing, R. F.: 1973b, in J. D. Fernie (ed.), *Variable Stars in Globular Clusters and in Related Systems*, D. Reidel Publ. Co., Dordrecht, p. 165.
Wing, R. F.: 1973c, in H. R. Johnson, J. P. Mutschlecner and B. F. Peery, Jr. (eds.), *Proc. Conf. on Red Giant Stars*, Indiana Univ., p. 52.
Wing, R. F. and Ford, W. K., Jr.: 1969, *Publ. Astron. Soc. Pacific* **81**, 527.
Wing, R. F. and Lockwood, G. W.: 1973, *Astrophys. J.* **184**, 873.
Wing, R. F. and Spinrad, H.: 1970, *Astrophys. J.* **159**, 973.
Wing, R. F. and Stock, J.: 1973, *Astrophys. J.* **186**, 979.
Wing, R. F. and Warner, J. W.: 1972, *Publ. Astron. Soc. Pacific* **84**, 646.
Wing, R. F., Baumert, J. H., Strom, S. E., and Strom, K. M.: 1972, *Publ. Astron. Soc. Pacific* **84**, 646.
Wing, R. F., Spinrad, H., and Kuhi, L. V.: 1967, *Astrophys. J.* **117**, 147.
Wing, R. F., Warner, J. W., and Smith, M. G.: 1973, *Astrophys. J.* **179**, 135.
Wyckoff, S.: 1970, *Astrophys. J.* **162**, 203.
Yamashita, Y.: 1967, *Publ. Dominion Astrophys. Obs.* **13**, No. 5.

DISCUSSION

Fujita: Have you any comments regarding the comparison of the abundances of ^{12}C and ^{13}C compared with other determinations?

Wing: No. I really cannot compare the relative accuracy of this technique of measuring carbon isotope ratios with spectroscopic measurement. I have not actually determined any numbers. We are hoping to be able to do that. It is a question of how realistic the synthetic spectrum predictions are which is difficult to evaluate. I can say that narrow band photometry is a much easier observation to make and can be applied equally to faint stars, but it will depend critically on accurate determinations from high resolution work in order to calibrate the photometric ratios.

Schwarzschild: May I say not just to the present speaker but certainly to all of this morning's speakers that the input you are making to the theories of stellar structure and evolution is rather larger than I think you realize, and it is not just the one point that was emphasized several times on the relative abundances both isotopic and otherwise. It is also that one feels implicitly underlying many of your comments that the structure of the outermost layers is after all only a boundary condition and a very nasty type of boundary condition for stellar models. I sort of hope that by and large the interests of at least several of you will switch to the dynamical state of the very layers that you are observing. There are several comments that came up. Firstly the microscopic turbulent velocity, then there was this beautiful doubling that is quite hair-raising dynamically where other types of dynamical problems come up. One point that I do not remember that I have heard this morning mentioned is that whenever you want velocity differences or temperature differences I would suggest one might not want to always stick in very stratified layers, you know everything constant on spheres, could it not very well be, particularly for the supergiants, that they really have a violently mottled appearance for all we know? And therefore whenever you get into trouble, instead of only looking into models with vertical stratification, might the time not be right to think (whenever you are in trouble, and only then) to make the surface the superposition of two different atmospheres and see whether that might be another potential possibility of getting out of trouble?

Townes: It clearly will get you out of trouble and one would be able to have more variables and fit things more, but then what do you know?

Schwarzschild: Sir, the astronomers always first find possibilities, and then try to sort out the true ones.

MEDIUM RESOLUTION STELLAR SPECTRA IN THE TWO-MICRON REGION

A. R. HYLAND

Mount Stromlo and Siding Spring Observatory, Australian National University, Australia

Abstract. The role of medium resolution spectroscopy in the two micron region is discussed. Examples of the spectral features amenable to observation are shown. These include the vibration rotation bands of the important molecules $^{12}C^{16}O$, $^{13}C^{16}O$, H_2O and CN, as well as the Bγ line of hydrogen. Band strength-colour and -luminosity relations for the CO and H_2O bands have been derived from observations of approximately 100 stars of spectral types later than G5. The interpretation of these and the outstanding problems (such as the colour dependence of the CO band strengths in carbon stars) are discussed.
Examples of the use of two-micron spectroscopy in the spectral classification of infrared sources, and the infrared components of symbiotic stars, etc., are shown. The extension of this kind of observation to the nuclei of galaxies is briefly discussed in relation to stellar population studies.

1. Introduction

Recent developments in high resolution Fourier transform spectroscopy in the infrared have led to exciting advances in the observation and interpretation of molecular absorption features in the atmospheres of late type stars. The high resolution obtainable is ideal for the investigation of stellar atomic and isotopic abundances (particularly for the important light elements C, N and O) and for detailing the physical processes operative within the photospheric layers.

In the present circumstances the role of medium resolution infrared spectroscopy (where we define medium resolution to cover the range 6–24 cm^{-1}), has been changed from its former use as a tool in the pioneering investigations of stellar molecular band structure in the 1–4 μ region (Kuiper 1962; Boyce and Sinton, 1964, 1965; Sinton, 1966; McCammon *et al.*, 1967). In my view, it should be used to provide quantitative information on a relatively small number of important observable quantities (such as the strengths of the CO and H_2O bands) for a statistically significant number of stars, and further to extend the use of infrared spectroscopy to the faintest available limits. Such a role parallels that of low resolution spectroscopy at optical and near infrared wavelengths and is typified by the near infrared results presented by Wing (1974).

Discussions of medium resolution infrared spectra in the literature have generally been qualitative in nature, aimed at the identification of molecular bands and the approximate temperature and luminosity dependence of these features, e.g. McCammon *et al.* (1967); Johnson and Méndez (1970). It is to be expected that future observations will concentrate on the quantitative measurement of the broad observable features and thus be amenable to explicit theoretical interpretations.

In this paper we shall discuss the astrophysical implications of available medium resolution spectra in the limited wavelength range 1.9–2.5 μ. Most of the discussion will be based on the extensive observations of Frogel and Hyland in that range (see Frogel, 1971; Frogel and Hyland, 1972; Hyland *et al.*, 1972), made with various telescopes of the Hale Observatories. The particular wavelength range was chosen because of its proximity to both the peak sensitivity of PbS detectors and the maximum photospheric emission in late type stars, and because it contains absorption bands of the important molecular species CO, H_2O and CN. The two major areas covered by this discussion will be:

(a) the observed quantitative empirical relationships between the colour and molecular band strengths for a variety of late type stars and the problems which arise in the interpretation of these relationships, including the determination of the $^{12}C/^{13}C$ isotopic abundance ratios from th CO bands; and

(b) the spectral classification of infrared sources, including the identification of the infrared components of symbiotic stars, spectroscopic binaries and novae.

2. Observational Techniques

While observations at medium resolution in the two micron region have been made with both Fourier transform techniques (e.g. Johnson and Méndez, 1970) and conventional grating spectroscopy, we shall be for the main part discussing observations made in the years 1968–1970 using the latter technique by J. Frogel and myself. The system comprised the 0.5-m Ebert-Fastie spectrometer described by McCammon *et al.* (1967) attached to the 60, 100 and 200-in. telescopes of the Hale Observatories. It was a dual beam system with star-sky chop and a monitor channel; the required wavelength range was covered by continuous rotation of the grating, and the output signal was digitized and recorded once a second. Using an entrance aperture of 4-mm provided an essentially slitless mode, while the exit aperture was set at either 0.5 or 1.0-mm to provide resolutions of 32.5 Å and 65 Å respectively (~ 6–$12 \, \text{cm}^{-1}$). The scan rate was adjusted so that the required resolution was achieved.

All spectrophotometric studies, but most especially those in the infrared meet two important calibration problems: (a) the correction of the spectra for terrestrial absorption features (CO_2, H_2O and CH_4 are the major contributors in two micron region); and (b) the calibration of the observed fluxes onto an absolute scale so that both the level and shape of the continuum will be correct.

By making the basic assumption that the spectrum of α Lyr is correctly represented by the most recent model atmosphere calculations, (Schild *et al.*, 1971), and that apart from Bγ no line absorption features are strong enough to affect the spectrum at the resolutions in question, both (a) and (b) could be achieved. This however necessitated the additional constraint that α Lyr and the program star be observed at the same air mass. In practice, because of the highly saturated nature of many of the terrestrial bands, this constraint could be relaxed, to a limited degree and also, other early type stars such as α CMa and α CMi were set up to be secondary standards. Although this method differs from the more rigorous approach of Johnson *et al.* (1968) (who used lunar spectra from ground based and air-borne observations to define the terrestrial absorption), the results have been found to be very similar. Examination of published spectra show that only in few cases is the terrestrial absorption poorly removed.

3. Features Observable at Medium Resolution

A detailed discussion of molecular and atomic features observed in the infrared spectra of stars has been given by Spinrad and Wing (1969) and will not be repeated here. We shall concentrate instead on those features which are observable and quantitatively measurable in the two micron region. This implies that a feature must necessarily have an equivalent width $\geqslant 6$ Å for it to be measurable with reasonable precision (say 10%). The exact value necessarily depends on the shape of the feature, i.e. whether it is a molecular band, a single atomic line or blend of atomic lines. There are a few features which fulfill these requirements, and these will be considered here.

3.1. Molecular bands

3.1.1. *Carbon Monoxide*

The first overtone ($\Delta v = 2$) vibration-rotation band sequence of $^{12}C^{16}O$ has six vibrational band heads between 2.29 and 2.45 μ. These are visible in all giant stars with spectral types later than $\sim$ G5, and have been seen by a large number of observers at both low and medium resolution, see e.g. Boyce and Sinton (1964, 1965); Mertz (1965); Moroz (1966); McCammon *et al.* (1967); Johnson *et al.* (1968). Typical spectra of M stars showing the sequence are given in Figure 1, where five band heads are easily visible. The bands are clearly amenable to precise measurement as long as the continuum shape can be reasonably defined.

In addition the isotopic molecule $^{13}C^{16}O$ has visible band heads in most giant stars later than K0, the first three of these are indicated in Figure 1 and are easily seen in the spectrum of μ Gem.

The importance of CO hinges on the fact that it is essentially totally associated in all cool stars and thus in principle the band strength provides a hold on the atomic abundance of the less abundant species. It is therefore an important link in the determination of C, N, O abundances in late type stars. The presence of the bands of $^{13}C^{16}O$ further provides an opportunity to derive the $^{12}C/^{13}C$ abundance ratio for comparison with theoretical predictions.

3.1.2. *Water*

The absorption bands of water vapour (H_2O) which have their peak absorption near 1.9 and 2.7 μ as in the terrestrial atmosphere, also occur in very cool stars. At stellar temperatures however, the bands are broadened by the increased strengths of lines from excited levels and extend well into the two micron window. The spectrum of U Ori (M8e) shown in Figure 1 provides a striking example of H_2O absorption between 2.0 and 2.2 μ. The same authors who observed CO have reported observations of the water vapour (steam) absorption bands in M type Mira variables, although

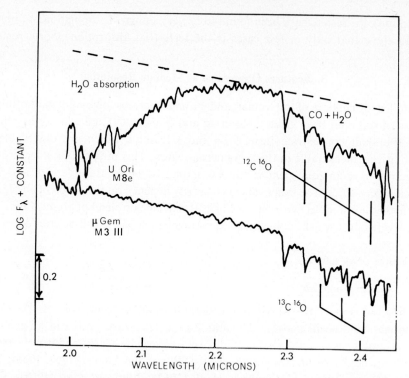

Fig. 1. Infrared spectra of μ Gem (M3 III) and the long period variable star U Ori (M8e) between 1.95 and 2.45 μ are shown. The resolution is 32.5 Å and the ordinate is the log of the flux per unit wavelength interval plus a constant. The prominent band heads of $^{12}C^{16}O$ and $^{13}C^{16}O$ are identified. Note the deep absorption of H_2O in U Ori at wavelengths shorter than 2.2 μ.

no quantitative estimates were made before the work of Frogel (1970). It is obvious from Figure 1 that at medium resolution the 1.9 μ band appears as a source of continuous opacity. While measurement of total absorption can be made from the available spectra, a water vapour index (WV) as defined by Frogel (1971) (See Section 4) appears to be a good measure of the strength of the 1.9 μ band, which may be compared with model atmosphere predictions (Auman, 1969).

3.1.3. *Cyanogen*

The extensive red system of cyanogen (CN) has several band heads in the two micron region. The R_{21}, R_{22}, R_{11} and Q_{11} band heads of the (2, 4) and (3, 5) ($\Delta v = -2$) systems were first positively identified in the spectra of carbon stars by Thompson and Schnopper (1970). The identification of the corresponding (1, 3) band heads of same system in carbon stars was made by Frogel and Hyland (1972). At medium resolution these CN bands are only visible in the spectra of carbon stars. Because of the large number of individual lines over a wide wavelength range the CN absorption behaves like a continuous opacity in a manner similar to H_2O. Considerable interest has recently been shown in the prediction of CN band strengths (Johnson *et al.*, 1972;

Alexander and Johnson, 1972) and the incorporation of CN opacities into realistic stellar atmosphere models. Thus although it is quantitatively more difficult to measure CN band strengths, the likely availability of synthetic spectra and the importance of CN in the atmospheres of carbon stars underline the necessity to consider CN seriously.

In Figure 2 the CN band heads are easily identified in the carbon star, U Hya (C7, 3) but are much less prominent in the long period variable carbon star V CrB (C6, 3). Very weak CN absorption is probably present in the S star AA Cyg, which has, on the other hand, among the strongest CO bands measured.

3.1.4. *Molecular Features not Seen*

Among other molecules, high excitation vibration-rotation bands of TiO, SiO are predicted to lie in the two micron region, but have not definitely been identified at the resolutions in question. Similar comments can be made regarding HCN and C_2H_2 whose predicted band heads are shown in Figure 2. Finally the predicted quadrupole

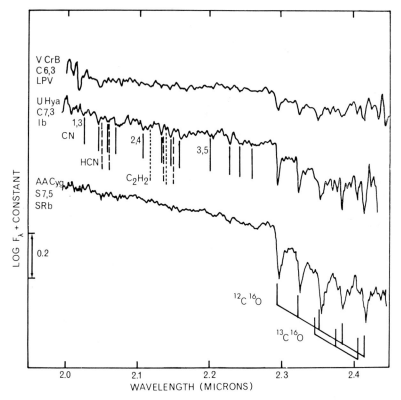

Fig. 2. Infrared spectra of the long period variable carbon star, V CrB, the irregular variable carbon star U Hya, and the S star AA Cyg are shown. In addition to the band heads of $^{12}C^{16}O$ and $^{13}C^{16}O$, the positions of the identified band heads of the red system of CN are shown, as are the predicted positions for bands of HCN and C_2H_2. The CO bands in the S star are among the strongest yet observed. Note the difference in the strengths of the CO bands in the two carbon stars.

line of H_2 (Spinrad and Wing, 1969) has been searched for unsuccessfully both at medium and high resolution.

3.2. Atomic features

In the remainder of this paper we shall be considering primarily measurements of the molecular species described above, largely because the majority of work has been done in that area. There are, however, several important atomic features for which some data is available, and which should be included in this discussion.

3.2.1. *Hydrogen*

The only measurable feature due to atomic hydrogen in the 2.0–2.5 μ region is the Brackett γ (B7) transition at 2.1655 μ. It is the major atomic feature in the region and a number of measurements have been reported in the literature. The strength of the line in absorption is such that it is measurable in stars with spectral types ranging from B to K, and its temperature dependence makes it a useful spectral type indicator for the earlier spectral classes.

The B7 line has also been observed in emission in the spectra of diffuse H II regions and planetary nebulae (Hilgeman, 1970), η Car (Neugebauer and Westphal, 1968) and in P Cyg and novae (Hyland, 1973). Knowledge of the emission strength of B7 is particularly helpful in deducing reddening estimates for these sources.

3.2.2. *Helium*

There is little available information on the occurrence of helium lines in the 2–2.5 μ region of the spectrum. No measurements of absorption lines have been reported. However, the $1s2s$–$1s2p$ transition of neutral helium at 2.05813 μ has been observed in emission in diffuse nebulae and planetary nebulae (Hilgeman, 1970) and is also present in the spectra of P Cyg and Nova Aql 1970 (Hyland, 1973). This line is also useful for reddening estimates and for the determination of He/H abundance ratios in these stars. It appears that there is considerable scope for wider investigations of these lines.

3.2.3. *Other Atomic Lines*

An investigation of the features observed in the spectra of K and M giants reveals several which may be identified with blends of closely spaced atomic lines of Si I, Mg I and Ti I (Frogel, 1971), given by Spinrad and Wing (1969). These show a systematic trend with spectral type, becoming stronger with increasing lateness of spectral type. The strengths are however all ≤ 3 Å and hence do not lend themselves to precise measurement such as we are discussing here.

4. Empirical Band Strength – Colour Relationships

In this section we shall discuss the measured CO and H_2O band strengths from observations of approximately 100 late type stars covering the spectral types G5–M8 and including the long period variable M stars, S stars and carbon stars. The large number of observations provides the opportunity for displaying the overall relations rather than the spectra of individual stars as is necessary for high resolution observations. The band strength of CO, $W(CO)$, is defined to be the equivalent width of the CO absorption between 2.29 and 2.39 μ, i.e. between the (2, 0) and (5, 3) band heads of $^{12}C^{16}O$. As can be seen in Figure 1 this is contaminated by the (2, 0) band of $^{13}C^{16}O$, although the effect is slight.

The strength of the 1.9 μ water vapour band is estimated by the use of the index WV, which is defined to be (after Frogel, 1971)

$$\log \frac{F_{cont}(\lambda = 2.10)}{F_\lambda(\lambda = 2.10)}$$

which differs slightly from the definition by Frogel (1970). $F_{cont}(\lambda=2.10)$ is the predicted continuum flux at 2.10 μ obtained from extrapolation of the continuum in the manner shown in Figure 1.

4.1. CARBON MONOXIDE IN G5–M8 GIANTS AND SUPERGIANTS

It is clear from the results of two micron spectroscopy over the last few years that the CO band strengths increase with increasing lateness of spectral type (e.g. Johnson and Méndez, 1970). There are several relevant questions which may be asked: (1) Are the CO band strengths sufficiently sensitive to be used for temperature determination, and how can luminosity effects be taken into account? (2) Is there evidence for abundance variations in the observed CO band strengths? This is of particular interest for the metal poor K giants and for the super metal rich stars, for abundance determinations may help define the relationship between the abundances of the light elements, C, N and O, and the heavy metals. (3) Can the observed strengths of the CO bands in the various species of late type star be explained theoretically? (4) What $^{12}C/^{13}C$ abundance ratios are derived from the $^{12}C^{16}O$ and $^{13}C^{16}O$ band strengths, and how certain are these.

In the following paragraphs and in Sections 4.2 and 4.3, we will consider how the observations bear on these questions.

In Figure 3 the $W(CO)$ strengths of G5–M8 giants are plotted as a function of $(J-L)$. The $(J-L)$ colour is known to be a good temperature indicator for cool stars when the interstellar reddening is small, as is the case for the bright stars in this discussion. The tight relationship of $W(CO)$ and $(J-L)$ may thus be interpreted in terms of the temperature dependence of $W(CO)$. The line in Figure 3 is a least square straight line fit to the data. Since CO is fully associated for all temperatures lower than ~ 4000 K, the increase in strength with decreasing temperature is not due to the increasing number density of CO molecules but rather to the decrease in the number density of H^- which provides the continuous opacity. Even so it has been necessary to invoke high microturbulent velocities (~ 10 km s^{-1}) to obtain theoretical fits to the CO band strengths. These velocities are considerably higher than those obtained by studies

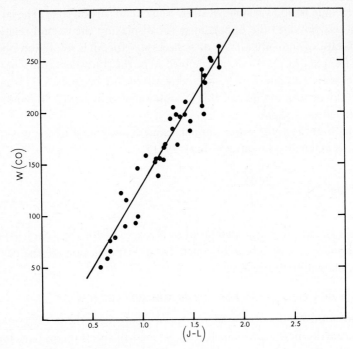

Fig. 3. The observed relation between $W(CO)$, as defined in the text, and the $(J-L)$ colour, for G5–M8 giants. The continuous line is the least square best fit straight line to the observations. The lines joining the dots refer to different spectroscopic observations of the same stars.

of optical spectra, and the relationship of the turbulence parameters in the optical and infrared are worthy of investigation at high resolution.

The equivalent observations for supergiant stars are shown in Figure 4, where the line is that derived for the giant stars. It is clear that at a given colour (i.e. temperature) the supergiants have stronger CO bands than the giants. The exact relationship is unfortunately clouded by the presence of strong interstellar and circumstellar absorption, for several of the supergiants. It appears that there is no easy two dimensional (luminosity, temperature) classification possible unless the colour is also known. The continuum slope obtained directly from the spectra can be used in place of $(J-L)$, and partly obviates the necessity for broad band measurements. Unfortunately the precision with which the slope can be measured is much lower than that obtainable with broad band measurements.

Qualitative inspection of giant and supergiant spectra reveals a fundamental difference in the shape of the CO absorption bands, however, no quantitative description of this difference has been proposed. Inspection of the CO bands in μ Gem (Figure 1), IRC +20439 (Figure 8) which are both M giants, and in the S star AA Cyg (Figure 2) (whose spectrum is identical to M supergiants) will illustrate the above point.

One further remark should be made about Figure 3. The spread in the observations at a given colour is larger in the early K stars $(J-L \sim 1.0)$ than would be expected from

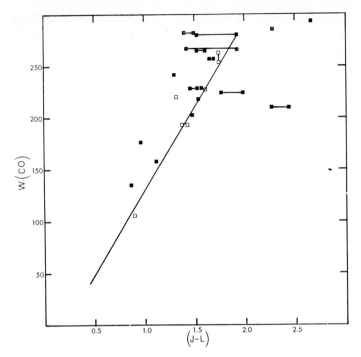

Fig. 4. The same as for Figure 3 for supergiant stars. The open squares are class II stars, while the filled squares refer to luminosity classes Ib, Iab and Ia. The straight line is the mean giant relation as shown in Figure 3. The straight lines joining the filled squares refer to independent measurements of the stellar colours.

the observational errors (± 15 Å). It is probable that these differences are due to real differences in the abundances in these stars. This is just that region of temperature in which the Cannon (1970) clump stars are found, and where spectral peculiarities such as strong CN become apparent. The correlation of CO band strength (which depends mainly on the abundance of carbon in these stars) with parameters such as CN or metal line strength may provide important clues as to the evolutionary status of early K giant stars.

4.2. Carbon monoxide band strengths in long period variable M stars, S stars and carbon stars

In Figure 5 the observed W (CO), $(J-L)$ relationships for all common species of late type giant star are presented. Omitted from this figure are the supergiants and the dwarf stars to minimise confusion. We note the following points:

(1) The long period variable M stars exhibit large cyclical variations in W (CO). The arrow indicates a typical excursion of the band strength. At maximum W(CO) the long period variable M stars an extension of the normal giant line to larger W (CO) and redder $(J-L)$.

(2) As a group the S stars have the strongest CO bands measured, being typical of,

and stronger than those of the M supergiants. This effect is probably due in part to the decrease in continuous opacity through the lack of H_2O absorption (see Section 4.4). Whether the S stars as a group have gas pressures more similar to supergiants than giants is not clearly established.

(3) Values of W (CO) in the semi-regular and irregular carbon stars show a definite decrease with increasing $(J-L)$ colour (Frogel and Hyland, 1972), from values typical of late M stars at $(J-L) \sim 2.0$ to values typical of K stars at $(J-L) \sim 2.6$. The long period variable carbon starts cover an even larger range of W (CO) values as illustrated in Figure 5, but are all much redder than the irregular and semi-regular carbon stars.

Of the above three remarks, the carbon star sequence is probably the most interesting, and the least understood. We shall therefore discuss in further detail the problems raised by the carbon star observations.

Several interesting correlations have been established by Frogel and Hyland (1972), the three most important being:

(1) the CO strength decreases with increasing colour, both in $(J-L)$ and the (3.5–8.4μ) colour;

(2) the CO strength decreases with increasing carbon parameter;

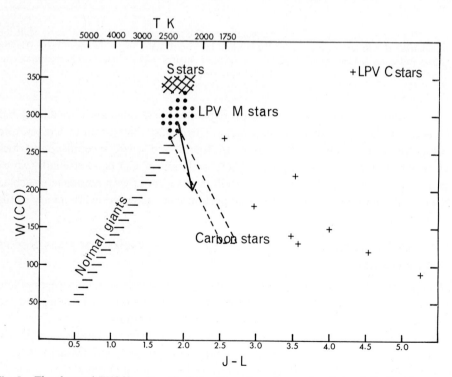

Fig. 5. The observed W(CO) $-(J-L)$ relation for various classes of cool stars including the normal giants, S stars, the long period variable (LPV) M stars, the irregular and semi-regular carbon stars, and the LPV carbon stars.

(3) the CO strength increases with decreasing temperature as determined from optical temperature parameters.

The last of these will be dealt with first. The temperature scale of the carbon stars is a rather controversial topic at the present time. From $(R-I)$ colours Eggen (1972) has proposed that all the carbon stars lie in the very restricted temperature range 3300–3500 K. Bessell and Lee (1972) from infrared stars of a similar group conclude that the blackbody continuum temperatures are lower than those of Eggen by ~ 500 K and that the spread of temperatures is somewhat larger. Unfortunately the broad band infrared colours (Mendoza and Johnson, 1967; Frogel and Hyland, 1972) only serve to confuse the issue because of the definite effects of circumstellar grain emission in many cases although the colour temperatures agree better with the scanner results. It is clear also that the effect of the atomic abundances on the atmospheric opacities is to alter the colour-temperature relation for the carbon stars. Until a definitive calibration of colour against temperature can be achieved via, for instance, lunar occultation measurements of carbon stars or realistic model atmospheres replace black body colours, this problem will remain one of the major obstacles in the study of carbon stars.

From (1) and (2) it follows that the $(J-L)$ colour also increases with increasing carbon parameter, which is an important clue as to the relevant mechanism involved. Two possibilities may be suggested to explain (1) and (2). Either (a) the abundance of oxygen is decreased as that of carbon is increased, while the major opacity source in the region of the CO bands is unchanged, or (b) as the carbon abundance is increased, the continuous opacity in the region of the CO bands increases, so that, although the number density of CO remains sensibly the same, the equivalent width of the CO bands decreases. The former suggestion is hard to reconcile with any process of mixing nuclear burning products to the surface, since it requires a decrease in the abundance of oxygen relative to H^- (which is assumed in this explanation to be the major opacity source).

The latter suggestion appears to be much more likely, but it requires that CN be the major opacity source in the vicinity of the CO bands, and that it increases with increasing abundance of carbon. Unfortunately the observations do not provide clear evidence of the predicted increase in the strength of the CN bands between 2.0 and 2.3 μ with increasing carbon parameter (see Frogel and Hyland, 1972). However, if these bands provide the major continuous opacity in that region also, such an effect would undoubtedly be masked.

The predictions of realistic model atmospheres could solve this problem. For example, if atmospheres with constant oxygen abundance and increased C/O have (a) redder broad band colours and (b) weaker apparent CO strengths, then the above suggestion would be a very plausible explanation. Certainly the explanation of the carbon star CO band strengths in terms of the absolute abundances of C and O poses crucial tests for the nuclear processes (e.g. the triple-α reaction or the CNO bi-cycle) proposed to explain the properties of carbon stars. Although some evidence supported the CNO bi-cycle hypothesis (Thompson *et al.*, 1971), recent more detailed analyses of molecular abundances in carbon stars appear to exclude it (Thompson, 1974). Thus the present emphasis lies in the comparison of the results of the triple-α process with the observations.

The position of the long period variable carbon stars relative to the irregular and semi-regular carbon stars in Figure 5 has been interpreted in terms of increased circumstellar dust shell characteristics in the long period variables (Frogel and Hyland, 1972) due to pulsationally coupled mass loss. This proposal remains an attractive and plausible explanation. The reason for carbon grain emission strongly affecting the two micron spectral region, whereas silicate emission does not (see Section 5.1) is that carbon grains can exist to much higher temperatures (i.e. 2000 K) than the silicates, and thus emit strongly at wavelengths down to 1 μ.

Many problems related to the evolutionary status and physical phenomena associated with carbon stars remain unsolved. We have seen that two micron spectroscopy is a valuable addition to the observational techniques available, which raises special problems of its own. However the recent upsurge of interest in carbon stars, both theoretical and observational, should result in major advances in our understanding of the subject over the next few years.

4.3. The isotopic $^{12}C/^{13}C$ abundance ratio

The presence of both $^{12}C^{16}O$ and $^{13}C^{16}O$ band heads in the 2.3–2.5 μ region allows one to obtain the $^{12}C/^{13}C$ ratio via analysis of the band strengths. The major problem in the derivation is the unknown degree of saturation of the $^{12}C^{16}O$ bands. While the degree of saturation becomes increasingly important at later spectral types, and is thought to be very high in the M supergiants (Thompson *et al.*, 1972), it appears, at least for early K stars, that it is low enough for realistic abundance ratios to be obtained. The derived values of $^{12}C/^{13}C$ for the early K giants (Frogel, 1971) reveal two interesting results. Firstly the majority of early K giant stars have values of $^{12}C/^{13}C$ in the range 5–10, which is markedly different from the solar and terrestrial value. Secondly, a few stars have values of $^{12}C/^{13}C$ much higher than the average. In particular the value for β Gem is >30, and for o' CMa >25.

We are forced to conclude that real differences in $^{12}C/^{13}C$ exist between stars of similar spectral type, and suggest that the exact ratio is critically dependent on the evolutionary history of the individual star. It should be emphasized again, that the early K giant region is where a large number of spectral peculiarities first become evident. For example the strong CN stars, the weak line stars, the SMR stars and extreme Population II stars all populate the early K giant region. The causal connection of these peculiarities to the 'clump', asymptotic branch stars has yet to be established. The correlation of the observed $^{12}C/^{13}C$ ratio with anyone of these peculiarities would be of considerable interest in this regard.

4.4. Water vapour band strengths in late-type stars

The importance of water vapour absorption in the 2.0–2.5 μ region in very cool stars can be judged from the spectrum of U Ori shown in Figure 1. A quantitative

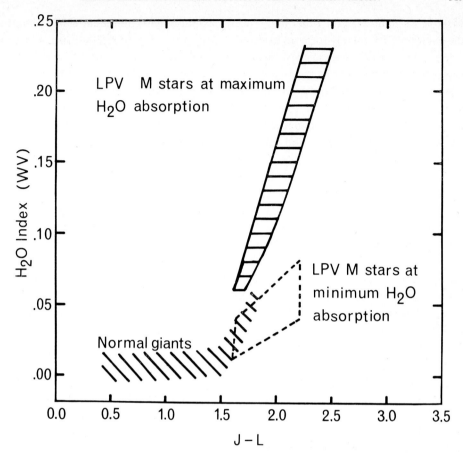

Fig. 6. The empirical colour − H₂O index (*WV*) relation for late type stars. The long period variable M stars at minimum *WV* form an extension of the normal non variable giant relation to redder colours. At maximum *WV* the long period variable M stars can be clearly distinguished by the strength of their H₂O absorption.

measure of the strength of the 1.9 μ band (WV) has been defined previously (Section 4). The observed relation of WV to the $(J-L)$ colour for both normal giants and long period variable M stars is shown in Figure 6. This shows that absorption due to the 1.9 μ band of H_2O becomes important only for stars with $(J-L) > 1.5$, i.e. with spectral types later than M6. Thus the presence of H_2O absorption in the spectrum of a star allows an upper limit to its temperature to be derived.

The position of M supergiants has not been shown in Figure 6, however, with the exception of the late supergiants VY CMa, NML Cyg, no H_2O absorption is seen, in agreement with molecular abundance calculations.

The H_2O index (WV) is greatest in the long period variable M stars and may be used to identify such sources. At minimum light, the H_2O index is larger for all long period variable M stars observed (Frogel, 1971) than in any of the irregular or non variable M giants. On the other hand, near maximum light (minimum H_2O strength) the values of WV merely form an extension of the normal giant relationship, albeit to larger values of $(J-L)$. One may define unambiguous spectral types in terms of the H_2O strength extending as far as M10 in similar manner to the TiO + VO relation of Wing (1967) however, there is no clear correlation of H_2O band strength with colour temperature. It appears that the band strength-temperature relation may in fact be multivalued. For a full understanding of the cyclical variations of the H_2O and CO band strengths in long period variable stars, the work of Frogel (1971) needs to be extended considerably.

Almost all of the S stars, both long period variables and non variables show no evidence for H_2O absorption, in agreement with the suggestion that $O/C = 1$ in S stars. One exception to this rule is W And, whose 2.0–2.5 μ spectrum is remarkably similar to that of the long period variable M star, U Ori. If W And has true S characteristics, in relation to the strength of its ZrO bands, it would appear that most S stars have been subject to two different processes, one which affects only the O/C ratio and the other which determines the abundance of the s-process elements. Furthermore, the existence of W And can be regarded as evidence that the s process alone has occurred. The possibility which this raises, that the abundance peculiarities in S stars are produced sequentially in two distinct phases, is worthy of full scale investigation.

5. Further Uses of Two-Micron Spectroscopy

In the previous sections we have discussed the observations and interpretation of the major molecular features observed between 1.9 μ and 2.5 μ in a variety of late type stars. By utilisation of the above results, however, two micron spectroscopy can make a valuable contribution to a number of other areas of investigation.

5.1. Spectral classification of infrared objects

Since the advent of the Two Micron Sky Survey (Neugebauer and Leighton, 1969) it has become clear that there are a large number of objects which radiate strongly in the infrared but which are either very faint or totally obscured visually. In the circumstances medium resolution infrared spectroscopy provides the means whereby the nature of the source can often be precisely ascertained. From the empirical band strength relationships discussed above it is possible to derive the abundance, temperature and luminosity characteristics of the source from the spectra alone. This method has proved extremely successful in the case of the infrared source NML Cyg (Johnson, 1968) and in the classification of the visually faint OH/IR stars (Hyland *et al.*, 1969, 1972). From the spectra NML Cyg was classified as an M supergiant, and the OH/IR sources as either M supergiants or long period variable M stars. Effective temperatures of these sources were determined by comparison with the temperatures of unreddened sources with similar two micron spectra. The validity of this procedure is dependent on the amount of distortion of the molecular CO and H_2O features by circumstellar dust emission. It has been shown that such an effect is only marginal

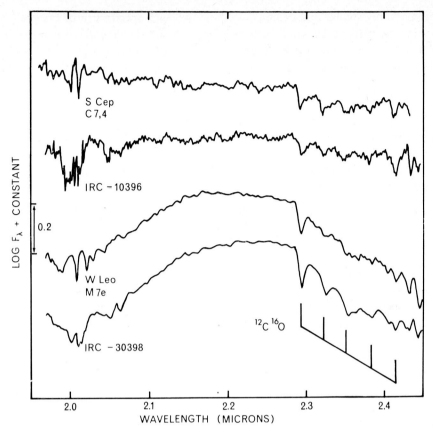

Fig. 7. Two micron spectra of the two infrared sources IRC −10396 and IRC −30398 are compared with spectra of the well known carbon star S Cep (C7, 4) and long period variable M star W Leo (M7e) respectively. Note the remarkable similarity of the spectra of the infrared sources to those of the comparison stars.

even in the case of an optically thick shell such as surrounds VY CMa (Hyland and Gingold, 1972) and hence that in the case of M stars the procedure is both valid and valuable. On the other hand it has been suggested as in 4.2 (Frogel and Hyland, 1972) that the weakening of the CO absorption in the reddest carbon stars is due to circumstellar emission from carbon grains which may exist at higher temperatures than the silicate like particles found in the shells surrounding M stars. Indeed the CO bands in IRC +10216, the most extreme carbon star known, are almost totally obliterated and the spectrum smooth and featureless (Becklin *et al.*, 1969).

As examples of spectral classification of infrared objects using two micron spectroscopy, the spectra of an oxygen rich M type long period variable W Leo (M7e), and the well known long period variable carbon star S Cep (C7, 4), are compared in Figure 7 with the observed spectra of two infrared objects IRC −30389 (OH/IR), and IRC −10396 respectively. The strength of the CO and H_2O bands in W Leo and IRC −30398, and the strength of the CO and CN bands in S Cep and IRC −10396 are remarkably similar and demonstrate how well such spectra can be matched with those of 'standard spectroscopic sources'. The only noticeable difference in each case lies in the slightly different continuum slope, which presumably results from the selective nature of circumstellar absorption and emission processes.

5.2. IDENTIFICATION OF THE INFRARED COMPONENTS OF SYMBIOTIC STARS, NOVAE, Be STARS, ETC.

There are several categories of stars which appear to consist of a hot and a cool component in close proximity. The problem in these cases is to identify the nature of the cool component, which dominates the infrared spectrum. In some cases clues as to the nature of the cool component are obtained from spectra in the red, where molecular band structure of TiO may be seen. The optical spectrum is of course dominated by the emission of the hot component.

Symbiotic stars are prototypes of this type of object, and in several cases the optical spectra have been sufficient for cool companions of spectral type $\sim$ M2 to be identified. There are however a number of cases for which the presence of a cool companion has been suspected but whose nature has not been established. It is in these instances that infrared spectra in the two micron region are particularly valuable, and allow for an unambiguous classification of the infrared continuum. This type of observation is also applicable to the infrared excesses observed in novae, P Cyg stars, Be stars and the supergiant F and G stars. We shall discuss here only the observations of four sources as examples of what is seen. The observed 2.0–2.5 μ spectra are shown in Figure 8.

5.2.1. *The Composite Star* IRC +20439

This source has been identified in the Two Micron Sky Survey with the eighth magnitude A2 star DM +22°3840. Close investigation has shown that the two sources are one and the same and spectra in the red (5500–7000 Å) confirm the presence of a late type companion whose spectral type is not unambiguously determined. The spectrum of IRC +20439 (in Figure 8) with W (CO) $\sim$ 250 Å allows us to identify the infrared component as an M7 giant with a temperature close to 2600 K. A report on similar observations of the infrared spectra of symbiotic stars is in preparation. Of particular interest is the unambiguous identification of the infrared component of XX Oph as a late M star.

5.2.2. Z CMa

This is an interesting early type star associated with nebulosity and apparently embedded in dust. It has been studied at infrared wavelengths out to 20 μ (Gillett and Stein, 1971) and discussed by Neugebauer *et al.* (1971). Without infrared spectra, the possibility that it is a composite star with a late M component could not be discounted. The two micron spectrum shown in Figure 8 (Racine *et al.*, 1971) shows that there is no evidence for the presence of a late type companion, but that the emission appears to be that of a smooth dust component. The peak in the continuous energy

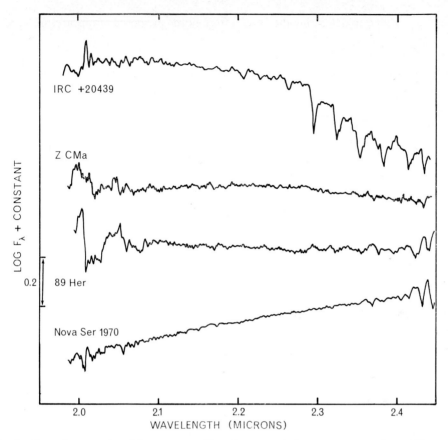

Fig. 8. Two micron spectra of four sources with early type optical components are shown. These are IRC +20439, Z CMa, 89 Her and FH Ser (Nova Ser 1970), and are discussed in the text.

distribution close to 2.2 μ indicates that there is a sizeable contribution from grains with temperatures ~1300 K. Neugebauer et al. (1971) have shown that a Larson (1969) type dust-cloud model where the dust opacity varies as λ^{-1} gives a reasonable fit to the overall energy distribution.

5.2.3. 89 Her

This is an interesting F supergiant star with a large infrared excess (Gillett et al., 1970) which has also been interpreted in terms of emission from a circumstellar dust cloud. The high rate of mass loss is consistent with the formation of the dust cloud from the expelled material. The 2.0–2.5 μ spectrum presented by Gillett et al. (1970) is shown in Figure 8, and also shows a smooth dust like continuum. The most interesting feature of both the broad-band infrared and spectral observations of 89 Her, is their remarkable similarity to those of the carbon rich variable R CrB. Unlike the infrared excesses found in the G supergiants (Humphreys et al., 1971) there is no evidence that the dust particles are composed silicates. Clearly further observations are needed to determine the exact nature of the proposed grains around 89 Her.

5.2.4. FH Ser (*Nova Ser 1970*)

Our final example is the nova which underwent a remarkable infrared phase some 40–60 days following its optical outburst (Hyland and Neugebauer, 1970). This infrared phase was interpreted in terms of dust emission from grains ($T \sim 900$ K) found at large distances from the star. The infrared spectrum shown is that of Hyland and Neugebauer (1970) after correction for terrestrial absorption, and in common with Z CMa shows the remarkably smooth and featureless continuum characteristic of blackbody emission from grains. Unfortunately, this region provides no knowledge as to the nature of the grains such as is available from narrow band photometry in the 8–14 μ region. Note that at the phase of observation no Bγ nor He $\lambda 2.058$ μ emission is seen in the spectrum of FH Ser.

The above four examples illustrate the diagnostic value of two micron spectroscopy for a variety of infrared sources.

5.3. Comparison of the Continuous Energy Distribution with Model Atmospheres

In the previous sections we have concentrated on the observations and interpretation of the molecular and atomic absorption features visible in the 2.0–2.5 μ wavelength region. Implicit throughout the discussion has been the possibility of comparing the overall observed energy distribution with the predictions of model atmospheres. This procedure has been generally overlooked although it should be invaluable for testing the predictions of realistic model atmospheres of late type stars as they become available (Alexander and Johnson, 1972; Querci, F. & M., 1974). It has been emphasized by Frogel (1971) that Auman's (1969) models including water vapour opacities do not agree with the observations, the predicted H_2O bands being too strong. Future observational and theoretical studies of medium resolution spectra should include detailed comparisons of the observed and predicted fluxes in the manner which has proved highly successful in the interpretation of visual scans of early type stars.

5.4. Stellar Population Studies in Galaxies and Globular Clusters

Knowledge of the spectral type – molecular band strength relations as discussed in Section 4 can be used in conjunction with observations to provide useful information on the cores of galaxies and globular clusters. Because these are generally faint in the two micron region it is necessary to reduce the resolution and/or to use a system of narrow band filters to define the molecular CO and H_2O band strengths. A pioneering study along these lines has been performed by Baldwin *et al.* (1973). Using narrow band filters centered at 2.10, 2.20 and 2.31 μ they measured CO and H_2O indices for the nuclear regions of the three galaxies M31, M81 and NGC 5195 in an effort to resolve the controversy regarding the luminosity function for cool stars. Comparison of the band strength with the standard relations for giants and dwarfs shows that all three galactic nuclei have band strengths similar to late type giant stars and precludes the existence of a dwarf enriched sequence ($M/L = 44$) such as has been proposed by

Spinrad and Taylor (1971). This result is in agreement with Whitford's (1972) inability to detect the Wing-Ford (9910 Å) band in the nucleus of M31.

Studies of this nature appear to be very promising for future investigations of the stellar population content of galactic nuclei, elliptical galaxies and globular clusters and emphasize the versatility of spectroscopic observations in the two micron region.

6. Conclusion

Throughout this paper I have tried to show that medium resolution spectroscopy in the two micron region has a vital role to play in a variety of astrophysical problems. We have seen that the molecular CO band strengths can be used as temperature and luminosity indicators, while also being useful under certain conditions for the study of light element abundances and the $^{12}C/^{13}C$ isotope ratios. The strength of the 1.9 μ band of H_2O appears to be intimately related to the LPV M stars, and has lead to interesting deductions regarding S stars. Two micron spectroscopy was also shown to be valuable in defining the properties of infrared components of various stars with large infrared excesses.

The future of observations of this nature seems assured with its extension to use as an indicator of stellar content in the nuclei of galaxies and globular clusters, and the advent of new high sensitivity detectors will ensure that even more exciting domains may be explored.

Acknowledgements

It is a pleasure to acknowledge the assistance of many of the Cal Tech Infrared Group in obtaining the spectra discussed here. The work at Cal Tech was supported in part by NASA Grants Nos. NGL 05-002-007, NGL 50-002-207 and NSF Grant GP 9527.

References

Alexander, D. R. and Johnson, H. R.: 1972, *Astrophys. J.* **176**, 629.
Auman, J. R.: 1969, *Astrophys. J.* **157**, 799.
Baldwin, J. R., Danziger, I. J., Frogel, J. A. and Persson, S. E.: 1973, *Astrophys. Letters* **14**, 1.
Becklin, E. E., Frogel, J. A., Hyland, A. R., Kristian, J., and Neugebauer, G.: 1969, *Astrophys. J. Letters* **158**, L133.
Bessell, M. S. and Lee, Y.: 1972, *Proc. Astron. Soc. Australia* **2**, 154.
Boyce, P. B. and Sinton, W. M.: 1964, *Astron. J.* **69**, 534.
Boyce, P. B. and Sinton, W. M.: 1965, *Sky Telesc.* **29**, 78.
Cannon, R. D.: 1970, *Monthly Notices Roy. Astron. Soc.* **150**, 111.
Eggen, O. J.: 1972, *Astrophys. J.* **174**, 45.
Frogel, J. A.: 1970, *Astrophys. J. Letters* **162**, L5.
Frogel, J. A.: 1971, Ph.D. Thesis, California Institute of Technology.
Frogel, J. A. and Hyland, A. R.: 1972, *Mem. Soc. Roy. Sci. Liège, 6th Ser.* **3**, 111.
Gillett, F. C. and Stein, W. A.: 1971, *Astrophys. J.* **164**, 77.
Gillett, F. C., Hyland, A. R., and Stein, W. A.: 1970, *Astrophys. J. Letters* **162**, L21.
Hilgeman, T. W.: 1970, Ph.D. Thesis, California Institute of Technology.
Humphreys, R. M., Strecker, D. W., and Ney, E. P.: 1971, *Astrophys. J. Letters* **167**, L35.
Hyland, A. R.: 1973, in preparation.
Hyland, A. R. and Gingold, R. A.: 1972, *Proc. Astron. Soc. Australia* **2**, 155.

Hyland, A. R. and Neugebauer, G.: 1970, *Astrophys. J. Letters* **160**, L177.
Hyland, A. R., Becklin, E. E., Neugebauer, G., and Wallerstein, G.: 1969. *Astrophys. J.* **158**, 619.
Hyland, A. R., Becklin, E. E., Frogel, J. A., and Neugebauer, G.: 1972, *Astron. Astrophys.* **16**, 204.
Johnson, H. L.: 1968, *Astrophys. J. Letters* **154**, L125.
Johnson, H. L. and Méndez, M. E.: 1970, *Astron. J.* **75**, 785.
Johnson, H. L., Coleman, I., Mitchell, R. I., and Steinmetz, D. L.: 1968, *Comm. Lunar Planet. Lab.* **7**, 83.
Johnson, H. R., Marenin, I., and Price, S. D.: 1972, *J. Quant. Spectrosc. Radiat. Transfer* **12**, 189.
Kuiper, G. P.: 1962, *Comm. Lunar Planet. Lab.* **1**, 1.
McCammon, D., Münch, G., and Neugebauer, G.: 1967, *Astrophys. J.* **147**, 575.
Mendoza, E. E. and Johnson, H. L.: 1965, *Astrophys. J.* **141**, 161.
Mertz, L.: 1965, *Astron. J.* **70**, 548.
Moroz, V. I.: 1966, *Soviet Astron.* **10**, 47.
Neugebauer, G. and Leighton, R. B.: 1968, *Two-Micron Sky Survey – A Preliminary Catalog*, NASA Sp-3047.
Neugebauer, G. and Westphal, J. A.: 1968, *Astrophys. J. Letters* **152**, L89.
Neugebauer, G., Becklin, E. E., and Hyland, A. R.: 1971, *Ann. Rev. Astron. Astrophys.* **9**, 67.
Querci, F. and Querci, M.: 1974, this volume, p. 341.
Racine, R., Becklin, E. E., Hyland, A. R., and Neugebauer, G.: 1971, data quoted by Neugebauer et al. (1971).
Schild, R. E. Peterson, D. M., and Oke, J. B.: 1971, *Astrophys. J.* **166**, 95.
Sinton, W. M.: 1966, *Astron. J.* **71**, 398.
Spinrad, H. and Wing, R. F.: 1969, *Ann. Rev. Astron. Astrophys.* **7**, 249.
Spinrad, H. and Taylor, B. J.: 1971, *Astrophys. J. Suppl.* **22**, 445.
Thompson, R. I.: 1974, this volume, p. 255.
Thompson, R. I. and Schnopper, H. W.: 1970, *Astrophys. J. Letters* **160**, L97.
Thompson, R. I., Schnopper, H. W., and Rose, W. K.: 1971, *Astrophys. J.* **163**, 533.
Thompson, R. I., Johnson, H. L., Forbes, F. F., and Steinmetz, D. L.: 1972, *Publ. Astron. Soc. Pacific* **84**, 779.
Whitford, A. E.: 1972, *Bull. Am. Astron. Soc.* **4**, 230.
Wing, R. F.: 1974, this volume, p. 285.
Wing, R. F.: 1967, Ph.D. Thesis, University of California, Berkeley.

DISCUSSION

Thompson: Could it be that as the temperature goes down it is simply just the association of C_2 and CN, that reduces the CO first overtone bands not so much an increase in carbon as you thought, but possibly just that we are associating more molecules.

Hyland: You mean that would increase the total opacity?

Thompson: Right.

Hyland: That is possible, but I am not sure that it would give the magnitude of the effect.

FOURIER TRANSFORM SPECTROPHOTOMETRY AND ITS APPLICATION TO THE STUDY OF K-GIANTS

S. T. RIDGWAY

Kitt Peak National Observatory, Tucson, Ariz.
and
State University of New York at Stony Brook, U.S.A.

Abstract. A Fourier Transform Spectrometer has been employed to obtain low and medium resolution spectrophotometry in the range 700 to 10000 cm^{-1}. The method is described briefly and typical data are shown. The technique is applied to a study of 15 K giants, and spectra of 4 and 16 cm^{-1} resolution in the range 4000 to 6500 cm^{-1} are discussed briefly. CN strong stars are found to be CO strong as well. Five K giants are found to have a relative ^{13}C enhancement over probable primordial abundances.

1. Introduction

For the last several years I have been applying Fourier spectroscopy to several planetary and stellar studies. Although high resolution has been obtained in some cases, much of the work has been done at low resolution employing techniques which might reasonably be called Fourier Transform Spectrophotometry. I will describe briefly the technique, then present typical data and discuss a number of results from a K-giant survey.

2. Fourier Transform Spectrophotometry

The scanning interferometer is a standard commercial model (Idealab, Inc., Maynard, Mass.). Maximum mirror travel is 1 cm and operation is in the rapid scanning mode first described by Mertz. A general discussion of the spectrometer system is available (Ridgway, 1972).

For the spectrometer to be useful beyond 3 μ, where thermal background emission is severe, it was especially important that the instrument has two extremely well balanced inputs for star + sky and sky. The foreoptics design is shown in Figure 1. Notice that the input optics are symmetric about the beamsplitter plane. Especially valuable for balancing the beams, two variable apertures are located in the prime focal plane. In addition, the 150 cm McMath solar telescope at Kitt Peak, which was used for this work, has a completely unobscured aperture, significantly reducing the thermal background problems. The two inputs are readily balanced to better than one part in 100, and moving the star from one aperture to the other between scans reduced the residual imbalance to below the detector noise for up to $\frac{1}{2}$ h integrations.

The detectors employed were PbS (Santa Barbara Research Corporation, Santa Barbara, California) with an NEP of 3×10^{-14} W Hz$^{-1/2}$ in the range 1 to 4 μ, and a Ge bolometer (Infrared Laboratories, Tucson, Arizona) with an NEP of 1×10^{-13} W Hz$^{-1/2}$ in the range 1 to 14 μ. It should be pointed out that detectors approximately 10 times more sensitive are currently available.

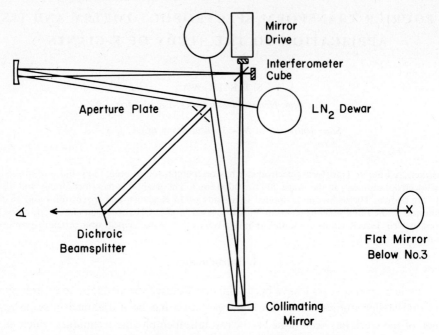

Fig. 1. Two-beam input foreoptics scheme for the FTS.

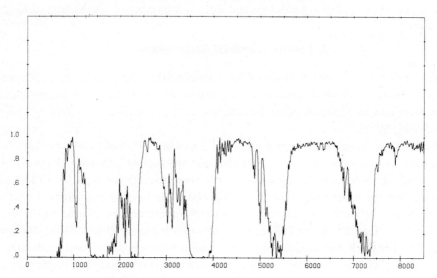

Fig. 2. Atmospheric transmission from Kitt Peak.

Figure 2 shows the atmospheric transmission in the range 700 to 9000 cm^{-1} for a typical good infrared night at Kitt Peak. Although the spectrometer has produced good spectra at up to 0.5 cm^{-1} resolution in each of these six atmospheric windows, I will restrict my discussion to the low resolution spectrophotometry.

The major non-instrumental difficulty to be surmounted was the development of

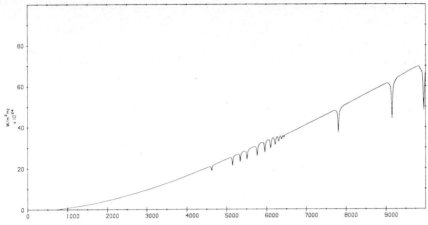

Fig. 3. Synthetic spectrum of Sirius.

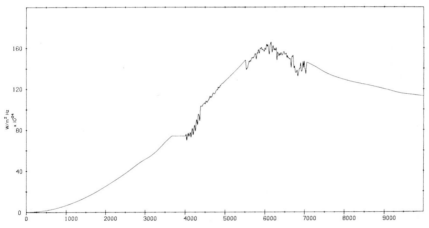

Fig. 4. Reference spectrum of Arcturus.

a suitable standard. In addition to having a known spectrum, the standard should have a high and reasonably uniform flux through the entire spectral region of interest. A computed synthetic spectrum of Sirius (Schild *et al.*, 1971; Peterson, 1971) is shown in Figure 3. Throughout this spectral range the synthetic spectrum is completely consistent with observational data. Short of about 2500 cm^{-1} Sirius is too faint to serve as a spectrophotometric standard. In order to cover the full range, 700 to 10000 cm^{-1} a kind of hybrid spectrum was devised for the bright K-giant Arcturus.

Sirius was used to determine atmospheric transmission, which in turn was used to obtain a corrected spectrum of Arcturus in the region 4000 to 6500 cm^{-1}. This was combined with the continuum flux predicted by model atmosphere studies, as shown in Figure 4. The spectral range 7000 to 10000 cm^{-1} has yet to be fitted with an observed spectrum. Also, the CO fundamental at 2000 cm^{-1} has not yet been taken into account. When this procedure has been completed, Arcturus will serve as a spectropho-

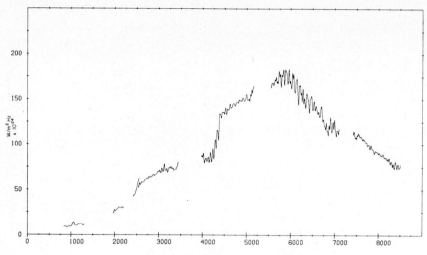

Fig. 5. The spectrum of α Her.

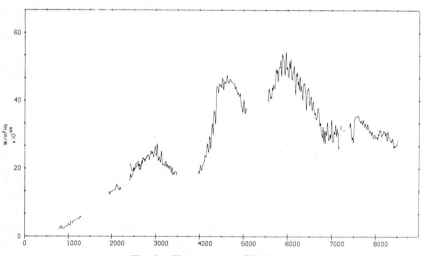

Fig. 6. The spectrum of RX Boo.

tometric standard over this very broad spectral range, ideal for reduction of the type of data described below.

In Figure 5, the spectrum of α Her from 800 to 8500 cm^{-1} is shown. This spectrum has been corrected for instrumental response and for telluric absorption using the Arcturus spectrum in its present form as described above. The range 800 to 3000 cm^{-1} was covered with a Ge on KBr beamsplitter, and the range 2500 to 8500 cm^{-1} with a Si on CaF$_2$ beamsplitter. The absolute flux scale is established by reference to broad band filter photometry at 4500 cm^{-1}. Throughout the rest of the spectrum fluxes are consistent with broadband photometry to within about 0.1 mag.

The spectrum of α Her is dominated by CO bands, atomic lines, and the H⁻ opacity minimum. Compare 3000 K α Her with 2500 K RX Boo in Figure 6. The luminosity classes are probably comparable, but the 500 K temperature difference leads to the sudden appearance of stellar water vapor absorption.

This very broad spectral bandwidth observational technique has been applied primarily to planets, and these examples of stellar spectra are merely illustrative. There are many potential applications, such as evaluation of blanketed model atmospheres (Querci *et al.*, 1973) and determination of effective temperatures (Dyck *et al.*, 1973).

3. The K-Giant Survey

I will now turn to a specific application of Fourier Transform Spectrophotometry – a study of K giants. The K giants are an especially fruitful subject for the infrared spectroscopist since they are relatively bright in the infrared, yet not so cool as to present the special complexities of polyatomic molecular formation. Also, an understanding of their composition will yield valuable information about the evolution of a star away from the main sequence.

Figure 7 shows the survey spectra of the six giants selected as normal standards with a comparison solar spectrum above. This spectral region is crowded with telluric absorption features, but these have been completely eliminated in the reduction

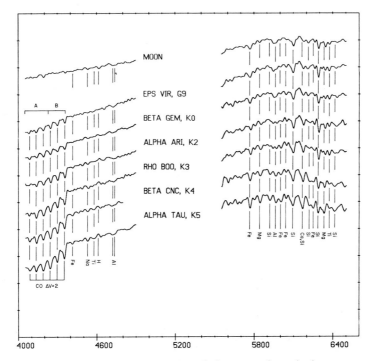

Fig. 7. K giants, 16 cm⁻¹ resolution, normal standards.

(Ridgway, 1972). The region 4900 to 5500 cm^{-1} is not plotted because of especially serious obscuration due to CO_2 and H_2O.

Instrumental response has been removed from the spectra using nightly photometric calibration. As a check of photometric quality, the infrared index $H-K$ can be computed from these spectra and shows better than 1% agreement with infrared photometry. The standard deviation computed point by point is $\frac{1}{2}$% or better, except near the edges of the atmospheric windows where the noise is up to 3 times greater. All of the spectra are normalized in the same way, then shifted vertically by increments of one unit for ease of comparison. The zero flux level of the lowest spectrum coincides with the base of the plot.

At this resolution all of the spectral lines are blended, but as a guide the strongest contributors to some of the atomic features are indicated (Montgomery *et al.*, 1969). The first overtone of CO is prominent starting with the 2–0 bandhead at 4360 cm^{-1}. The second overtone cannot be identified at this resolution. The general strengthening of features with later spectral type is due to increased abundances of neutral species and decreasing continuous opacity. Note the shift of the flux distribution to smaller wavenumber for the cooler stars. The distortion of the continuum due to the H$^-$ opacity minimum at 6000 cm^{-1} is also prominent.

Survey spectra of four stars identified as SMR (Spinrad and Taylor, 1969) are shown in Figure 8. These stars are also known to have unusually strong CN bands (McClure, 1970). Figure 9 includes the rest of the survey spectra. This is a mixed group in-

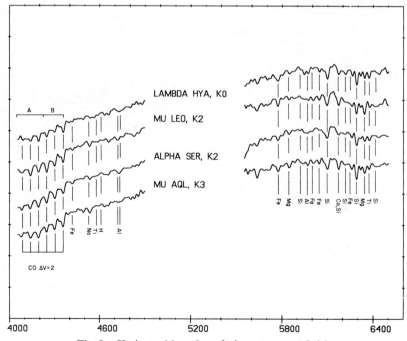

Fig. 8. K giants, 16 cm^{-1} resolution, super metal rich.

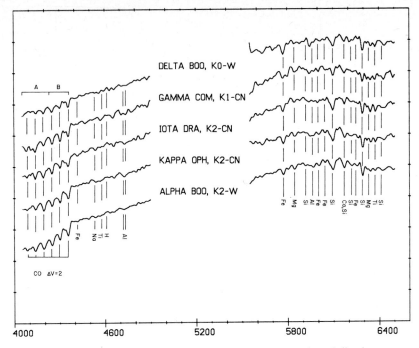

Fig. 9. K giants, 16 cm^{-1} resolution, CN strong and weak lined.

cluding CN strong stars which are not presently considered SMR, and two weak lined stars.

Because of the high photometric quality it is possible to mathematically 'synthesize' narrow and medium bandwidth filters to explore small differences among the spectra. As a first example, it is possible to define a temperature dependent color in the vicinity of the H$^-$ opacity minimum. This index can be compared with an index T measured by Spinrad and Taylor near 7000 Å. Since the continuum opacity at 7000 Å is much greater than at the minimum, the relationship between the two indices should be sensitive to stellar atmospheric temperature structure differences.

In Figure 10, $F(5875)/F(4525)$ is plotted against T for the K giants studied. The dashed line through the normal stars represents the behavior expected from model computations. There is a clear dispersion in the sense that metal poor stars appear to have a relatively steeper temperature gradient with optical depth and the CN strong stars a relatively shallower temperature gradient. This is precisely opposite the effect we would expect if high metallicity and CN strength are associated with line blanketing. Differential blanketing in the filters might overwhelm any temperature structure effects but there is no evidence that this is the case here. Any other pair of filters in the range 4400 to 6800 cm^{-1} leads to a similar result.

By employing this filter 'synthesis' technique, it is possible to investigate the CO band strength in some detail. For the purpose of distinguishing the causitive mechanisms, we can study separately the integrated CO absorption for two regions. Region A,

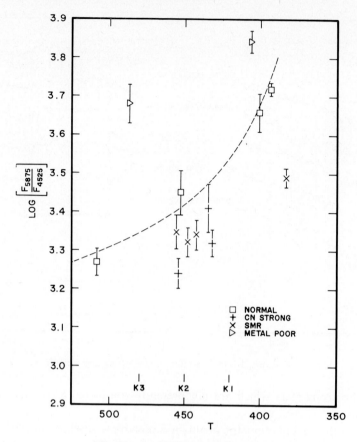

Fig. 10. log [F(5875)/F(4525)] vs T.

4000 to 4230 cm^{-1}, is dominated by lines on the linear part of the curve of growth for giants earlier than about K4. Region B, 4230 to 4360 cm^{-1}, is dominated by a relatively small number of very strong lines which are saturated for giants later than about K0.

A photometric study of CO band strength by Baldwin *et al.* (1973) employed a filter corresponding roughly to the region B. Since several stars were common to both studies, measures for these stars can be compared. In Figure 11 corresponding indices are plotted. The agreement is generally consistent with assigned observational error. The slope of the plotted points is consistent with the fact that Baldwin *et al.* used an adjacent rather than a local continuum.

Careful study of the spectroscopic data of this survey, and of the more extensive photometric survey of Baldwin *et al.* fails to reveal any systematic peculiarities in the CO band strength of the saturated region B for weak lined or CN strong stars.

Turning now to the unsaturated section of the CO bands, the integrated blocking fraction of region A is plotted against the temperature index T in Figure 12. The smooth line is a theoretically predicted curve due to Schadee (1968) fitted to the normal K-giants. Clearly all CN strong and SMR stars show excessive CO strengths in region

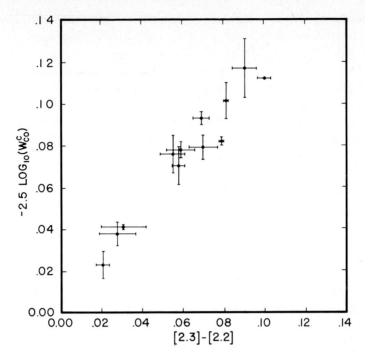

Fig. 11. $W_{CO}{}^B$ vs (2.3)–(2.2).

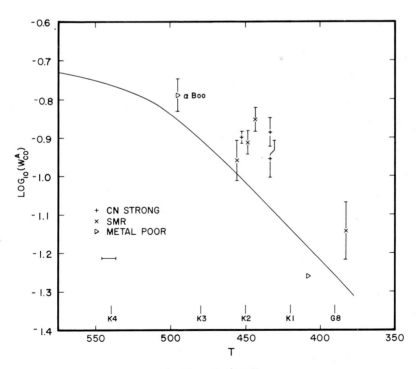

Fig. 12. $W_{CO}{}^A$ vs T.

A. This might be due to differences in luminosity, microturbulent velocity, line blanketing, systematic errors in the T index, or abundance differences.

On the basis of Schadee's work and the surface gravities tabulated by Williams (1971, 1972), it can be verified that the results of Figure 12 are unaffected by luminosity differences, except for α Boo which appears above the normal curve due to its substantially greater luminosity.

If the CO excesses were due to differences in microturbulent velocity, line blanketing with consequent surface cooling, or systematic errors in the T index, we would expect saturated region B of the CO bands to exhibit an equal or greater variation in CO strength. In fact, as noted above, it does not. This is consistent with a genuine abundance difference, since saturated region B would be less sensitive to abundance differences than unsaturated region A. By studying the details of the correlation between CO and CN strengths, it may be possible to isolate the abundance peculiarities responsible.

4. The $^{12}C/^{13}C$ Ratio

As a final topic I would like to describe briefly an application of spectral synthesis techniques to the evaluation of the $^{12}C/^{13}C$ ratio from medium resolution spectra. The synthesis method was implemented by Carbon (1972), based on the model stellar atmosphere program SOURCE (Gingerich and Carbon, 1973). From a model atmosphere and a line list, the program solves the equations of state (under the assumption of LTE) at each specified depth to determine the molecular number densities. Then for each wavelength the source function is computed at each depth. The source function is then integrated over depth to find the emergent flux. The line absorption profile is assumed to be described by a Voigt function.

Of course, the greatest potential of this technique will be realized in the study of the highest resolution spectra; but it is also the best approach to the analysis of medium resolution spectra such as those in Figures 13 and 14. Here, at 4 cm^{-1} resolution, the $^{13}C^{16}O$ bandheads are clearly resolved, yet each resolution element contains many unresolved lines.

As a guide to the identification of atomic features, the positions of strong lines in α Boo (Montgomery et al., 1969) are indicated. Resolution was not sufficient to identify many comparably strong lines in both earlier and later spectral types. The CO $\Delta v = 3$ sequence, although labeled, is not clearly distinguishable. The study of a spectrum involves the selection of suitable effective temperature, surface gravity and composition; generation of a model atmosphere; selection of microturbulent velocity and $^{12}C/^{13}C$ ratio; and synthesis of an 'infinite' resolution spectrum. The synthetic spectrum is then convolved with an appropriate line profile and compared with the observed spectrum. Model parameters can then be adjusted to obtain a satisfactory fit. The technique has been applied to 0.5cm^{-1} resolution carbon star spectra (Carbon et al., 1973) and will be applied shortly to similar spectra of a number of K and M giants. I have recently used this technique to obtain $^{12}C/^{13}C$ ratios for these spectra.

Note first that α Ser shows weak or absent $^{13}C^{16}O$ bands at this resolution.

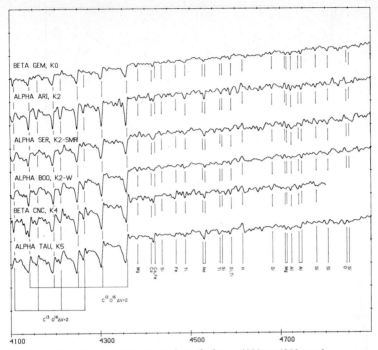

Fig. 13. K giants, 4 cm^{-1} resolution – 4100 to 4900 cm^{-1}.

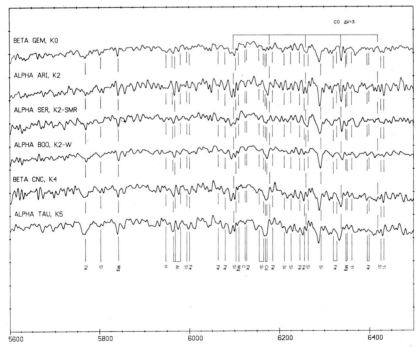

Fig. 14. K giants, 4 cm^{-1} resolution – 5600 to 6500 cm^{-1}.

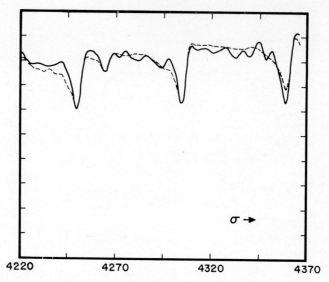

Fig. 15. α Ari observation, solid line, and synthetic spectrum, dashed line.

For the rest of the stars, a $^{12}C/^{13}C$ ratio of 10 yields a satisfactory fit to the observations. A typical example is shown in Figure 15 for α Ari. The dashed line is the synthetic spectrum. The fit is generally consistent with the S/N of 100 and the presence of several strong atomic lines in the stellar spectrum.

A $^{12}C/^{13}C$ ratio of 10 is roughly consistent with several studies of α Boo especially by Lambert and Dearborn (1972) at higher resolution and with other molecules. Since evidence now indicates that a terrestrial $^{12}C/^{13}C$ ratio is probably typical throughout the solar system and the local interstellar medium (Ridgway, 1972), it is reasonably certain that these stars have substantial ^{13}C enhancement over the primordial abundances.

References

Baldwin, J. R., Frogel, J. A., and Persson, S. E.: 1973, *Astrophys. J.* **184**, 427.
Carbon, D. F.: 1972, unpublished Ph.D. dissertation, Harvard University.
Carbon, D. F., Thompson, R. I., and Ridgway, S. T.: 1973, in preparation.
Dyck, H. M., Lockwood, G. W., and Capps, R. W.: 1973, to be published.
Gingerich, O. and Carbon, D. F.: 1973, Smithsonian Observatory Special Report No. 300, in preparation.
Lambert, D. L. and Dearborn, D. S.: 1972, *Mem. Soc. Roy. Sci. Liège, 6th Ser.* **3**, 147.
McClure, R. D.: 1970, *Astron. J.* **75**, 41.
Montgomery, E. F., Connes, P., Connes, J., and Edmonds, F. N.: 1969, *Astrophys. J. Suppl.* **19**, 1.
Peterson, D. M.: 1971, private communication.
Querci, F., Querci, M., and Tsuji, T.: 1973, to be published.
Ridgway, S. T.: 1972, Ph.D. dissertation, SUNY at Stony Brook, University Microfilms, Ann Arbor, Michigan.
Schadee, A.: 1968, *Astrophys. J.*: **151**, 239.
Schild, R., Peterson, D. M., and Oke, J. B.: 1971, *Astrophys. J.* **166**, 95.
Spinrad, H. and Taylor, B. J.: 1969, *Astrophys. J.* **157**, 1279.
Williams, P. M.: 1971, *Monthly Notices Roy. Astron. Soc.* **153**, 171.
Williams, P. M.: 1972, *Monthly Notices Roy. Astron. Soc.* **158**, 361.

DISCUSSION

Townes: You have given a ratio of 10 for ^{12}C and ^{13}C and I wonder if you could say what the general range of variability would allow in your analysis?

Ridgway: The uncertainties are mostly hidden in the sense that they are not observational errors. My value of 10 compared with Lambert's results of 6 or 7 certainly is not inconsistent in any way. 5 to 20 would be fair enough.

Williams: I am most interested in your comparison of your carbon monoxide excess with your cyanogen excess of your super metal rich and cyanogen strong stars. The error bars were large, but there did seem to be a slight systematic difference between them, in the sense that the super metal rich stars were of a slightly stronger cyanogen for their carbon monoxide. This suggests that whereas the strong CN stars may be mostly carbon strong, the super metal rich stars are also nitrogen strong. And I think you will see CO photometry seems to be an extremely valuable way of finding out just what is happening among these stars.

INTERPRETATION OF CARBON STARS SPECTRA FROM MODEL ATMOSPHERES COMPUTATIONS*

F. QUERCI and M. QUERCI

Laboratoire du Télescope Infrarouge, D.E.P.E.G., Observatoire de Paris, Meudon, France

1. Introduction

The first part of this paper presents a grid of model atmospheres for carbon stars in which the most striking feature is the inclusion of the molecular line blanketing effect of CO, CN and C_2 through opacity probability distribution functions. The techniques and the main results have been fully discussed in previous papers (Querci, 1972; Querci *et al.*, 1972 and 1973), and consequently we will give just a rapid description asking the reader to refer to these papers for details. However, the grid has been extended here to lower temperatures.

The second part presents original results from the M. Querci's 'Thèse d'Etat' (1974), and concerns the computation of synthetic spectra. A kind of grid of synthetic spectra has been computed from each model atmosphere, that is to say for each particular effective temperature and gravity, over a selected spectral range. The influence of the gravity and of the effective temperature on the computed spectra and the problem of the line cutting procedure are considered. Finally, comparisons with a high resolution infrared spectrum of a carbon star are made.

The conclusion summarizes the experimental progress that must be made in order to be able to make meaningful comparisons between predictions and observations.

2. The Grid of Model Atmospheres

In the construction of the model atmospheres, in order to take into account the blanketing by about 10^6 lines of the three following diatomic molecules,
  CN (red and violet systems),
  CO (fundamental, first and second overtones),
  C_2 (Swan, Phillips and Ballik-Ramsay systems),
for all the isotopic species observed in carbon stars, we have adapted the statistical method proposed by Ström and Kurucz (1966).

As explained in Querci *et al.* (1972, 1974) and Querci (1972), for each subinterval of the primary spectral intervals dividing the total spectral range, we build up and put on a magnetic disk a table of opacity probability distribution functions consisting of a grid in temperature T and gas pressure P_g, in which the values required during the run of the model are obtained by interpolation.

The lines used for building the OPDF tables are put on a magnetic tape, the ATLAS

---

* Presented by F.Q.

tape. The selection of the lines is such that, for a given pair of values of temperature and gas pressure, $T_1 - P_{g_1}$, which furnish a maximum number of lines for a given molecular system, we keep those lines for which: $l_{v_0} \geqslant 0.1\, k_{v_{min}}(T_1, P_{g_1})$, where l_{v_0} is the monochromatic absorption coefficient at the line center and $k_{v_{min}}$ is the minimum continuous absorption coefficient over the spectral range from 0.2 to 20 μ. Of the lines previously stored on the ATLAS tape, within each primary interval, Δv, we keep the lines such that $l_{v_0} \geqslant 0.1\, k_v(T, P_g)$. The line wings are also cut when $l_v < 0.1\, k_v$, where k_v is the continuous absorption coefficient per gramme of stellar material in Δv.

Sixty primary intervals Δv divide the total spectral range and the molecular absorption is smoothed within each interval. The subintervals of Δv have equal statistical weights.

The OPDF tables have been established for:

$$T = 840, 1006, 1260, 1680, 2100, 2800, 4200, 6300, 8400 \text{ K},$$

and

$$\log P_g = -3, -1, +1, +3, +5.$$

The abundances have been taken to be:

$$\text{C/H} = 4.1 \times 10^{-5}, \quad \text{N/C} = 37.0, \quad \text{C/O} = 3.2,$$
$$\text{C}^{12}/\text{C}^{13} = 1.11 \times 10, \quad \text{N}^{15}/\text{N}^{14} = 4.1 \times 10^{-4}, \quad \text{O}^{16}/\text{O}^{18} = 2.0 \times 10^{-3}.$$

In solving the ionisation and dissociation equilibria, 26 atoms (neutral and once ionized) and 200 molecules, respectively, have been included (from Tsuji, 1973).

The microturbulence velocity has a single value of 5 km s^{-1}.

The model atmospheres presented are characterized by the effective temperatures:

$$T_e = 4500, 4200, 3800, 3400, 3000, 2600, 2200 \text{ K},$$

and the gravities:

$$g = 0.1, 1.0, 10.0.$$

The lower temperature model atmospheres have to be considered as preliminary to more realistic models in which polyatomic opacities are included.

The first value of the standard optical depth has been taken to be $\log \tau_{std} = -6.0$ and the standard wavelength to be 0.8 μ.

The models are calculated with 80 optical depth points.

The chemical composition is the same as in the computation of the table of opacity distribution functions.

The total pressure used in this work consists of gas, radiation and turbulent pressures. The equation of hydrostatic equilibrium is integrated by Hamming's method.

The radiative flux F through a layer is computed by summation of the fluxes F_v corresponding to each spectral interval Δv:

$$F_v = \sum_{i=1}^{n} W_i F_v^i,$$

where F_v^i are the elementary fluxes, W_i are the weights and n is the number of sub-intervals in Δv.

The elementary fluxes imply the source function, that is calculated by the Feautrier difference equation method.

I have selected some of the results for illustration of the grid. Others are discussed in Querci (1972) and Querci *et al.* (1974).

Figure 1 shows the $T(\tau)$ laws of the grid of model atmospheres. Figures 2 and 3 present the run of the physical quantities with the standard optical depth for effective temperatures corresponding to very cool atmospheres: $T_e = 2600$ K and $T_e = 2200$ K with $g = 1.0$ and 10. The run of the $T(\tau)$ laws and of the logarithm of the density ϱ are also presented. The variation of the temperatures is well correlated with that of the partial pressures, and we see how large variation in the absorption in the molecular lines can modify the blanketing effect.

To illustrate the importance of blanketing in the coolest stars, model atmospheres for $T_e = 4500$, 3400 and 2600 K and $g = 1$ have been run with and without line blanketing by CO, CN and C_2. The curves in Figure 4 also show the importance of surface cooling and deep backwarming. In the deepest layers of the star, where the temperature is so high that no lines can be formed, the unblanketed and the blanketed models are obviously very similar.

Figure 5 presents the predicted emergent fluxes for 3 typical effective temperatures, 4500, 3400 and 2600 K, for the same gravity, $g = 10$.

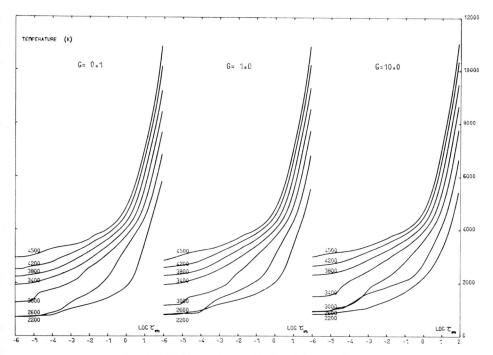

Fig. 1. $T(\tau)$ laws of the grid of model atmospheres.

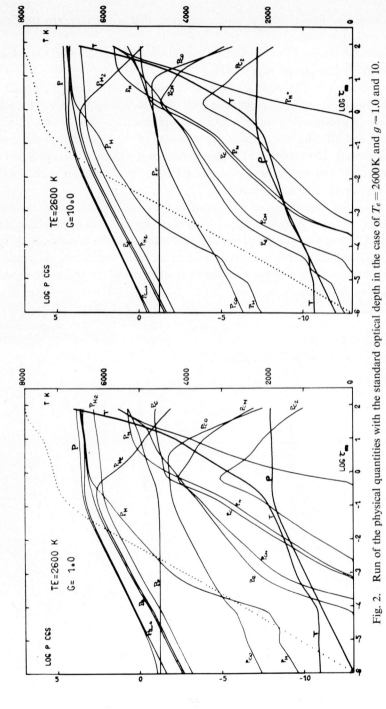

Fig. 2. Run of the physical quantities with the standard optical depth in the case of $T_e = 2600$ K and $g = 1.0$ and 10.

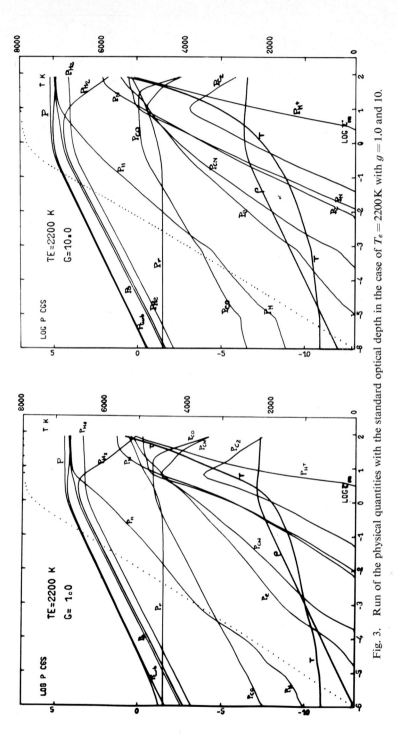

Fig. 3. Run of the physical quantities with the standard optical depth in the case of $T_e = 2200$ K with $g = 1.0$ and 10.

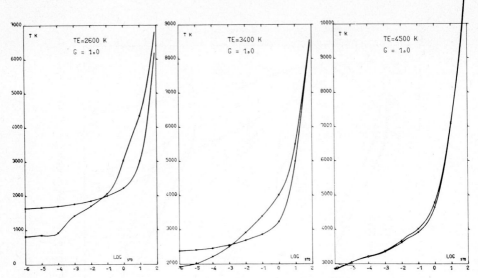

Fig. 4. $T(\tau)$ laws from the models with $T_e = 4500\,\text{K}$, $3400\,\text{K}$ and $2600\,\text{K}$, and $g = 1$, including and without including the molecular line blanketing.

While for the atmosphere with $T_e = 4500\,\text{K}$ the emitted flux stays near the black-body curve with only the most important bands of CN and CO appearing, for $Te = 3400\,\text{K}$ the flux at $1.8\,\mu$ to the right of the H^- minimum absorption is heavily absorbed by CN and CO. In the $2600\,\text{K}$ atmosphere, the flux emitted at $1.8\,\mu$ is enhanced and the molecular features appear clearly:
- the violet CN at $0.39\,\mu$ (0 sequence), $0.42\,\mu$ (-1 sequence), $0.46\,\mu$ (-2 sequence),
- the sequences 5 to -4 of the red CN,
- the C_2 Swan system at $0.47\,\mu$, $0.51\,\mu$ and $0.62\,\mu$,
- the sequences 0, -1, -2 of the C_2 Ballik-Ramsay system,
- the fundamental, the first and second overtones of CO.

By following the variation with the effective temperature, it is easy to see the influence of the molecular line blanketing by the 3 molecules CO, CN and C_2 on the fluxes emitted by carbon stars.

3. The Synthetic Spectra

At each optical depth point in the atmosphere, the program uses previously computed values of τ_{std}, $T(\tau)$, $\log P_g$, and the standard total continuous absorption coefficients $(k_\nu + \sigma_\nu)_{\text{std}}$. It also uses the assumed H/C/N/O abundance ratios and the isotopic abundance ratios. The number of lines considered is the number used to compute the opacity distribution functions. These lines are picked up from the ATLAS tape over the spectral region for which synthetic spectra are being computed.

For each τ_{std}, various quantities are computed: the abundance of CO, CN and C_2, the continuous absorptions, the scattering and the induced absorptions. The program

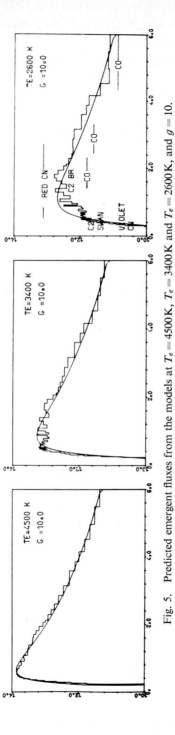

Fig. 5. Predicted emergent fluxes from the models at $T_e = 4500$ K, $T_e = 3400$ K and $T_e = 2600$ K, and $g = 10$.

checks the width of the biggest line of the spectrum; this quantity is used to compute the contribution of all the line profiles to the line absorption coefficient $\sum l_v^i$ at the frequency v.

The frequency step length is set equal to the distance between two primary points of the observed spectrum ($\simeq 0.08$ cm^{-1}).

The transfer equation at each frequency point is solved by the Feautrier difference equation method with the second order upper boundary condition from Auer (1967).

The emerging continuous flux F_c is computed once for each spectral interval. The emerging flux F_v in the line and the ratio $R_v = F_v/F_c$ are computed at each frequency point.

The computed spectra are finally convolved with the same instrumental and apodized function as the observed spectra, and the synthetic and observed spectra can then be directly compared.

To compute synthetic spectra, spectral intervals are chosen which contain characteristic molecular features. We present here computations for a spectral interval around the head of the (2, 0) band of the CO first overtone. In this spectral range, from 4319 to 4400 cm^{-1}, the CO vibration-rotation spectrum and the electronic red

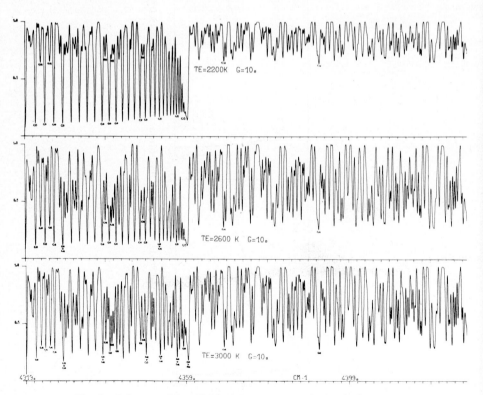

Fig. 6. Influence of the effective temperature on the synthetic spectra: $T_e = 2200$K, 2600K and 3000K for $g = 10$.

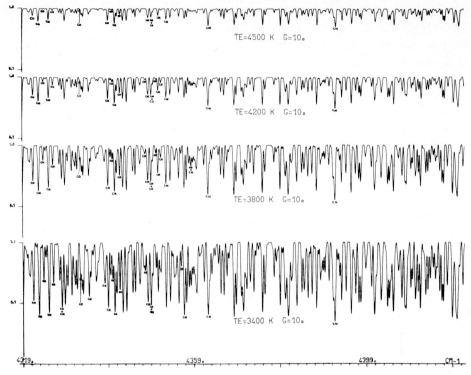

Fig. 7. Influence of the effective temperature on the synthetic spectra: $T_e = 3400\,\mathrm{K}$, $3800\,\mathrm{K}$, $4200\,\mathrm{K}$ and $4500\,\mathrm{K}$ for $g = 10$.

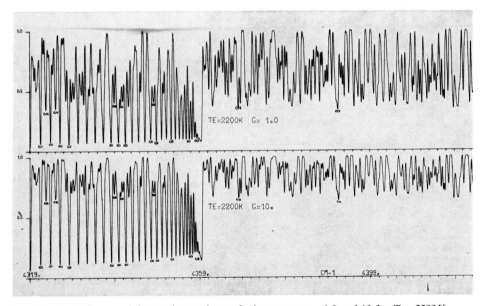

Fig. 8. Influence of the gravity on the synthetic spectra: $g = 1.0$ and 10 for $T_e = 2200\,\mathrm{K}$.

system of CN are very efficient. Moreover, these sysems are the two most important absorbers in the atmospheres considered.

We note that the line wavenumbers have been corrected for the radial velocity peculiar to the molecular system considered.

First, we look at the influence of the effective temperature on the synthetic spectra.

Figure 6 shows the case of low temperatures, $T_e = 2200, 2600, 3000$ K, for $g = 10$. In the atmosphere with $T_e = 2200$ K, the CO lines are very strong, while the CN lines are very weak. For $T_e = 2600$ K, the CN lines become more important; the intensity of the CO lines remains the same. For $T_e = 3000$ K, the CN lines and the CO lines have about the same strength and CO begins to weaken.

Figure 7 groups the following temperatures: $T_e = 3400, 3800, 4200, 4500$ K, for $g = 10$. For $T_e = 3400$ K, the CO bandhead is weaker than the CN lines, while for $T_e = 4500$ K, the CO lines have nearly disappeared.

We now look at the influence of the gravity on the synthetic spectra.

It is striking that, for $T_e = 2200$ K (Figure 8), the gravity has a strong influence on CN; the effect is also noticeable for CO.

In Figure 9 ($T_e = 3400$ K and $g = 0.1, 1.0$ and 10), we see that the CN system is not so much influenced by the gravity. On the other hand, the strengths of the bandhead of CO or of other lines are fairly sensitive to the gravity.

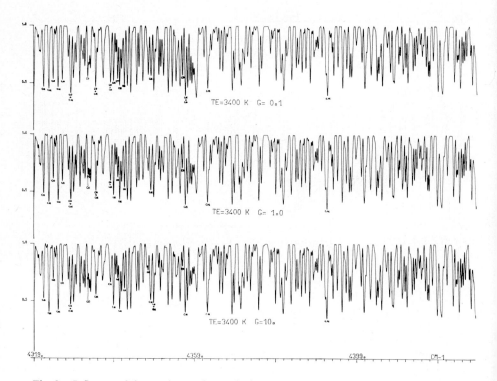

Fig. 9. Influence of the gravity on the synthetic spectra: $g = 0.1, 1.0$ and 10 for $T_e = 3400$ K.

For $T_e = 4200$ K and $g = 0.1$, 1.0 and 10 (Figure 10), the high effective temperature is not favourable to the formation of molecules. Atomic absorption should be included in the opacity distribution functions. However, we see that the CO spectrum is stronger for a weak gravity ($g = 0.1$ corresponding to supergiant C stars), while for $g = 10$, the CN spectrum is the stronger.

We may make several remarks:

– All the synthetic spectra have been computed with the same H/C/N/O abundance ratio. However, the molecular features of CO and CN vary strongly and differently with the effective temperature and the gravity. Consequently, we have to be very careful when we determine abundances from slab models of cool stars.

– To explain the behaviour of the two molecular systems, it is necessary first to compare the synthetic spectra with the $T(\tau)$ laws and the run of the partial pressure through the atmospheres, in order to see the variation of the abundances of the two molecules with depth in the atmosphere; secondly, it is necessary to study how these molecular bands evolve with the temperature.

– Some synthetic spectra computed over correctly chosen spectral ranges, are good tools for the determination of effective temperatures and gravities.

It must be pointed out that when studying line blanketing in cool atmospheric layers, the effect of the cut off for weak lines has to be considered very carefully before making comparisons between computed and observed stellar spectra. It is

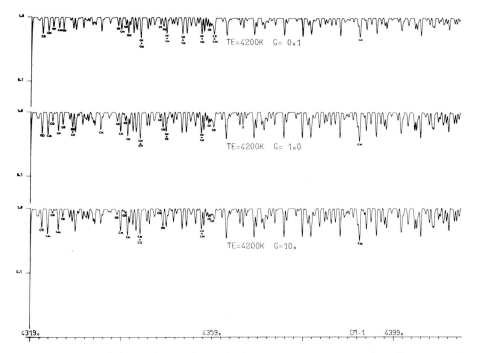

Fig. 10. Influence of the gravity on the synthetic spectra: $g = 0.1$, 1.0 and 10 for $T_e = 4200$ K.

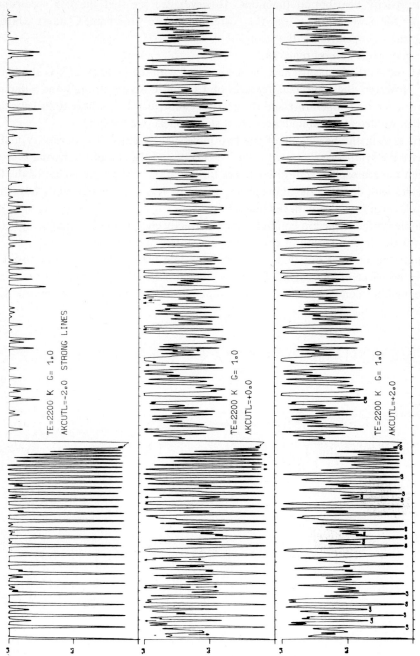

Fig. 11. Influence of the line cutting procedure illustrated by the atmosphere with $T_e = 2200$ K and $g = 1.0$.

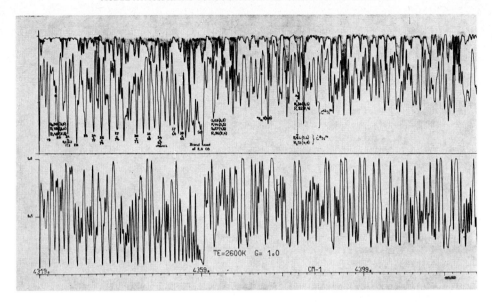

Fig. 12. At the top, solar spectrum; in the middle, spectrum of the carbon star UU Aur; at the bottom, synthetic spectrum computed with $T_e = 2600$ K and $g = 1$.

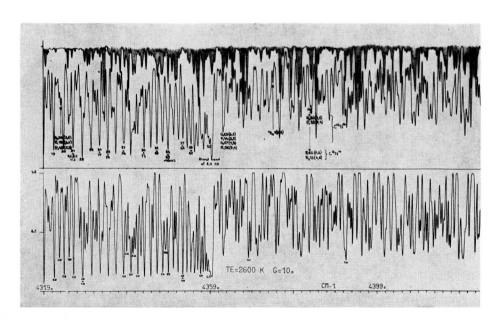

Fig. 13. *Idem quod* Figure 12, except that the synthetic spectrum has been computed with $T_e = 2600$ K and $g = 10$.

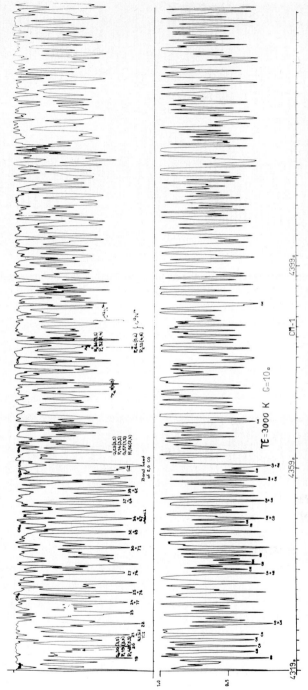

Fig. 14. *Idem quod* Figure 12, except that the synthetic spectrum has been computed with $T_e = 3000$ K and $g = 10$.

useful to show the effects of the line cutting procedure ($l_v < k_v \times 10^{AKCUTL}$) for the weak lines and for the wings of strong lines.

Figure 11 has the following characteristics: $T_e = 2200$ K, $g = 1.0$ and AKCUTL = $= -2, 0, +2$. For this temperature, the cutting procedure has little influence on the strong lines but a strong influence on the weak lines, when AKCUTL is positive. This is a further reason for thinking it necessary to include, in the coolest model atmospheres, the weak lines furnished by polyatomic molecules. It is evident that the appearance of the spectrum varies with the cut adopted.

In the next three figures, we compare the spectrum of the carbon star UU Aur, obtained at a resolution of 0.3 cm^{-1} by Maillard in 1972, with some synthetic spectra. In these figures, we show, at the top, the solar spectrum, the lines of which blend the stellar spectrum, then the stellar spectrum itself, and finally the synthetic spectrum for $T_e = 2600$ K and $g = 1$ (Figure 12), for $T_e = 2600$ K and $g = 10$ (Figure 13) and for $T_e = 3000$ K and $g = 10$ (Figure 14).

From the comparison of the most prominent features of the observed and synthetic spectra, we suggest that UU Aur has $T_e = 2600$ K and $g = 10$. Baumert (1972) obtains a similar value by narrow band photometry. We may refine these values of the effective temperature and the gravity by studying bands of less strong molecular system that could be more sensitive to these parameters, such as the C_2 Ballik-Ramsay system.

4. Conclusion

We note that for the high temperature model atmospheres, we should include the visible and the infrared atomic lines, and for the cooler ones we should include polyatomic molecules and grains. Departures from LTE ought to be examined, as was shown by Querci (1972) and emphasized by Thompson (1973).

Attempts in these directions are being made at Meudon. The identification of the infrared atomic lines on high resolution cool star spectra is being currently pursued (see, for example, Chauville et al., 1970). We have recently identified some titanium lines arising from two electron jumps, the interpretation of which is not going to be easy. The computation of atomic oscillator strengths are made following Decker's formulation (1969). Many more or less sophisticated methods are currently available for the computation of the atomic radial wavefunctions.

As to the molecules, everyone working in this field realises that we lack a great deal of data on wavelengths and oscillator strengths for the molecules of astrophysical interest, especially in the infrared region. In our laboratory, we are studying the C_2 Ballik-Ramsay spectrum by Fourier Transform Spectroscopy with a high resolution interferometer devoted to laboratory spectra. We plan to extend this work to other carbon and oxygen molecules, and similarly to atomic spectra.

Acknowledgements

We would like to thank Dr T. Tsuji for his contribution to the construction of model atmospheres and for many helpful discussions. Mrs F. Gautier from CIRCE (Orsay) is greatly thanked for her help in the drawing of the spectra.

References

Auer, L.: 1967, *Astrophys. J. Letters* **150**, L53.
Baumert, J. H.: 1972, Ph.D. Thesis, The Ohio State University.
Chauville, J., Querci, F., Connes, J., and Connes, P.: 1970, *Astron. Astrophys. Suppl.* **2**, 181.
Decker, E.: 1969, *Astron. Astrophys.* **1**, 72.
Querci, F.: 1972, Thèse d'Etat, Université de Paris.
Querci, F., Querci, M., and Tsuji, T.: 1974, *Astron. Astrophys.*, in press.
Querci, M., Querci, F., and Tsuji, T.: 1972, *Mem. Soc. Roy. Sci. Liège, 6th Ser.* **3**, 179.
Ström, S. E. and Kurucz, R. L.: 1966, *J. Quant. Spectrosc. Radiat. Transfer* **6**, 591.
Thompson, R. I.: 1973, *Astrophys. J.* **181**, 1039.
Tsuji, T.: 1973, *Astron. Astrophys.* **23**, 411.

DISCUSSION

Hyland: I am very interested in the model atmospheres and emergent fluxes that you derive, in particular from the point of view that the temperature scale of carbon stars is just not known, and it is a very important one, and therefore the difference is that your model atmospheres compared with Alexander and Johnsons – it is very important to clear up why there is such a difference. Do you have any ideas on this?

Querci: For me narrow band photometry is good, but I prefer to see a problem with high resolution spectra. We try to pick up some better spectral intervals. I have one computed curve in terms of UBV and so on, and sometimes between two spectral intervals. You can see this in the diagram.

Hyland: A lot of us use further out, like JHKL bands, and the fact that yours and Johnson's differ does not depend on our observations. We would like to know which ones are right, and why.

Querci: But I think this is not a good model for making comparisons. When the temperature decreases all the bands of the visual are very important.

STELLAR SPECTROSCOPY AT 1.1 μ

H. ZIRIN

Big Bear Solar Observatory, Hale Observatories, Carnegie Institution of Washington, California Institute of Technology, Pasadena, Calif., U.S.A.

This is a progress report on the program of stellar spectroscopy at 1.1 μ reported on by Vaughan and Zirin (1968) and by Zirin (1971). About 450 plates of about 200 stars have been obtained, using a magnetically focused RCA image converter with the 144″ and 72″ Palomar coudé cameras, giving dispersion of 8 and 17 Å mm^{-1} respectively. Most of the plates were taken at the latter dispersion and cover about 300 Å. In good seeing a star with $J=3$ is obtained in about 90 min. A tube with a fibre optic backplate is now in use. Cooling such a tube is difficult because the cathode is held at high voltage. By completely insulating the cold box, we may use dry ice cooling, and 2 h exposures, reaching $J=5$, may be made with this ITT tube.

Our principal interest in this region of the spectrum is the He I 10830 line, but there are other features of interest such as Paschen γ and a series of CN bands. λ10830 is of particular interest because it is produced at relatively high temperatures and thus its presence in stars later than B5 can only be explained by the existence of a high temperature corona or chromosphere or of a strong ultraviolet source, which amounts to the same thing. Vaughan and Zirin found the presence of such chromospheres in a number of stars, and I have pursued the matter since then to determine the existence of chromospheric variability. The 10830 absorption in the Sun is very small from the normal chromosphere and is principally due to the presence of active regions. Emission is only found in large flares. Thus cyclic variations in Sun-like stars could be attributed to sunspot cycles. Unfortunately, main sequence stars are near the limit of our observations. Although definite 10830 absorption has been found in such main sequence stars as 61 Cyg A and B, 70 Oph br, K Cet and ε Eri, the absorption is just barely detectable and variation from plate to plate could be explained by variable quality.

On the other hand, definite variation, possibly cyclic, has been observed in giant stars, in particular θ Her and ε Gem. In both these stars the He line has been seen in emission, absorption and absence. In θ Her the following sequence has been observed

May, 1966	str emission
May, 1968	wk emission
Sept., 1969	nothing
Oct., 1970	some abs
June, 1970	mod emission
May, 1972	nothing
February, 1973	str emission

Thus there is irregular variation, and the determination of periodicity if any awaits more accurate measurements. ε Gem was in emission in 1965 but has only shown irregularly varying absorption since then. Whatever periodicities exist in these stars are only a few years long.

Continuing observation has confirmed the original conclusion of Vaughan and Zirin that strong K emission was a necessary but not sufficient condition for the presence of 10830, and that close double stars had a strong tendency to show 10830. Emission has only been found in giant stars and late type variables.

Various CN band heads are found in this region of the spectrum and we have measured the band intensities in all our plates, but there seems no definite advantage to observing these bands instead of visible wavelengths.

We have obtained spectra of a number of late type stars because they are so easy. He emission and Pγ emission is found in R Aqr, R Hya and R And, but not in any other late stars. In R Aqr, I found emission in the La II line at 11012 Å, the furthest emission line I have found in the IR. He emission was also found in the RV Tau variable R Scu, but disappeared later. Some of these spectra appear in the papers cited, and a new comprehensive report will appear soon.

The following stars show moderate to strong 10830 emission or absorption. The list is presented in the hope that someone will look at these stars in the ultraviolet, where they should have strong emission lines:

TABLE I

Emission		Absorption	
R Sct	η Aql	θ Tau	β Dra (str)
ε Crv	α Cet	β Gem	58 Per
θ Her		61 Cyg A and B	λ And
12 Peg (str)		ε Lep	ε Gem
θ Cet		σ Gem (str)	α Tri
ξ Cyg		HD 131977	ζ And (str)
μ Her		β Cam (str)	ε Leo (str)
μ Per		β Cet	31 Mon (str)
R Hya		β Sct	HD 88236
R And		α Aur	70 Oph br
R Aqr		ϱ Cas (str)	

Acknowledgements

This research was supported by NASA under NGR 05-002-034.

References

Vaughan, A. H. and Zirin, H.: 1968, *Astrophys. J.* **152**, 123.
Zirin, H.: 1971, *Phil. Trans. Roy. Soc. London* **A270**, 183.

DISCUSSION

Vardya: Have you measured the radial velocities of this line?

Zirin: Normally the line is very broad. In a few cases one gets this sort of P Cyg type variation with emission with an absorption on the blue side, and that typically is shifted by about 1 Å or so, which would be one part in 10000 or whatever that is, 30 km s^{-1}. But in cases of absorption the absorption is normally well centred on the line.

OPEN DISCUSSION

Chairman of the morning session: A. R. Hyland
Chairman of the afternoon session: R. I. Thompson

Thompson: I have persuaded Dr Townes to say a few words about the infrared work he is doing. He has been mainly concerned with a project involving spatial interferometry in the infrared. However, the technique does have some applications in infrared spectroscopy.

Townes: Let us talk about high spectral resolutions work using heterodyne spectroscopy. Of course that kind of spectroscopy has been used in the radio region a long time and obviously gives immediately almost infinite resolution because one simply needs to mix with a local oscillator the incoming signal and then filter out the resulting signal with various radio frequencies filters which one can make as narrow as he pleases. The problem is not the resolution then but poor signal to noise. A heterodyne detector has basically a noise present due to the uncertainty principle. One can get rid of a noise temperature of hv/k when v is the frequency and k is the Boltzmann constant. This means that at $10\,\mu$ for example, the noise temperature of the detector is necessarily high. It is about 15 000 deg. And so one is looking at a background of 15 000 deg and trying to determine some stellar spectrum against it. Now that is not impossible, but it does mean that about $10\,\mu$ is as high a frequency as heterodyne techniques are likely to be useful. I think, one might go another factor of two or so. Of course if we go down to the radio region then this noise temperature is of the order of one degree or so. I have given you only the theoretical limit. In fact our present detectors miss that by about one order of magnitude. Detectors in the laboratory have done better and miss that theoretical limit by maybe a factor of three only, and certainly one can get such detectors in the field. (This is the theoretical result I say one order of magnitude.) Actually there are two kinds of detector, a photoconductor has a noise temperature twice that, and we miss that by a factor of 10 so we have $20\,hv/k$, and in the lab this has gotten down to maybe about three or four times hv/k and that certainly can be achieved in the field. The result is that when we look at a star we can get a ten to one signal of noise in the band width which we have which is 13 000 mHz and get ten to one signal to noise in something like five to ten minutes for the brighter stars. This means that one can divide that band width up then with filters into practically maybe 10 or 20 units and with reasonable observation times can get a signal. We are using a 30-in. telescope. Obviously a 100-in. telescope would give you 10 times more signal to noise, so that for the brighter stars this a perfectly practical kind of technique. One might ask do you need that resolution because most spectral lines in the $10\,\mu$ region are going to be as broad as $1/20$ of a wave number probably or something like that, so that most work can be done with other kinds of interferometry. On the other hand in cool stellar shells there is a need at least there for higher resolution. We also hope to use this for planetary work for the planetary atmospheric lines where high resolution is needed. And so if one wants a really high resolution this is a way of doing it. On Mars, for example, we get a signal to noise of 10 in about five seconds. Mars because of its broader solid angle is much more favourable than stars and also a good deal cooler, so that in the atmosphere of Mars one can look at CO_2 lines in enormous detail getting 1/1000 of a wave number easily, if that's what resolution is needed, and it is just about needed down to that level for some of these cases. For stars the time required is longer but for the brighter stars one certainly can detect a spectrum. This study has been carried about by two students Betts and Jansen aided to a lesser extent by two other students, Gailhouse and Petersen, and Petersen, Betts and Jansen in particular hope to look at some spectrum of Mars before long. We have been headed in a different direction and have not really produced any spectra yet but they hope to do that before long. Now I must warn you of another technical difficulty. This is a difficult technical field. Now we can get this resolution you look at one time at a band width of maybe 15000 mHz, but that is 1/20 of a wave number. So your total field of view is 1/20 of a wave number, and you can break that up as finely as you wish. Now, you say as in the radio frequency range let us tune our local oscillator around and we move this 1/20 of a wave number around, and we can then get a complete spectrum and look in enormous detail at any reasonable spectral region we wish. The trouble with that is that the local oscillators and the infrared lasers are not all that tunable and flexible. You will have heard of tunable solid state lasers in this region, and they work, but they do not work consistently at the power levels required for the presently available detectors. They do in certain laboratories and for certain limited times work at power levels adequate for certain special detectors, and none of these things are commercially available, but within the course of perhaps two or three years they ought to be commercially available, and sooner than that on a personalized kind of basis. With that one can then tune through the spectrum, tune the local oscillator getting this exceedingly high resolution obviously needed for only very special problems, but they are undoubtedly giving us a great deal of information. You may possibly have seen such spectra taken in the laboratory where the problem is easy. In the laboratory from an absorption cell you can have a tunable local oscillator, a tunable laser, simply a receiver on the other end, and then the resolution, not by heterodyne techniques, but simply because of the high spectral purity of the laser, is just fantastic. And people get down to resolutions of anything you want but normally down to about 1/30000 of a wave number. With nice heavy gases with lots of rotational structures this is sometimes needed.

Schwarzschild: I would certainly like to emphasize how dreadfully important this type of work is for any stellar evolution in the advanced phases. Practically everything interesting happens exactly in the stars that you have been discussing all day. I am a little more squeamish being a theoretician of the use of our theoretical work that some of you are doing, because I think it is less certain than you think, but that is a good state of affairs. Obviously in spite of my very great admiration of the work that you are attempting there are additional points that I hope by and by you will come to.

One of the points, for example, I will suspect will come home to you and to us is the question does one have to worry already in the photosphere when one is building stellar models as some of you have been doing about the formation of grains adding to the opacity or vice versa. Can one in this range, which is not the range in which the grains radiate by themselves, possibly detect the existence of a more or less neutral absorber? I mean that for constructing for using your model atmosphere as boundary conditions for our stellar interiors will become important.

However, I would like to come back once more and emphasize anything you can do to learn about the hydrodynamics of the atmospheres in addition of course to the abundances is of fundamental importance to the interior.

Thompson: I think before this session draws to a close we ought to give a very hearty round of applause for Dr Fujita for working so hard to bring this together (applause).

III. KINEMATICS AND AGES OF STARS NEAR THE SUN

(Edited by L. Perek)

Organizing Committee

L. Perek (Chairman), T. A. Agekjan, O. J. Eggen, Ch. Fehrenbach, W. Gliese, K. F. Ogorodnikov, H. F. Weaver, R. Woolley

PREFACE

L. PEREK

Astronomical Institute, Czechoslovak Academy of Sciences, Prague, Czechoslovakia

An unprecedented number of Commissions manifested their interest in the above topic proposed by Commission 33. They were: 24, 25, 27, 29, 30, 33, 37, and 45. The Organizing Committee, composed of: L. Perek (Chairman), T. A. Agekjan, O. J. Eggen, Ch. Fehrenbach, W. Gliese, K. F. Ogorodnikov, H. F. Weaver, and R. Woolley, had the difficult task to select talks and speakers who would give an adequate picture of the present state of this intricate problem.

Held one year after the IAU Colloquium No. 17 on Stellar Ages, which provided a solid basis in one aspect of the topic, it prominently showed the importance of nearby stars as a sample with the best available knowledge of selection effects and the most precise motions.

The prosaic solar neighbourhood covers a negligibly small part of the Galaxy. It contains, however, information on general questions of structure, dynamics, stability, and evolution of the entire galactic system.

THE AGES OF STARS IN THE NEIGHBOURHOOD OF THE SUN

G. CAYREL DE STROBEL

Laboratoire d'Astrophysique, Observatoire de Paris-Meudon, France

1. Introduction

This introductory paper is divided into two sections.

In the first section a short summary is given of the IAU Colloquium No. 17 on Stellar Ages held in Paris-Meudon Observatory in September 1972. The summary will concern particularly those subjects relevant to this Joint Discussion.

In the second part the ages of nearby stars located in the subgiant region of the $(\log T_{\text{eff}}, M_{\text{bol}})$ diagram are discussed.

2. IAU Colloquium No. 17: Stellar Ages

In planning the Colloquium we divided the stellar ages contributions into four arbitrary groups. Stellar ages as understood:
 (a) from an evolutionary point of view,
 (b) from kinematic criteria,
 (c) from nucleosynthesis aspects,
 (d) from spectrophotometric criteria.
We shall now report briefly on these four points.

2.1. Age from the HR diagram

The main method of determining stellar ages is still to comput a grid of isochrones in the $(\log T_{\text{eff}}, M_{\text{bol}})$ plane from a set of evolutionary tracks. Two isochrone grids have been presented by Hearnshaw (1972; see Figures 2 and 3, paper XLI, IAU Colloq. No. 17). They are the two sets of isochrones interpolated by Sandage and Eggen (1969) from the evolutionary tracks of both, Iben (1967) and Aizenman *et al.* (1969).

Other isochrone sets have been presented by Hejlesen *et al.* (1972; Figure 2, paper XVII, IAU Colloq. No. 17) and Hejlesen (1972; Figures 3 and 4, paper XVIII), and have been calculated by the Danish school.

Despite of all the uncertainties remaining in the physics of stellar structure these sets of isochrones agree with each other within 25% in age, and are suitable for fixing the proper stellar chronology with a possible error in the total age of the Galaxy.

In the practical use of isochrones very important limitations do occur:

(1) The theoretical Zero Age Main Sequence (thereafter ZAMS) is not accurate enough to fit without corrections to the empirical ZAMS.

(2) In some parts of the HR diagram three different isochrones cross at a given

point and therefore there is no unique solution to the age of a star of given effective temperature and luminosity. Fortunately, when that happens one solution corresponds to a very fast stage of evolution (phase of gravitational contraction of an exhausted convective core) and has a nearly zero probability of being observed, and the ages of the isochrones corresponding to the other two solutions do not differ by more than 20 to 30% (see Figure 3, paper XVII, IAU Colloq. No. 17).

(3) In some parts of the HR diagram the isochrones are so close to each other that observational errors result in very large uncertainty in stellar ages determinations. The accuracy of the age determinations are largely dependent on the accuracy of the effective temperatures and the parallaxes. Great uncertainties occur in particular in the vicinity of the main sequence and along the yellow red giant branch, close to the Hayashi limit. In practice one is left, for accurate dating, with subgiant stars and with early-type stars which cannot be old anyway.

(4) Strictly speaking, in the determination of the age of a star its helium content and its metal content should be taken into account, as well as its rotation.

The metal content is usually known from a metallicity index or a spectral analysis. The helium content is usually unknown, but for stars like the subgiants which have not yet undergone the mixing process, recent works favor a rather uniform value of the helium/hydrogene ratio (Demarque, 1972).

About the rotation of a star, it has been shown that rotation does not affect the life time on the main sequence by more than 5% (Maeder, 1972).

2.2. Age and kinematical properties

As stellar ages were not readily available most former works use spectral class groups instead of age groups. That has of course the inconvenience that if early spectral

TABLE I

Mean motion and velocity dispersion of nearby dwarfs

Sp	Mean motion			Dispersion		
	$\langle U \rangle$	$\langle V \rangle$	$\langle W \rangle$	σ_U	σ_V	σ_W
B0	+10	−15	−7	10	9	6
A0	+ 7	−14	−7	15	9	9
F0	+ 9	−10	−7	3	3	13
G0	+15	−21	−6	26	18	20
K0	+11	−15	−7	28	16	11
M0	+ 6	−15	−7	32	21	19

types are really a selection of young stars, more advanced types are older in average but contain a mixture of old and young objects. It was found very early that the dispersion of the velocities increases steadily from the early types to the late types. This can be seen in Table I (Delhaye, 1965), in which the first column contains six spectral groups of nearby dwarfs, the next three columns the mean space motions for each spectral group, and the last three columns their velocity dispersions. Table I shows that the young early-type stars have a much smaller velocity dispersion than the more advanced spectral types which are a mixture of stars of different ages.

It was also found that stars of intermediate age, of the order of one galactic year (~ 200 m. y.) have a very asymmetrical velocity distribution with respect to the galactic coordinates. It is now believed that this phenomenon, known as the asymmetric drift, has something to do with the spiral structure of the Galaxy and we shall hear more on this subject later on in this Joint Discussion.

The lion's share in the study of 'kinematical properties vs age' comes from Sir Richard Woolley and his Herstmonceux school (Woolley, 1970).

If one defines a pseudo-eccentricity for the galactic orbit of each star in the solar neighbourhood, there is a strong correlation between each class of eccentricity and the age of the group. Figure 1 shows this correlation: three M_v vs $B-V$ diagrams are reproduced from a paper by Woolley (1971). The first diagram shows the position

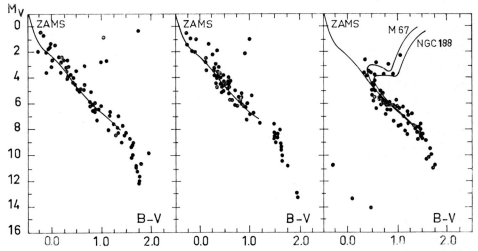

Fig. 1. Empirical HR diagrams after Woolley (1972) for three samples of nearby stars having different eccentricities ($e \leq 0.05$, $0.075 \leq e \leq 0.100$, $0.150 \leq e \leq 0.200$) and the same box angle ($i \leq 0.05$).

of nearby stars having an eccentricity of their galactic orbit smaller than 0.05: $e \leq 0.05$ and $i < 0.05$, i being the box angle of the orbit of a star, defined as the maximum elongation from the galactic plane, with the center of the Galaxy as origin. The second and the third diagrams show the position of nearby stars having an eccentricity within: $0.075 \leq e \leq 0.100$ and $0.150 \leq e \leq 0.200$ respectively and with the same condition on i. On the two first diagrams the upper main sequence is well populated, whereas on the third one there is a turn-off point corresponding to an age of the old galactic clusters: M 67 and NGC 188. Woolley also found that the perigalactic distance is well correlated with the ages of the stars.

2.3. AGE FROM NUCLEOSYNTHESIS ASPECTS

Reeves (1972) discussed the chronology of the chemical elements resulting from nuclear interactions taking place at various times and in various locations of the Galaxy.

Furthermore Reeves spoke about the relative abundance of long-lived radioisotopes. The abundance of such isotopes, observed directly through their fossil remnants in the Earth and in the meteorites can give us some information on various events and their epoch of occurence during the lifetime of the Galaxy.

Lithium as a stellar age indicator has been discussed by Vauclair (1972). The lithium abundance is correlated with age in F and G dwarfs.

2.4. AGE FROM SPECTROPHOTOMETRIC CRITERIA

2.4.1. *Age from Emission Lines*

Wilson and Skumanich (1964) and Skumanich (1965, 1972) have derived an empirical relationship between the intensity of the emission in H and K lines in dwarf stars and the age of the star. They have found that:

$$\tfrac{1}{2}(F_\text{H} + F_\text{K}) \sim (\text{age})^{-1/2}.$$

Kuhi (1972) discussed this relation which is not expected to be valid without an important dispersion but is nevertheless useful for F, G, and K dwarfs for which no evolutionary age is available.

It is interesting to note that as early as 1953 Delhaye (1953) found that M dwarf stars with strong emission lines have a velocity distribution resembling the velocity distribution of relatively young stars and not the velocity distribution of ordinary non-emission late dwarfs.

2.4.2. *Age from Chemical Composition*

The systematic enhancement of heavy elements with time in the Galaxy was supposed fifteen years ago to be a main clue to the time of birth of a star.

Recent works of Powell (1972) and Clegg and Bell (1973) suggest that a slight enrichment does occur, but is unfortunately of the order of magnitude of the scatter at any time in the history of the Galaxy. In particular Janes and McClure (1972) find that CN stars, which are believed to be young stars, include some significant old population.

The only safe statement is that strongly metal deficient stars, by more than an order of half a magnitude in respect to normal solar composition, are all belonging to the halo population and are therefore as old as the Galaxy. The reciprocal is not true: one can find some old stars having normal chemical composition.

3. Ages of Nearby Subgiants

I shall present now some results that I have just obtained concerning the nearby subgiants contained in the catalogues of Gliese (1969) and Woolley (1970).

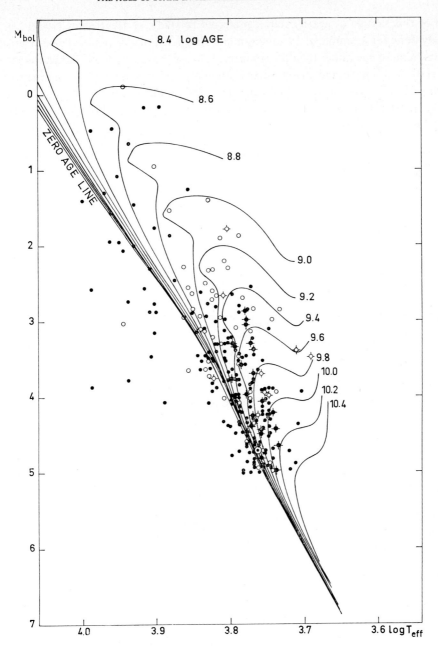

Fig. 2. Grid of isochrones in the HR diagram derived for $(X, Z) = 0.7, 0.03$. The full circles in the diagram are the loci occupied by stars classified in Woolley's Catalogue as dwarfs, the open circles are stars classified as subgiants; full circles with crosses and open circles with crosses are stars for which a detailed analysis is available in the literature.

For this sample of nearby stars I preferred to study, as Hearnshaw (1972) has already done for a sample of 19 southern late subgiants, the kinematical characteristics in function of age rather than the contrary, because the kinematical characteristics are of statistical nature and therefore any attempt to select stars with given kinematical characteristics necessarily includes the tail distribution of other kinematical groups.

To do so, it is necessary to group the nearby stars by ages and to plot their velocity distribution by age groups.

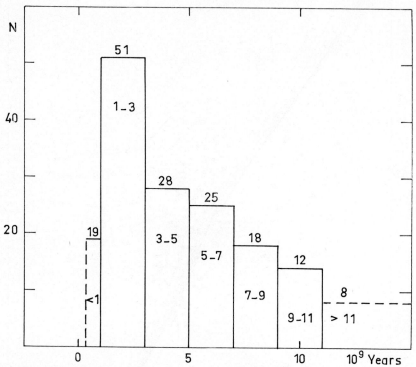

Fig. 3. Histogram of the age of 161 nearby subgiants falling in the ($\log T_{\rm eff}$, $M_{\rm bol}$) diagram region within the limits $\log T_{\rm eff} = 3.68$ and 4.0 and $M_{\rm bol}$ within 0 and 5. The numbers above the top of each lane are the number of stars with an age in billions of years in the range given below the top.

The difficulty is that out of 1566 stars in Woolley's catalogue with trigonometric parallaxes larger than 0".040, only 10% show enough evolution to be dated according theoretical evolutionary tracks. The 90% of the remaining stars are so close to the main sequence that only an upper limit to their age can be found.*

In Figure 2 nearby dwarfs and subgiants are plotted on the isochrone set computed by Hejlesen et al. (1972). These isochrones are the most complete set computed up to date. The isochrones have been derived for $(X, Z) = 0.7, 0.03$. Hejlesen et al. attempted to make the models of internal structure as realistic as possible within the limitations

* Dr Gliese advised me during the Joint Discussion to use only absolute magnitudes derived from trigonometric parallaxes; therefore I worked chiefly with Woolley's catalogue.

imposed by present uncertainties in the theory of convective flux and available opacity tables.

On the isochrones of Figure 2 only stars with effective temperatures between 4000 K and 10000 K and bolometric magnitudes between 0 and 5 have been plotted. The inclusion of stars cooler or fainter does not allow to date any other object with some accuracy in view of the closeness of the isochrones in such a region.

The full circles in the diagram are stars classified as dwarfs, the open circles stars classified as subgiants; full circles with crosses and open circles with crosses are stars for which a detailed analysis is available in the literature. The Sun is represented as an open dotted circle.

To get the bolometric magnitude of the stars we have used Johnson (1966) bolometric correction, corrected by 0.07 to have the standard definition of M_{bol}. The effective temperature of the stars has been mainly derived from $(R-I)$ colour indices and from high dispersion analysis.

Figure 2 shows that stars classified as dwarfs are already evolved and that stars classified as subgiants are not yet evolved.

In total 161 evolved stars have been dated. Figure 3 shows a histogram of the age of these stars. This histogram is not representative of an unbiased sample of the nearby stars because of the selection procedure. Nevertheless it suggests that the rate of star formation one or two billion years ago was very substantial, compared to the average rate of star formation during the lifetime of the Galaxy. This conclusion can be drawn because the lifetime in the subgiant stage of the stars falling in the one to three billion lane is shorter than the lifetime of older subgiants.

We grouped the stars by age and we plotted the groups on a Böttlinger diagram.

Figure 4 shows this diagram. Filled dots stand for stars younger than 3×10^9 yr, open circles for stars with an age between 3×10^9 and 7×10^9 yr, and crosses for stars older than 7×10^9 yr. The 'kidney-like-lines' are iso-eccentricity lines in the U' and V' plane, in which the direction of U is taken opposite to the galactic center, and V in the direction of the galactic rotation.

The Böttlinger diagram of nearby subgiants shows that the youngest stars have

TABLE II

Mean motion and velocity dispersion of nearby subgiants

N	Age	Mean motion			Dispersion			e
		$\langle U \rangle$	$\langle V \rangle$	$\langle W \rangle$	σ_U	σ_V	σ_W	
19	$<10^9$	+9.1	−7.2	−8.1	23.0	10.1	9.7	0.09
51	$1-3 \times 10^9$	+12.5	−16.4	−10.3	30.3	16.7	12.9	0.12
28	$3-5 \times 10^9$	+9.4	−12.8	−13.1	30.6	24.2	22.6	0.15
25	$5-7 \times 10^9$	+17.2	−22.0	−3.8	39.5	24.6	18.1	0.18
18	$7-9 \times 10^9$	+20.4	−36.3	−7.7	54.6	20.9	28.9	0.20
12	$9-11 \times 10^9$	+15.2	−22.3	−4.7	33.9	22.4	19.0	0.14
8	$>11.10^9$	+27.1	−31.9	+9.5	61.1	32.4	31.5	0.23

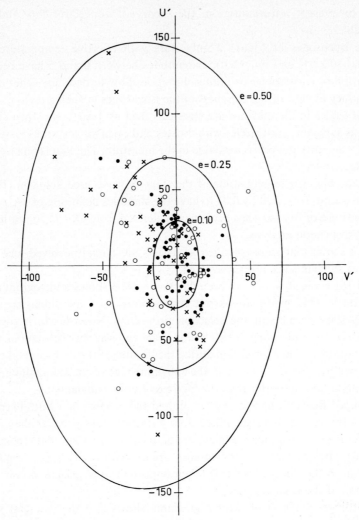

Fig. 4. Böttlinger diagram of the sample stars. Filled dots stand for stars younger than 3×10^9 yr, open circles for stars with an age between 3×10^9 and 7×10^9 yr, crosses for stars older than 7×10^9 yr.

the more circular orbits and they show a significant asymmetric drift; stars between three and seven billion years show a larger dispersion and stars older than 7 b.y. have a significant drag with respect to the galactic rotation. On the diagram we can see that old stars do not avoid small eccentricity orbits.

Table II shows the mean space motions and the velocity dispersions for seven groups of ages.

The velocity dispersions increase from younger to older groups. The mean space motion in the direction of galactic rotation $\langle V' \rangle$ is significantly higher than for the average mean motion of G and K dwarfs in Table I. Here the drag in the galactic rotation is visible.

4. Conclusion

In the first section of this paper we learned something about the actual tools which permit a star to be dated.

In the second section we applied one of the dating techniques to a sample of nearby stars. Some of the results of this exercise are:

(a) The delimitation between class V and class IV luminosities from spectral classification is poorly representative of actual luminosities.

(b) The peak we found on the histogram of Figure 3 suggesting a substantial rate of star formation two billion years ago is in agreement with a statement of Woolley (1970, p. 139). Nevertheless Clegg and Bell (1973) who have recently studied the abundance and age distribution of 500 F stars in the solar neighbourhood have not found this peak. In view of difficulties involved by selection effects in our sample we are not going to discuss this point further.

(c) The fact that the velocity dispersion increases with age (Table II) is of course not a new result. But it is satisfactory to find that this result can be obtained directly starting from stars which have been grouped according to their physical age.

References

Aizenman, M. L., Demarque, P., and Miller, R. H.: 1969, *Astrophys. J.* **158**, 669.
Clegg, R. E. S. and Bell, R. A.: 1973, *Monthly Notices Roy. Astron. Soc.* **163**, 13.
Delhaye, J.: 1953, *Compt. Rend. Acad. Sci. Paris* **237**, 294.
Delhaye, J.: 1965, *Stars and Stellar Systems* **5**, 61.
Demarque, P.: 1972, in G. Cayrel de Strobel and A. M. Delplace (eds.), 'L'âge des étoiles', *IAU Colloq.* **17**, Paper XXI.
Gliese, W.: 1969, Catalogue of Nearby Stars, *Veröffentl. Astron. Rechen-Inst. Heidelberg* No. 22.
Hearnshaw, J. B.: 1972, in G. Cayrel de Strobel and A. M. Delplace (eds.), 'L'âge des étoiles', *IAU Colloq.* **17**, Paper XLI.
Hejlesen, P. M.: 1972, in G. Cayrel de Strobel and A. M. Delplace (eds.), 'L'âge des étoiles', *IAU Colloq.* **17**, Paper XVIII.
Hejlesen, P. M., Jørgensen, H. E., Otzen Petersen, J., and Rømcke, L.: 1972, in G. Cayrel de Strobel and A. M. Delplace (eds.), 'L'âge des étoiles', *IAU Colloq.* **17**, Paper XVII.
Iben, I.: 1967, *Astrophys. J.* **147**, 624.
Iben, I.: 1967, *Ann. Rev. Astron. Astrophys.* **5**, 571.
Janes, K. A. and McClure, R. D.: 1972, in G. Cayrel de Strobel and A. M. Delplace (eds.), 'L'âge des étoiles', *IAU Colloq.* **17**, Paper XXVIII.
Johnson, H. C.: 1966, *Ann. Rev. Astron. Astrophys.* **4**, 193.
Kuhi, L. V.: 1972, in G. Cayrel de Strobel and A. M. Delplace (eds.), 'L'âge des étoiles', *IAU Colloq.* **17**, Paper XLIII.
Maeder, A.: 1972, in G. Cayrel de Strobel and A. M. Delplace (eds.), 'L'âge des étoiles', *IAU Colloq.* **17**, Paper VII.
Powell, A. L. T.: 1972, *Monthly Notices Roy. Astron. Soc.* **155**, 483.
Reeves, H.: 1972, in G. Cayrel de Strobel and A. M. Delplace (eds.), 'L'âge des étoiles', *IAU Colloq.* **17**, Paper XXXII.
Sandage, A. and Eggen, O. J.: 1969, *Astrophys. J.* **158**, 685.
Skumanich, A.: 1965, *Astron. J.* **70**, 692.
Skumanich, A.: 1972, *Astrophys. J.* **171**, 565.
Vauclair, S.: 1972, in G. Cayrel de Strobel and A. M. Delplace (eds.), 'L'âge des étoiles', *IAU Colloq.* **17**, Paper XXXVIII.
Wilson, O. C. and Skumanich, A.: 1964, *Astrophys. J.* **140**, 1401.

Woolley, R.: 1970, in H.-Y. Chiu and A. Muriel (eds.), *Galactic Astronomy*, Gordon and Breach, N.York, Vol. 2, p. 95.
Woolley, R., Epps, E. A., Penston, M. J., and Pocock, S. B.: 1970, *Roy. Obs. Ann.* No. 5.
Woolley, R., Pocock, S. B., Epps, E. A., and Flinn, R.: 1971, *Roy. Obs. Bull.* No. 166.

DISCUSSION

Gliese: I would warn against using the resulting parallaxes instead of the trigonometric data in the 'Catalogue of Nearby Stars'. Resulting parallaxes include spectroscopic and/or photometric parallax data. The latter have been derived by setting the stars exactly on a mean sequence.

Further, the collection of parallax data in this catalogue, naturally, has preferred objects with positive errors in trigonometric parallax measurements. Therefore, on an average, the mean distance of the star groups will be too small. In the F star region this effect causes a systematic error in the mean absolute magnitudes of about 0.15, even if we restrict ourselves to parallaxes with small accidental errors (p.e. smaller than 10%).

Cayrel de Strobel: I used trigonometric parallaxes of subgiants wherever they were available and resulting parallaxes in all other cases.

Gliese: If you would have taken only stars with trigonometric parallaxes your sample of subgiants would have been smaller.

Cayrel de Strobel: Yes, therefore I took also the resulting parallaxes but that did not matter because the probable error of the resulting absolute visual magnitudes does not exceed 0.2. And with this error in magnitude the conclusion on the ages of the F and G subgiants can still be true.

Iwanowska: To summarize what we know about the lifetimes of stars located in different parts of the HR diagram from the evolutionary isochrones we show Figure 5.

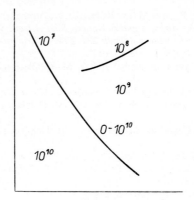

Fig. 5.

Further, from our investigations on stellar motions in connection with stellar populations we are led to believe that the velocity dispersion as well as eccentricities and inclinations of stellar galactic orbits depend not only on the lifetimes but also on the masses of stars. The arguments to this we see in the fact that e.g. kinematical parameters for very old white dwarfs are not greater in the mean than those of younger late-type dwarfs. Also, young dMe stars show these characteristics larger than young stars of large mass. I do not claim that our Galaxy is in a state of equipartition of energy but it seems to have made some steps in this direction.

Cayrel de Strobel: I certainly agree with you if you want to compare ages and velocity distributions of MeV and AV stars. I have not considered mass groups for the results presented in my table. Anyway the mass spectrum of F and G subgiants is very limited.

Cayrel: I just wish to point out a recent work now submitted to *The Astrophysical Journal* by Ann Merchant Boesgaard and Wendy Hagen on the age of our neighbour α Cen A. From high dispersion spectra taken at Mauna Kea with 1.7 and 3.4 Å mm^{-1}, the authors have derived an age for α Cen A using the three criteria: Li 6707 line strength, rotation of the star, and the H and K emission and the decay curves published by Skumanich in 1972. Li strength gives 3.4×10^9 yr, rotation 3.6×10^9 yr, and H and K emission gives only a lower limit of 2×10^9 yr. The authors conclude that the star is 3.5×10^9 yr old, i.e., 10^9 yr younger than the Sun. The fact that Proxima Centauri, a member of a triple system, is a flare star, is interesting because the age of flare stars is generally assumed to be 10^9 yr at the most.

Buscombe: As a practising spectroscopist, what criteria do you think should be used for classifying late-type subgiants, which might yield absolute magnitudes more closely fitting the isochrones?

Cayrel de Strobel: (1) Trigonometric parallaxes up to 0″035, (2) High dispersion detailed analyses, knowing in advance the temperature of the star from a colour index, e.g. $(R-I)$, (3) Some good line ratios.

GOULD'S BELT

P. O. LINDBLAD

Stockholm Observatory, Sweden

The zone of bright stars inclined 20° to the galactic plane, pointed out by John Herschel (1847) and by Gould (1874, 1879), is generally referred to as Gould's Belt. This General Assembly is well timed. With the splendour of the centre of the Milky Way in the zenith in the early hours of the night anyone who comes out of the city-lights of Sydney can try to convince himself of the reality of this phenomenon.

Shapley and Cannon (1922, 1924) showed that the concentration towards Gould's Belt is very pronounced for the brighter B stars, while the fainter ones are highly concentrated towards the galactic plane. Their results indicated the radius of this local system to be of the order of 500 pc. The motions of the B stars brighter than a magnitude of about $5^{m}\!.5$ are characterized by the K-effect, which thus seems to be a property of this local Gould Belt system.

Numerous investigations of the distribution and motion of the bright B stars have been published. A very thorough analysis has been carried out by Mrs Lesh (1972b),

TABLE I

Velocity gradients as observed for nearby early type stars together with theoretical values for an expanding group and for pure differential rotation. Unit 1 km s^{-1} kpc^{-1}

Gradient	Observed	Exp. model age 60×10^6 yr	Diff. rot.
$\dfrac{\delta U}{\delta X}$	$+23 \pm 2$	$+26$	0
$\dfrac{\delta U}{\delta Y}$	-16	-8	-25
$\dfrac{\delta V}{\delta X}$	$+18$	-8	-5
$\dfrac{\delta V}{\delta Y}$	$+10$	$+7$	0

who refined a method due to Blaauw and Bonneau. Table I shows the observed velocity gradient matrix as derived by Mrs Lesh (where X and U respectively represent position and velocity in the direction of the galactic anticentre, and Y and V corresponding quantities in the direction of galactic rotation). Assuming an expanding system in the gravitational field of the Galaxy, values for these gradients can be predicted for various expansion ages. The table gives these values for an expansion age of 60×10^6 yr, which seems to be close to a best fit. The table also gives expected gradients for pure differential rotation without expansion. $\delta U/\delta X$ and $\delta V/\delta Y$ agree well with the expanding model, while $\delta U/\delta Y$ is somewhat undecided and $\delta V/\delta X$ completely off for both models.

At the Stockholm Observatory we have been investigating the velocity-distribution of the very local neutral hydrogen as well as formaldehyde velocities of dark clouds along Gould's Belt (Lindblad *et al.*, 1973). Dr Weaver will discuss the local gas this afternoon although he will give a slightly different interpretation. In Figure 1 the filled circles indicate velocities for a narrow H I component with very wide extent in galactic latitude. The open circles represent another, broader and more irregular, component with less extent in latitude, in some regions identical with what is called the Orion arm. These two components together constitute the local gas. Crosses represent formaldehyde observations, where crosses in boxes indicate clouds definitely associated with the inclined Gould's Belt.

It now proves that the velocity relation for the filled circles can be reproduced by the model of a slowly expanding cloud – or rather a doughnut – following the equations of motion given by Blaauw. The theoretical relation is shown by the full-drawn curve in Figure 1. The expansion age of the cloud as given by the best fit is 60×10^6 yr and thus happens to coincide with the recent value given by Mrs Lesh. Figure 2 shows the position of the model dough-nut and the projection on the galactic plane of Mrs

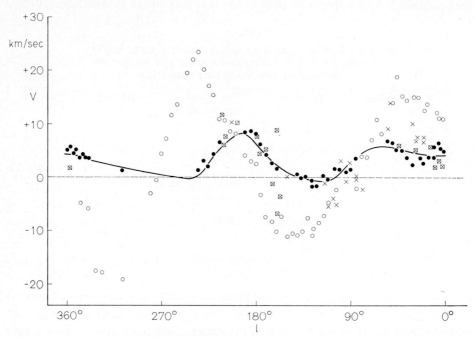

Fig. 1. Velocity-longitude diagram for the two components of local neutral hydrogen gas (filled and open circles). Crosses mark formaldehyde velocities, crosses within squares representing clouds falling in the Gould Belt region when separated from the galactic plane. The full-drawn line shows the theoretical relation for the expanding doughnut shaped cloud represented by the ellipse in Figure 2. (Figures 1 and 2 have been reproduced, with the kind permission of Springer Verlag, from: *Proceedings of the First European Astronomical Meeting*, Vol. 2, 'Stars and the Milky Way System', ed. L. N. Mavridis, Springer, Berlin-Heidelberg-New York, 1974.)

Lesh's stars. The centre of the cloud falls in the neighbourhood of the Pleiades, and the original velocity of expansion is 3.6 km s^{-1}.

The similarity of kinematic behaviour between the neutral hydrogen, the dark clouds along Gould's Belt and the Gould Belt stars, would indicate that this local hydrogen component is related to the expanding Gould Belt system, and that this system was very much more concentrated 60×10^6 yr ago. One may then speculate that the original compression occured in a spiral shock. The expansion age and a knowledge of the present position of the shock then could give a valuable check on the spiral pattern velocity.

However, once we introduce the density wave theory we are led to examine carefully if this gas cloud and local system really exists as such, as expansion is a general phenomenon in inter-shock regions. The effect to consider is velocity-crowding and can be exemplified by an investigation by Roberts (1972). In Figure 3 of Roberts' paper we see how in a spiral shockwave pattern the gas density and velocity may vary along the line of sight at a certain longitude. This gives rise to a spurious component

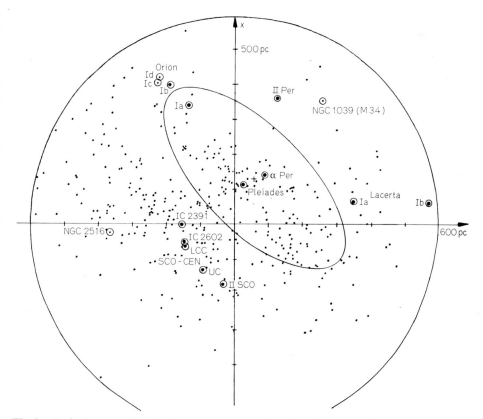

Fig. 2. Projections on the galactic plane of stars earlier than B5 with well determined distances smaller than 600 pc according to Lesh (1968, 1972a). The centre of the coordinate system marks the position of the Sun and the positive X-axis points towards the galactic anticentre. The ellipse with its centre marked by a cross represents the ring model for the neutral hydrogen described in the text.

close to zero velocity in the resulting line-profile, due to the tenuous inter-arm gas between us and the shock. Could the hydrogen component dealt with here be due to velocity crowding in an otherwise featureless inter-arm gas? This possibility is at present being investigated in Stockholm applying Roberts' non-linear shock wave theory. Burton, who presently is applying the linear density wave theory on this problem, reports positive results (see the discussion below).

It is then of great importance to establish whether the stars form a separate local system or not. The observations we would like to suggest may be illustrated by a figure due to Dixon (1970). Dixon plotted radial velocities vs distance for stars of two age-groups in the direction towards the galactic centre and anticentre in order to determine $\delta U/\delta X$. However, when examining Dixon's figure it is interesting to note that the velocity gradients would be 26 km s^{-1} kpc^{-1} for an expanding group of age 60×10^6 yr and would be 0 for pure differential rotation. Dixon's diagram may suggest a high velocity gradient in the solar neighbourhood which shifts back to zero at larger distances indicating a local expanding system. First order density wave theory, on the other hand, would give a sine-wave with minima at the density maxima of the spiral arms.

If we could establish, at longitudes where the gradients of different models differ greatly, a clear discontinuity in the radial velocities as a function of distance from the Sun, we could determine the extent of the expanding Gould Belt system in different directions and perhaps determine an upper limit for the age of its stellar population.

References

Dixon, M. E.: 1970, *Monthly Notices Roy. Astron. Soc.* **151**, 87.
Gould, B. A.: 1874, *Proc. Am. Assoc. Adv. Sci.* **1874**, 115.
Gould, B. A.: 1879, *Uranometria Argentina. Result. Obs. Nac. Argentina Cordoba* **1**, 355.
Herschel, J. F. W.: 1847, Results of Astronomical Observations Made During the Years 1834, 5, 6, 7, 8, at the Cape of Good Hope, Smith, Elder and Co., London, p. 385.
Hobbs, L. M.: 1971, *Astrophys. J.* **166**, 333.
Lesh, J. R.: 1968, *Astrophys. J. Suppl.* **17**, 371.
Lesh, J. R.: 1972a, *Astron. Astrophys. Suppl.* **5**, 129.
Lesh, J. R.: 1972b, in G. Cayrel de Strobel and A. M. Delplace (eds.), 'L'âge des étoiles', *IAU Colloq.* **17**, Paper XXIII.
Lindblad, P. O., Grape, K., Sandqvist, Aa., and Schober, J.: 1973, *Astron. Astrophys.* **24**, 309.
Roberts, W. W.: 1972, *Astrophys. J.* **173**, 259.
Shapley, H. and Cannon, A. J.: 1922, *Harvard Circ.* 239.
Shapley, H. and Cannon, A. J.: 1924, *Harvard Repr.* 6.

DISCUSSION

Burton: There is evidence on a large scale that the motions of the neutral hydrogen in the Galaxy are ordered according to the general predictions of the linear density-wave theory. It is reasonable to examine the effects of those motions on a small scale (say, less than 0.5 kpc) near the Sun. T. M. Bania and I at NRAO are doing this; our results show that most of the observational characteristics of Lindblad's feature A can be accounted for by the sort of flow pattern observed on a large scale also being present near the Sun. This explanation of course does not require special local mechanisms. Specifically, we found the flow parameters required by the hydrogen-line profile cut-offs which are contributed locally (positive cut-off at $70° < l < 170°$, negative velocity cut-off at $190° < l < 290°$).

These cut-offs contain unambiguous velocity information and are easier to isolate than features deeper in the profiles. The derived flow parameters are consistent with those found to hold on a large scale. When model hydrogen profiles are calculated using these parameters, a feature results which shows approximately the same velocity-longitude relationship as found by Lindblad for feature A. Our preliminary results suggest that some of the kinematic characteristics (projected to $b = 0°$) of the Gould Belt stars can also be accounted for by this simple model.

Lindblad: This is the investigation I was referring to. I think the results are very interesting.

Toomre: Does either Dr Lindblad or Dr Burton have any theoretical idea why that Belt should be tilted?

Lindblad and Burton: We have no idea.

Buscombe: The high-resolution optical interstellar gas velocities, e.g. by Hobbs, should indicate which components are caused closer to the Sun than the stars. How do they fit with the H I observations you have mentioned?

Lindblad: One interesting observation which has to be explained is that Hobbs' (1971) interstellar lines in the spectra of the Pleiades stars already show the positive velocities which our model would ascribe to gas further away.

RED VARIABLES AND THEIR MAIN SEQUENCE PROGENITORS

O. J. EGGEN

Mount Stromlo Observatory, Canberra, Australia

The papers upon which this paper was based are all currently published or in the press of the *Astrophysical Journal* and the *Publications of the Astronomical Society of the Pacific*.

DISCUSSION

Rountree Lesh: Has the calibration of M_v vs β and $[u-b]$ been published? I have been working with the $M_v - \beta$ calibration, and I agree that another term is needed to account for evolution away from the main sequence.

Eggen: The paper has been submitted to *The Astrophysical Journal*.

LOW LUMINOSITY STARS AND WHITE DWARFS

W. J. LUYTEN

University of Minnesota, Minneapolis, Minn., U.S.A.

(1) When discussing properties of stars of low luminosity we should first define what we mean by that term. My own feeling is that a bolometric luminosity of 0.001 of that of the Sun appears to be a reasonable upper limit for 'low luminosity'. Further, it is obvious that, in order to find any numbers of them, one must go down to the very faintest objects, near the limit of what our present telescopes can show. This, in turn, means that virtually no accurate magnitudes, colors, or parallaxes will be available, and that our observational data must be obtained almost exclusively from proper motion surveys for very faint stars. Thus we end up by saying that only the 48-in. Palomar Schmidt telescope can produce large numbers of them. Such a statement is, of course, an oversimplification, but how much of one can be judged from the fact that when three years ago I published a catalogue of 1055 such stars only 36 had come from other sources, and 1019, or $96\frac{1}{2}\%$ from the Palomar Schmidt plates. We have now processed more than 400 pairs of plates with our automated-computerized blink machine, and we have at least another 1000 new, low-luminosity stars, and the Palomar Schmidt contribution reaches at least 98%.

In getting back to the data, accepting $L=0.001\odot$ or $M_{bol}=12.2$ as our upper limit, and realizing that most of these stars will probably be very red, it means we have to look for stars with M_{pg} larger than $+16.5$. Since no parallaxes are known, we must finally identify these objects statistically from their proper motions. If we assume that, on the average, they have tangential velocities of 75 km s^{-1}, this means that the quantity $H=m+5+5\log\mu=M+5\log T$ must be larger than 22.5 pg. At every step of this derivation large uncertainties enter – still, I believe this is a workable way of finding low-luminosity stars. I have also tried to extend this same line of reasoning to stars less red, but there one almost certainly runs into white dwarfs and degenerates, and objects will be included in the list because of extraordinary large velocities up to almost ten times what I have assumed here, which, in turn, would mean stars with luminosities one hundred times greater than the limit assumed.

In order to reach a value of H larger than 22.5 pg a star must have the minimum proper motion as shown below, for various apparent photographic magnitudes. It is not surprising that no red star brighter than the fourteenth photographic magni-

m_{pg}	μ_{min}
14	5
16	2
18	0.8
20	0.32
21.5	0.16

tude can make it – Barnard's star, Proxima Centauri and the components of L 726-8 do not make the grade.

When we turn to the statistical discussion of the properties of these stars we must remember that virtually *all* our information comes from proper motion surveys, hence we can expect that, selectively, our stars will be overwhelmingly high-velocity stars. If there are stars fainter than, say, $m=18$ pg (apparent), with small enough tangential velocities, and therefore small proper motions to belong to our group of low-luminosity stars, we have no means of discovering them at present.

The frequency in space of these low-luminosity stars is important because they provide the bulk of the material near, and below the maximum of the luminosity function. Ever since Kapteyn made his first determination of the Luminosity Function in 1918 and found the maximum to lie at $M=+7.7$ (vis) each subsequent determination pushed the maximum fainter. When I analyzed the Bruce data in 1937, I derived $M=+14.5$ pg but thirty years later when the first results from the Palomar Survey became available, I went down to $M=+15.7$ pg. A few months ago La Bonte and I completed the Blink-machine analysis of a region of some 3000 sq deg near the South Galactic Pole, down to apparent magnitude 21 pg, for motions larger than $0''.18$ annually. The first preliminary analysis indicates that the maximum lies at $M=+15.4$ – so for the first time we are receding a little, and maybe we really have reached the maximum this time.

But all of this is still preliminary, for we have only 7000 stars for our analysis, so of course, we cannot be as certain as all these recent critiques of my luminosity function and determinations of star density at high galactic latitudes by Gliese, who had some 80 stars, Jones who had a hundred or so, or Murray and Sanduleak who had all of 21 stars 7 of which had no proper motion. The latter found that my star density near the North Galactic Pole must be multiplied by a factor of five, but a more recent article by Jones dealing similarly with red stars near the South Galactic Pole, comes to the conclusion that Murray's numbers must be divided by five.

Recently it has become fashionable to populate the region of the North and South Galactic Poles with very large numbers of red dwarfs, most of which have very small tangential velocities, and therefore also small proper motions. Now in 1960 I made counts on three-image color plates taken with the Palomar 48-in. Schmidt of a region near the South Galactic Pole and published the results for 4000 stars down to $m=$ $=19$ pg. I found that there were far *fewer* red stars than expected. Now it is quite possible that I was, and am wrong in that conclusion. I applied all the tests I could think of to check my colors, galaxies, faint asteroids, white dwarfs, etc. – but even so, there is always the possibility that my data are subject to some serious but unknown systematic error. However, no one has ever bothered to analyze or discuss my data. My conclusion, you see, was then, and is now unpopular and so most people just ignore it – and this is considered modern science.

We have now finished a similar survey of proper motion stars near the North Galactic Pole, and in a year or so I expect to have the discussion available on some 10000 proper motion stars brighter than apparent magnitude 21 pg and with motions larger than 0″.18 annually in an area of 4400 sq deg surrounding the North Galactic Pole. Again, these results will be preliminary for when we have finished the entire Palomar Proper Motion Survey we expect to have available similar data – though without colors – for half a million stars brighter than 21 pg and with motions larger than 0″.09 annually.

(2) Since I am to talk also on the space distribution and the kinematics of white dwarfs – but not on their ages, as that is outside my province – I can appropriately discuss them here, for the degenerate stars also belong to the group of low-luminosity stars. Before we start on any discussion of their properties we should first investigate how we find them, i.e. what selection effects are involved. There are three main techniques for finding white dwarfs; (1) from Proper Motions, (2) from Faint Blue Star Surveys, (3) from objective Prism Spectral Surveys. By far the most efficient is the first, and in the course of the Bruce and Palomar Proper Motion Surveys I have by now found and published more than 4500 probable, and possible white dwarfs. When an apparently faint star with a sizeable proper motion proves to be white, or perhaps we should say, much less red than expected, for its value of $H = m + 5 + 5 \log \mu$ it is a pretty safe conclusion that it is a degenerate object. In identifying faint blue stars one does not have that certainty. I have published more than 20 000 faint blue stars but I think I can say fairly definitely that most of them are *not* white dwarfs. The luminosities among them range all the way from $+10$ – a reasonably bright white dwarf – through $+3$ or $+4$, the typical halo Population II 'intermediate' through 0 or $+1$, a typical horizontal branch star to perhaps a few blue main-sequence stars at -2 or so, to eventually even quasars at -24 or so. Extremely accurate three-color photometry might help, but is rarely conclusive; proper motions are conclusive only for stars brighter than 16-17 because a very blue white dwarf fainter than this would have only an immeasurably small proper motion, and spectra for such faint stars are just non-existent. Spectroscopic surveys could be very useful but obviously these are slow, and restricted to bright stars – I doubt whether even a dozen white dwarfs have been found this way. And here, one must point to the fact that mere spectroscopic evidence alone is obviously not reliable. At the White Dwarf Conference in St. Andrews much was made of a new very bright white dwarf in the southern hemisphere, with, apparently, little or no proper motion. If this were really a white dwarf, it should have a parallax of nearly 1″ but nothing has been heard of it since, so I presume it has fizzled. I might also point to the number of alleged white dwarfs announced by the spectroscopists from among faint blue stars – but a very large percentage of these have evaporated again, all of these objects had relatively small proper motions which I classified as 'intermediates' with M around $+4$ or so, instead of around $+10$. The most extreme case is that of Ton 202: Greenstein now proudly claims that he was the first person ever to obtain a spectrogram of a quasar – Ton 202 – at the time he classified it as a white dwarf and thus – by his own figures – underestimated the luminosity by a factor of 10^{14} or more, and the distance by a factor of 10^7. Now the spectroscopists are very fond of saying that parallaxes estimated by proper motion people are not reliable, and are not to be compared with spectro-

scopic determinations which are always completely accurate. I am continuously making estimates of parallax from proper motions and I have often been wrong – but never by a factor of more than ten. That one guess of Greenstein's will raise the mean error of all spectroscopic determinations of distance to a larger value than that of all proper-motion estimates for the next century.

To come back to the white dwarfs proper: we must accept the fact that the vast majority of those now known were picked up in proper motion surveys and can thus be expected to have large tangential velocities. I am therefore much puzzled by Greenstein's remark some years ago that now for the first time he had identified some Population II white dwarfs. One of the first two white dwarfs discovered – o_2 Eridani B is a high-velocity star, so are most of those found in proper motion surveys hence they must abound in Population II stars. Population I white dwarfs, with small tangential velocities may exist – even in large numbers – but as yet we have no foolproof method for finding them. Another statement of Greenstein's I take issue with is that he claims that yellow degenerates are much less frequent than previously supposed. At the St. Andrews Conference Greenstein stated that he had observed a large number of proper motion stars of yellow color but that he had found only one or two real yellow degenerates, while the vast majority proved to be G or K 'subdwarfs', i.e. high-velocity stars. Now Greenstein picked his observing list from my proper motion catalogue and chose mainly yellow stars around $m=14$ (apparent) with large proper motions, but these were mostly stars I had not designated as white dwarfs. Yellow degenerates of this color are expected to have an absolute magnitude around $M = +14$ so if one looks around apparent magnitude 14 one is searching for stars at distances of 10 pc – and how many yellow degenerates does one expect to find. But in the Palomar Survey I have found many stars between $m=18$ and $m=20$ (pg) with colors of g or k, and large proper motions. A typical star of this kind, with $m=19$ pg, color $g-k$, and a proper motion of 0″3 annually, would, if it were merely a Population II high-velocity star of $M = +9$ or brighter, have a tangential velocity of 1500 km s^{-1}, so it is pretty certain to be a degenerate of $M = +14$ or fainter.

Summing up, it is again the same story as for the low-luminosity stars – the large majority of white dwarfs now known have been found from proper motion surveys and can therefore be expected to behave the same way, kinematically, as high-velocity stars. Again, if low-velocity white dwarfs exist we have not been able to find them in large numbers, nor do we have efficient techniques for doing so.

Whether degenerate stars of the color and temperature of a main-sequence M dwarf exist we do not yet know. A few have been announced by spectroscopists, but it is fairly plain that their luminosities are not anywhere near low enough. Many years ago I pointed out that in wide binaries with one degenerate component, the main sequence star is the more luminous one, bolometrically, in an overwhelming number of cases. Hence, as I suggested about four years ago, perhaps the best prospects for M-type degenerates are the fainter components of common proper motion pairs where the primary is a yellow degenerate. We now have about a dozen of those, some with a primary of known parallax, and hence absolute magnitude around +13, and of color g. If the secondary in such a case is five or more magnitudes fainter, and of color m, it would seem to be a likely prospect. From the Palomar Proper Motion Survey alone I now have about 250 such wide proper motion pairs with at least one degenerate component. I should like to emphasize again that these are the only degenerate objects for which we can really determine the masses by observing the orbital motion, but this, of course, means that we should begin now taking first epoch plates with large reflectors, and not simply wait for someone else to do this – maybe during the next century.

DISCUSSION

Buscombe: Prof. Hynek hopes to determine UBV data for stars of 16th to 20th mag. with image enhancement on the Corralitos 60 cm reflector near Las Cruces, New Mexico. Although initially stars in Kapteyn selected areas will be observed, he may be interested in finding-charts for a few of your tricky candidates.

Luyten: I am very glad to hear this and shall be glad to cooperate in any way I can, but I am afraid I have to add that since by now we have some 250000 new proper motions I could not begin to supply finding charts except for some very few extremely interesting stars.

Irwin: How many stars have you found with $\mu > 3''$ yr^{-1} for example?

Luyten: None. The machine stops searching at $\mu = 2.5''$ yr^{-1}.

Irwin: How many $\mu > 2.5''$ yr^{-1} stars (on your plates) do you estimate you have missed?

Luyten: If I have to stick my neck out on this, I would say, almost zero.

Gliese: We know nothing about the real velocity dispersion of the red low-luminosity stars. We know about the differences in velocity dispersion of dM and dMe stars. Among the McCormick stars the ratio dM/dMe is about 2:1. Is anything known whether this ratio varies with absolute magnitude when going to the low luminosities?

Luyten: I am afraid nothing is really known about this, since for stars fainter than $m_{pg} = 15$ we have only very crude colors, and no spectra, radial velocities, or parallaxes.

Gliese: A preliminary investigation of the number of proper motion M stars ($\mu \leq 0\rlap{.}''2$ yr^{-1}) with m_{pg} between 17 and 21 given by Prof. Luyten near the South Galactic Pole shows no evidence of a number of low-luminosity objects larger than that given by Luyten's luminosity function. In so far this count does not agree with some estimates made near the North Galactic Pole.

Cayrel de Strobel: I am very interested in your discovery of G and K stars, but I would like to know how sure you can be that these are really G and K?

Luyten: The data were taken from the Palomar 48-in. Schmidt plates exposed with Haro's three image method. Our exposures were calibrated on SA 68 at declination $+15°$, hence we would not expect that our stars at $-20°$ and $-30°$ would appear too blue. Moreover, I made three further checks – against white dwarfs, asteroids, and elliptical galaxies, and all appeared to have the right colors. Yet when getting down to 17, 18 and 19 m_{pg} we found some red M stars but an unexpectedly large number of F, G and K stars and I concluded therefore that these are probably halo stars.

THE KINEMATICS AND AGES OF STARS IN GLIESE'S CATALOGUE

R. WIELEN

Astronomisches Rechen-Institut, Heidelberg, F.R.G.

1. Introduction

This contribution gives some results on the kinematics and ages of stars near the Sun. These results are mainly based on the catalogue of nearby stars compiled by Gliese (1957, 1969 and minor recent modifications). Table I shows the number of objects under consideration. While the old catalogue (1957) contained only stars with distances r up to 20 pc, the new edition (1969) includes many stars with slightly larger distances. In Table I, a 'system' is either a single star or a binary or a multiple system. The number of systems with known space velocities nearer than 20 pc has increased by about 30% from 1957 to 1969. The first edition of Gliese's catalogue (1957) has been analyzed in detail by Gliese (1956) and von Hoerner (1960).

TABLE I

Gliese's Catalogue of Nearby Stars

Number	Edition 1969		1957
	All	$r \leqslant 20$ pc	$r \leqslant 20$ pc
Stars	1890	1277	1095
Systems	1529	1036	916
Systems with known space velocity	1131	770	598

Although Gliese's catalogue is the most complete collection of stars within 20 pc, this sample of stars is severely biased by selection effects and is not fully representative for all the nearby stars. Only the stars brighter than $M_v \sim +7$ are almost completely known within 20 pc. For the fainter stars, the following selection effects occur: (a) The southern sky is deficient in detected faint nearby stars; (b) Since most of the faint nearby stars are found by their high proper motions, the sample is deficient in stars with small tangential components of their space velocities (measured with respect to the Sun); (c) Due to incomplete detection, the apparent space density of faint stars decreases rapidly with increasing distance r, and this effect becomes stronger with increasing M_v. From the luminosity function derived in the following section, we predict that there are at least 3600 stars nearer than 20 pc. Only about 1300 of them have been detected up till now. Hence the majority ($>65\%$) of stars within 20 pc are still undetected.

2. Luminosity Function

We derive the luminosity function φ for nearby stars by counting Gliese's stars in appropriate volumes of space (all declinations and $r \leqslant 20$ pc for $M_v < 7.5$; $\delta > -30°$ and $r \leqslant 20$ pc, 10 pc, 5 pc for $M_v = 7.5$ to 9.5, 9.5 to 11.5, $\geqslant 11.5$). For $M_v > 13.5$, only lower limits for φ can be obtained. Our unit of φ is 'stars per unit magnitude interval in a complete sphere of radius $r = 20$ pc' (Table II and Figure 1). The resulting luminosity function is rather flat in the range $5 < M_v < 9$ (see also Arakelyan, 1968; Mazzitelli, 1972). From the data of Gliese's catalogue, it remains uncertain where the maximum of φ occurs, but $M_{v,\max} \geqslant 13$ is indicated. From the predicted total number of stars within 20 pc, namely at least 3600, the majority are stars on or near the main sequence (about $3300 = 91\%$). 27 giants (0.7%) are observed and about 300 white dwarfs (8%) are predicted for $r \leqslant 20$ pc. The local stellar number density is at least 0.11 stars pc^{-3}.

The luminosity function derived from Gliese's catalogue is in rather good agreement with the results obtained by Luyten (1968). This is remarkable, since the methods are quite different. While we directly count Gliese's stars according to their individually known absolute magnitudes, Luyten uses proper motion surveys and needs therefore additional statistical assumptions about the velocities of the stars. Luyten found the maximum of φ at $M_v \sim 13.9$ (1968) and more recently at $M_v \sim 13.6$ (1974). This means that our luminosity function may have actually reached the maximum, and that the decrease of our values of φ for $M_v > 14$ may be partly real. At the bright end, our luminosity function agrees well with the data derived by McCuskey (1966) from a larger volume of space (LF regions).

For the local stellar mass density, we find $\varrho_s = 0.046\ \mathcal{M}_\odot$ pc^{-3} by using our data (Table III). If we use Luyten's luminosity function for $M_v \geqslant 13.5$, the value of ϱ_s increases only from 0.046 to 0.049 $\mathcal{M}_\odot$ pc^{-3}. The value of ϱ_s is much smaller than the total mass density in the solar neighbourhood, $\varrho_{\text{tot}} = 0.15\ \mathcal{M}_\odot$ pc^{-3}, determined dynamically (Oort, 1965). The local density of the observed interstellar matter is about 0.02 $\mathcal{M}_\odot$ pc^{-3}. Hence the problem of the missing local mass remains as severe as it was before, if we do not invoke many faint M dwarfs of uncommonly low space velocities, as proposed by Murray and Sanduleak (1972) and Weistrop (1972).

Since we know for many stars in Gliese's catalogue both the absolute magnitude and the space velocity individually, we can investigate whether the luminosity function is correlated with the kinematical behaviour and hence with the ages of the stars.

TABLE II

Luminosity function

M_v	−1	0	1	2	3	4	5	6	7	8	9
φ	1	4	14	24	43	78	108	121	102	132	159

M_v	10	11	12	13	14	15	16	17	18	19
φ	245	341	512	597	>341		>213		$\geqslant 16$	

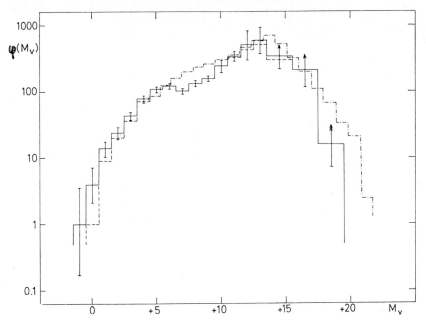

Fig. 1. Luminosity function $\varphi(M_v)$. Full line: total φ; dashed line: φ for stars on or near the main sequence; dashed-dotted line: Luyten's φ (1968).

TABLE III

Stellar mass density near the Sun

$\varrho\,[\mathcal{M}_\odot\,\text{pc}^{-3}]$		Based on
Stars on or near the main sequence:		
$M_v < 9.5$	0.020 ⎫	776 stars within
$9.5 \leqslant M_v < 13.5$	0.014 ⎬ 0.038	various volumes
$13.5 \leqslant M_v$	0.004 ⎭	depending on M_v
Giants	0.001	27 stars within 20 pc
White dwarfs	0.007	5 stars within 5 pc
Sum	0.046	

The luminosity of stars brighter than $M_v \sim 5$ is affected by evolutionary effects which are difficult to eliminate. Hence we shall consider stars fainter than $M_v \sim 5$. For those stars, the luminosity function is essentially determined by the spectrum of stellar masses at the time of formation. In Table IV, we have grouped the stars according to their space motions (W is the velocity perpendicular to the galactic plane, S is the total space velocity, both referred to the adopted circular velocity). Because of the strong correlation between the space velocities and the ages of the stars, these groups have rather different mean ages. The quantity shown in Table IV is the number of stars as a function of M_v and $|W|$ or S, normalized in the range $5.5 \leqslant M_v <$

TABLE IV
Relative luminosity functions for kinematical groups

M_v	$|W|$ [km s^{-1}]				Total space velocity S			
	0–5	5–10	10–20	>20	0–20	20–35	35–50	>50
+ 5	1.6	0.8	0.8	0.9	1.9	0.8	1.4	0.7
6	0.9	1.0	1.1	1.1	0.7	1.1	1.1	1.1
7	1.1	1.0	0.9	0.9	1.2	0.9	0.9	0.9
8	0.9	1.1	1.0	0.8	1.1	0.9	1.3	0.7
9	0.9	0.8	0.8	0.8	0.8	1.0	0.9	0.7
10	0.9	1.0	0.9	1.2	0.5	0.6	0.9	1.7
+11	0.8	0.9	0.9	1.2	0.6	1.0	1.5	0.7

<7.5 and given relative to the total luminosity function at each value of M_v. If the luminosity function does not vary with age, then we expect the value 1.0 everywhere in Table IV. Except in the line for $M_v=5$, which may be slightly biased by evolutionary effects, there does not occur any statistically significant variation of the luminosity function among the different kinematical groups. We conclude that the distribution of stellar masses at birth does not significantly depend on the time of formation, at least when averaged over longer periods of time and larger regions in space in the covered range of masses.

3. Age Groups in the Colour-Magnitude Diagram

The ages of Gliese's stars may be estimated either from the position in the colour-magnitude diagram or from the intensity of the Ca II emission (Section 4). If we wish to determine individual ages by fitting theoretical isochrones, we can use only those absolute magnitudes which are derived from accurate trigonometric parallaxes, since Gliese's photometric or spectroscopic parallaxes already assume M_v. Unfortunately, the number of evolved stars with accurate trigonometrical values for M_v is too small for a sound statistical treatment. This is obvious from Figure 2, which shows the colour-magnitude diagram for the stars of Gliese's catalogue with values of M_v based on trigonometric parallaxes having a probable error not larger than 10%. Therefore, we form groups in the colour-magnitude diagram and use the average age of each group for our kinematical studies. The groups are shown in Figure 2. Groups 1–5 and 6a–d are stars on or near the main sequence in various intervals of $B-V$. The mean age τ is assumed to be roughly half the lifetime of a main sequence star at the mean $B-V$. Group 7 consists of old giants.

Figure 3 shows the velocity distribution of Gliese's stars on or near the main sequence with $B-V<0.50$ (spectral types $\leqslant$F6) and $r\leqslant 20$ pc. The U-axis points towards the galactic center and the V-axis in the direction of galactic rotation. The space velocities are corrected for the solar motion, $U_0=+9$ km s^{-1} and $V_0=+12$ km s^{-1} (Delhaye, 1965). The most prominent feature in Figure 3 is the vertex deviation.

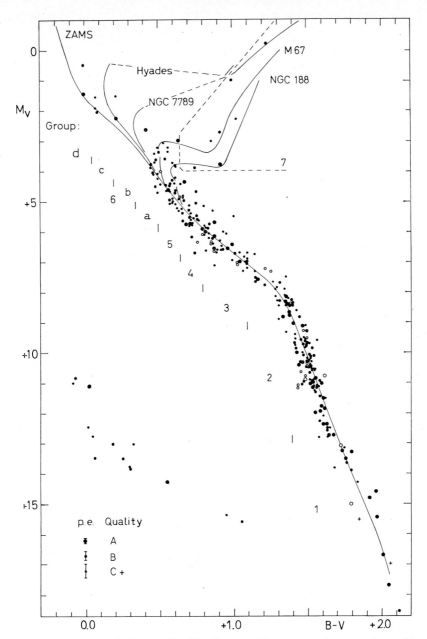

Fig. 2. Colour-magnitude diagram for Gliese's stars with accurate trigonometric parallaxes

No obvious streams or moving groups (Eggen, 1965) show up in Figure 3. In Table V, we present the results for the velocity dispersions σ and for the asymmetric drift of the stars in our groups. We include the McCormick K+M dwarfs contained in Gliese's catalogue, since they should be representative for the majority of the common

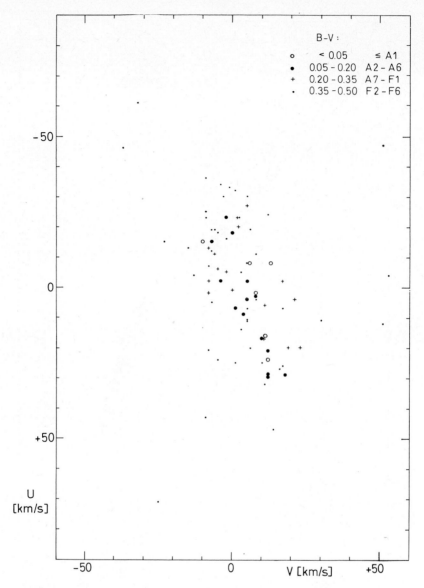

Fig. 3. Velocity distribution of Gliese's stars with $B-V < 0.50$ and $r \leqslant 20$ pc.

nearby stars (see Section 6). The quantity $V_\odot$, listed in Table V, is the difference between the V-component of the Sun and of the mean motion of the stars in the group. Both the observed velocity dispersions and the asymmetric drift increase rather monotonically with the mean age of a group. The kinematical behaviour of the old giants and of the white dwarfs (see also Gliese, 1971) is quite similar to that of the McCormick K+M dwarfs. From Figure 4, we conclude that the relative increase of the velocity dispersions σ with age τ is roughly the same in the com-

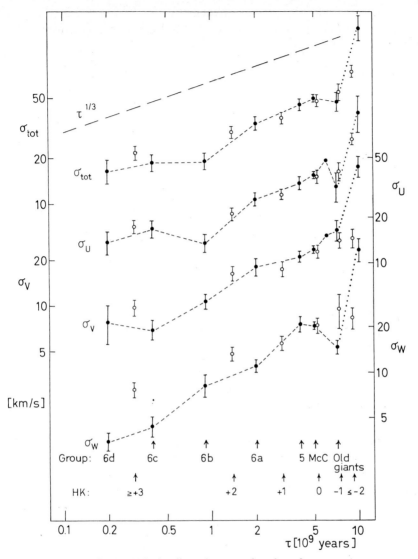

Fig. 4. Velocity dispersions as a function of age.

ponents parallel to the galactic plane, U and V, and perpendicular to the plane, W. The accumulating effect of encounters of the field stars with large complexes of stars and interstellar matter would lead to $\sigma \propto \tau^{1/3}$ (Spitzer and Schwarzschild, 1953), which is in quite good agreement with the observed behaviour (Figure 4). But other mechanisms may also explain the increase of σ with τ (Barbanis and Woltjer, 1967). In numerical experiments, we have studied the increase of σ_U and σ_V for stars born with small velocity dispersions, by the gravitational field of a stationary density wave of Lin's type. Preliminary results indicate that at least a significant fraction of the increase in σ_U and σ_V can be quantitatively explained by such a mechanism.

TABLE V
Velocity dispersion for groups in the colour-magnitude diagram

Group	N	σ_{tot}	σ_U	σ_V	σ_W	$V_\odot$	Mean age $\langle\tau\rangle$ [10^9 yr]
				[km s^{-1}]			
6d	6[a]	16	14	8	4	+5	0.2
6c	14[a]	19	17	7	4	+7	0.4
6b	16[a]	19	14	11	8	+7	0.9
6a	47[a]	34	27	18	11	+10	2
5	76[a]	44	34	21	21	+17	4
4	84[a]	48	34	27	20	+27	5
3	126[a]	49	36	28	17	+19	5
McCormick K+M dwarfs	317	50	39	23	20	+19	5
Old giants	25[b]	47	32	31	15	+31	7
White dwarfs	13[b]	50	42	22	18	+15	
Subdwarfs	8[b]	145	101	82	65	+104	10

[a] $r \leqslant 20$ pc, without classified subdwarfs.
[b] $r \leqslant 20$ pc.

4. Age Classification According to the Ca II Emission

It has long been known that the different kinematical behaviour of dwarfs with and without emission lines (dMe and dM stars) indicates an age difference for these groups. However, significant progress has been recently achieved by the observational work of O. C. Wilson. Wilson classifies the Ca II emission intensity at the H and K lines, using a visual scale ranging from about +8 (very strong emission) to −5 (extremely weak or no emission). Wilson's estimates (Wilson and Woolley, 1970) of the Ca II emission intensity, HK, are available for many McCormick K+M dwarfs in Gliese's catalogue, which should be representative of the common nearby stars (Figure 5). In Table VI, we give the velocity dispersions for these stars as a function of their Ca II emission intensity HK. The monotonic increase in σ with decreasing HK proves that the Ca II emission intensity is strongly correlated with the age of a star. Hence, the Ca II emission intensity may be used as a rather accurate indicator for the ages of the unevolved late-type dwarfs.

It is, however, difficult to convert the relative ages, obtained from Wilson's HK estimates, into absolute ages. Since the basic astrophysical mechanism for the decay of the Ca II emission is not accurately known, we may use either the relation between σ and τ from Section 3 to obtain a 'kinematical' calibration of HK(τ), or we may use the cumulative frequency of stars as a function of HK and an assumed rate of star formation for estimating τ. In Table VII we derive the mean ages of the various HK groups by assuming a constant rate of star formation over 10^{10} yr in a cylinder perpendicular to the galactic plane. The relative number of stars in such a cylinder is approximately obtained by multiplying the number N of stars in a volume near the plane by $|W|$ (see also Section 6). Although the absolute ages obtained should

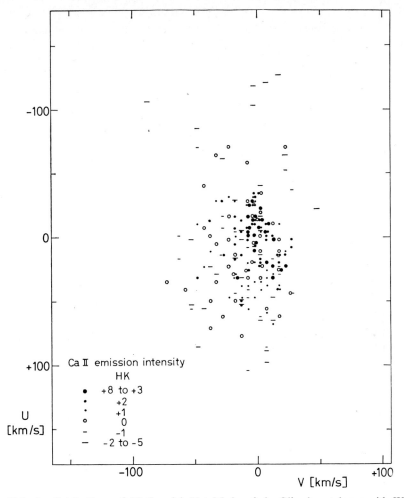

Fig. 5. Velocity distributions of McCormick K+M dwarfs in Gliese's catalogue with Wilson's classification of the Ca II emission intensity.

TABLE VI

Velocity dispersion of McCormick K+M dwarfs as a function of Ca II emission intensity

HK	N	σ_{tot}	σ_U	σ_V	σ_W	$V_\odot$
				[km s^{-1}]		
+8 to +3	23	22	18	10	8	+11
+2	40	30	21	16	13	+15
+1	36	37	29	17	15	+18
0	41	48	38	23	20	+25
−1	24	55	40	27	26	+21
−2 to −5	31	75	66	27	23	+25
Hα emission	18	34	25	19	14	+18

TABLE VII

Age estimates for McCormick K+M dwarfs assuming a constant rate of formation

Ca II emission intensity HK	N	$\langle|W|\rangle$ [km s^{-1}]	$N \cdot \langle|W|\rangle$	$\langle\tau\rangle$ [10^9 yr]
+8 to +3	23 = 12%	7	158 = 6%	0.3
+2	40 21%	11	420 16%	1.4
+1	36 18%	13	464 17%	3.0
0	41 21%	16	665 25%	5.2
−1	24 12%	17	399 15%	7.2
−2 to −5	31 16%	18	547 21%	9.0

TABLE VIII

Vertex deviation ψ

for stars with $\sqrt{U^2+V^2} \leq 40$ km s^{-1} and $|W| \leq 15$ km s^{-1}

Group	N	ψ	m.e.	McCormick K+M dwarfs HK	N	ψ	m.e.
On or near the main sequence:				+8 to +3	21	+23°	± 6°
				+2	30	+35	20
6d	6	+23°	±13°	+1	17	+ 2	6
6c	14	+20	4	0	19	+37	54
6b	15	+29	13	−1	10	+37	16
6a	34	+18	5	−2 to −5	4	(+57	40)
5	45	+22	6				
4	36	+ 4	14	Hα emission	12	+18	6
3	68	+ 5	6				
McCormick K+M dwarfs	159	+17	5	McCormick K+M dwarfs $|W|$	N	ψ	m.e.
				0−5	67	+16°	6°
Other stars:				5−10	52	+18	10
Old giants	11	−70	26	10−20	58	+13	12
				>20	30	+14	16
White dwarfs	3	(+4	4)	All	159	+17	5
				All, but with weight $|W|$		+20	

be very crude, they fit astonishingly well with the relation between σ and τ derived in Section 3 (Figure 4). Furthermore, for the Hyades, the observed Ca II emission intensity, HK = +3 (Wilson, 1964, 1966), would lead to an age similar to the evolutionary cluster age of about 0.9×10^9 yr. The ages derived in Table VII indicate that the total decay time of the Ca II emission seems to be quite long ($\gtrsim 10^{10}$ yr) for K and M dwarfs.

5. Vertex Deviation

The question how the vertex deviation depends on the age of the stars, has recently been studied from an observational point of view by Woolley (1970), Wilson and Woolley (1970), Woolley *et al.* (1971), and Mayor (1972). They all find that the vertex deviation ψ is large for very young stars, small for stars of intermediate ages and practically zero for old stars. Unfortunately, however, the method used by these authors is biased in favour of such a result: They consider stars with an orbital excentricity e smaller than some chosen value, $e \leqslant 0.15$ for example. The region $e \leqslant 0.15$ in the velocity space is, however, roughly an ellipse with a major axis pointing exactly towards the galactic center. For a group of young stars, with a very small velocity dispersion, the bias towards $\psi \sim 0$ is insignificant. But for old stars, with a large velocity dispersion, the sampling volume in the velocity space defined by $e \leqslant 0.15$ favours small values of ψ, even if the real vertex deviation should be as large as for young stars. In order to avoid such a bias, but still to restrict the sample to low-velocity stars, we have chosen a circular limit in the UV-plane. The resulting vertex deviations ψ for our groups, using $(U^2 + V^2)^{1/2} \leqslant 40$ km s^{-1} and $|W| \leqslant 15$ km s^{-1}, are given in Table VIII. The large mean errors of ψ which occur are mainly due to the circular sampling volume in the UV-plane. While the young stars show a significant vertex deviation of about $+20°$, it is not obvious whether the vertex deviation for low-velocity stars decreases with increasing age. The old stars with high peculiar space velocities (not listed in Table VIII) do not show a significant vertex deviation.

6. Representative Common Nearby Stars

For many dynamical purposes, such as the construction of dynamical models of the Galaxy or the discussion of the gravitational stability of the Galaxy, one is interested in the total velocity distribution of all nearby stars. As pointed out in the introduction, the sample of all Gliese stars is not adequate for that purpose because of selection effects. However, the subgroup of the McCormick K + M dwarfs in Gliese's catalogue should be a rather representative sample for the velocity distribution of all nearby stars. The McCormick stars are free from the kinematical selection effects, since they are discovered on objective prism plates (Vyssotsky, 1963), and they should be a representative mixture of stellar ages. The velocity dispersions and the asymmetrical drift of the McCormick K + M dwarfs in Gliese's catalogue are listed in Table V. The velocity distribution of these stars, not reproduced here, shows the well-known asymmetry in the velocity component V, and a pronounced vertex deviation for the low-velocity stars.

The solar neighbourhood is a sampling volume situated close to the galactic plane. While the velocity distribution of the McCormick stars is representative for nearby stars, now at $z \sim 0$, this local distribution is not representative for all stars in a cylinder perpendicular to the galactic plane, because the velocity distribution varies with the distance z from the plane (Vandervoort, 1970). This can be shown by grouping the

TABLE IX

Velocity dispersion of McCormick K+M dwarfs grouped according to $|W|$

$|W|$	N	σ_U	σ_V	$V_\odot$	Mean height $\langle|z|\rangle$
		[km s^{-1}]			[pc]
0–5	86	31	19	+15	28
5–10	66	31	20	+17	55
10–20	91	37	21	+20	100
>20	74	52	31	+25	300
All	317	39	23	+19	
All, but each star weighted by $|W|$		48	29	+23	200

nearby stars according to their vertical space velocity, $|W|$. Table IX shows the remarkable increase of the velocity dispersions parallel to the galactic plane, σ_U and σ_V, with $|W|$. We have also indicated the mean height $\langle|z|\rangle$, averaged over an oscillation period in z, of stars which pass now through the plane with a vertical velocity $|W|$. The increase of σ_U and σ_V with $|W|$ is an effect of the different ages of the stars. Stars with higher vertical velocities $|W|$ are on the average older and move therefore with higher velocity dispersions in U and V, since σ_U, σ_V grow together with σ_W (Section 3).

How can we derive a velocity distribution which is representative for all stars in a cylinder perpendicular to the galactic plane? Let us assume that the z-motions of the stars are well-mixed (stationary state) and decoupled from the motions parallel to the plane, and that the oscillation period T_z does not depend on the amplitude of the z-motion (approximately fulfilled for $|z|<500$ pc). The probability p of finding a star in a small element Δz at $z=0$ is given by the time for crossing Δz, T_c, divided by T_z. Since T_c is inversely proportional to $|W|$ and $T_z \sim$ const., we get $p \propto |W|^{-1}$. In order to derive a velocity distribution representative for the cylinder, we have to weight each star in our sampling volume at $z=0$ by a weight p^{-1} or just by $|W|$. The weight $|W|$ is a conservative lower limit in so far that T_z actually increases slowly with $|W|$ (see also Woolley et al., 1971). In the last line of Table IX, we give the resulting velocity dispersions representative for the cylinder. The velocity dispersions σ_U and σ_V integrated over z, are higher than those at $z=0$ by about 25%. The z-averaged value $\sigma_U = 48$ km s^{-1}, compared with 39 km s^{-1} at $z=0$, has some interesting implications: the stability of our Galaxy against a local gravitational collapse is improved, but the propagation of a density wave of Lin's type may be rendered more difficult. These topics are treated in detail by Toomre in his contribution to this Joint Discussion.

Acknowledgements

This contribution rests heavily on the results of a broader statistical analysis of Gliese's catalogue which has been carried out by Mr H. Jahreiss (1974) and myself. We would like to thank Dr W. Gliese very much for many stimulating discussions and helpful comments.

References

Arakelian, M. A.: 1968, *Astrofizika* **4**, 617.
Barbanis, B. and Woltjer, L.: 1967, *Astrophys. J.* **150**, 461.
Delhaye, J.: 1965, *Stars and Stellar Systems* **5**, 61.
Eggen, O. J.: 1965, *Stars and Stellar Systems* **5**, 111.
Gliese, W.: 1956, *Z. Astrophys.* **39**, 1.
Gliese, W.: 1957, *Mitt. Astron. Rechen-Inst. Heidelberg Serie A* No. 8.
Gliese, W.: 1969, *Veröffentl. Astron. Rechen-Inst. Heidelberg* No. 22.
Gliese, W.: 1971, in W. J. Luyten (ed.), 'White Dwarfs', *IAU Symp.* **42**, 35.
Hoerner, S. von: 1960, *Fortschritte der Physik* **8**, 191.
Jahreiss, H.: 1974, unpublished Thesis, Heidelberg University.
Luyten, W. J.: 1968, *Monthly Notices Roy. Astron. Soc.* **139**, 221.
Luyten, W. J.: 1974, this volume, p. 389.
Mayor, M.: 1972, *Astron. Astrophys.* **18**, 97.
Mazzitelli, I.: 1972, *Astrophys. Space Sci.* **17**, 378.
McCuskey, S. W.: 1966, *Vistas in Astronomy* **7**, 141.
Murray, C. A. and Sanduleak, N.: 1971, *Monthly Notices Roy. Astron. Soc.* **155**, 463.
Oort, J. H.: 1965, *Stars and Stellar Systems* **5**, 455.
Spitzer, L. and Schwarzschild, M.: 1953, *Astrophys. J.* **118**, 106.
Vandervoort, P. O.: 1970, *Astrophys. J.* **162**, 453.
Vyssotsky, A. N.: 1963, *Stars and Stellar Systems* **3**, 192.
Weistrop, D.: 1972, *Astron. J.* **77**, 849.
Wilson, O. C.: 1964, *Publ. Astron. Soc. Pacific* **76**, 28.
Wilson, O. C.: 1966, *Science* **151**, 1487.
Wilson, O. and Woolley, R.: 1970, *Monthly Notices Roy. Astron. Soc.* **148**, 463.
Woolley, R.: 1970, in W. Becker and G. Contopoulos (ed.), 'The Spiral Structure of Our Galaxy', *IAU Symp.* **38**, 423.
Woolley, R., Pocock, S. B., Epps, E. A., and Flinn, R.: 1971, *Roy. Obs. Bull.* No. 166.

DISCUSSION

Buscombe: The claim of completeness of bright southern stars is incorrect. A 5th mag. F dwarf, γ Cir B, is the closest bright member of an optical double. In my opinion, the dependence on trigonometrical parallaxes to delineate groups of stars has been over-emphasized.

Wielen: The completeness of bright stars within 20 pc is meant in a statistical sense. Of course, due to changes in the adopted values of the parallaxes, there will be always some borderline cases for any chosen limit in r. In the present investigation, we have used the 'resulting parallaxes' given in Gliese's catalogue which are a combination of trigonometric, spectroscopic and photometric distance determinations. However, only the trigonometric parallaxes should be used for deriving individual ages of evolved stars from isochrones, since the spectroscopic and photometric parallaxes derived by Gliese are based on *assumed* absolute magnitudes.

Irwin: Do you implicitly assume, when you calculate incompleteness factors for stars from 10 to 20 pc distant, that the stars are uniformly distributed in space? Or do you allow for any thinning out of the stars in the z direction?

Wielen: For deriving the luminosity function, I assume that the stars would be uniformly distributed within $r \leqslant 20$ pc, if there were no selection effects. I do not use incompleteness factors. The luminosity function for fainter stars is derived from smaller volumes for which the data indicate a uniform distribution within the chosen limit of r.

Murray: The kinematics of the stars in Wilson's classes $+8$ to $+3$ are very similar to those of the stars in the north galactic cap which I discussed. Is there any selection effect operating against the inclusion of these stars in Gliese's catalogue?

Wielen: The McCormick K + M dwarfs in Gliese's catalogue for which Wilson gives Ca II emission intensities, should be a representative sample of dwarfs in the spectral range covered (mainly from K8 to M2). Considering Gliese's catalogue *as a whole*, the stars in Wilson's classes $+8$ to $+3$ should be underrepresented, because they have small space velocities and are therefore affected by the kinematical selection effect.

THE THIRD AND FOURTH MOMENTS OF THE LOCAL STELLAR VELOCITY DISTRIBUTION

R. H. MILLER

University of Chicago, U.S.A.

Abstract. The velocity moments through the fourth order were calculated for a subset of 870 astrophysically selected stars from Gliese's catalog. An essential part of this program was the determination of the covariances of estimates of the moments; both sampling and observational error contributions were taken into account in evaluating the total covariance matrix. In applications to the study of galactic structure, the Gliese stars are regarded as tracers of the galactic potential. The moments found are consistent with conventional assumptions about symmetries of the Galaxy. Certain galactic parameters, such as the curvature in the law of rotation and the asymmetric drift, were studied with the help of the values obtained for the moments.

The higher moments of the local stellar velocity distribution can be of use in several ways in the study of the properties of the Galaxy. Three problems were addressed: (1) Can the third and fourth moments be determined at all? Contrary to the folklore, fourth order moments can be determined with a 'signal-to-noise ratio' of about 5–8. (2) Are the resulting moments consistent with the assumed symmetries of the Galaxy? Except for vertex deviation, they are. (3) What do the moments tell about the Galaxy? This work was largely carried out by a student, R. R. Erickson; his discussion of questions (2) and (3) follows a hydrodynamical development by P. O. Vandervoort.

Gliese's (1969) catalog provided the data used for this study; the data were accepted uncritically. This sample of stars has widely appreciated properties; both the strengths and shortcomings of the catalog are well known. This is advantageous for a study of the kind described here in which the principal scientific contribution is the demonstration that a certain kind of problem can feasibly be attacked. The statistical methods used have been checked by this example, and shown to yield usable results.

The sample provided by Gliese's catalog was ruthlessly trimmed to obtain a subset that could be used in the required manner, but all criteria were based on data contained in the catalog. Any entry that was incomplete or uncertain was thrown out; this included any system for which radial velocity determination was based on a white dwarf. Multiple systems were treated as a single entry. Among the acceptance criteria, that most likely to be controversial was the rejection of subdwarfs and of stars listed as possible subdwarfs. This was necessary because there are too few subdwarfs to provide a statistically usable sample. The final sample contained 870 stars.

In describing the statistical method, many important details must be omitted. Only a few of the most important features will be described. The appropriate method for estimating moment-like functions is a tensor generalization of Fisher's k-statistic, for which a sampling theory is available (Kaplan, 1952). Error estimates are the essence of the problem since the question of whether the higher moments can be determined must be decided on the relative sizes of the moments and of their estimated errors (signal-to-noise ratio or coefficient of variation). The final covariance is the sum of two parts: the propagated observational errors (available from Gliese's catalog) and the sampling errors. Sampling errors are familiar as the term the yields the usual $1/\sqrt{N}$ relative accuracy. Determination of the sampling covariances between two fourth-order moments requires the calculation of an eighth-order moment. There are 31 non-trivial moments, so the covariance matrix is 31×31 real symmetric positive definite. The full set of moments and the full covariance matrix, along with the details of the calculation, will be given elsewhere.

TABLE I

Values of the product $\Delta \mathbf{x}^t C^{-1} \Delta \mathbf{x} = \Delta \chi^2$

Subsample	Number of systems	$\Delta \chi^2$
Half of sample (first, third, fifth, ... systems)	435	24.6
Half of sample (second, fourth, sixth, ... systems)	435	26.0
G0 to M9	718	10.9
G0 to K9	466	20.6
F5 to K9	556	13.5
Parallax $<0\rlap{.}''075$	601	10.1
Parallax $\geqslant 0\rlap{.}''075$	269	43.8

The comparisons are made to the basic sample of 870 stellar systems.

The feature of this statistical exercise that yields the most convincing demonstration of its validity is that the values obtained are quite stable when the moments are re-computed using various subsets of the basic 870-system sample. The results for some of these sectionings are shown in Table I; there the square of the 'distance' between the solution obtained from each of these subsets and that obtained from the full sample of 870 stars is shown, calculated using the covariance matrix belonging to the subset. For a random scatter of independent values, the distances should be χ^2-distributed on 31 deg of freedom (expectation value about 31). With due regard for the lack of independence, the sample seems remarkably stable under these sectionings.

The contributions of observational and of sampling error to the total covariance are of the same order of magnitude for the fourth-order moments, with the sampling term almost always larger. Thus this sample is nearly optimal for present purposes in the sense that an improvement in observational techniques, leading to a reduction of the observational errors, would not improve our knowledge of the moments significantly without a substantial increase in the number of stars in the sample; but neither would an increase in the number of stars help much without a corresponding improvement in observational techniques.

We are now ready to examine the values obtained for the moments from the standpoint of the assumed symmetries before applying them to the study of galactic structure. The stars of the Gliese catalog are regarded as tracers of the galactic potential in the applications. Completeness of the sample is not required – only that the sample be representative of stars that have lived long enough to explore the galactic potential. Stars that have undergone the violence of early galactic collapse would prejudice the results undesirably. The stars actually used should represent the disk population; they explore a toroidal region of the Galaxy that extends outward and inward by a kpc or so. That region contains so many stars that the population from which the sample is drawn is effectively infinite.

Under the usual galactic symmetry assumptions (rotational symmetry, symmetry about a plane, the only differential motion allowed is a differential rotation, and the system is in equilibrium), all moments that contain an odd number of π (galactic radial) or Z (normal to the galactic plane) indices should vanish. This is true within the statistical accuracy for all the estimated second, third, and fourth order moments except for two related to vertex deviation: the $\langle \pi \Theta_p \rangle$ and $\langle \pi\pi\pi\Theta_p \rangle$ moments. The fourth-order moments that should survive stand up at about 5–8 standard deviations.

The next exercise was to legislate that those moments that should vanish by symmetry are exactly zero: the solution is constrained so that those moments must be zero. There are 12 allowed nonzero moments. The subsequent results are quoted for 4 solutions: the full set of 31 moments (none forced to zero), the set of 12 moments with all forbidden moments forced to zero, a set of 13 that leaves the $\langle \pi \Theta_p \rangle$ moment free in addition to the allowed 12, and a set of 14 that also leaves the $\langle \pi^3 \Theta_p \rangle$ moment free. All of these are admissible solutions from a purely statistical point of view, although statistical criteria must be strained a bit to suppress the $\langle \pi \Theta_p \rangle$ term. Even if we could not see beyond 20 pc, we could conclude that the Galaxy has the assumed symmetries. The well-known non-normal character of the velocity distribution appears as well (fourth moments too large, implying too many stars in the high-velocity wings of the distribution).

The stage is now set to use the values of these moments to investigate properties of the Galaxy. The square of the axis ratio of the velocity ellipse, $B/(B-A)$, appears as the ratio of several combinations of moments; consistent values are obtained, as shown in Table II. These estimates are not independent; a 'best value' for the ratio is not a simple weighted mean, but rather is a single value consistent with all the covariances among the moments. This 'weighted mean' is quite close to the value 0.4 that results from the standard values for the Oort constants. In principle, the

TABLE II

Values of $B/(B-A)$ and their standard errors

	Set of 31 moments	Set of 14 moments	Set of 13 moments	Set of 12 moments
$\langle \Theta_p^2 \rangle / \langle \pi^2 \rangle$	0.45 ± 0.04	0.46 ± 0.03	0.45 ± 0.03	0.44 ± 0.04
$\langle \Theta_p^4 \rangle / \langle \pi^2 \Theta_p^2 \rangle$	0.39 0.06	0.39 0.05	0.39 0.05	0.39 0.06
$3 \langle \pi^2 \Theta_p^2 \rangle / \langle \pi^4 \rangle$	0.66 0.12	0.71 0.11	0.65 0.11	0.60 0.11
$\langle \Theta_p^2 Z^2 \rangle / \langle \pi^2 Z^2 \rangle$	0.41 0.09	0.42 0.07	0.41 0.08	0.35 0.07
Resultant $B/(B-A)$	0.43 0.03	0.41 0.03	0.42 0.03	0.42 0.03

number of independent parameters could be further reduced by eliminating 4 of these moments in favor of the ratio $B/(B-A)$ to leave a 9-parameter solution, but we have not yet done this. The hydrodynamical equations do not permit separate solutions for B and A.

Some of the galactic parameters obtainable through these arguments are shown in Table III. There, l_v is the longitude of the vertex, which vanishes for the 12-moment solution. The quantity $D = \partial^2 (\ln \Theta_0)/\partial (\ln \varpi)^2$, is essentially the curvature in the law of galactic rotation. A mixture of the radial 'scale heights' for N, the number density, and for $\langle \pi^2 \rangle$, the velocity dispersion in the radial direction, appears as Λ; the scale height is around 5 kpc. The two quantities that contribute to Λ cannot be separated without additional assumptions; two 'reasonable' assumptions appear as $\partial (\ln Q)/\partial \varpi = 0$ or 0.1; these assumptions concern arguments relating to the stability of the Galaxy (itself a slippery argument that need not be entered upon here). The scale-heights obtained are fairly insensitive to the value assumed. Once a value is assumed, it is possible to calculate the asymmetric drift – it is remarkable that the asymmetric drift can be determined from a sample such as this. The quantity, $\partial (\ln \langle \pi^2 \rangle)/\partial \varpi$ is sometimes regarded as too small to admit observational determination (Oort, 1965). This cannot be so: a value of $\partial (\ln \langle \pi^2 \rangle)/\partial \varpi = 0$ leads to values for $\partial (\ln N)/\partial \varpi = -1.8 \pm 0.5$ kpc^{-1}, or to a radial scale-height in the number density of 550 pc, a value more appropriate to vertical scale-heights.

It has not been possible, in this report, to present more than a brief synopsis of this work. It is presented with the hope of stimulating your interest by indicating the usefulness of results of this kind in the study of galactic structure. Papers containing the full details of the statistical work, the development of the equations of

TABLE III

Galactic parameters and their standard errors

	Set of 31 moments		Set of 14 moments		Set of 13 moments		Set of 12 moments	
l_v	$9° \pm 3°$		$9° \pm 2°$		$7° \pm 2°$		suppressed	
D	0.6	1.4	0.2	1.4	0.2	1.5	-0.1	± 1.5
$\partial^2 \Theta_0/\partial \varpi^2$ (km s^{-1} kpc^{-2})	2.2	3.6	1.2	3.4	1.0	3.6	0.5	3.7
Λ (kpc)	4.6	0.9	4.3	0.9	4.6	1.0	5.1	1.1
For $\partial \ln Q^2/\partial \varpi = 0.0$ kpc^{-1}								
$\partial \ln N/\partial \varpi$ (kpc^{-1})	-0.20	0.07	-0.22	0.07	-0.22	0.07	-0.22	0.07
$\partial \ln \langle \pi^2 \rangle/\partial \varpi$ (kpc^{-1})	-0.22	0.05	-0.23	0.06	-0.21	0.06	-0.19	0.06
$\Theta_0 - \Theta_c$ (km s^{-1})	-9	3	-9	4	-9	3	-8	3
For $\partial \ln Q^2/\partial \varpi = 0.1$ kpc^{-1}:								
$\partial \ln N/\partial \varpi$ (kpc^{-1})	-0.25	0.07	-0.26	0.07	-0.27	0.07	-0.27	0.07
$\partial \ln \langle \pi^2 \rangle/\partial \varpi$ (kpc^{-1})	-0.21	0.05	-0.23	0.06	-0.21	0.06	-0.19	0.06
$\Theta_0 - \Theta_c$ (km s^{-1})	-9	3	-10	4	-9	3	-9	3

Where needed in calculating these parameters, the 'standard' self-consistent set of values $\{A = 15$ km s^{-1}, $B = 10$ km s^{-1}, $\Theta_0 = 250$ km s^{-1}, and $\varpi = 10$ kpc$\}$ were used. The standard errors (equal to the probable errors divided by 0.6745) that are listed reflect only the propagation of the observational and sampling errors of the moments and do not include any uncertainties in the 'standard' values.

stellar hydrodynamics to include fourth-order moments, and the application to galactic structure, have been submitted to *The Astrophysical Journal* by Erickson and by Vandervoort.

References

Gliese, W.: 1969, *Veröffentl. Astron. Rechen-Inst. Heidelberg,* No. 22.
Kaplan, E. L.: 1952, *Biometrika* **39**, 319.
Oort, J. H.: 1965, *Stars and Stellar Systems* **5**, 455.

THE ABUNDANCE AND AGE DISTRIBUTION OF 500 F STARS IN THE SOLAR NEIGHBOURHOOD

R. E. S. CLEGG and R. A. BELL
University of Maryland, U.S.A.

A short account of the paper published in *Monthly Notices Roy. Astron. Soc.* **163**, 13, 1973 was presented by R. A. Bell.

THE SPACE DENSITY OF FAINT M-DWARFS

J. H. OORT

The Observatory, Leiden, The Netherlands

Donna Weistrop, a student of Maarten Schmidt's, has found evidence for an unexpectedly high space density of very faint M-dwarfs. This was confirmed by Murray and Sanduleak. The space density inferred from these investigations is comparable to, or possibly higher than, the local gas density. Because there is evidence that the stars concerned are young this appears to present a problem.

As the investigation by Murray and Sanduleak is the simpler and more direct of the two I confine my discussion to this. The authors determined proper motions for 21 M dwarfs down to 17^m found in an objective-prism survey at the Warner and Swasey Observatory in regions close to the North Galactic Pole. The distance of the stars can be inferred from the reflection of the solar motion. Assuming the 'basic' solar motion they found $\langle \pi \rangle = 0''.021$ and a space density of 0.23 pc^{-3} for these stars, corresponding with a mass density of about 0.04 $M_\odot$ pc^{-3}. This is exceedingly high, in view of the fact that the average gas density near $z=0$ is only 0.03 $M_\odot$ pc^{-3}. The problem is aggravated by the fact that the stars have low velocities, the dispersion in velocity in one coordinate coming out ± 10 km s^{-1}. This is considerably lower than the velocity dispersion of A-type stars, and shows that the stars are young, perhaps about 10^8 yr. How is it then that the gas has not been completely used up long ago?

I believe one cannot discard this conclusion on the ground of inaccuracy of the proper motions; on the average the motions are considerably higher than the probable errors. This is further confirmed by independent data derived by Luyten. Moreover, the entirely different investigation by Donna Weistrop led to very similar, partly even *higher* densities.

The difficulty can be *somewhat* relieved by using the 'standard' solar motion instead of the 'basic' solar motion used by Murray and Sanduleak. If the stars are quite young they may be thought to follow the motion of the interstellar medium which gives a higher solar velocity. If we do this, we find a density of 0.016 $M_\odot$ pc^{-3}, therefore about half the present gas density. The velocity dispersion is raised to ± 13 km s^{-1}.

But even such a density seems inacceptable for a general average density. A possible way out of this dilemma is that we are seeing a region where the density is considerably higher than the average. This might be the case if we happen to be inside an interstellar cloud.

DISCUSSION

Murray: I must thank Prof. Oort for this very clear exposition of the work by Sanduleak and myself. I fully agree that the choice of the basic solar motion for estimating the mean parallax was quite arbitrary; if we had chosen a larger motion, the mean parallax, and hence the density, would have been smaller. But whatever solar motion one adopts the transverse velocity dispersions appear to be relatively small. Although we only have kinematic data for some twenty stars, Sanduleak actually found 1200 stars in 120 sq deg near the North Galactic Pole; this is not an insignificant sample. It appears, however, that there may be differences between the two galactic polar caps; thus we may be seeing a localized phenomenon. Recent photoelectric photometry in four colours by D. H. P. Jones confirms that probably all the stars in the kinematic sample are in fact dwarfs.

Luyten: In connection with the Murray-Sanduleak stars I might add that whereas their distances were derived from kinematics, I remeasured all their motions and found good agreement except for the seven very small motions. I also redetermined their magnitudes and found them 0.43^m fainter than those of Murray and Sanduleak. Then all you need is to shift the spectra 0.06 of a spectral class earlier, i.e. from M3 to M2.4 and there is no (?) more discrepancy with my previous data. I would again emphasize the danger of drawing such far-reaching conclusion from such minuscule samples.

Oort: In my communication I made no use of the magnitudes to estimate the mean distances. This was derived from the reflection of the solar motion.

THE CORRELATION BETWEEN KINEMATICAL PROPERTIES AND AGES OF STELLAR POPULATIONS

J. EINASTO

W. Struve Astrophysical Observatory, Tôravere, Estonia, U.S.S.R.

The spatial and kinematical properties of galactic populations are very conservative for time changes. Therefore the study of these properties gives us certain information on the past dynamical evolution of the Galaxy, in particular on the evolution of star generating medium (interstellar gas, as generally accepted). The detailed study of spatial structure of stellar populations in our Galaxy is possible in exceptional cases only. But the study of kinematical properties is possible practically for all populations, which makes these studies very useful for cosmogonic purpose.

In order to obtain adequate quantitative information for the study of dynamical history of the Galaxy the statistical data on stellar velocities are to satisfy the following requirements: populations under study must be physically homogeneous; statistical samples of stars must be free from selection effects, especially from velocity selection; information on r.m.s. errors of observed quantities must be known in order to correct the results for accidental observational errors; the data for the determination of the age of the sample must be available.

We have collected published data on stellar velocities and determinations of kinematical parameters back to the fundamental work by Parenago (1951). A critical analysis of these data shows, however, that only a part of the available data can be used for our purpose. For all populations mean velocity dispersion $\sigma = \frac{1}{3}(\sigma_R^2 + \sigma_\theta^2 + \sigma_z^2)^{1/2}$ and mean heliocentric centroid velocity in rotational direction, $\bar{V}_\theta$, have been calculated (σ_R, σ_θ, σ_z are the velocity dispersions in galactic cylindrical coordinates). The velocity dispersions have been corrected for observational errors using a method, proposed by us (Einasto, 1955). The age of populations has been determined from Iben's evolutionary tracks. For halo populations the individual age determinations coincide within possible errors. The relative age of these populations has been estimated theoretically, adopting for oldest halo population the age of the Galaxy, 10^{10} yr, and for other halo populations an age, needed for the population considered to collapse with free fall acceleration to its observed dimensions.

The Strömberg diagram for populations studied is given in Figure 1. Populations with metal deficit are represented by open circles, populations with normal metal content by points, the interstellar gas by a cross. The smooth curve shows the mean dependence between σ and $\bar{V}_\theta$ of populations of different ages; the latter is indicated in 10^9 yr, starting from the formation of oldest galactic populations known.

The main results of this study may be formulated as follows.

(1) If we attribute all metal deficient subpopulations to the halo, then it appears that the halo is rather heterogeneous in its kinematical properties: it contains all subpopulations with velocity dispersion $\sigma \geq 50$ km s^{-1}. The corresponding axial ratio

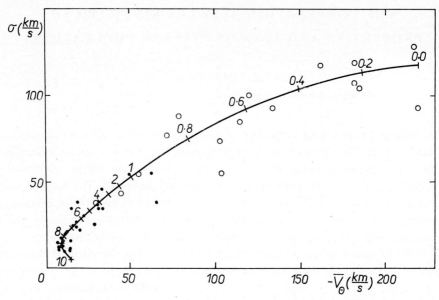

Fig. 1. The Strömberg diagram for populations. The numbers give the birthdates in 10^9 yr starting from the formation of the oldest populations.

ε of equidensity ellipsoids, calculated from our recent model of the Galaxy (Einasto, 1970), is equal to or larger than 0.10. Studying the structure of the Andromeda galaxy M31 we also came to the conclusion that its halo consists of a mixture of subpopulations with $\varepsilon \geqslant 0.10$ (Einasto, 1972).

These results show that intermediate subsystems of the Galaxy according to Kukarkin (1949) also belong to the halo.

(2) Direct age determinations of stellar populations are too inaccurate to estimate the duration of the initial galactic collapse. There exists, however, indirect observational (Sandage, 1969) and theoretical (Eggen *et al.*, 1962) evidence that the collapse proceeded in a short time scale compared with the age of the Galaxy.

(3) The populations of the galactic disc have $15 \leqslant \sigma \leqslant 50$ km s^{-1} and, respectively $0.02 \leqslant \varepsilon \leqslant 0.10$ (Einasto, 1970). The age dependence of spatial and kinematical properties of these populations may be caused by the action of irregular gravitational forces (Spitzer and Schwarzschild, 1953; Kuzmin, 1961).

(4) The subsystems of interstellar gas and young stars rotate with a velocity, smaller than the circular one. Therefore the young stellar subsystems are nonsteady and time is needed for them to obtain steady structure. This result supports the recent discovery of the non-stationary state of young populations by Dixon (1967a, b, 1968) and Jôeveer (1968).

References

Dixon, M. E.: 1967a, *Monthly Notices Roy. Astron. Soc.* **137**, 337.
Dixon, M. E.: 1967b, *Astron. J.* **72**, 429.
Dixon, M. E.: 1968, *Monthly Notices Roy. Astron. Soc.* **140**, 287.

Eggen, O. J., Lynden-Bell, D., and Sandage, A. R.: 1962, *Astrophys. J.* **136**, 748.
Einasto, J.: 1955, *Tartu Astron. Obs. Publ.* **33**, 35.
Einasto, J.: 1970, *Tartu Astron. Obs. Teated* **26**, 1.
Einasto, J.: 1972, *Tartu Astron. Obs. Teated* **40**, 3.
Jôeveer, M.: 1968, *Tartu Astron. Obs. Publ.* **36**, 84.
Kukarkin, B. V.: 1949, *The Investigation of the Structure and Evolution of Stellar Systems on the Basis of the Study of Variable Stars*, Gosud. Izd. Tekhniko-Theoreticheskoj Lit., Moscow.
Kuzmin, G. G.: 1961, *Tartu Astron. Obs. Publ.* **33**, 351.
Parenago, P. P.: 1951, *Trudy Astron. Inst. Sternberg* **20**, 26.
Sandage, A.: 1969, *Astrophys. J.* **157**, 515.
Spitzer, L. and Schwarzschild, M.: 1953, *Astrophys. J.* **118**, 106.

SPACE DISTRIBUTION AND MOTION
OF THE LOCAL H I GAS

H. WEAVER

*Dept. of Astronomy and Radio Astronomy Laboratory, University of California,
Berkeley, Calif., U.S.A.*

In a Joint Discussion devoted to the ages and kinematics of the local stars, inclusion of a paper on the local gas may seem anomalous. There is, however, strong justification for considering such a topic. Newly formed stars retain many properties of the gas from which they originated. To understand the spatial and kinematic properties of local young stars, we must understand the spatial distribution and the kinematics of the local gas from which they were formed.

A cloud within the interstellar gas can collapse gravitationally if its mass, density, and temperature satisfy the Jeans Criterion. Collapse is favored by low temperature and high density.

Various investigators have pointed out that in the Lin Spiral Density Wave Theory a shock must occur on the inner edge of a spiral arm. Such a shock compresses the gas and hence promotes cloud formation with subsequent gravitational collapse to form stars. Shu and several collaborators have shown that such a shock is very effective in triggering cloud formation in a two-phase interstellar medium of the type discussed by Field *et al.* (1969), and it is widely believed that this is the principal step in the process of forming stars.

Observationally, however, it is found that the majority of *local* dark interstellar clouds, H I regions, and young stars are located *not* in the galactic plane, as one would expect from galactic spiral shock theory, but in Gould's Belt, the irregular quasi-planar structure inclined at approximately 20° to the galactic plane and spoken of earlier in this Joint Discussion by Dr Lindblad. Such observations suggest that in the *local* region some process other than, or in addition to, the galactic spiral shock process is operative in initiating cloud and star formation. The key to understanding this situation is, I believe, knowledge of how Gould's Belt is related to the local gas out of which clouds and stars form.

In Figure 1, I have employed observations which David Williams and I made to show the column density of local neutral hydrogen over a large portion of the sky extending from longitude 10° to 250° within the latitude range +30° to −30°. On the diagram blackness indicates the column density of local neutral hydrogen. The distribution of local gas is decidedly wispy and filamentary in character. It is quite irregular in latitude extent in different longitudes, with a large extension to latitude −45° in the longitude range 150° to 210°. There are many interesting details worthy of comment on this picture, but our purpose here is to examine the overall structure of the local gas. On that scale, the local gas clearly maps onto the sky in the roughly

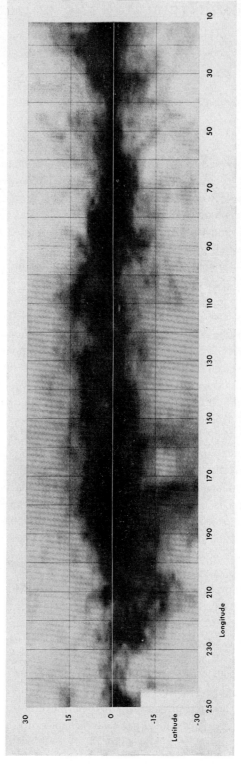

Fig. 1. The distribution of column density of local neutral hydrogen over the sky. On this picture blackness is proportional to column density.

sinusoidal pattern that characterizes Gould's Belt. For example, at $l=200°$, the column density at $b=-15°$ is five times as great as it is at $b=+15°$.

It has long been known that a high degree of correlation exists between the sky distribution we see on this diagram and the sky distribution of dark clouds, H I regions, nearby OB supergiants, and young stars in general. All local young objects are closely associated in space within the irregular gas structure we see illustrated in this slide. Since these young objects form from the gas, it is this gas structure that is the basis of Gould's Belt. The distribution of local young stars, for example is *strongly* associated with the great southward extension of the gas in the longitude range 150° to 210°.

Lindblad (1967) has shown that in the local neighborhood what he describes as a ring of gas is expanding in the region of the galactic plane. Hughes and Routledge (1972) provided additional evidence for such an expanding ring. Lindblad *et al.* (1973) have recently improved the numerical parameters originally used by Lindblad to describe the ring. All investigators have associated this expanding ring with Gould's Belt and with the expansion of the local group of early-type stars discussed by Blaauw (1956), by Bonneau (1964), and by Lesh (1968).

Figure 2a, taken from the paper by Lindblad and his collaborators, illustrates the nature of the Lindblad hypothesis. The plane of the diagram is the galactic plane; the direction to the galactic center is along the negative X-axis; galactic rotation is towards the right. From the origin of the coordinate system a ring of gas extends uniformly at a speed of, say, 4 km s^{-1}. After 60 m.y., the ring of gas occupies the position shown, with the sun in the interior of the ring at the position marked by the cross. The ring is elliptical in form because of the combination of expansion and differential galactic rotation operative on the gas.

To an observer at the Sun, the velocity of the ring would, as a function of longitude, appear as shown on Figure 2b as a solid curve. The observational points derived by Lindblad and his collaborators are shown on the diagram. From these points they locate the center of expansion at a distance of 140 pc in the direction $l=150°$. The expansion velocity is 3.6 km s^{-1}; the expansion age is 60 m.y. They find a small motion of the center with respect to the local standards of rest.

The extensive Berkeley observations of neutral hydrogen which David Williams and I have made provide the material for investigating the local hydrogen in considerable detail. Figure 3 shows as a function of longitude the velocity distribution of all neutral hydrogen encompassed in the galactic latitude range $-20°$ to $-30°$. This particular rather high latitude range was chosen for the illustration because (i) it is substantially free from contamination by any non-local gas, and (ii) it crosses a strong section of Gould's Belt at $l=200°$.

We immediately note from this diagram that the average velocity of the gas is positive and, in particular, in the vicinity of $l=0°$ and 180° the velocity of the gas is positive. Both these aspects of the velocity distribution denote expansion of the gas.

In Figure 4 mean velocity points derived from this distribution are superimposed on the theoretical curve computed by Lindblad and his collaborators. The agreement

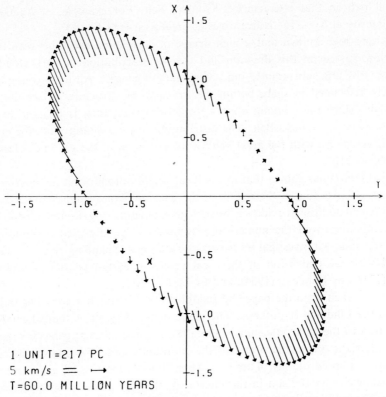

Fig. 2a. The velocity distribution investigated by Lindblad. The gas expanded uniformly from the origin of coordinates. The position of the Sun is marked by the cross. The direction to the galactic center is along the negative X-axis. See text for further explanation.

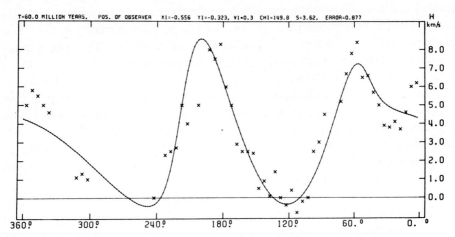

Fig. 2b. The theoretical velocity-longitude diagram for the velocity distribution shown in Figure 2a is shown by the solid line. Observational values of velocity derived by Lindblad *et al.* from the neutral hydrogen data are shown as crosses.

SPACE DISTRIBUTION AND MOTION OF THE LOCAL H I GAS 427

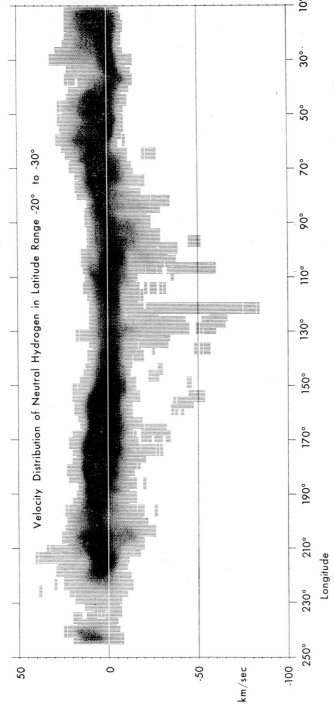

Fig. 3. The observed velocity-longitude distribution for neutral hydrogen in the latitude range −20° to −30°.

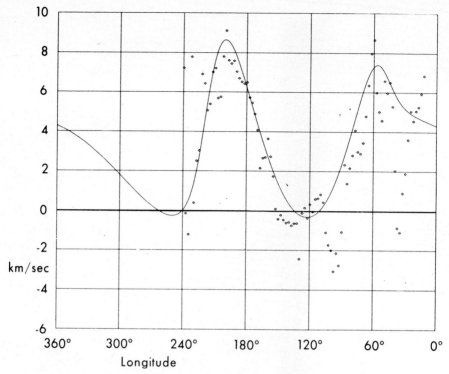

Fig. 4. Mean velocity points for the velocity-longitude distribution shown in Figure 3, superimposed on the theoretical velocity-longitude relation derived by Lindblad *et al.*

is impressive particularly in the range $l = 120°$ to $240°$, and must be taken as strong confirmation of the basic Lindblad hypothesis. In the longitude range $30°$ to $120°$ there are local perturbations in the gas. The one from $30°$ to $60°$ is connected with the north galactic spur and is particularly marked.

In Figure 5 we examine, as representative of lower latitude gas, the velocity distribution of the neutral hydrogen encompassed in the latitude range $+10°$ to $+20°$. Here there is some contamination from gas lying over distant spiral arms at high z-distances. In the diagram this contamination shows as the light background (denoting low intensity) against which the dark central part of the distribution is seen. The local gas (in which we are interested) shows as the dark low-velocity concentration in this figure. Darkness, of course, here respects radiation brightness or column density.

The end regions of this distribution are very similar to the end regions of the distribution we just saw for the latitude range $-20°$ to $-30°$. But the central part of the distribution illustrated in Figure 5 shows a strong representation of negative velocities in the longitude range $90°$ to $200°$. The Lindblad hypothesis as originally proposed does not predict such negative velocities.

The suggested explanation of this phenomenon follows from the model shown in

SPACE DISTRIBUTION AND MOTION OF THE LOCAL H I GAS

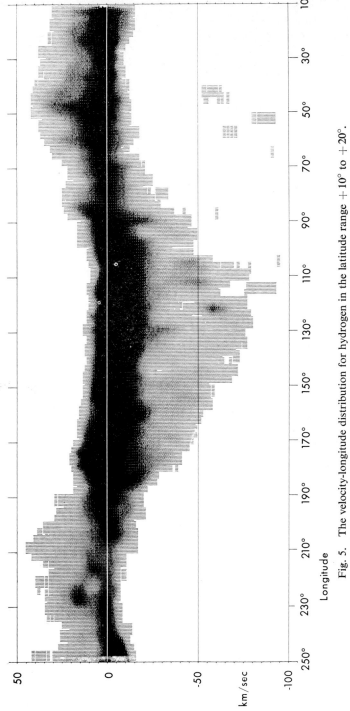

Fig. 5. The velocity-longitude distribution for hydrogen in the latitude range $+10°$ to $+20°$.

Figure 6a. At the origin of coordinates there is a source from which gas flows uniformly and continuously. After the passage of 12, 24,... 60 m.y. and so on, the gas reaches the boundaries indicated. The density of crosses in any region indicates the relative density of gas at that point. Seen from the sun, gas at each of these loci representing different ages would show velocities as indicated in Figure 6b.

If the flow of gas is continuous, then the velocity distribution we will observe is the envelope of these separate curves. Within this envelope the expected density of points in the longitude-velocity distribution will be proportional to the areal density of crosses on the velocity curves. Taking into account the density of crosses, we see that the predicted velocity-longitude distribution for a continuous flow is strikingly similar to that illustrated for the gas in the latitude range $+10°$ to $+20°$ shown in Figure 5. The observed distribution shows a somewhat greater extent in the negative velocity direction than the simple uniform continuous model predicts. The real gas flow may not be entirely uniform in all directions. For the present, however, we

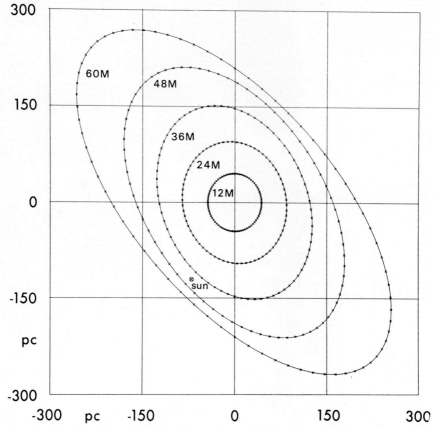

Fig. 6a. Results from a model calculation. Loci occupied by material flowing from the origin of coordinates after periods of 12×10^6, 24×10^6,..., 60×10^6 yr. Differential galactic rotation is present as well as uniform outflow from the origin. The position of the Sun is indicated. The direction towards the galactic center is as in Figure 2a. See text for explanation of the crosses on the various loci.

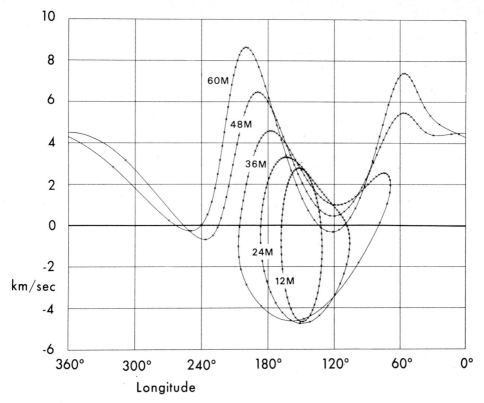

Fig. 6b. Velocity-longitude diagrams for the loci shown in Figure 6a.

ignore this slight irregularity in discussing a theoretical model. We adopt as our model of the gas flow in Gould's Belt a uniform continuous flow from a source located at a distance of approximately 140 pc in longitude 150°. When we look in the latitude range $-120°$ to $-30°$, we look above much of the continuous flow; we see the periphery of the flow, which might be described as an expanding ring. At lower latitudes our line of sight encounters the continuous flow pictured in the model.

If such a continuous flow of gas exists in Gould's Belt, and generally near the galactic plane, other objects clearly associated with the gas in the Belt must show that flow also.

The local dark clouds have long been known to be closely associated with Gould's Belt. We should thus expect the velocity distribution of these clouds to lie within the envelope of the velocity distribution representative of the local gas. The velocities of the clouds need not, of course, uniformly fill that envelope. There is no reason to expect that the clouds will be distributed uniformly throughout the gas.

Dieter (1973) has observed velocities for more than 100 dark clouds from their formaldehyde absorption lines at 4830 MHz. The velocity distribution of the dark clouds is shown in Figure 7 as a function of longitude. The dark cloud velocities are

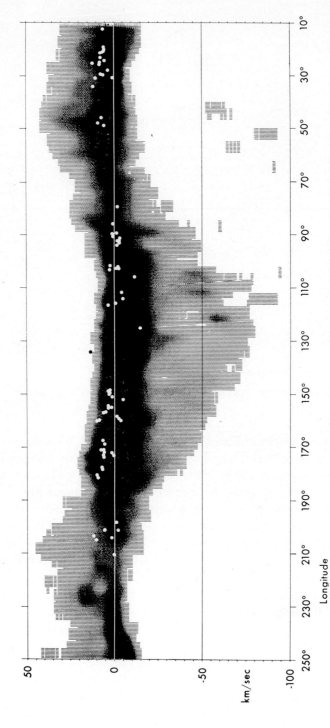

Fig. 7. Velocity-longitude distribution of dark clouds superimposed on the velocity-longitude distribution of neutral hydrogen in the latitude range $+10°$ to $+20°$. (Repeated from Figure 5.)

SPACE DISTRIBUTION AND MOTION OF THE LOCAL H I GAS   433

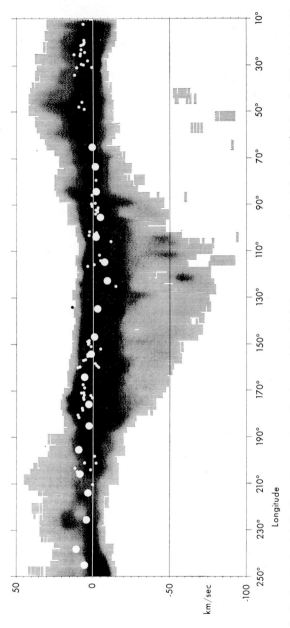

Fig. 8. Velocity-longitude distribution of young stars (the large white dots) shown as mean values taken in 10° intervals of longitude. The large white dots are superimposed on a copy of Figure 7.

shown as dots against the background of the velocity distribution of the local gas. The dark clouds show a velocity distribution characteristic of an expanding group. There are, for example, positive velocities in the regions $l=0°$ and $180°$. The velocity distribution of the dark clouds lies within the envelope of the velocity distribution of the local gas as we might have anticipated.

Young stars are closely associated with the gas and dark clouds in Gould's Belt. Lesh (1968) has discussed the kinematics of young stars, B5 and earlier, within 600 pc from the Sun. The population she studied spatially overlaps the gas in which we are interested between longitudes $60°$ and $250°$. We limit distances to 400 pc. In Figure 8 we see represented as very large dots the mean velocities of these young stars taken in $10°$ intervals over the longitude range of interest. The stellar mean velocities – the large dots – lie closely among the dark cloud velocities and well within the gas velocity distribution envelope. The young stars in Gould's Belt also form an expanding group like the dark clouds and reflect the kinematic properties of the local gas.

The local dark clouds and the local young stars are closely associated in Gould's Belt with the local gas from which they must have originated. They all show a common kinematic pattern of outflow from a specified region of the galactic plane.

What is the source of this outflow, which appears to have been surprisingly constant over a time scale at least as great as the ages of the clouds and stars we have considered? For an answer, we examine the gas at higher latitudes.

An example of the character of the velocity distribution of local neutral hydrogen at high galactic latitudes is seen in Figure 9, which shows the velocity distribution for all the gas in a great circle through the galactic poles, perpendicular to the galactic plane. The left side of each figure refers to longitude $49°.5$; the right hand side of each figure refers to longitude $229°.5$. The left figure shows the directly observed velocity distribution of the hydrogen gas; the right-hand figure displays the mean velocity of the hydrogen as a function of latitude. Negative velocity is towards the center of the circle, positive velocity is outwards. Stationary gas, that is, gas showing no radial motion with respect to the local standard of rest, would lie along the zero-velocity circle. Note that the gas is not stationary. In the region of the galactic plane, the average velocity is positive, denoting outflow from a source as we have already found. As we go towards higher latitudes in both the northern and southern hemispheres, the mean velocity goes through zero and becomes negative, denoting downflow. The mean downflow is at least as great as 10 km s^{-1}.

The phenomenon of universal negative velocity of the gas in the polar regions has long been known. The velocity pattern of downflow from the polar regions in both hemispheres, and outflow at low latitudes, is completely characteristic of the local region of the Galaxy. A major gas circulation pattern exists in the solar neighborhood.

A model that accounts for all of the observed phenomenons is shown in Figure 10.

There is flow from above the galactic plane in both hemispheres. The two approximately equal streams collide. Gas flows outward, away from the collision region. The fact that the colliding streams are not precisely equal or precisely centered

SPACE DISTRIBUTION AND MOTION OF THE LOCAL H I GAS 435

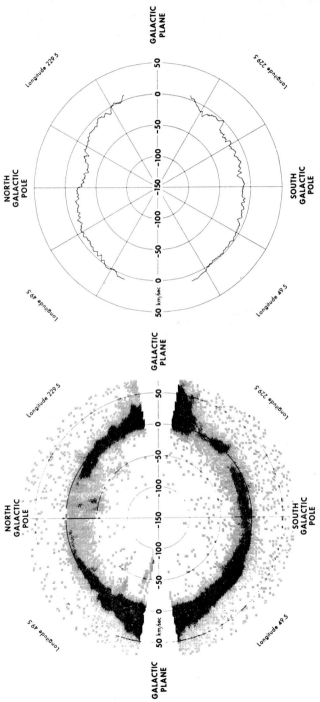

Fig. 9. Velocity-latitude distribution of neutral hydrogen in a plane perpendicular to the galactic plane. See text for complete explanation.

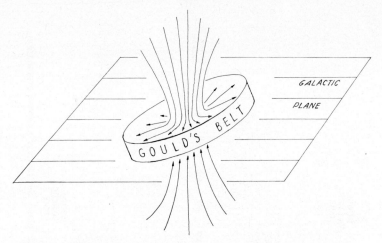

Fig. 10. Model showing schematically the explanation of Gould's Belt.

accounts for the roughly 20° tilt of the principal plane of the flow which is, of course, Gould's Belt. Irregularities of outflow exist in both velocity and direction as, for example, the great southward extension of the local gas in the longitude range 150° to 210°. The average half-thickness of the flow is ~ 200 pc, slightly more than one scale height, but the thickness is quite variable from one longitude to another. The outflow is ~ 4 km s^{-1} but shows variability with direction and location. The downflow velocity is at least 10 km s^{-1}, and probably rather higher. About 10^6 solar masses of hydrogen are involved in the flow; the energy involved is of the order 10^{50}–10^{51} ergs. The downflowing gas appears to cover approximately 40% of the area of the expanding gas region. About 1.7×10^{-2} $M_\odot$ yr^{-1} flows down onto the galactic plane. The gas density per cm^3 is approximately twice as great in the downflow region of the plane as it is in the region near the edge of the flow.

Most importantly, the collision of the two streams causes an increase in pressure of at least a factor of two in the colliding streams of gas, and possibly by a much greater factor depending on the exact character of the downflow and collision. Such overpressure strongly stimulates cloud formation and subsequent star formation in the outward flowing gas.

The age of the flow may be estimated from the age of the Lindblad expanding ring, which is here identified with the edge of the outflow of gas, 60 m.y., or from the expansion ages of objects participating in the flow, 45–90 m.y. We take as an estimate of the age a value 45–60 m.y., of the order $\frac{1}{4}$ to $\frac{1}{3}$ of a galactic rotation period.

On the galactic scale, Gould's Belt may be thought of as a small transitory gas eddy in the local spiral feature. What initiates the downflow that starts the eddy?

Observations show that the regions above spiral arms are filled with substantial masses of gas at considerable z-distances above the axis of the arm. Kepner (1970) and other observers have found concentrations of gas having masses as great as 0.5–1×10^6 $M_\odot$ at z distances greater than 1 kpc.

SPACE DISTRIBUTION AND MOTION OF THE LOCAL H I GAS   437

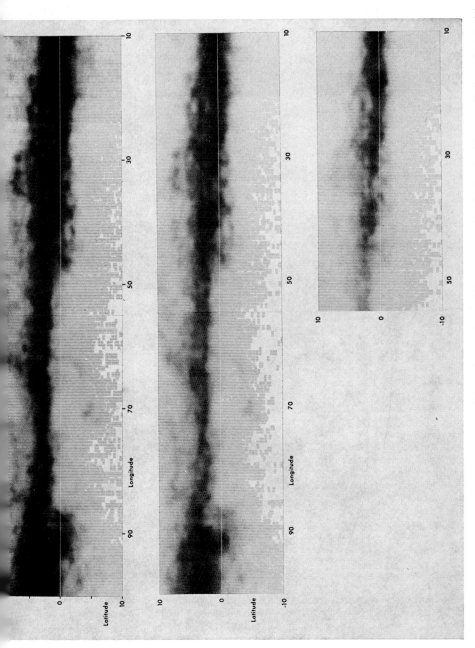

Fig. 11. A computer-produced picture of the outer arm of the galaxy as it would appear on the sky in the 21-cm radiation of neutral hydrogen. Three pictures of the outer arm are shown. These are analogous to optical photographs taken on plates of different speeds and contrasts. The pictures shown were produced by using in the computer program three different intensity-blackening relations, the analogs of different HD curves in the case of photographic plates.

Figure 11 shows a computer-produced picture of the outer arm of the Galaxy as it appears on the sky in neutral hydrogen radiation. The large amount of gas present in the form of overlying cloud-like structures is evident. Many features are a kpc above the axis of the arm; a great many are at heights of 400 to 500 pc. Masses of many of these features are $\sim 10^6 \, M_\odot$. Such features above the arms cannot be permanent. They must be buoyed up from the arm for a short time and then return to the plane, guided, perhaps, by the buoyed-up or inflated magnetic lines as in a Parker-type instability. The velocity of downfall we observe in the solar neighborhood corresponds to free-fall from z-heights of the order of 300–400 pc.

One may suggest, then, the following schematic picture of the gas in the solar neighborhood. Masses of gas at moderate z-distances above the plane become unstable and fall back to the plane. Two streams of descending gas, one from above the plane, the other from below, flow onto the plane, collide, and form the Gould's Belt gas structure we observe. The tilt of the gas structure and irregularities within it are caused by density and velocity irregularities in the colliding streams of gas. The collision of the down flowing gas streams compresses the gas and promotes the formation of clouds with subsequent gravitational collapse to form stars. The stars formed from the gas reflect the spatial and kinematic properties of the gas. The gas flow has existed for only a small fraction – perhaps $\frac{1}{4}$ to $\frac{1}{3}$ – of a galactic rotation period. We may expect that in a further small fraction of a rotation period, this small local eddy in the gas will damp out and disappear, leaving behind, for a much longer time, the group of aging stars to which it gave birth.

References

Blaauw, A.: 1956, *Astrophys. J.* **123**, 408.
Bonneau, M.: 1964, *J. Obs.* **47**, 251.
Dieter, N. H.: 1973, *Astrophys. J.* **183**, 449.
Field, G. B., Goldsmith, D. W., and Habing, H. J.: 1969, *Astrophys. J. Letters* **155**, L149. (See also Goldsmith, D. W., Habing, H. J., and Field, G. B.: 1969, *Astrophys. J.* **158**, 173.)
Hughes, V. A. and Routledge, D.: 1972, *Astron. J.* **77**, 210.
Kepner, M.: 1970, *Astron. Astrophys.* **5**, 444.
Lesh, J. R.: 1968, *Astrophys. J. Suppl. Ser.* **17**, 371.
Lindblad, P. O.: 1967, *Bull Astron. Inst. Neth.* **19**, 34.
Lindblad, P. O., Grape, K., Sandqvist, Aa., and Schrober, J.: 1973, *Astron. Astrophys.* **24**, 309.

DISCUSSION

Van Woerden: Dr Weaver's paper is highly interesting and thought-provoking, but I wish to make two remarks.

(1) There is, in the low-velocity hydrogen, another striking feature in addition to Gould's Belt. Fejes and Wesselius (*Astron. Astrophys.* **24**, 1, 1973) have pointed out two ridges in opposite locations, together outlining a tilted disk ('Scheve Schijf') inclined by 45° to the galactic plane and with its pole at $l \sim 120°$, $b = 45°$. This disk may be genetically related to the intermediate-negative-velocity gas discussed by Wesselius and Fejes (*Astron. Astrophys.* **24**, 15, 1973) and to the high velocity clouds.

(2) Weaver's picture of two gas streams flowing in from high latitudes, $|b| \sim 60°$ to 70°, towards the centre of Gould's Belt appears overly schematic. At high negative latitudes his stream may well be present. At high positive latitudes, however, the scene is dominated (cf. Blaauw and Tolbert,

Bull. Astron. Inst. Neth. **18**, 405, 1966; Wesselius and Fejes, *Astron. Astrophys*. **24**, 15, 1973) by a big hole in the low-velocity hydrogen *and*, closely over lapping, a big complex of hydrogen with average velocity -42 km s^{-1}. Wesselius and Fejes show in detail that the hole and the complex are probably related, and that the complex is likely to come, at a speed of 70 km s^{-1}, from the direction $l = 120°$, $b = +40°$. They estimate the distance of this complex at 70 pc, its mass at 1800 $M_\odot$, its energy at 0.9×10^{50} erg. The origin of this complex is probably the same as that of the high-velocity clouds, which come from the same direction.

Weaver: Dr van Woerden has introduced the topic that would have been the second part of my talk at this Joint Discussion if I had been allocated another ten or fifteen minutes. I will comment on his second remark first.

Before the appearance of the paper by Wesselius and Fejes, completely independent analysis of the Berkeley Neutral Hydrogen Survey data made without any knowledge of their work, led to precisely the same picture of the hole and the intermediate-negative-velocity gas derived by Wesselius and Fejes. My estimates of distance, mass, and so forth are in excellent – indeed, in several instances, exact – agreement with theirs.

What Dr van Woerden terms a 'complex of hydrogen' I would describe rather more specifically and simply as a mass of hydrogen consisting of $\sim 2 \times 10^3 \, M_\odot$ moving in a non-circular galactic orbit with small z-motion. This mass of gas is now approximately 70 pc north of the galactic plane, and is colliding with the local gas. It probably originated in the next outer arm of the Galaxy, perhaps as much as 2×10^7 yr ago, when its orbit was changed from circular or near circular to quite elliptical by some energetic event such as a supernova explosion. While internal motions have dispersed the mass over several hundred parsecs during this time interval, it appears to have a fairly compact core that is the most evident colliding mass.

Since the orbit of this gas is non-circular, the gas has a radial component of motion inward towards the galactic center, and it has a lower speed in the direction of galactic rotation than the local gas. We are running into the mass of gas. From our point of view, however, and with respect to the local standard of rest, the mass of gas is like a jet stream moving just over our heads at a distance of 70 pc with a velocity of possibly 70 km s^{-1}. It has seriously perturbed a portion of the down-flowing gas; it has punched a hole in a portion of that gas. But the duration of the collision is short, a few percent of the duration of the down flow, and the mass of the colliding gas is very small compared to the mass in the down-flowing stream. The collision now in progress will certainly affect the future development of the evolving structure we call Gould's Belt, but it does not completely obscure the stream of down-flowing gas in the northern hemisphere as Dr van Woerden's remark implies. The down-flowing stream is easily visible.

I regret that there is not more time to discuss this interesting topic on which a great deal of information is now available.

In regard to Dr van Woerden's first comment I would remark only that the intriguing 'Scheve Schijf' found by Fejes and Wesselius may well be related to the collision phenomenon I have just been discussing. Clearly, however, the Scheve Schijf is a minor feature on the scale of Gould's Belt; its total mass is a few percent of that of Gould's Belt.

Tolbert: Why should the streams of gas from either side of the galaxy come inward at the same time? The gas distributions away from the plane are very different on either side, and it's not clear to me why these two streams should arise at the same time and in the same place near the Sun.

Weaver: There is no immediate and satisfying answer to Dr. Tolbert's question. I have discussed what is observed and I have tried to synthesize what is observed in the form of an understandable and physically consistent model.

There is at present no clear understanding of the physics of high-z clouds of the sort we saw so clearly over the spiral arm shown on the last slide. We do not understand how such masses of gas rise above the plane, or how their flow back to the plane is initiated.

In view of my discussion of Dr van Woerden's remarks, it should be clear that what may appear as a very different distribution of gas on the two sides of the plane now, may be no more than a momentary perturbation on the time scale of Gould's Belt, which is the time scale of interest in these considerations. What we see now was not the case over the time period of significance in the formation of Gould's Belt.

Murray: It seems remarkable that the suspected high concentration of young M stars in the north galactic cap is just where your gas density is least; whereas in the south, the gas density is more dense and there may not be as many low velocity M dwarfs.

Weaver: The gas density in the north galactic cap is normal over the z-range 0 to approximately 70 pc. The young M dwarfs to which Dr Murray refers are, I believe, closer to the Sun than 70 pc on at least no farther away than 70 pc. I see no suggestion of a discrepancy that the stars are found in a region of low gas density as implied by Dr Murray's question.

We should also keep in mind that 'young' is a relative term. Gould's Belt is young, perhaps 60×10^6 yr old, the collision (and the resultant low gas density above the plane) discussed earlier is young, perhaps 2×10^6 yr old. How old are the M dwarfs?

Clube: Would you like to comment on the fact that A stars are moving towards us with the same velocity as the HI? It seems to me that the idea of gas streams contained by magnetic fields may be difficult to uphold.

Weaver: It is not my recollection that the A stars are moving with such velocity. The down-flow velocity of the gas is 10 or more km s^{-1}. The radial velocities of A stars in the north galactic polar cap were observed by Perry (*Astron. J.* **74**, 139, 1969). His results indicate for A stars within approximately 25° of the NGP a mean radial velocity of essentially zero. He did find, in the course of his investigation, that the A stars appear to form two populations which have different kinematics properties. The physical meaning of this fact remains to be clarified, but it does not appear to indicate a discrepancy of the sort mentioned by Dr Clube.

We do not understand the physics of the high-z concentrations of gas observed to lie above the spiral arms – how the gas gets there or what triggers the process by which it falls back to the plane. A suggestion made some time ago by Parker in a different context is that gas in the spiral arms may be buoyed up by the pressure of cosmic rays and may slide back to the plane along magnetic lines of force in the arms. I do not propose this as *the* mechanism operative here; I suggest that it is one possibility that is well established in the literature and that we should consider in the present context.

INFLUENCE OF A SPIRAL GRAVITATIONAL FIELD ON THE OBSERVATIONAL DETERMINATION OF GALACTIC STRUCTURE

C. C. LIN
Massachusetts Institute of Technology, U.S.A.

C. YUAN
City College of New York, U.S.A.

and

W. W. ROBERTS
University of Virginia, U.S.A.

(Presented by C. Yuan)

1. The Plan of Investigation

To investigate the nature and the extent of the influence of a spiral gravitational field on the observational determination of galactic structure, we examine several models of the solar vicinity, including a spiral gravitational field and the accompanying shocks. (These models are described in the Appendix.) We calculate the radial motions and the proper motions that would be observed from the solar vicinity. *We then treat these theoretically calculated data by the conventional methods used for analyzing observational data.* In this way, we hope to recover the galactic parameters originally adopted in our models. The accuracy of this process is then a measure of the uncertainty in the existing determinations of galactic constants, because of the influence of the spiral gravitational field.

2. Observational Justification of this Approach

Before this procedure can be justified, we must examine whether the field of flow calculated is qualitatively of the same nature as the field of stellar motions observed. As it turned out, the outstanding features of the two fields of motion appear to be in general agreement for the model already adopted (Lin *et al.*, 1969), on the basis of a number of other observational data. In this model, the Sun is placed approximately midway between two major spiral arms. The comparison of these fields will be presented below in this section.

In contrast, if the Sun were placed near a major spiral arm, there would have been rather peculiar features noticeable in the field of motion (see Section 4 below), in strong disagreement with observations.

Figure 1 shows the theoretical line-of-sight velocity v_{ls}, divided by radial distance r, plotted against galactic longitude for three radial distances, $r = 1, 2, 3$ kpc respectively.

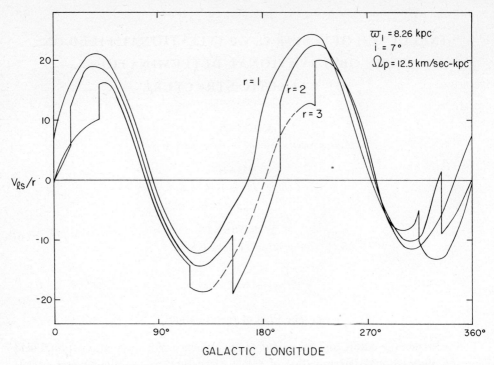

Fig. 1. The plot of v_{ls}/r against galactic longitude, where v_{ls} is the light-of-sight velocity, and r is the radial distance of the star (theory).

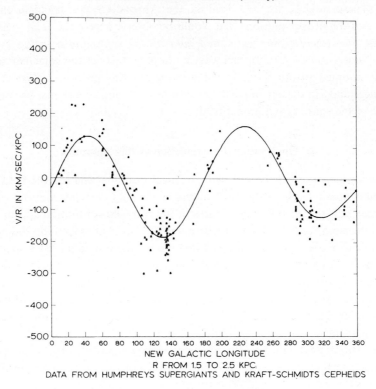

Fig. 2. Same as Figure 1, observation.

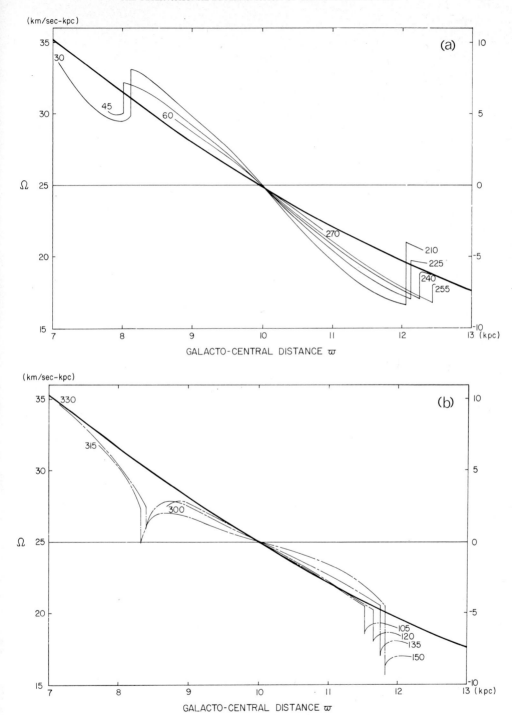

Fig. 3. The rotation curve of the Galaxy and the theoretically expected deviations. (a) First and third quadrants ($l = 0°–90°$, $180°–270°$). (b) Second and fourth quadrants ($l = 90°–180°$, $270°–360°$).

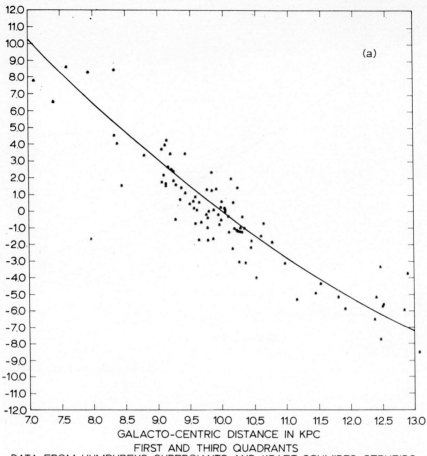

Fig. 4a.

We also prepared a similar plot (Figure 2) of the combined data for early-type stars, cepheids, and supergiants (cf. papers by Kraft and Schmidt, 1963; Rubin *et al.*, 1962, 1964; Humphreys, 1970). Similar characteristics can be noted.

A more clear-cut demonstration of the similarity of the flow fields is shown by Figure 3, where $v_{ls}/\varpi_0 \sin l$ is plotted against $\varpi - \varpi_0$ ($\varpi =$ galacto-centric distance, whose value ϖ_0 for the Sun is assumed to be 10 kpc). The slightly curved thick line is the Schmidt curve. The other curves indicate where the stellar data may be located. From the figure, one can clearly expect the data to drop below the Schmidt curve near $\varpi = 12$ and then rise above it at a somewhat larger distance. This is a prominent characteristic noted by the observers, and attributed by them to the peculiar behavior in the Perseus arm. (See Figures 4a and 4b, which are reproduced from papers by Kraft and Schmidt, and by Humphreys.)

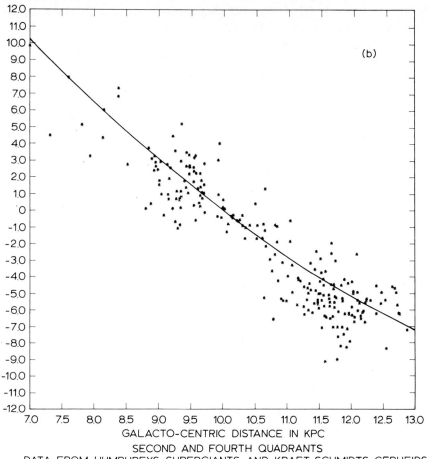

Fig. 4b.

Figs. 4a–b. The rotation curve of the Galaxy and the observed deviations. (a) First and third quadrants. (b) Second and fourth quadrants.

3. Results

A summary of the principal results is presented in the following tables. We note that the Oort constants are recovered with good accuracy in the models S-1 and E-1. (These models are described in the appendix).

Especially the determination of

$$\frac{\Theta}{\varpi} = A - B$$

is blessed with extremely good accuracy. There is a good theoretical reason for this fact. Thus, the influence of the effects we have studied here appears to be below that of the inaccuracy in the observational determination of proper motions.

We have not yet calculated the curvature term from the theoretical data shown in part in Figure 3. But it is clear from the plot that we would obtain values similar to those obtained by earlier authors. The data show clear upper limits to the curvature term in the rotation curve, but not a clear lower limit.

We have also adopted a method of analysis somewhat more general than the customary ones. We make *Fourier analyses of the radial and proper motions at several radial distances from the Sun.*

We write

$$\frac{v_{ls}}{r} = \alpha_0 + \sum_{n=1}^{\infty} (\alpha_n \cos nl + \beta_n \sin nl).$$

Two samples of such Fourier coefficients are shown. Clearly, the values of β_2 give good approximations to the Oort constant A. The coefficient α_0 is K, and is indeed

TABLE I
Tabulation of numerical values of A, B, $A-B$, $A+B$

1. Oort constant A; assumed value = 15 km s^{-1} kpc^{-1}

	$r=1$	$r=2$	$r=3$
S	14.7	14.3	13.9
E-1	15.2	15.1	14.7
E-2	17.2	17.0	14.8
S-1	16.6	16.3	15.1

2. Oort constant B; assumed value = -10 km s^{-1} kpc^{-1}

	$r=1$	$r=2$	$r=3$
S	-10.3	-10.6	-10.9
E-1	-8.8	-9.5	-11.1
E-2	-7.0	-9.3	-11.7
S-1	-7.9	-9.9	-11.1

3. Angular velocity $A-B$; assumed value = 25 km s^{-1} kpc^{-1}

	$r=1$	$r=2$	$r=3$
S	25.0	24.9	24.8
E-1	24.0	24.6	25.8
E-2	24.2	26.3	24.5
S-1	24.5	26.2	26.2

4. Radial gradient of circular velocity $A+B$; assumed value = 5 km s^{-1} kpc^{-1}

	$r=1$	$r=2$	$r=3$
S	4.4	3.7	3.0
E-1	6.4	5.6	3.6
E-2	10.2	7.7	3.1
S-1	8.7	6.4	4.0

TABLE IIA
Fourier coefficients for theoretical values of v_{ls}/r(E-1)

Distance (kpc)	1	2	3
α_0	4.42	4.03	−0.12
α_1	−2.79	−0.34	0.36
β_1	−1.05	−2.37	−3.66
α_2	4.10	4.06	−1.81
β_2	15.24	15.08	14.69
α_3	−0.60	1.17	0.27
β_3	0.48	0.39	0.90

TABLE IIB
Fourier coefficients for theoretical values of v_r/r(S-1)

Distance (kpc)	1	2	3
α_0	5.00	1.81	−0.16
α_1	−0.96	1.88	0.31
β_1	−1.53	−3.33	−4.10
α_2	4.26	−0.78	−1.13
β_2	16.57	16.13	15.05
α_3	0.17	1.46	−0.38
β_3	0.01	−0.50	0.88

found to agree with the value of K calculated by following density changes along a stream line. The value of K is positive, consistent with our theoretical picture of an expanding motion between spiral arms (cf. Figure 5). Reasons can be given why a smaller value (or even a negative value) is obtained by earlier authors from observational data. Briefly, it is associated with a bias of observational data because the young objects are found predominantly in the spiral arms. This effect can be detected, when Figure 3 is compared in detail with similar plots based on observational data. Fortunately, when the same method of Fourier analysis is applied to observational data, the value of the Oort constant A obtained still agrees approximately with those obtained from the theoretical curves.

4. Discussion

To summarize, we have made the following two points:

(1) The systematic deviation of the observed stellar motions from the mean rotation curve shows features similar to those of the streaming motions due to the presence of a spiral gravitational field, provided that we adopt a model similar to that used by Lin *et al.* (1969).

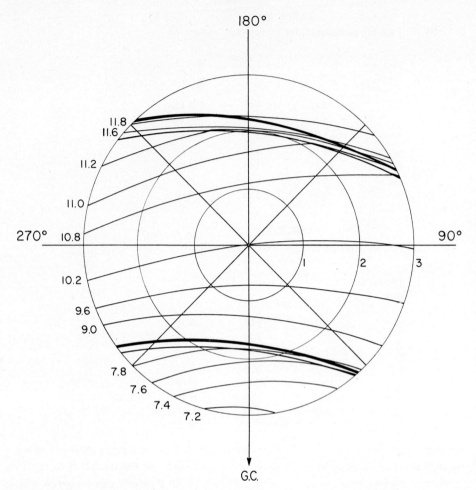

Fig. 5. The flow field in the solar vicinity. Divergence of streamlines indicates that there is expansion in the solar vicinity.

(2) The conventional methods of analysis of stellar motions yield a quite accurate model of the rotation curve of the Galaxy. The determination of Oort constants is not much influenced by the spiral gravitational field.

To gain a better overall perspective, we must ask ourselves what consequences would follow if a different model were adopted. We know that the Sun is located at the inner side of the Orion arm. However, we have insisted that the Orion arm is not a major spiral arm because there is a paucity of important H II regions. Similarly for the Carina arm. Let us see what the consequences are if the Sun were close to a major arm. If the center of a major arm were placed at 10.3 kpc, the diagrams in Figures 1 and 3 would look like Figures 6 and 7. Especially in Figure 6, we note that the double-sine nature of the curve is completely lost, in sharp contrast with well-known obser-

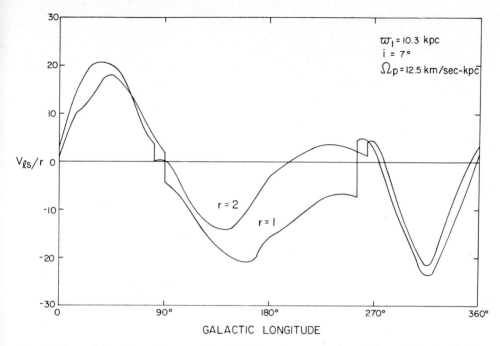

Fig. 6. Same plot as Figure 1 for a model with the Sun close to a major spiral arm. The double-sine character of the curve is suppressed.

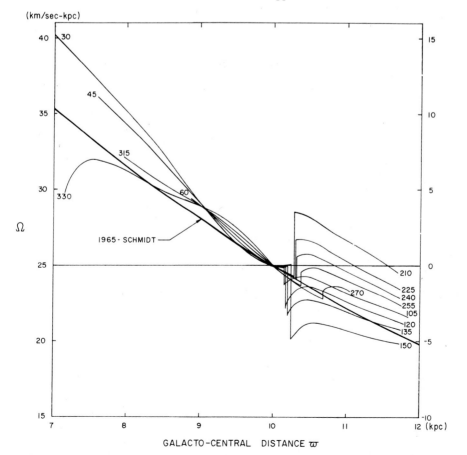

Fig. 7. Same plot as Figure 3 for the model of Figure 6. Note the difference with Figure 3.

vational data. Thus, the good agreement between the theoretical and observed streaming motions obtained in the present study also lends general support to the model adopted in earlier studies.

Appendix – The Models

Five theoretical models were investigated. These are designated S-1, S-2, S-3 and E-1, E-2. The spirals in the S series have a pitch angle of seven degrees; the E series, eight degrees. The location of the Sagittarius and the Perseus arms are given in the following table

	Sagittarius arm (in kpc, from G.C.)	Perseus arm (in kpc, from G.C.)
S-1	8.26	12.3
S-2	8.5	12.55
S-3	7.0	10.3
E-1	7.9	12.4
E-2	7.4	11.5

Acknowledgement

This work is supported in part by the National Science Foundation.

References

Humphreys, Roberta M.: 1970, *Astron. J.* **75**, 602.
Kraft, R. P. and Schmidt, M.: 1963, *Astrophys. J.* **137**, 249.
Lin, C. C., Yuan, C., and Shu, F. H.: 1969, *Astrophys. J.* **155**, 721.
Rubin, V. C., Burley, J., Kiasatpoor, A., Klock, B., Pease, G., Rutscheidt, E., and Smith, C.: 1962, *Astron. J.* **67**, 491.
Rubin, V. C. and Burley, J.: 1964, *Astron. J.* **69**, 80.

DISCUSSION

Buscombe: (1) A new finding list of 1569 supergiants is available from me at the Northwestern University. Colour excesses and radial velocities are still needed for nearly half of them. (2) Formally, I have shown (*Observatory*, 1971) that the kinematics of young luminous galactic clusters is fitted by algebraically larger A and B than the older, less luminous open clusters.

Yuan: Of course, one has to look at the actual distribution of these young galactic clusters in their longitudes and distances before making any fair remark. The selection effect is not very strong here. From our theoretical study, young stars clustered near the shock region indeed tend to give a relatively higher value of A. This certainly is consistent with your findings.

SOLAR NEIGHBOURHOOD AS THE LOCAL MACROSCOPIC VOLUME ELEMENT WITHIN THE GALAXY

T. A. AGEKJAN and K. F. OGORODNIKOV
Dept. of Astronomy, University of Leningrad, U.S.S.R.

The notion of the local Macroscopic Volume Element (MVE) is introduced in order to facilitate the application of hydro-dynamical methods of treating internal motions in stellar systems and in particular in our Galaxy. The MVE must characterize the properties of the stellar system in a given point within the system. Hence the diameter of the MVE must be taken as small as possible; it must, however, contain a sufficiently large number of stars in order that the laws of statistics might be applied. The greater is the local star density the smaller can be made the size of the MVE.

Modern catalogues of nearby stars by Gliese (1969) and Woolley *et al.* (1970) afford a good collection of data for a determination of local characteristics of the Galaxy in the immediate neighbourhood of the Sun. In the first section we discuss such characteristics as the local stellar density, the local luminosity function and the influence of observational selection upon them. In the second section we consider the local solar motion and the effect of the asymmetry of stellar motions. The responsibility for the content of paragraph 1 lies primarily upon T. A. Agekjan while that for paragraph 2 upon K. F. Ogorodnikov. We both are indebted to G. V. Ishkhanov and to L. V. Chuvicova for substantial help in numerical computations.

(1) The stellar density, the proportion of double and multiple stars and the luminosity function of stars were determined using the catalogue of nearest stars by Woolley *et al.* (1970).

The method of extrapolation to zero distance was applied. In this method several overlapping concentric spherical layers within the sphere around the Sun are considered and the mean density of registered catalogue objects $D^*(r_i)$ in each spherical layer is considered.

r_i is the mean distance of the ith spherical layer from the center.

The solution of the equations

$$a + br_i + cr_i^2 = D^*(r_i)$$

by the method of least squares gives

$$a = D^*(0) = D$$

the true density of objects in the neighbourhood of the Sun.

In our calculations 22 overlapping spherical layers were considered in the sphere of radius 25 pc, the common part of two adjacent layers being 3 pc thick.

The results are:

(a) The mean stellar density in the neighbourhood of the Sun is equal to

$$D = 0.138 \pm 0.009 \text{ stars pc}^{-3}.$$

(b) The proportion of stars which are components of double and multiple systems among all stars around the Sun is equal to

0.73 ± 0.04.

(c) All the stars were divided in eight groups according to their absolute magnitudes. The mean density of the stars of each individual group was calculated by means of Equation (1). Table I shows the proportions of stars in each interval of absolute magnitude among all stars population.

TABLE I

Luminosity function

M	Δ
$-1.0 \leqslant M \leqslant +4.6$	0.0052 ± 0.0008
$+4.5 < M \leqslant +6.5$	0.0107 ± 0.0008
$+6.5 < M \leqslant +8.5$	0.0088 ± 0.0007
$+8.5 < M \leqslant +10.0$	0.0079 ± 0.0008
$+10.0 < M \leqslant +11.5$	0.0237 ± 0.0011
$+11.5 < M \leqslant +13.5$	0.0254 ± 0.0029
$+13.5 < M \leqslant +15.5$	0.0112 ± 0.0012
$+15.5 < M \leqslant +19.5$	0.0071 ± 0.0009

The corresponding luminosity function has the appearance shown in Figure 1.

(2) The solar motion ordinarily determined from motions of stars brighter than some given visual magnitude has little sense, if any, from the point of view of stellar dynamics since it is based practically upon stars of large absolute brightness only and thus disregards the multitude of dwarf stars which really constitute the main content of the solar surroundings, i.e. the MVE. In order to make first steps in clearing up the situation we determined the linear galactic components u_0, v_0, w_0 of the solar motion. These quantities were computed for 25 concentric spherical volumes with radii ranging from 10 to 25 pc. The results are shown in Figure 2. The total number of stars used is 1361 (for $r = 25$ pc). The smallest of the spheres has the radius equal to 10 pc and contains 202 stars. The intermediate radii in the average differ from the adjacent ones by approximately 0.6 pc but in fact the radii were allowed to vary in order that the increase of the number of stars in adjacent regions would be more or less uniform. It appears from Figure 2 that with increasing r the values of u_0 and v_0 are steadily increasing while those of w_0 are decreasing.

This phenomenon is not easy to explain. It may be suggested that it is due to the asymmetry of stellar motions. It is now generally accepted beginning with classical works of Strömberg and Oort that high velocity stars are those which move along very elongated galactic orbits and belong to Population II. This effect is more clearly seen in the v_0 and w_0 components. The larger is the spherical volume the greater is the percentage of Population II stars within it since, roughly speaking, the latter rises proportionally to r^3 while the number of the 'flat' Population I stars rises only as r^2.

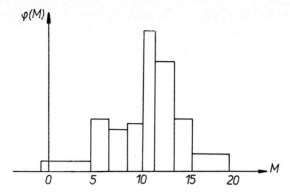

Fig. 1. The luminosity function.

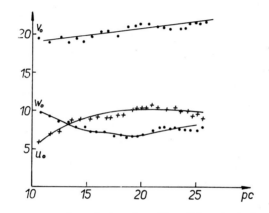

Fig. 2. Velocity components dependent on the radius of the spherical volume considered.

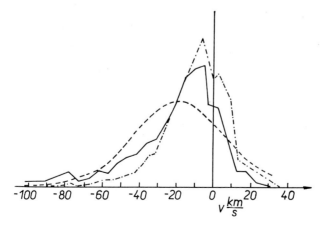

Fig. 3. Distribution of the v-component. Observed (full), best fitting Gaussian (dash), and the weighted distribution (dot-dash).

In the v_0 component the Population II stars lag behind the Population I and thus increase the solar velocity. This we shall consider in some detail later on. On the contrary, the surplus of Population II stars makes the w_0 component smaller since the elongated orbits which happen to pass through the solar neighbourhood must have preferentially orbits with small inclinations towards the galactic plane, i.e. small w-components. The increase in the u_0 component perhaps is due to a minor asymmetry of stellar motions in projection to this direction. This should have been expected apriori since the axis of Strömberg's asymmetry as determined by himself makes an angle of 6°.5 with the adopted v-direction. Of course all these remarks are of qualitative character only and require a more detailed consideration.

The stellar velocity components most strongly affected by the asymmetry are the v-components. This is clearly seen in Figure 3 where the distribution curve of v-components of 1355 stars of the catalogue by Woolley *et al.* is plotted (full line). It is seen that at the positive end the curve sharply drops to zero at $v = 35$–40 km s^{-1} while toward the negative side the curve streches up to -125 km s^{-1} and there are a few stars (12 in number) much farther off, up to more than -400 km s^{-1}, while at the positive side there is not a single star. This is quite in accordance with Oort's high velocity star phenomenon.

The ordinary procedure with assigning equal weights to all stars, except 12 'discordant' ones which were rejected, gives $v_0 = 19.6$ km s^{-1} and $\sigma = 24.4$ km s^{-1}. The dashed smooth curve represents the best fitting Gaussian with the same values of the three parameters as the observed (full) curve.

The ordinary procedure of deriving the solar motion (this time v) is equivalent to assigning equal weights $p = 1$ to all stars except the discordant ones to whom we assign the weight $p = 0$. The difference between the observed curve and the Gaussian is marked. The observed curve shows, besides the asymmetry, a substantial positive excess which is an indication of a statistical non-homogeneity of the stellar ensemble. A more rational method of assigning weights may be the following one (s.f. Ogorodnikov, 1928).

We put the observed distribution function of $F(v)$ to be a superposition of Gaussians with different centers and dispersions in the form

$$F(v) = \frac{N}{\pi^{1/2}} \int_0^\infty f(h) \, h e^{-h^2(v - v_h)^2} \, dh, \tag{1}$$

where $h^2 = \tfrac{1}{2}\sigma^{-2}$ and v_h is a partial centroid of the particular Gaussian which depends upon its value of h. $f(h)$ is an unknown normalized distribution function of the h-values. We then easily have that

$$\varphi(v) = -\frac{F'(v)}{2F(v)} = \overline{h^2(v)} \left[v - \overline{v_h(v)} \right], \tag{2}$$

where

$$\overline{h^2(v)} = \frac{\int_0^\infty f(h) h^3 \exp[h^2(v-v_h)^2] \, dh}{\int_0^\infty f(h) h \exp[h^2(v-v_h)^2] \, dh};$$

$$\overline{v_h(v)} = \frac{\int_0^\infty f(h) v_h h^3 \exp[h^2(v-v_h)^2] \, dh}{\int_0^\infty f(h) h^3 \exp[h^2(v-v_h)^2] \, dh}$$
(3)

are respectively the average weight and the weighted mean value of v_h corresponding to all stars of the ensemble which have the same velocity v.

But following Strömberg (1923) we may write

$$v_h = ah^{-2} + b,$$
(4)

where $a = -0.0096$ s km^{-1}; $b = -10.0$ km s^{-1} are Strömberg's constants. Substituting (4) into the formula for $\overline{v_h(v)}$ we get

$$\overline{v_h(v)} = a[\overline{h^2(v)}]^{-1} + b$$
(5)

and then using (2)

$$\overline{h^2(v)} = \frac{a - \varphi(v)}{v - b},$$

$$\overline{v_h(v)} = \frac{av - b\varphi(v)}{a - \varphi(v)}.$$
(6)

From (4) it is seen that for $h = \infty$ $v_h = b$. But $h = \infty$ means that σ the dispersion of peculiar stellar velocities in this particular subgroup is zero. Thus b is equal to velocity relatively to the Sun of mass points. From the second Equation (6) we see that $\overline{v_h(b)} = b$. It seems that $v = b$ is a point of discontinuity for $\overline{h^2(v)}$. But this is not so since it is easily shown that $\varphi(b) = a$ and the indetermination is avoided by using de l'Hospital rule. The above weights are decreasing rather rapidly for large v. For instance they drop to less than 0.05 of the maximum value for $v = -60$ and $v = +30$ km s^{-1}. In Figure 3 the dot-dash curve represents the weighted distribution of the v-component velocities subject to the condition that the total number of stars remains the same, i.e. $N = 1355$. The weighted solar velocity appears to be $v_0 = 11.7$ and $\sigma = 18.7$ km s^{-1}, i.e. both are substantially smaller than with equal weights.

References

Gliese, W.: 1969, *Veröffentl. Astron. Rechen-Inst. Heidelberg*, Nr. 22.
Ogorodnikov, K. F.: 1928, *Astron. Zh.* **1**, 1.
Strömberg, G.: 1923, *Astrophys. J.* **59**, 229.
Woolley, R., Epps, E. A., Penston, M. J., and Pocock, S. B.: 1970, *Roy. Obs. Ann.* No 5.

DISCUSSION

Perek: How does the derived value of the density of 0.138 compare with the value derived by you, Dr Wielen?

Wielen: For stars in Gliese's Catalogue it would be about 0.11, that is in a good agreement.

Gliese: The region around 25pc is sensitive to the accuracy of parallaxes. We must keep in mind that positive errors are preferred. I assume that the authors have tried to eliminate this effect.

HOW CAN IT ALL BE STABLE?

ALAR TOOMRE

Massachusetts Institute of Technology, Cambridge, Mass., U.S.A.

My contribution to this Joint Discussion can hardly be termed very novel or original. Rather, I just want to call your attention here to two headaches – one perhaps only a hangover, but the other a real migraine. Sooner or later these two headaches beset anyone who persists in asking how the random motions of stars in the only vicinity where we tolerably observe them can possibly jibe with the sensible presumption that not only our neighborhood but this entire Galaxy should by now be reasonably stable.

1. Local Stability

There is of course nothing strictly 'local' about any problem involving the far-reaching gravity. Yet if some conceivable instability of our collection of nearby stars comes even close to deserving such a label, it is surely the Jeans instability or tendency toward gravitational collapse. After all it is not difficult to estimate (e.g., Toomre, 1964) that the most troublesome of such incipient clumpings ought to have significant dimensions (such as half-wavelengths) of the order of 3 or 4 kpc. Even the latter scale is probably not too large a fraction of our distance from the galactic center for 'local' analyses in the WKBJ spirit to remain coarsely trustworthy.

It seems just as clear, however, that any Jeans instability here can really be honored only *in absentia*. The reason is that any actual clumping tendencies in our disk of stars would exhibit e-fold growth times about as short as 10^8 yr. Hence any such troubles now could scarcely have been the first. Yet if similar instabilities indeed did arise in the past, it seems a safe bet that already they would rapidly have increased the random motions of the stars. And that in turn should soon have cured the problem – for there seems little question that all simple Jeans instabilities are avoided once the typical random speeds of stars exceed some relatively modest minima.

With much of this already in mind when I first considered such matters quantitatively in my 1964 paper, I remember being very struck that our locally observed random velocities seemed barely adequate even for that humble purpose. This is not to say that there seemed any real danger of contradiction, but only that the observed and required motions appeared identical within their uncertainties. However, in retrospect I have often wished that I had not been so captivated by that seeming agreement. For one thing it caused me to overlook one significant correction (to be discussed below) which tends to increase our margin of local stability. Far more important, it also lulled me – and perhaps others – into a false sense of security that various other kinds of collective instability of stars in this Galaxy were probably not much harder to suppress. We know today that isn't so. But let me not race ahead yet to that second and more serious headache. What I want to stress first is simply the frustration that despite the lapse of a decade, the uncertainties in the observed data remain such as to leave our actual factor of safety against even the local clumpings still quite poorly known.

To review those numbers, let us recall my old local criterion for the stability of a supposedly razor-thin disk of stars with a Schwarzschild distribution of horizontal velocities: It asserted that no short axisymmetric disturbances remain unstable if the rms radial speed σ_u exceeds

$$\sigma_{u,\min} = 3.36 \, G\mu/\kappa, \tag{1}$$

where G is the gravitational constant, μ is the projected mass density, and $\kappa = = [-4B(A-B)]^{1/2}$ is the so-called epicyclic frequency. Within its limited sphere of competence, that criterion itself seems to have fared well enough over the years. For instance, Goldreich and Lynden-Bell (1965) partly corroborated it through their analogous local findings for gaseous disks. Graham (1967) verified for star disks with several other sensible velocity distributions that the stability condition is likewise very similar. Julian and Toomre (1966) even reckoned that the criterion is not ruined by *short* non-axisymmetric disturbances. And Hohl (1971) showed among other things that when his full-disk n-body experiments were constrained to be axisymmetric, the speeds prescribed by Equation (1) were indeed about adequate for overall stability.

Yet some systematic correction for the finite thickness of a disk is clearly in order: Though I had guessed that such a reduction of $\sigma_{u,\min}$ might amount to perhaps 15 or 20%, first Shu (1968) and then Vandervoort (1970a) determined much more reliably that the numerical factor in Equation (1) had better be replaced by about 2.6 for an assumed ratio of vertical-to-radial motions $\sigma_w/\sigma_u \cong 0.6$ like observed. Finally, if some small fraction of the total projected density μ were to consist of cold gas instead of moving stars, its mere presence would *raise* the required stellar speeds by almost the same ratio; assuming 10% gas hereabouts, we thus arrive at the adjusted minimum

$$\sigma_{u,\min} \cong 2.9 G\mu/\kappa \tag{2}$$

to be used below.

Unfortunately this simple local formula remains distinctly more certain than the set of 'observed' quantities σ_u, μ and κ which it invites us to intercompare. The trouble is that none of those three quantities is obtained very directly. Probably best known is the epicyclic frequency $\kappa = 32$ km s^{-1} kpc^{-1} that is consistent with the conventional values $A = 15$, $B = -10$ for the local Oort constants; however, even it must be deemed uncertain by at least 10% and conceivably 20. As for the projected density of matter in our *disk* here, I prefer to think that $\mu = 75 \pm 15 M_\odot$ pc^{-2}. This estimate rests largely on the well-known deductions of the vertical force by Oort and others, most recently by Lacarrieu (1971). Though the end product of such studies has usually been a local volume density, the method in fact determines primarily a mean or projected density of all matter contained within modest heights z of the order of 200 or 500 pc above and below the Sun. Hence Oort's (1960) implied 62, 71 and 75 $M_\odot$ pc^{-2} within $z = 400$, 600 and 800 pc, respectively, seem almost as significant as his $\varrho(z=0) \cong$

$\cong 0.15 M_\odot$ pc^{-3} that is usually cited – and they are surely more accurate than if that volume density had simply been multiplied by some assumed equivalent thickness. As it happens, those particular values must now be increased to about 66, 77 and 83$M_\odot$ pc^{-2} (cf. Vandervoort, 1970b) to account for non-local contributions to K_z implied by $A=15$, $B=-10$ instead of $A=19.5$, $B=-6.9$. The extrapolated total thus comes to perhaps $\mu=90$, but it seems fairest to reduce it slightly both to exclude various transient halo-like stars, and for one other reason: If indeed, as now seems likely from the work of Plaut, Clube and others, our measured galactocentric distance $R_0 \cong 10$ kpc is overdue for downward revision, such independent estimates as $\mu(R_0)=$ $=114$ obtained from the 1965 Schmidt model seem themselves destined to shrink to roughly 90 if $R_0=9$ and to perhaps only $70 M_\odot$ pc^{-2} if $R_0=8$ kpc (cf. Toomre, 1972).

If in fact we adopt $\kappa=32$ and $\mu=75$, Equation (2) claims $\sigma_{u,\min}=30$ km s^{-1} to be the least rms radial speed that can be presumed adequate. However the alternatives $\mu=60$ or 90 would of course have yielded $\sigma_{u,\min}=24$ or 36 km s^{-1}. And taking into account also the uncertainty involving κ, we observe that just the required random velocity can conceivably still range between extremes differing by a factor of two.

To be compared with this elusive target is the data on the actual random velocities of nearby stars. That topic has already been reviewed at this Joint Discussion by Wielen, and here I simply wish to reemphasize three things he told us.

The first is that it seems about as true as ever that not only do K and M dwarfs constitute the bulk of the *known* nearby stellar density, but also their own galactocentric radial motions *in this vicinity* seem best characterized by $\sigma_u=$ mid-30 km s^{-1}.

The second point concerns that oversight of mine to which I have already alluded. As first stressed by Vandervoort (1970b), any likely positive correlation between the vertical and horizontal motions of stars means that our $z \cong 0$ mid-layer sampling volume tends to be deficient even in radial velocities compared with stars found at various heights in this disk. Indeed Vandervoort has already suggested that the nearby σ_u's should for this reason be multiplied by $(\pi/2)^{1/2} \cong \frac{5}{4}$ to become truly representative – but his example was somewhat arbitrary and hence inconclusive. However Wielen has now come along and done us a related service that is both beautiful and convincing in its simplicity: He merely weights the observed radial motions of the McCormick selected-area K and M dwarfs with their likewise observed vertical speeds (though even this presumably still under-compensates for those 'polling errors'), and thereby he judges the correct rms speed to be no less than 48 km s^{-1}. To be sure, to exclude the undue influence upon such statistics of a few passing halo stars, I am inclined to diminish Wielen's estimate slightly. Yet even so it seems hard to dislodge this new impression that the typical radial dispersion of the most common known disk stars is $\sigma_u=$ low-to-mid 40 km s^{-1}.

The third and last point is actually the most frustrating: As Wielen has already implied, even the inclusion of various earlier main-sequence stars, and of the giants, the known white dwarfs, and presumed dark companions still means that of the three-quarters or so of Oort's $0.15 M_\odot$ pc^{-3} that cannot comfortably be attributed to the local volume density of interstellar material, roughly half remains to this day totally unidentified. Perhaps one should not even speculate on the motions of such 'missing' stars. However, whether or not that other half turns out to consist of many yet fainter M dwarfs and/or white dwarfs and/or some yet more exotic objects, it seems to me historically implausible that their present random motions could on the whole average less than those of the dK and dM stars just cited. On the contrary, their average motions (like their ages) seem apt to be even greater.

(In case anyone wonders, I remain skeptical of the reality of the dense local layer of faint, low-velocity M stars that has been suggested by the recent work of Weistrop (1972) and Murray and Sanduleak (1972)). At issue here is chiefly the mean distance of the stars detected by these workers: If one accepts fully the reasoning of Murray et al. based on transverse motions, the nearby mass density of such stars emerges as an impressive 0.05 $M_\odot \text{pc}^{-3}$, but the indicated less-than-10 km s^{-1} spatial motion in any one coordinate is even more startling. I find it incredible that so small a velocity dispersion could have survived random gravitational forces from gas concentrations, various spiral wave sloshings, and perhaps even Jeans instabilities for the presumed large age of those stars; moreover, even if one accepts that such stars provide all of Oort's 'missing' mass in the *volume* near the Sun, the small implied thickness of their disk means that they still cannot account for the bulk of the *projected* missing mass.)

To conclude, the above evidence suggests that the best single estimate possible nowadays for the ratio Q of the existing peculiar motions to the minimum required ones in the galactic disk near the Sun must be roughly 1.5. We have seen however that this simple factor of safety could easily range between 1.2 and 2.0, and it may possibly climb yet higher.

Thus the Jeans instabilities now seem certifiably impossible. But I would not have inflicted all this numerology upon you only to establish something so 'obvious'. I confess I had another motive as well: Though many of you may not have realized it, the ratio Q also plays a surprisingly sensitive role in the ability or willingness of disks of stars to carry density waves of the sort envisaged by Lin and Shu. With $Q=1.0$, the region of conceivable waves extends all the way from the so-called inner Lindblad radius to the outer. However, already when $Q=1.5$, the Lin-Shu-Kalnajs dispersion relation no longer admits any waves at all within a fairly extensive intermediate annulus where a certain relative frequency $|v|<0.6$ (cf. Figure 1 of Toomre, 1969). And when $Q \cong 2$, the remaining tightly-wrapped wave picture has become so cramped as to be practically valueless.

One may of course, if one wishes, reverse this reasoning and argue that just the likely existence of waves in this Galaxy implies that $Q \cong 1$. That is not my aim here. I merely wanted to caution explicitly that – contrary to a misimpression which I am afraid I helped begin – it is *not* the observations of σ_u or μ or κ which compel one to adopt $Q \cong 1.0$ for spiral waves or any other purpose.

2. Overall Stability

All these nitpicking details, however, pale by comparison with a near-scandal in our understanding of galactic structure that has surfaced unmistakably only during the past year or two. In its potential impact, this particular difficulty reminds me already of the solar neutrino embarrassment from another area of astronomy: It raises doubts even about fundamental assumptions, it seems unlikely to vanish overnight, and I can here do no more than describe it briefly.

The difficulty in short is that at least four independent analyses or numerical experiments have now converged to testify that disks of stars remain very susceptible to large-scale or 'global' instabilities of a non-axisymmetric sort, even when endowed with random velocities well in excess of those estimated to suppress the simple local clumpings. Though their exact nature is still unclear, these growing disturbances are not Jeans instabilities by any reasonable standard. Rather, they seem to represent a strong, if perhaps only transient tendency of the central regions of the model disks to develop bar-like structures, often accompanied by wide-open but temporary spiral structures or even waves farther out.

What is startling about these large-scale instabilities is not their occurrence as such (since local studies could logically neither predict nor refute them) or even their frequent bar-likeness (which is vaguely reassuring in view of the many barred spirals found in the sky). The real surprise is that such troubles seem unavoidable in all thin disks of stars examined, unless the kinetic energy of their random motions is more than two and a half times the kinetic energy of their systemic *rotation* itself!

To my knowledge, something like this alarm was first raised by Miller, Prendergast, and Quirk (1970), and by Miller (1971). Those authors remarked that their 10^5-body "calculations typically produce 'hot' systems that are largely pressure-supported", with "velocity dispersions... considerably greater than those needed to stabilize" in the local sense. Unfortunately no one knew at the time just how seriously that warning was to be taken: For one thing, Miller *et al.* had quickly added an inelastic or 'gas' component to their stellar disks. Though intended for greater realism, that addition made it unclear how much of the increase was to be blamed on the stars themselves and not on the evident Jeans instabilities and clumpings of the 'gas'. Likewise uncertain were the effects of possible numerical errors, inasmuch as both the integration steps and the potential mesh used by Miller *et al.* were purposely quite coarse. And then, too, an early report of related *n*-body calculations by Hockney and Hohl (1969) had claimed no such further instabilities.

The plot thickened, however, with the fine analytical study by Kalnajs (1972, and earlier) of the various linear modes of certain very thin disks of stars possessing a uniform angular velocity of rotation. Kalnajs again found strong indications that "disks which are hot enough to avoid axisymmetric instabilities can still evolve rapidly in a nonaxisymmetric manner". Yet a skeptic could have retained some mild reservations, this time notably about the strange velocity distributions required in those disks to begin with.

A third investigation deserved even fewer such quibbles: This one, by Hohl (1971), in a sense only continued the work of Miller *et al.* using roughly as many mass points; however, it distinctly excelled in that it consisted of numerous separate experiments, was much smoother in its numerical treatment, and above all dealt only with imagined stars. (It also retracted as premature the earlier contrary claims of Hockney and Hohl.) Again the bulk of the evidence was that "disks of stars are considerably more difficult to stabilize than indicated by local analyses". Hohl also guarded against the suspicion that the disks might have become excessively hot by having begun too unstable; he did so by 'cooling' one such stable final disk, and by finding that fresh instabilities soon reappeared. Finally, even possible relaxation effects as a cause of excessive 'heating' seem now to have been largely exonerated by Hohl (1973).

Most impressive is the fourth and latest chapter of this unfolding story. It is impressive not because the few-hundred-body calculations which Ostriker and Peebles (1973) undertook were themselves very remarkable. Rather it is so because it occurred to Ostriker (1973) to propose and to test numerically a new stability criterion that turned out to unify all four of the investigations: He wondered if it is as true of the thin galaxy models as it seems true of many models of differentially rotating and inhomogeneous stars (cf. Ostriker and Bodenheimer, 1973) that the criterion for the avoidance of bar-making practically matches that for the non-bifurcation of the Jacobi ellipsoids from the classical Maclaurin spheroids. In that classical setting, secular instability toward triaxial or bar-like forms occurs whenever the total kinetic energy of rotation, T_{rot}, exceeds a meager 13.8% of the absolute value of the potential energy W of the system. Practically the same critical value was found for the star models just cited. And now in a similar vein, Ostriker and Peebles report that the approximate criterion

$$T_{\text{rot}}/|W| \cong 0.14 \pm \text{perhaps } 0.02 \tag{3}$$

characterizes not only their own n-body calculations but also the 'coolest' of the largely pressure-supported stable disks achieved in every one of the three previous investigations!

Though of course it is no proof that every conceivable model disk must be as hot to be fully stable, this astonishing numerical agreement means at the very least that the Ostriker-Peebles criterion is an excellent rule of thumb summarizing all the available evidence. Thanks to the virial theorem, an equivalent summary would have been that the portions T_{rand} and T_{rot} of the total kinetic energy associated with random and mean rotational motions, respectively, must satisfy

$$T_{\text{rand}}/T_{\text{rot}} \gtrsim 36/14 \cong 2.6 \tag{4}$$

for a completely stable equilibrium.

For various reasons including our locally-observed stellar motions, such a ratio seems about the reciprocal of what one might intuitively have expected to find in a real spiral galaxy. Hence it raises all sorts of questions and worries about the whereabouts of so many high-velocity stars and/or other mass points not only in our Galaxy but also in others which have heretofore been thought relatively cool and disk-like. Possibly the answer lies in major stellar halos, such as Ostriker and Peebles have already suggested tentatively, and within which relatively cool disks might still be embedded stably – but it is hard to believe, for instance, that any disk instabilities themselves would ever have propelled stars to great heights in the z-direction.

The only sure bet seems to be that, from now until it has been resolved, this second of the headaches that I wanted to complain about will seriously plague *all* large-scale galactic dynamics.

Acknowledgements

I want to thank especially Drs F. Hohl, A. J. Kalnajs, R. H. Miller, J. P. Ostriker and R. Wielen for their frank and generous communications long before publication. This review was supported in part by the (U.S.) National Science Foundation.

References

Goldreich, P. and Lynden-Bell, D.: 1965, *Monthly Notices Roy. Astron. Soc.* **130**, 97.
Graham, R.: 1967, *Monthly Notices Roy. Astron. Soc.* **137**, 25.
Hockney, R. W. and Hohl, F.: 1969, *Astron. J.* **74**, 1102.
Hohl, F.: 1971, *Astrophys. J.* **168**, 343.
Hohl, F.: 1973, *Astrophys. J.* **184**, 353.
Julian, W. H. and Toomre, A.: 1966, *Astrophys. J.* **146**, 810.
Kalnajs, A. J.: 1972, *Astrophys. J.* **175**, 63.
Lacarrieu, C. T.: 1971, *Astron. Astrophys.* **14**, 95.
Miller, R. H.: 1971, *Astrophys. Space Sci.* **14**, 73.
Miller, R. H., Prendergast, K. H., and Quirk, W. J.: 1970, *Astrophys. J.* **161**, 903.
Murray, C. A. and Sanduleak, N.: 1972, *Monthly Notices Roy. Astron. Soc.* **157**, 273.
Oort, J. H.: 1960, *Bull. Astron. Inst. Neth.* **15**, 45.
Ostriker, J. P.: 1973, A.A.S. Warner Prize lecture (delivered at the January meeting in Las Cruces, N.M.).
Ostriker, J. P. and Bodenheimer, P.: 1973, *Astrophys. J.* **180**, 171.
Ostriker, J. P. and Peebles, P. J. E.: 1973, *Astrophys. J.*, **186**, 467.
Shu, F. H.: 1968, Ph. D. Thesis, Harvard University.
Toomre, A.: 1964, *Astrophys. J.* **139**, 1217.
Toomre, A.: 1969, *Astrophys. J.* **158**, 899.
Toomre, A.: 1972, *Quart. J. Roy. Astron. Soc.* **13**, 241.
Vandervoort, P. O.: 1970a, *Astrophys. J.* **161**, 87.
Vandervoort, P. O.: 1970b, *Astrophys. J.* **162**, 453.
Weistrop, D.: 1972, *Astron. J.* **77**, 849.

DISCUSSION

Van Woerden: I believe there is no real problem in assuming two thirds of our Galaxy to have random motions of about 200 km s^{-1}. Halo objects have such velocities and can (I think) account for 70% of the mass of our Galaxy.

Oort: I do not find a real difficulty with the supposition that a large proportion of the mass of our Galaxy is contained in a halo and 'central bulge'. One has to admit only a small density of late-type subdwarfs in the vicinity of the Sun. In an article in Vol. 5 of *Stars and Stellar Systems* I estimated that a density of 0.005 solar masses per cubic pc near the Sun would produce a halo having a mass equal to the entire mass of the galactic system if one makes the reasonable assumption that the space density of the halo is proportional to the inverse third power of the distance from the centre. The density mentioned is only about 10% of the 'missing' mass density, and could in my opinion be easily furnished by subdwarfs which are intrinsically fainter than the faintest known subdwarfs.

I should also like to draw attention to the dynamical model of the Galaxy worked out several years ago by Woltjer and Ng. This was, I believe, the first model of the entire Galaxy in which full account was taken of Poisson's law. Ng obtained likewise the result that the velocity dispersion of the bulk of the mass must be very high.

Toomre: I largely agree with both of you. Yet it also seems to me that the really tough nut to crack is not whether a massive halo is plausible, but to establish that it indeed *must* exist. I am afraid that Ng's theoretical models offer no such proof whatever; his method of construction all but guaranteed that they would be very hot.

Freeman: We really have little reliable data yet on the local density of halo stars or on the distribution of density with radius in the galactic bulge, so it is difficult to decide whether or not the galactic bulge is capable of stabilizing the disk in the way Ostriker and Peebles suggest. On the other hand, there are plenty of disk galaxies (spirals and S0) with very weak bulge components (see the Hubble Atlas). The bulge components must be much less massive than the disks in these systems, unless their M/L is extremely large, so it seems unlikely that disk stabilization by the bulge is a general feature.

Toomre: That is certainly one frustration with postulating major halos. Those can indeed remain blissfully unknown to us if M/L is large enough, but like you I would be happier if the average edge-on spiral gave some decent positive evidence of such halos.

Innanen: Unless one makes the proposed massive halo with a radius smaller than R_0, it seems to me that placing $10^{11} M_\odot$ in the halo unavoidably contributes significantly to $\mu(R_0)$ and further aggravates Toomre's pleas for lower values of μ. Even in Innanen's 1966 model cited by Toomre in the discussion (*Astrophys. J.* **143**, 163) a halo of $0.4 \times 10^{11} M_\odot$ increased μ from 74 to $98 M_\odot$ pc^{-2}. In more recent models (Innanen, 1973, *Astrophys. Space Sci.*, in press) a more modest halo of 1 to $2 \times 10^{10} M_\odot$ already contributes at least 10% to μ so it is not really easy to hide it.

Toomre: You are here talking of the total projected density, from the halo and disk together. In my discussion of the local stability I tried to include only the material from the most evident disk.

Miller: The Ostriker-Peebles and your earlier criterion both refer to initially axisymmetrical configurations. It may be possible to stabilize a non-axisymmetrical system more easily.

Toomre: I wish we could be sure that is so!

Mestel: Does the rapid heating of the disk occur whatever the details of the physical processes by which stellar energy is tapped by the gaseous component and radiated away? For example, if one pretended that random velocities are killed if they exceed a modest value (much less than the circular velocity), wouldn't the disk-like structure be maintained? Doesn't the Ostriker-Peebles result depend on the fact that star-gas interactions are much less effective for high relative motions?

Miller: Our model galaxies had two components, one of which was cooled and one of which was not. The cooled component would end up being cool enough, but the un-cooled component was very hot – as in the models that Toomre has talked about. We also tried some models in which the entire population was cooled – but these collapsed catastrophically.

CLOSING REMARKS

F. K. EDMONDSON
Indiana University, U.S.A.

My presence here as summarizer of this joint discussion is an example of how persuasive the president of Commission 33, Dr McCuskey, can be. I feel somewhat overwhelmed by this responsibility, since I have been an onlooker, not a practitioner, in this field for the past decade.

The outstanding feature in this joint discussion, as I see it, is the marriage of galactic astronomy and astrophysics, with clear benefits to both fields. The correlations between ages of stars and their kinematical properties, as discussed by Mrs Cayrel, represents major progress in our understanding of the history of the Galaxy. The rehabilitation of Gould's Belt by Mrs Lesh and Dr Lindblad is another important development.

Repetition of everything you have heard today in these closing remarks would serve no useful purpose, so let me comment on only a few additional matters. The discussions of the disk and halo stars and the importance of the local gas were interesting and important contributions. The relationships between velocity and age, and vertex deviation and age were also valuable contributions.

I find that I am still bothered by the way many people speak of different values of the solar motion with respect to different classes of objects. As I see it, *the* solar motion (in the Milne sense) can have only one value, and this should be the solar motion relative to the circular velocity in the Sun's neighborhood. Differences from this are due to different group motions of the different classes of objects and not to the Sun's motion. *The* solar motion should always be identified with the reduction to the local standard of rest. It is misleading to talk about the solar motion with respect to the globular clusters, for example.

The density wave theory and other theoretical discussions have had their rightful place in this program, and have served to show how the interaction between observation and theory are necessary for real progress.

This has been a stimulating and informative day. Our thanks are due to the organizers and participants.

IV. ORIGINS OF THE MOON AND SATELLITES

(Edited by G. Contopoulos)

Organizing Committee

S. K. Runcorn (Chairman), E. Anders, A. Dollfus, G. H. Pettengill,
J. Ring, H. Wood

ON THE GROWTH OF THE EARTH-MOON SYSTEM

F. L. WHIPPLE

Smithsonian Astrophysical Observatory and
Harvard College Observatory, Cambridge, Mass., U.S.A.

This paper elaborates the postulate that the Earth and Moon became a binary system during their accretional development and that the Moon's growth was essentially completed before the assumed solar nebula dissipated. The solar nebula was still hot enough at the formation of the two bodies that both consisted largely of the refractory and relatively low-density minerals now characteristic of the Moon. During the subsequent condensation, agglomeration and accretion of siderophile and more volatile higher density minerals, the Earth grew very much faster than the Moon because of (a) its much greater gravitational capture area coupled with retention by a sizeable atmosphere and (b) the Moon's velocity, with respect to the solar nebula, which produced a wind that aerodynamically blew away volatiles and smaller debris resulting from hypervelocity impacts of larger planetesimals. This 'impact differentiation' process favored the retention of the refractory minerals on the Moon (Figure 1). The Moon's surprisingly high moment of inertia follows naturally from the basic postulate.

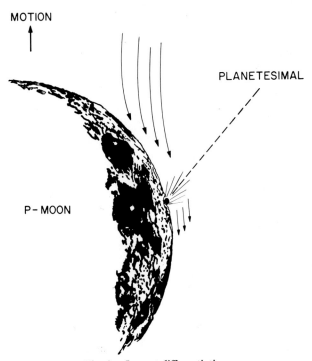

Fig. 1. Impact differentiation.

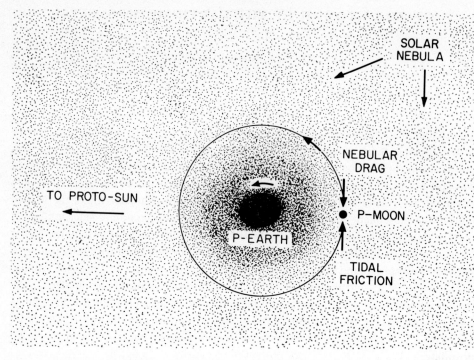

Fig. 2. Proto-Earth and Moon in Solar nebula.

A sizeable velocity of the Moon with respect to the solar nebula may have arisen in two ways, which could have been sequential. If the proto-Moon formed near the proto-Earth's orbit, the perturbations by the Earth's mass would have forced it into a rather eccentric and inclined orbit before subsequent capture. (See detailed discussion of such motions by Safronov, 1969.) After capture, possibly by collision with the Earth as discussed by Öpik, its orbital velocity was significant as will be discussed below. The proto-Moon may conceivably have been formed in a ring-system about the proto-Earth while the solar nebula was still hot.

Regardless of the early history, the proto-Moon, once in orbit about the proto-Earth, was subject to drag by the solar nebula causing it to spiral in towards the proto-Earth. The opposed force of 'tidal friction,' however, produced a quasi stable orbit in which drag and tidal-friction forces were balanced (Figure 2).

The drag acceleration on the proto-Moon (mass, M_m; density, ϱ_m; radius, R_m) at distance, a, and velocity V_m with respect to the proto-Earth (M_e, ϱ_e, R_e) in a solar nebula of density ϱ, would be of the classical Newtonian form:

$$\text{Acceleration} = -\frac{\varrho v_m^2}{6\varrho_m R_m} = -\frac{\mu \varrho}{6a\varrho_m R_m}, \tag{1}$$

where $\mu = G(M_e + M_m)$ neglecting the effect of lunar gravity on the density and velocity of the nebular gas.

Hence the rate of change of a in a nearly circular orbit became

$$\frac{da}{dt} = -\frac{\mu^{1/2}\varrho a^{1/2}}{3\varrho_m R_m}, \qquad (2)$$

neglecting terms in M_m/M_e.

The tidal acceleration by the proto-Earth with polar axis at some moderate angle to the orbital pole and rotation period much shorter than the orbital period would (e.g. MacDonald, 1964) produce an acceleration in a given by

$$\frac{da}{dt} = 0.9 \frac{\mu^{1/2}}{a^{11/2}} \frac{M_m}{M_e} R_e^5 \sin 2\delta, \qquad (3)$$

for Love Number $=0.30$, where δ is the assumed tidal lag and terms in M_m/M_e are neglected.

By equating the nebular drag to the tidal friction as given by Equations (2) and (3) respectively, we solve for the orbital parameter, a_s, in the case of quasi-stable equilibrium:

$$a_s^6 = 2.7 \frac{\varrho_m^2}{\varrho \varrho_e} R_m^4 R_e^2 \sin 2\delta. \qquad (4)$$

The tidal lag, $\sin 2\delta$, in Equation (4) could alternatively be expressed as an energy dissipation term, but neither expression for tidal friction can be confidently evaluated under the assumed physical circumstances. The present-day value of δ is $2°.16$, which as MacDonald shows, leads back in time to the Roche limit in 1.7×10^9 yr, much shorter than the age of the Earth and Moon. With little uncertainty in a_s ($a_s \sim \delta^{1/6}$) we may reasonably adopt a tidal friction rate for the proto-Earth and Moon one tenth the present rate so that the dimensionless term $2.7\sin 2\delta$ in Equation (4) becomes 0.020.

The condition that the proto-Moon cannot exist for a_s within the Roche limit of the proto-Earth provides an upper limit to R_e/R_m in Equation (4) shortly after capture or at formation of the binary system. Let us adopt $a_s > 2.455(\varrho_e/\varrho_m)^{1/3}R_e$ for the Roche limit and apply it to Equation (4). Then

$$\left(\frac{R_e}{R_m}\right)^4 < \frac{0.020}{(2.455)^6} \frac{\varrho_m^4}{\varrho \varrho_e^3}, \qquad (5)$$

represents an upper limit to the ratio R_e/R_m for the quasi-equilibrium condition stated by Equation (4).

Basic to any numerical calculations is the density, ϱ, to be assumed for the solar nebula in the neighborhood of the proto-Earth. Abundances of the elements by Urey (1972) lead to the mass distribution and molecular weights given in Table I for materials divided into the classes *gaseous* (H, He, noble gases), *icy* (hydrides of C, N, O) and *earthy* (oxides of heavier elements). At the proto-Earth, following Larimer and Anders (1967), the temperature is taken as 550 K. At this temperature the gas plus ice mixture would be gaseous with an adopted mean molecular weight of 2.37 (Table I).

The assumption that the Earth-Moon system was essentially complete at the time of dissipation of the solar nebula leads to a minimum density nebula, if we assume a central mass about equal to the present Sun. Thus the total surface density across the solar nebular disk at the Earth's distance is the order of 6×10^3 gm cm^{-2}. at $T=550$ K during most of the accumulation of the Earth if the mass fraction of *Earth* in the nebula is 0.004. This corresponds to an Earth mass in the zone half-way to Venus and to Mars. The space density of the *gas* and *ice* mixture at the proto-Earth's distance then becomes $\varrho = 3.7 \times 10^{-9}$ gm cm^{-3}.

TABLE I

Material	Solar abundances	
	Mass	Mol. Wt.
Gaseous	0.976	2.33
Icy	0.020	17.2
Earthy	0.004	45.0
Gaseous and icy	0.996	2.37

Application of Equation (5) with the value of ϱ leads to a minimum ratio of the radii, proto-Earth to proto-Moon, at the earliest stage of their binary formation, $R_e/R_m < 13.3$. (Assumed: $\varrho_e = 4$, $\varrho_m = 3$, gm cm^{-3}). Smaller bodies in orbit about the proto-Earth would spiral in rather rapidly because of drag by the solar nebula.

For a spherical body of radius s, density ϱ_s and in a circular orbit of radius a_0 about a proto-Earth of mass M_e the spiral time, t, to the proto Earth is given by

$$t = \frac{6\varrho_s s}{\varrho G^{1/2} M_e^{1/2}} \left[a_0^{1/2} - (R_e + s)^{1/2} \right]. \tag{6}$$

For a proto-Earth mass of one-tenth the Earth's mass, a body of $s = 10$ km, $\varrho_s = 3$ gm cm^{-3} would spiral to the proto-Earth from $a_0 = 100000$ km in 2000 yr because of the nebular drag. It is difficult to see how a ring system could develop or persist in a nebula, at least about a terrestrial planetary mass.

With the density now assumed for the solar nebula we can apply Equation (4) to determine the quasi-stable separation of the proto-Moon and proto-Earth under the opposed forces of nebular drag and tidal friction. Table II lists values of this quantity distance, a_s, the orbital period, the circular velocity and the velocity of escape from the proto-Moon, all for various values of proto-Earth and proto-Moon masses and densities.

The resultant separations of the binary in Table II are the order of 30–40000 km or 6–9 of the various proto-Earth radii over a small range in proto-Moon masses, consistent with commonly favored early distances.

The calculations from Equations (4), (5), and (6) neglect gravitation effects of the proto-Earth and proto-Moon that will increase the density of the solar nebula near the proto-Earth. The drag force on the proto-Moon will be increased both by this increased density and by the increased drag coefficient induced by the gravitational attraction of the proto-Moon in its passage through the gas. The ability of the nebular wind about the proto-Moon to carry away impact debris from planetesimal collisions will be increased by the increased velocity and particularly by the increased density near the proto-Moon caused by these gravitational effects. Correspondingly the impact velocities will also be increased, producing more impact debris to be blown away.

TABLE II
Calculations for assumed conditions of
proto-Earth and proto-Moon (Equation (4))

M_e/M_e (present)	1.0	0.5	0.25	0.1
ϱ_e gm cm^{-3}	5.5	5.0	4.5	4.0
M_m/M_m (present)	1.0	1.0	0.8	0.5
ϱ_m gm cm^{-3}	3.3	3.3	3.3	3.0
a_s (10^3 km)	40	38	35	29
a_s/R_e	6.3	7.3	8.1	8.8
Period (hr.)	22	29	35	42
v (circ.) km s^{-1}	3.1	2.3	1.7	1.1
v_m (escape) km s^{-1}	2.4	2.4	2.2	1.8

If we neglect these gravitational effects the present Moon moving about the present Earth in the assumed solar nebula would appear, from crude theory, to lose rather than gain mass by planetesimal accretion. About an appreciably less massive proto-Earth the effect would be reversed were it not for the above-mentioned gravitational processes. Hence it appears worthwhile to continue this research into the more difficult theoretical area of gravitating spheres moving through compressible nebular gas of significant density. Quite possibly the Moon has not gained or lost a significant fraction of mass since the binary system developed, either in the nebular stage or subsequently. The Earth may have gained enormously in mass during the same interval.

References

Larimer, W. J. and Anders, E.: 1967, *Geochem. Cosmochim. Acta* **31**, 1239.
MacDonald, G. T. F.: 1967, *Rev. Geophys.* **2**, 467.
Öpik, E. J.: 1972, *Irish Astron. J.* **10**, 190.
Safronov, V. S.: 1969, *Evolution of the Protoplanetary Cloud and Formation of the Earth and Planets*, Ch. 7, Moscow, Transl. NASA.
Urey, H. C.: 1972, *Ann. N.Y. Acad. Sci.* **194**, 35.

EVIDENCE FOR LUNAR-TYPE OBJECTS IN THE EARLY SOLAR SYSTEM

H. C. UREY

Chem. Dept., University of California, San Diego, La Jolla, Calif. 92037, U.S.A.

Objects of the solar system, in addition to the Sun, can be classified into four groups – the planets, objects of lunar mass, smaller objects of variable mass and the comets.

If the solar proportion of gases relative to non-volatile compounds of the variety in the terrestrial planets, namely about 300 times the mass of these elements, were added to the terrestrial planets, they would have masses comparable to those of the major planets. Mercury is low in mass but has a high density, indicating that it has lost several times its mass of silicate materials relative to high density metallic iron. If this were restored and then the component of gases were added, it would also fall into the group rather naturally. Mars appears to be rather small. Uranus and Neptune have rather high densities indicating some loss of gases, probably hydrogen and helium. When we attempt to estimate the mass of primitive solar material from which the planets were evolved, we conclude that they evolved from very similar masses. Later, I shall argue that the process was a very inefficient one.

Schmidt (1944) suggested that the planets accumulated from many small objects, and, in this case, one might reasonably suggest that the axes of rotation might be oriented in a vertical direction to the invariant plane. Only for Jupiter is this true, and this may indicate that large objects such as of lunar mass were present, and were part of this accumulation process. Table I gives the orientation of planetary axes. Tidal effects have probably slowed the rotations of Mercury and Venus, though the reverse rotation of Venus can hardly be explained in this way. As Singer (1970) has pointed out, the reverse rotation could be explained by a collision of an object of lunar mass with the planet in which the relative angular momentum of planet and lunar object could be sufficient to reverse the direction of rotation. Also, the somewhat erratic orientation of the axes of the other planets could be explained by similar processes.

A second group of objects in the solar system consists of seven satellites having the mass of the terrestrial Moon within a factor of 2. Table II gives the masses of these satellites in units of the Moon's mass, the densities of the satellites and the fraction of lunar type material, assuming that they consist of material of this type and a material of density 1.0 g cm^{-3}, presumably water, and the masses of lunar type material so calculated in units of lunar mass. If it is assumed that low densities are due to admixtures of water, the masses become even more constant. A third group of objects consists of the smaller satellites of the planets and the asteroids, the largest of which have masses of a few percent of a lunar mass. It is interesting and probably informative that a group of seven objects of nearly lunar size exists in the solar system, and that a distinct discontinuity in masses exists within the planetary satellites.

It is interesting that Triton moves in a very nearly circular orbit with an axis tilted 160° to the ecliptic plane. If Pluto is an escaped moon of Neptune, as suggested by Lyttleton, this large tilt of the axis is explained, but is it not surprising that the orbit was left with a very small eccentricity? Possibly Triton was captured in a circular orbit lying in the plane of the equator of Neptune, and a lunar object was subsequently captured into the body of Neptune, thus tilting its axes but leaving Triton approximately in its original orbit. Possibly Pluto is a lunar object which has not been captured. Possibly other such objects exist and have not been observed as yet. A fourth group of solar objects are the comets which may be the source of some meteorites.

If this evidence for a substantial number of objects of lunar mass is to be taken as valid, it is necessary to assume that these objects were captured by the planets, and, hence, originated in the solar nebula as independent objects. This is a popular assumption in regard to the origin of the terrestrial satellite, mostly because of chemical arguments. Urey (1958, 1966) has postulated such an origin and has proposed that gravitational instability in a gaseous solar nebula was a suitable physical mechanism for producing such objects, and that the Moon was one of these. The formulae for such masses have been given in previous publications (Urey, 1958, 1966). Using the theory of Chandrasekhar (1955), it was postulated that early in the history of the solar system, there existed a nebula in quiet rotation about the Sun with variable

TABLE I

Inclination of axes

Mercury	–
Venus	180°
Earth	23°27′
Mars	25°12′
Jupiter	3°7′
Saturn	26°45′
Uranus	98°
Neptune	29°

TABLE II

	Mass	Radius	Density	Lunar solid %	Mass
Io	0.985	1.00	3.22	98.4	0.969
Europa	0.641	0.89	3.02	95.5	0.612
Ganymede	2.112	1.60	1.73	60.2	1.271
Callisto	1.316	1.44	1.48	46.3	0.609
Titan	1.87	1.40	2.31	80.9	1.513
Triton	1.85	1.08±0.37	$4.8^{+12}_{-2.8}$		
Pluto	?	?	?		
Moon	1.00	1.00	3.34	100	1.00
				Av.	0.994

temperatures as a function of distance from the Sun, but constant temperatures in directions perpendicular to the nebula plane. In this case, the variation of density vertical to the plane is given by:

$$\varrho = \varrho_0 \operatorname{sech}^2\left(\frac{x}{H}\right), \quad H = \left(\frac{RT}{2\pi G\mu\varrho_0}\right)^{1/2},$$

where ϱ_0 is the density at the median plane, x is the distance vertical to this plane, μ is the mean molecular weight of the gas and here assumed to be 2.4. The mass per unit is found to be $2\varrho_0 H$. Chandrasekhar's formula for the minimum unstable wavelength in a direction perpendicular to the axis of rotation is:

$$\lambda = \left(\frac{\pi\gamma RT}{G\mu\varrho}\right)^{1/2} \left(1 - \frac{\Omega^2}{\pi G\varrho}\right)^{-1/2},$$

where γ is the ratio of heat capacities, C_p/C_v, and Ω is the angular velocity of rotation. Using the mass of the present Sun to calculate Ω and assuming that the value of ϱ to be used is $\varrho_0/2$, the quantity in the second parenthesis is 0.819 for all distances. If the field due to the nebula is included, a massive nebula would decrease this somewhat. The minimum unstable mass is:

$$m = 2\varrho_0 \left(\frac{RT}{2\pi G\mu\varrho_0}\right)^{1/2} \left(\frac{2\pi\gamma RT}{G\mu\varrho_0}\right) \left(1 - \frac{2\Omega^2}{\pi G\varrho_0}\right)^{-1}.$$

We use the Roche density for ϱ_0, namely

$$\varrho_0 = \frac{M_\odot}{2\pi R^3 \times 0.04503} = \frac{2.10 \times 10^{-6}}{c^3},$$

where c is the radial distance in astronomical units.

These formulae are very approximate when applied to a real solar nebula instead of an idealized gas of uniform density in all directions. One does not believe that cubes of gas, in the three dimensional case, nor that square areas of gas, in the two dimensional case, collapse to spheres. Also, the presence of solids, which may have been of various sizes and might have settled to the median plane of the nebula, will surely modify the results deduced from these formulae. In particular, the temperatures that have been calculated in previous papers are probably not realistic as has been noted previously. Radioactive substances will be present in the solar nebula, and, hence, some electrical conductivity will be present and magnetic fields may be trapped and considerably modify the behaviour of gases.

If solids accumulate in the median plane of the nebula, one could expect a development of variable densities of solid masses. Such variable accumulations would seem to be most probable, and, in fact, these may constitute the way that planetary growth occurred. If such variable masses occurred and by chance became more concentrated in some area, gravitational instability of the gases would be promoted and a gas sphere with solids accumulating at the center would probably be promoted. Hence, higher temperatures and lower densities than those required by the theory of gases would be required.

In past discussions of this problem, it was assumed that objects of lunar mass, plus the quota of gases based on estimated solar abundances, were formed throughout the solar system. With this hypothesis and the assumption of the Roche density at the median plane, the temperature and densities at the median plane can be calculated. These are presented in Table III. It is immediately evident that the temperatures are very low. In fact, at Uranus and Neptune, only helium would remain in the gaseous state. The assumption of lunar masses was justified on the assumption that all satellites of approximately lunar mass throughout the solar system accumulated in this way, and that they were captured by their planets. The table also shows the value of ϱ_0 for the different planetary distances. Kusaka *et al.* (1970) have recently discussed the accumulation of the planets from a solar nebula, and their values for the temperatures and densities are given in the last two columns. These data apply to the stage of solar development when the Sun had a luminosity of 10 $L_\odot$ at which time the solar nebula was present. Their temperatures fall off approximately with $c^{-1/2}$, and these temperatures apply to the outer surface of the nebula which are illuminated by the high temperature Sun. Table III, in the last two colums, gives the nebular masses for each planetary region and the total masses. Our total mass for the nebula of 0.6 $M_\odot$ is only three times the value of 0.2 $M_\odot$ preferred by Herbig (1971) in order to account for the dust in space. The Kusaka *et al.* (1970) nebula has about one fourth of Herbig's preferred mass. On either model a high temperature process occurred as the nebular material left the contracting solar mass followed by the cooling of this nebula to lower temperatures. Probably the fine dust to produce the T-Tauri stage was formed at this time. This was followed by other processes which formed the planets, here assumed to have occurred through gravitationally unstable masses. Table III also shows the densities per unit area of the nebula to secure lunar masses of 2.2×10^{28} g for the lunar gas mass and the values assumed by Kusaka *et al.* (1970). These latter masses are based on the assumption that the solar nebula had exactly the masses needed to produce the planets with no excess planetary matter lost to space. It seems reasonable to suppose that a contracting mass of gas with excess angular momentum would throw off a nebula in a rather continuous way without such gross minimum and maximum amounts of material at neighbouring radii, and that the apparently erratic varying masses of planetary bodies are due to methods of accumulation of these bodies.

If such gas spheres were formed, the solids would settle to their center. If the solids were present as very small particles, they would settle slowly, and if as larger objects, more rapidly. The energy of accumulating a lunar object would be absorbed by the great heat capacity of the gas so that the lunar object would be formed at a low temperature. If the energy of accumulation were distributed uniformly through the gas sphere, its temperature would be raised by less than one degree. If the gas sphere lost energy by radiation, it would contract and temperatures would increase at the

TABLE III

	AU	$m = 2.2 \times 10^{28}$ g					$m = 2.2 \times 10^{28}$ g	
					K N H		Urey	K N H
		ϱ_0, g cm^{-2}		H km			Neb. masses	Neb. masses
		T	$(\times 10^{-3})$	$(\times 10^{-5})$	T	g cm^{-2}	$(\times 10^{-32}$ g$)$	$(\times 10^{-30}$ g$)$
Mercury	0.3871	106.15	1127.0	1.56	405	1600	1.3543	0.2159
Venus	0.7233	56.81	323.0	2.91	274	11000	1.1240	3.434
Earth	1.	41.09	169.0	4.02	225	7200	0.8793	4.024
Mars	1.5237	25.97	72.7	6.12	176	2200	1.2272	0.4245
Asteroid	2.7673	14.85	22.0	11.13	130	0.22	1.4695	0.00161
Jupiter	5.2028	7.84	6.24	20.92	97	1500	1.4547	37.08
Saturn	9.5388	4.31	1.85	38.36	73	130	1.5587	12.19
Uranus	19.191	2.14	0.458	77.18	54	36	1.3692	10.02
Neptune	30.07	1.39	0.196	120.93	45	28	0.8704	23.69
Pluto					40	1		0.6372
						Total	11.3078×10^{32}	91.172×10^{30}
							$\sim 0.6\, M_\odot$	$\simeq 0.45\, M_\odot$

center. Thus, one could expect that the surface might melt and a rather thick layer of liquid would form if the surface were agitated by massive storms which would be expected. The formulae for the central temperature and pressure for a sphere of mass M and radius R as calculated from Emden gas spheres (1929) are:

$$T_0 = 1.16 \times 10^{-15} \frac{M}{R}$$

$$P_0 = 1.09 \times 10^{-7} \frac{M^2}{R^4} \text{ dyn cm}^{-2},$$

providing the mean molecular weight is 2.4 and the value of $\gamma (= C_p/C_v)$ is 1.5. If the mass is 2.2×10^{28} g and $R = 8.5 \times 10^{10}$, $T_0 = 300$ K and $P_0 = 1$ bar, and if $R = 2 \times 10^{10}$, $T_0 = 1276$ K and 330 bar. More exact calculations have been made by Janet Bainbridge (1962). Thus, objects could be accumulated at low temperatures and be melted later at the surface. Later, as the gas sphere was dissipated by a high temperature Sun, the surfaces of such lunar objects would be subjected to a thermal history appropriate to their distance from the Sun. It is to be expected that all such objects at some particular distance from the Sun's center would not be of the same mass and would not proceed through the same thermal history. We should note that the gas spheres would be on the verge of instability due to the solar field, but they must not have too great stability for the gases might not be removed when the high temperature Sun with its sweeping magnetic field became effective.

Again, it should be noted that the theory is a very approximate one. It assumes that square areas in the nebula became gas spheres of equal mass and that the suspended solids had no effect on the process. Not all masses accumulated at the same time, all did not have the same mass, and they did not go through the same identical history.

Some became hot at their surfaces, some remained at low temperatures and some may have become melted throughout. The solid objects collided with each other and produced fragments of metal and silicate which fell on other objects. The great complications of the meteorites attest to such complications.

If these seven objects of approximate lunar mass were captured by the Earth and other planets, there is the serious problem of the mechanism of capture. Several suggestions in regard to the capture of the Earth's Moon have been made, i.e. dissipation of energy through tidal effects, collision with many smaller objects previously captured, and dissipation of energy in a terrestrial gaseous nebula, and possibly others will be proposed. Capture by interaction with other objects or with gaseous nebula might be applied to the satellites of other planets, but tidal effects could hardly be a general mechanism.

There have been serious difficulties with the suggestion that the Moon was accumulated in a primitive gas sphere because of its low density and the general belief that the carbonaceous chondrites represent the solar abundance of iron. This abundance of about 0.9 relative to silicon in numbers of atoms would lead one to expect a density of the non-volatile fraction of solar material considerably higher than that of the Moon, and would require a considerable amount of water in the lunar interior. Since the lunar rocks are remarkably free of water, this has not seemed probable. The high abundance of iron in the Sun, as determined a few years ago based upon Whaling's (1970) oscillator strengths, depends upon solar data on strong lines which are markedly affected by damping in collisions with hydrogen atoms in the Sun's atmosphere. Brueckner (1971) has shown that damping constants are larger than previously estimated, and, hence, that the abundances reported have been too high. Gilbert *et al.* (1974) and his associates have determined the oscillator strengths of weak lines, and these criticisms do not apply to abundances determined using these data. These latter determinations give a solar ratio of iron to silicone of about $\frac{1}{3}$, and the density of the solar material of low volatility agrees rather closely with that of the Moon. Thus, after considerable uncertainty, this particular objection to the Moon being a primitive object may possibly be resolved.

The difference in chemical composition of the Moon and terrestrial planets is a serious problem. The density of Mars is about 5.5, indicating that it is about 65% iron by mass while the Earth and the Moon must consist of about 30% and 15% of iron respectively. Some fractionation of this density element relative to magnesium and silicon and other elements must have occurred. Of course, we do not know whether other satellites of similar masses to that of the Moon have the composition of the Moon relative to these low volatile elements.

The hypothesis that there were many objects of lunar mass in the early solar system was advanced some years ago in an attempt to understand the differences in chemical composition of the Moon and Earth. The capture of the Moon by the Earth appears to be improbable, and this improbability seems to be relieved if many Moons were present. Also, the tilt of the axes of the planets is readily explained by this hypothesis, and other features of meteorites, variation in densities of the terrestrial planets, etc., are explained by this hypothesis. If many moons were not present in the early solar system, this student of the subject would prefer to believe that the Moon escaped from the Earth and would prefer to try to explain the chemical differences between the Earth and Moon rather than to assume that capture of one lone Moon by the Earth occurred, and would ascribe the irregular tilt of planetary axes to some unknown process. The 'Many Moon' hypothesis is viewed with disfavor by many students of solar origin, but it seems to me as being no less probable than the many asteroid hypothesis, and, in fact, I have assumed that many objects of both kinds were present.

One of the major problems of this 'Many Moon Theory' is the formation of regularly spaced satellite orbits. However, it seems probable that Mars had a nebula (Urey, 1972), and, if so, it seems likely that the other planets also had nebulae and that the capture and spacing of satellites were determined by the action of such nebulae. Describing such processes in detail is a very difficult problem. In discussing such complicated problems, it is desirable to be able to have observational data. This paper is an attempt to apply observational data to the problem.

References

Bainbridge, J.: 1962, *Astrophys. J.* **136**, 202.
Brueckner, K. A.: 1971, *Astrophys. J.* **169**, 621.
Chandrasekhar, S.: 1956, *Vistas in Astronomy* **1**, 344.
Gilbert, A., Sulzmann, K. G. P., and Penner, S. S.: 1974, *J. Quant. Spectrosc. Radiat. Transfer*, in press.
Herbig, G. H.: 1971, private communication.
Kusaka, T., Nakano, T., and Hayashi, C.: 1970, *Prog. Theoret. Phys.* **48**, 1580.
Schmidt, O. Yu.: 1944, *Dokl. Akad. Nauk* **45**, 245.
Singer, S. F.: 1970, *Science* **170**, 1196.
Urey, H. C.: 1958, *Proc. Chem. Soc. London* **67**.
Urey, H. C.: 1972, in Centre National de la Recherche Scientifique (ed.), *Symp. on Origin of the Solar System*, Nice, France, p. 206.
Whaling, W.: 1970, *Nucl. Instr. Methods* **10**, 363.

THE MOVEMENT OF SMALL PARTICULATE MATTER IN THE EARLY SOLAR SYSTEM AND THE FORMATION OF SATELLITES

T. GOLD

Center for Radiophysics and Space Research, Cornell University, Ithaca, N.Y., U.S.A.

(Read by E. E. Salpeter)

Satellites are a common feature in the solar system, and all planets on which satellite orbits would be stable possess them. (For Mercury the solar perturbation is too large, and the retrograde spin of Venus would cause satellites to spiral in to the planet through tidal friction.) An explanation of the formation of satellites must hence be one which makes the phenomenon exceedingly probable at some stage in the solar system formation processes, and very improbable processes like a capture cannot be the answer in most cases.

Small particulate matter must have been very abundant in the early solar nebula. Such particulate matter must have existed both from the first condensation of the low vapor pressure components of the gas in the first round, and it must also have been composed of material scattered from impacts after some major bodies had begun to form, frequently finding themselves no doubt on collision orbits.

In general, small particulate matter will not follow the same orbits as large bodies would, due to the action of drag forces. In the early solar system such drag must have been present within the original gaseous disc, and at later stages there continued to be a slight drag due to the Poynting-Robertson effect. The gas drag, depending on the mass and temperature distribution of the gas in the early solar nebula, could have acted to supply a force on small particles, either in the forward direction of planetary orbits, or in the retrograde direction. Particles could thus be caused to spiral either outwards or inwards, as a result of such forces.

Such material will frequently not continue to move on a spiral orbit, for it will frequently happen that, as the mean period gradually changes, a particle comes into resonance with a perturbing force arising from the motion of major bodies. Such resonances can be of two kinds: one causing stability for that particular period, the other removing material even more quickly from orbits of that period than the spiral motion would have done. In the one location in the solar system where we still see small particulate matter on long-lived orbits, we indeed see the material organized into well-defined rings, namely the rings of Saturn. It may well be that even at the present time there exists a similar banded structure for particles of a certain size range in the plane of the solar system, with some bands as a result of planetary perturbations having permanent stability, despite the Poynting-Robertson effect. The zodiacal light may be due to a set of bands that would make the solar system look like a faint version of Saturn's rings when viewed from outside the central plane.

Such bands will have been a most important feature in the condensation processes in earlier times. It is in these circumstances, where a nonconservative force has been active, that Liouville's theorem is not satisfied, and indeed particles will be driven into much higher densities than the densities at which they were originally supplied. It is important to realize that the process will also sort particles for a certain size range, in the sense that the ones that are too small (for the strength of any particular celestial mechanics resonance) cannot be arrested against the drag force that causes spiraling, while particles that are too large may spiral so slowly as not to reach a resonant condition. A resonant band will thus first be supplied with the smallest particles that can be stable in that resonance and will later gradually acquire larger and larger objects. The relative velocities, and thus the erosion rates through mutual collisions, will become very small, and the circumstances will be favorable for snowballing, to make larger particles with the help of any surface stiction. The concentration of asteroids into the Trojan orbits cannot readily be understood without such an action. It is also important to realize that such lanes may in some cases have been supplied principally from original condensate, but in other cases from debris of collisions of earlier bodies. Different lanes may thus be chemically and mineralogically distinctive. In the course of long times, the stability of individual planetary bands may be lost as a result of various changes. Major perturbations among the planets, further accretion, or major collision, could so change the celestial mechanics situation as to destroy the resonance with a particular band, and it would then again begin to spiral. Now, however, the starting point would be an enormously more concentrated, narrow lane of highly collimated orbits, rather than a diffuse distribution in the whole disc. The densities of particulate matter in such a band may well have risen above the mean by a factor of the order of 10^4. We must then visualize that there will have been many events of such highly concentrated bands becoming unlocked from their resonance and therefore spiraling in the solar system. When such material reaches the vicinity of a planet each particle may suffer one of three possible fates. Firstly it may impact the planet and thus contribute to its growth. Secondly it may, after a period of large perturbations, cross safely to the other side of the sphere of influence of that planet and then continue with its spiral. Thirdly, it may be placed on a satellitic orbit to that planet, a probability which would be greatly increased by the presence of a frictional medium such as the main solar nebula, or its temporary concentrations in the vicinity of the planets. It is extremely difficult to make any estimate of the relative probability of the three types of encounter. Even if the setting up of a satellite orbit were many times less probable than the other two possibilities, it would still suffice to lead to the eventual formation of the satellites that we know, if the process occurred early enough so that a large proportion could still be added to make the planets grow.

The step from circumplanetary rings attenuated by gas friction to the formation of satellite bodies is not a difficult one and has been discussed on many occasions. If in the first place a number of separate satellites form, as may well be the rule, it will depend on their size distribution, and the tidal friction with the planet, whether they can be maintained as separate bodies. If tidal friction is sufficient and there is a forward spinning planet, and if the innermost satellite formed is more massive than the others, then they will all be swept up into one body finally. This is because the most massive satellite will spiral out the fastest, and there is no possibility in this case of one body crossing the lane of the next without colliding with it. Perhaps this is the set of events that occurred to make our rather large Moon, while in the case of the major planets the evolution through tidal friction has been somewhat slower. The material that now forms the surface of the Moon is perhaps the last addition, and the soil that is found there is material acquired directly in its present form from orbit. The fact that the material has suffered chemical differentiation on a planetary body in its past does not argue against such a theory. As we have said, many of the bands will be debris from collisions, and the last material acquired by the Moon may be one of those. Whatever finely divided material fell into the Moon in the late phases would tend to make a set of layers that may be chemically distinctive. Any of the larger impacts, such as those that caused the mare basins, must then rework the layered structure into one that makes for regional differences in composition. Much detail that is now known about the lunar soil is difficult if not impossible to account for within the view that this soil resulted from the grinding up by meteorite bombardment of solid lunar rocks. In particular, the high intensity and remarkable uniformity of the cosmic ray exposure of all the soil seems to accord much better with a picture in which this soil was in diffuse form in orbit and fell in to make the last addition to the Moon. (The impact on collision of each grain with the lunar surface would not heat the particles enough to eradicate the cosmic ray tracks if infall was from other Earth satellite orbits only.)

The recent radar observations from the Jet Propulsion Laboratory at Goldstone of Saturn's rings suggest that they are made largely of metallic particles. (The alternative theory that they are made of dielectric material in accurately spherical form – cat's eyes – is considered unlikely, since we know of no mechanism that would tend to assemble meter-sized spheres as would be required.) This observation emphasizes the view that bands of particular composition can be formed and become placed on satellite orbits. The rings of Saturn are too close to the planet to form further satellites, but the greatly varying composition of the satellites suggests that similar processes had occurred there but with differently sorted out, second generation debris material.

Acknowledgement

Work on lunar studies is carried out under NASA Grant NGL-33-010-005.

GRAVITATIONAL COLLAPSE AND THE FORMATION OF THE SOLAR NEBULA

R. B. LARSON

Dept. of Astronomy, Yale University, New Haven, Conn., U.S.A.

Abstract. Most theories of the origin of the solar system begin by assuming that, at the time of its formation or shortly afterwards, the Sun somehow acquired a disc-like 'solar nebula' in which the planets later formed by accretion. The presently available gravitational collapse calculations, while not yet sufficiently sophisticated to delineate in detail the processes involved, at least support the plausibility of this idea and suggest that the solar nebula may have formed by the accretion or capture of leftover protostellar material into orbit around the early Sun. Infall of matter into the disc probably continued for roughly 10^5 to 10^6 yr, and may have produced transient strong heating effects through the thermalization of the infall energy in shock fronts. It seems likely that the satellite systems formed in a basically similar way through the capture of residual solid matter into orbit around the planets.

A more complete discussion of gravitational collapse and the formation of the solar system has been given in *The Origin of the Solar System*, ed. H. Reeves, p. 142. (Paris: CNRS).

THE PRINCIPLE OF LEAST INTERACTION ACTION

M. W. OVENDEN

Dept. of Geophysics and Astronomy, and Institute of Astronomy and Space Science, University of British Columbia, Vancouver 8, B.C., Canada

Abstract. The intuitive notion that a satellite system will change its configuration rapidly when the satellites come close together, and slowly when they are far apart, is generalized to 'The Principle of Least Interaction Action', viz. that such a system will most often be found in a configuration for which the time-mean of the action associated with the mutual interaction of the satellites is a minimum. The principle has been confirmed by numerical integration of simulated systems with large relative masses. The principle lead to the correct prediction of the preference, in the solar system, for nearly-commensurable periods. Approximate methods for calculating the evolution of an actual satellite system over periods $\sim 10^9$ yr show that the satellite system of Uranus, the five major satellites of Jupiter, and the five planets of Barnard's star recently discovered, are all found very close to their respective minimum interaction distributions. Applied to the planetary system of the Sun, the principle requires that there was once a planet of mass $\sim 90\ M_\oplus$ in the asteroid belt, which 'disappeared' relatively recently in the history of the solar system.

References

Ovenden, M. W.: 1973, in B. D. Tapley and V. Szebehely (eds.) *Recent Advances in Dynamical Astronomy*, D. Reidel Publ. Co., Dordrecht, p. 319.
Ovenden, M. W.: 1974, 'Bode's Law – Truth or Consequences', *Vistas in Astronomy* **16**, in press.
Ovenden, M. W., Feagin, T., and Graf, O.: 1974, *Celes. Mech.* **8**, 455.

V. JOVIAN RADIO BURSTS AND PULSARS

(Edited by F. G. Smith)

Organizing Committee

F. G. Smith (Chairman), M. M. Komesaroff, V. K. Prokof'ev, M. J. Rees, B. Warner

SUMMARY

Professor G. R. A. Ellis reviewed the wide range of radio emission from Jupiter. At centimetric wavelengths the thermal radiation corresponds to a blackbody at 130 K. Between 2 m and 10 cm wavelength there is a powerful component of synchrotron radiation from the electrons trapped in the radiation belts. At longer wavelengths there is a great variety of impulsive radio emission from coherent plasma oscillations.

The magnetic field of Jupiter is known from the polarisation of the synchrotron radiation to be situated centrally (within one tenth of the radius) and inclined at 10° to the rotation axis. The radiating electrons have energies of the order of 10 MeV, and a density of 10^{-3} cm^{-3}, much greater than in the case of the Earth's radiation belts.

The decametric radiation varies with the rotation of Jupiter, possibly analogously to pulsar radiation. Bursts at around 4 MHz reach very high brightness temperatures, exceeding 10^{17} K. The occurrence of these strong bursts is closely related to the position of the Jovian satellite Io, which must have an interaction with the main magnetic field.

The bursts are usually almost completely circularly polarised, varying in detail with radio frequency. There are many different time scales of fluctuation, from hours to microseconds: again this is reminiscent of pulsar radiation. The bursts drift rapidly in frequency, as in some types of solar radio bursts. The drift rate takes characteristic values at different frequencies. Drifts can occur both upward and downward in frequency.

Jovian radio bursts may be closely related to some radio emission from the terrestrial ionosphere, such as the 'hiss', 'dawn chorus', and 'whistlers'. These are stimulated by energetic particles from the Sun. Artificial stimulation by terrestrial radio sources has also been deomonstrated.

Prof. Ellis gave a brief review of the theory of Jovian radio bursts, and referred the audience to the following key references:

References

Carr, T. D. and Gulkis, S.: 1969, *Ann. Rev. Astron. Astrophys.* **7**, 577.
Ellis, G. R. A.: 1965, *Radio Sci.* **69D**, 1513.
Goldreich, P. and Lynden-Bell, D.: 1969, *Astrophys. J.* **156**, 59.

Dr J. Ables presented a review of the radio observations of pulsars, concentrating on those observations which are particularly important in understanding the radiation mechanism.

The radio pulses may be described by an integrated pulse envelope extending over a few percent of the period, obtained by the superposition of some hundreds of pulses.

Individual pulses contain narrower components known as subpulses; these may contain variations on a shorter timescale, known as the microstructure. The subpulses often appear at steadily changing times in successive pulses, so that they 'drift' through the integrated pulse profile. In some cases, the drifting occurs clearly at the beginning and end of the profile, but not at the centre.

The pulse energy is very variable from pulse to pulse. The statistical distribution of energy in the Crab Pulsar may be consistent with Poisson statistics, in which pulses are made up of a randomly varying number of individual spikes. There are also random variations in intensity on timescales of months and years. The radio spectral index is generally between -1 and -3.5, but it often becomes even steeper at the highest radio frequencies, and also becomes flatter or reverses in slope at low frequencies.

The polarisation of the radio pulses gives an important clue to the radiation process. Integrated profiles may be made by adding the Stokes parameters of a sequence of pulses. These integrated profiles show polarisation which is predominantly linear, often with a monotonic swing of position angle by up to 180°. A circularly polarised component is also seen for some pulsars near the centre of the integrated profile.

The degree of polarisation generally falls at higher radio frequencies. The subpulses are typically very highly polarised, showing changing forms of elliptical polarisation within each subpulse as it drifts.

Theories of the radiation mechanism fall into three classes according to the location of the emitting region. The pulsar is a rotating neutron star, with a strong magnetic field which forces any ionised magnetosphere into co-rotation with the star. The three classes of theory refer to an origin in a magnetic polar region close to the surface, or in a region close to the velocity of light cylinder where co-rotation would produce speeds approaching the speed of light, or further out beyond the velocity of light cylinder.

The polar cap theories involve a radiation mechanism (curvature radiation) which beams radiation along the polar magnetic field lines. They provide in particular an explanation of the absence of 'interpulses' in most pulsars, since the radiation from the opposite pole is not observed unless the angle between the magnetic and rotation axes is near 90°. The 'velocity of light cylinder' theories are derived from the observed independence of pulse width on frequency. The width is determined by geometrical considerations only, through relativistic beaming.

A third theory, proposed by Lerche, involves a source beyond the velocity of light cylinder which is excited by the anisotropic magnetic pressure of the rotating field.

Prof. P. A. Sturrock gave an account of the 'polar-cap' model, following the lines of an early article (P. A. Sturrock, *Astrophys. J.* **164**, 529, 1971), modified in accordance with a more recent article (D. H. Roberts and P. A. Sturrock, *Astrophys. J.* **181**, 161, 1973).

Intense electric fields develop near the magnetic polar caps which accelerate both ions and electrons to high energies. The electrons radiate high-energy gamma rays which may annihilate in the magnetic field to produce electron-positron pairs. If this occurs, the flow becomes unstable, giving rise to bunching and coherent radio emission. If the transition from closed to open field lines occurs at the 'force-balance radius', rather than the light cylinder, one obtains good agreement with the observed braking index, pulse widths, and period-age distribution. The pair production cascade also explains the high particle flux from the Crab pulsar into the nebula. X-ray radiation is attributed to synchrotron radiation from the secondary electrons and positrons, but the optical radiation is believed to be due to coherent radiation from the electron-positron bunches resulting from the cascade as these bunches moved along the curved magnetic field lines.

Prof. F. G. Smith showed that the relativistic beaming theory resulted from a more detailed study of the observed characteristics of the pulses, and particularly of the width and polarisation of the sub-pulses. The emitting region appeared to be situated typically at 0.7 to 0.9 of the radial distance to the velocity of light circle, and located in a particular region of the magnetic field pattern. The radiation mechanism appeared to be coherent cyclotron radiation, in which bunches of high-energy electrons moved in approximately circular orbits. This would be narrow-band radiation; the wide spectrum must be generated by an assembly of such sources. Optical and X-ray radiation from the Crab Pulsar would then be incoherent synchrotron radiation from the same particles.

DISCUSSION

In a general discussion the following points offered possibilities of distinguishing between the theories:

Cole: The maintenance of sub-pulse structure over many pulsar rotations must be explained as a long-lived moving electron configuration.

Rees: The fine frequency structure from narrowband sources might appear differently across a relativistically compressed pulse, due to the varying Doppler effect.

Ables: The lack of an interpulse in most pulsars is difficult to explain by the relativistic beaming model for co-rotating sources since it implies only a single such emitting region. If emitting regions are associated with magnetic poles, it is necessary to explain why this theory would not predict at least two pulses per rotation.

Lyne: The smooth and very rapid changes of polarisation within the sub-pulses favours the theory of beam compression.

Komesaroff: On Professor Smith's model the cyclotron frequency is emitted. This being longitude dependent would result in a relative delay of different frequencies which is not observed.

Sturrock: Estimates of the particle flux from the pulsar into the Crab nebula, of 10^{41} electrons or positrons per second, seem not to be compatible with the model proposed by Smith. Also, Smith's estimate of the magnetic field strengths required by long period pulsars leads to estimates of the 'age' which are much shorter than those found observationally.

Smith: The two theories approach the problems in different ways. Although the relativistic beaming theory does not attempt to explain the plasma physics, it is directly related to the observations and provides a description of the location and nature of the emitter which seems to be inescapable. The model involving flat sheets of charge cannot match the detailed description of the pulse structure.

Sturrock: I agree that the earlier model involving flat sheets of charge is incompatible with observational data. It now seems that the current pattern at the polar caps must have highly variable small-scale structure.

Prof. F. D. Kahn discussed the propagation of the electromagnetic wave associated with pulsar rotation. According to some theories a pulsar emits most of its energy in the form of very low frequency electromagnetic waves. The period of such a wave is the same as that of the pulsar. Its amplitude can be very large indeed: for example, in the case of the Crab pulsar, the vector potential would have a magnitude A of order 10^{14} (in Gaussian units) at the speed of light cylinder. In the presence of such a wave an electron or a proton will have a mass of order $eA/c^2 = 5 \times 10^{-17}$ gm. This is several orders of magnitude larger than the rest mass of either particle. The transmission of the vlf waves through a plasma therefore depends very much on their amplitude, and simple propagation properties can be expected only for circularly polarized waves with a single frequency.

Various interesting suggestions have been made about the manner in which vlf waves actually force their way through a diffuse plasma. One notable proposal, due to Rees, is that the waves clear channels for themselves through the plasma, within which their vector potential is large enough and along which they can propagate.

But in order to be able to use any such theory one needs to know whether a vlf wave of pure frequency is stable. Just recently a stability calculation has in fact been undertaken (Claire Ellen Max, 1973, preprint). It is found that a circularly polarised wave of large amplitude, with a given frequency and wavenumber, is unstable to a break-up into circularly polarised waves of neighbouring frequencies and wavenumbers. So far only a linearized calculation is available. But the probable inference is that no circularly polarised wave of pure frequency can continue to propagate. Instead it will break up into a mixture of waves with different frequencies and different handedness of polarisation. The mass eA/c^2 of the charged particles will then fluctuate rapidly, and this will cause equally rapid changes in the refractive index of the medium for waves of any frequency. The propagation of any wave will therefore be impeded. Very probably the energy of vlf wave is thus converted into random energy of the charged particles quite close to the source of the radiation.

Prof. L. Mestel presented the theory of Force-Free Pulsar Magnetospheres.

The standard model of an obliquely rotating neutron star is adopted for the pulsar: the aim is to formulate and solve the equations to the structure of the external magnetic field and associated electric field. Since only a small fraction of the rotational energy being lost from the Crab pulsar is tapped to supply the radio, X-ray and optical pulses, it is a plausible approximation to ignore such losses and treat the magnetospheric field as simply coupling the pulsar with the surrounding nebula *via* the light-cylinder. The theory should predict a breaking index; it could also conceivably shed light on the pulsar radiation mechanism, e.g. if the field could be shown necessarily to have current singularities.

The field is assumed steady in the rotating frame, so that

$$\frac{\partial}{\partial t} = -\Omega \frac{\partial}{\partial \theta}, \tag{1}$$

where $\Omega \mathbf{k}$ is the angular velocity vector and θ the azimuthal angle about $\mathbf{k}$. The postulate that the star is surrounded by a strict vacuum – with zero charge-current density – then yields the classical Maxwell-Deutsch wave of frequency Ω, discussed e.g. by

Pacini (1967, 1968) and Ostriker and Gunn (1969). Well within the light-cylinder this wave has an electric field component $E_\parallel$ along B of the same order as $E_\perp$, and this can act as a powerful accelerator on individual charges. However, in the presence of a very small plasma density, the component $E_\parallel$ causes a charge-separation which is likely to reduce $|E_\parallel|$ to a value much below $|E_\perp|$. To a first approximation, the vacuum conditions $\varrho_e = 0, j = 0$, are now replaced by the plasma condition

$$\mathbf{E} + \frac{(\Omega \mathbf{k} \times \mathbf{r}) \times \mathbf{B}}{c} = 0, \qquad (2)$$

this being equivalent to the more general $\mathbf{E} + (\mathbf{v} \times \mathbf{B})/c = 0$ when the velocity $\mathbf{v}$ is the sum of the co-rotation velocity $\Omega \mathbf{k} \times \mathbf{r}$ plus a component parallel to $\mathbf{B}$. (For (2) to hold beyond the light-cylinder we require that the velocity components along the field-lines are directed so as to keep the total velocity below c). Associated with $\mathbf{E}$ from (2) is the charge density

$$\varrho_e = \frac{\nabla \cdot \mathbf{E}}{4\pi} = -\frac{\Omega \mathbf{k}}{2\pi c} \cdot \{\mathbf{B} - \tfrac{1}{2}\mathbf{r} \times (\nabla \times \mathbf{B})\}. \qquad (3)$$

Goldreich and Julian (1969) argued that the pulsar itself would supply the plasma: $E_\parallel$ acting at the pulsar surface would extract charges of one sign or the other. Ruderman (1971) estimated that the work-function resisting electronic emission is quite modest, but that ions would probably remain bound, implying that any currents into and out of the pulsar probably consist of electrons only. We postulate that at each point where Equation (3) requires ϱ_e to be positive, the magnetosphere has in fact acquired a sufficient ion density (e.g. via accretion). A strictly *charge-separated* magnetosphere is not consistent with the steady-state condition (Goldreich and Julian, 1969; Okamoto, 1973); however, even if the net charge-density is a good deal smaller than the density of either sign, the associated mass-density ϱ is still extremely small, in the sense that $B^2/8\pi\varrho c^2 \gg 1$ (Mestel, 1971). The equations of motion of the plasma then reduce to the *relativistic force-free condition*

$$0 = 4\pi\varrho_e \mathbf{E} + \frac{4\pi \mathbf{j}}{c} \times \mathbf{B} = (\nabla \mathbf{E})\mathbf{E} + \left(\nabla \times \mathbf{B} - \frac{1}{2}\frac{\partial \mathbf{E}}{\partial t}\right) \times \mathbf{B}. \qquad (4)$$

(The vacuum solutions are clearly a sub-class, in which the two terms in (4) vanish separately). Equations (2), (3) and (4) and condition (1) jointly yield (Mestel, 1973; Endean, 1973)

$$\nabla \times \tilde{\mathbf{B}} = \psi \mathbf{B}, \quad \mathbf{B} \cdot \nabla \psi = 0, \qquad (5)$$

where

$$\tilde{\mathbf{B}} = \left\{B_r\left(1 - \frac{\Omega^2 r^2}{c^2}\right), B_\theta, B_z\left(1 - \frac{\Omega^2 r^2}{c^2}\right)\right\} \qquad (6)$$

in cylindrical polar coordinates (v, θ, z). Equations (5) are generalizations – to non-axially symmetric systems that are steady in the rotating frame – of the axisymmetric equations given by Michel (1973) and Scharlemann and Wagoner (1973).

The adoption of the 'magnetohydrodynamic approximation' (2) does *not* imply at all that $\mathbf{E}_\parallel$ is assumed to be strictly zero, so that electrical acceleration parallel to $\mathbf{B}$ is ruled out. For the solutions of Equations (5) to be the zero-order terms in a self-consistent iterative scheme, we merely require $|\mathbf{E}_\parallel| \ll |\mathbf{E}_\perp| \simeq \Omega r B/c$. There will in general be currents flowing parallel to $\mathbf{B}$, and $\mathbf{E}_\parallel$ can be found from the component along $\mathbf{B}$ of the equation of motion of the electrons. The approximation (2) for constructing a zero-order field model is justified if $|\mathbf{E}_\parallel|$ so determined is much below $|\mathbf{E}_\perp|$. The most important correction to (2) may very well come not from the relativistic inertial terms near the light-cylinder but from resistance due to two-steam instability. The essential point is that $\mathbf{E}_\parallel$ – so important in theories of radio emission such as Prof. Sturrock's – should be *determined* rather than settled by 'fiat' (Ruderman, 1972).

Well within the light-cylinder $(r \ll r_c = c/\Omega)$, Equations (5) can be approximated by the curl-free condition, whereas when $r \simeq r_c$ both particle and displacement currents are significant. When $r \gg r_c$ only the displacement current is important, so that for those field-lines which get beyond r_c the wave is essentially a Maxwell vacuum wave. A crucial question, which solution of Equation (5) will answer, is: how much magnetic flux will pass through the singularity at r_c? For the only simple case – cylindrical symmetry, with $\partial/\partial z = 0$, but $B_z \neq 0$ – the answer is rather dramatic: all the field-lines leaving the pulsar are trapped within r_c, yielding a zero radial Poynting vector at r_c. Thus the presence of the charge density (3) kills the cylindrical wave that would otherwise carry energy to infinity.

This result cannot be generalized to more realistic geometry, but it is plausible that the field structure near the light-cylinder will usually be significantly affected by the particle current. As noted by Dr Claire Max (private communication), there is a partial similarity between the pulsar magnetosphere problem and that of the propagation of *plane* strong waves in dense plasmas. The critical electron density for such waves to be stifled is comparable with the charge density (3) at the light-cylinder. It is therefore not surprising that the pulsar wave is killed in one simple geometry, and that it will probably be significantly modified in general. Further results – in particular, the prediction of a braking index – must await solution of Equations (5).

References

Endean, V. G.: 1973, *Astrophys. J.*, in press.
Goldreich, P. and Julian, W. H.: 1969, *Astrophys. J.* **157**, 869.
Mestel, L.: 1971, *Nature Phys. Sci.* **233**, 149.
Mestel, L.: 1973, *Astrophys. Space Sci.* **24**, 289.
Michel, F. C.: 1973, *Astrophys. J.* **180**, L133.
Okamoto, I.: 1973, in preparation.
Ostriker, J. P. and Gunn, J. E.: 1969, *Astrophys. J.* **157**, 1395.
Pacini, F.: 1967, *Nature* **216**, 567.
Pacini, F.: 1968, *Nature* **219**, 145.
Ruderman, M.: 1971, *Phys. Rev. Letters* **27**, 1306.
Ruderman, M.: 1972, *Ann. Rev. Astron. Astrophys.* **10**, 427.
Scharemann, E. T. and Wagoner, R. V.: 1973, *Astrophys. J.* **182**, 951.

VI. THE OUTER LAYERS OF NOVAE AND SUPERNOVAE

(Edited by C. de Jager)

Organizing Committee

E. R. Mustel (Chairman), C. de Jager

THE CHEMICAL COMPOSITION OF
THE ENVELOPES OF NOVAE

L. I. ANTIPOVA

Astronomical Council, USSR Academy of Sciences, Moscow, U.S.S.R.

Abstract. There are two methods to study the chemical composition of the envelopes ejected by novae: (a) the analysis of the absorption spectra of novae; (b) the analysis of emission lines during the nebular stage.

1. The Absorption Method

It is known that the spectra of novae contain usually absorption and emission components. The strength of the emission is smallest at light maximum: sometimes there is even practically no emission in the spectra of novae at the very moment of light maximum. Therefore, only for this moment we may obtain the most reliable results on the chemical composition.

In order to obtain quantitative results from the investigation of the absorption spectrum, we have to know a mechanism of formation of the line profiles.

Before light maximum the outer layers of the nova undergo an expansion with a relatively high velocity. Therefore, it is expected that this expansion may significantly influence the line profiles. However the size of the envelope of the nova at light maximum is usually much larger than the size of the 'photosphere' at this moment (Mustel, 1945). For example the 'photospheric' radius of DQ Her 1934 at light maximum was about 75 $R_\odot$, whereas the radius of the envelope was about 500 $R_\odot$ (Beer, 1937). Furthermore, calculations show (Mustel, 1945), that in this case the influence of expansion of the envelope on the profiles of absorption lines is very small.

Now we may expect a certain velocity gradient to exist in the expanding envelope of a nova close to light maximum. This velocity gradient may also influence the absorption lines. There are many papers on this problem (Abhyankar, 1964a, b, 1965; Kubikowski and Ciurla, 1965; Ciurla, 1966). The principal conclusions from these papers are the following: (1) The lines, widened by the velocity gradient are asymmetric. (2) Strong and weak absorption lines have different displacements. (3) The equivalent widths of absorption lines is increased, those of the weaker lines are increased more strongly.

At the same time different observers find that at light maximum the absorption lines of spectra of novae are often relatively narrow and sharp. Moreover, we found that there is no asymmetry in the tracings of DQ Her 1934 and HR Del 1967. Then, the equivalent widths W_λ of the lines of DQ Her have been compared with the values of W_λ for the same lines in the spectra of ε Aur, the supergiant of the same spectral type (Mustel and Baranova, 1965). It has been found, that the lines of DQ Her are systematically weaker than those in the spectrum of ε Aur, the weaker lines being weakened more strongly, whereas we might expect the opposite effect for the case with a velocity gradient. Thus, it can be concluded that, at least in some novae, the velocity gradient in the envelope is not too large and does not influence appreciably the profiles of absorption lines. This, of course, does not mean that generally the velocity gradient in these envelopes is absent or very small. It seems that the absorption lines are formed mainly inside some effective level, which has a small thickness in comparison with the thickness of the whole envelope and the velocity dispersion in this particular layer is relatively small.

Thus we may assume that the turbulent motions and the radiation damping (or collisional damping) are the main broadening factors of the absorption lines in the spectra of those novae, where the absorption lines have symmetric profiles.

In such cases the usual method of the curve of growth can be used for the determination of the chemical composition. First of all, it is necessary to choose some model for the atmosphere of the star. Since the novae have a very extended 'reversing layer', we may accept the Schwarzschild-Schuster's model. According to this model the upper layers produce mainly absorption lines whereas the continuous radiation comes from a 'photosphere'.

The first results of the analysis of the chemical composition of the envelopes of novae by means of the curve of growth method at light maximum are published by Mustel and Boyarchuk (1959). The spectrograms of DQ Her with a dispersion of about 36 A mm^{-1}, obtained by G. A. Shajn at the Simeiz Observatory were used for this analysis. It has been found that the relative abundances of C, N and O in comparison to the metals are considerably higher in the envelope of DQ Her 1934, than in the atmospheres of 'normal' stars. At the same time the relative abundances of the metals in the envelope of DQ Her and in the atmospheres of 'normal' stars are prac-

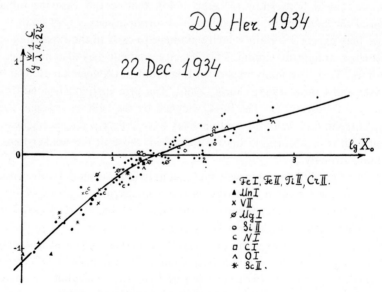

Fig. 1. Curve of growth for DQ Her at light maximum.

tically identical. Taking into account the importance of the problem of the chemical composition of novae, this analysis was repeated in papers of Mustel and Baranova (1965) and Mustel and Antipova (1971). Improved data on oscillator strengths were used in these papers. Figure 1 shows the curve of growth for DQ Her at light maximum (22 December, 1934). We see, that the scattering of the individual points is small in this figure. This is a direct independent argument in favour of the conclusion about the possibility to apply the method of the curve of growth for the determination of the chemical composition to the envelopes of novae at light maximum.

It is very important to know in such investigations the excitation temperature for different atomic transitions. The usual method here is to plot the dependence of the shift of the multiplets vs the excitation potential. This dependence is normally a straight line and the slope of the line gives the excitation temperature. However, in the case of DQ Her this dependence is not a straight line and we observe the effect of an over-

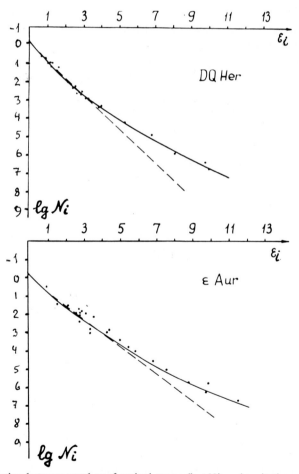

Fig. 2. Relation between number of excited atoms (log N_i) and excitation potential (ε_i) for DQ Her and ε Aur.

excitation of the levels with high excitation energies (Figure 2, Mustel and Baranova, 1965). It is possible, that this fact has the following explanation. Since the lines with different excitation potentials are formed in different layers of a very extended envelope with a temperature varying with height, the excitation temperature for lines with different excitation potentials has to be different. In particular, this temperature may be relatively high in relatively deep and relatively hot layers. This leads to the effect shown in Figure 2. It is not excluded also that we deal here with some high-temperature cells inside the extended envelope.

It is very important to point out that the dependence shown for DQ Her in Figure 2 is very similar to the one obtained for the supergiant ε Aur, see again Figure 2. Both stars (DQ Her at light maximum and ε Aur) have approximately the same spectral class. This last fact (similarity between spectral classes) was already noted by McLaughlin (1937). This similarity between the two objects is illustrated also by Table I which shows the physical parameters determined by means of the method of the curve of growth for DQ Her and ε Aur. These parameters are approximately identical. Hence the conditions in the envelope of DQ Her were more or less identical to these in the atmosphere of supergiant a of the F-type.

A comparison of the chemical composition of the envelope of DQ Her found by Mustel and Antipova (1971) with the chemical composition for 'normal' stars is shown in Figure 3. We see that the relative chemical composition of the envelope of DQ Her for the metals coincides practically with the 'normal' one. However the abundances of C, N and O in the envelope of DQ Her are two orders higher than the 'normal' ones.

TABLE I

Comparison of physical properties of the envelope of DQ Her and the atmosphere of ε Aur

	DQ Her 1934	ε Aur
v_t	19 km s^{-1}	17.8 km s^{-1}
T_{ex}(0–3 eV)	4500°	4800°
T_{ex}(7 eV)	6300°	6200°
T_i	7000°	6800°
lg N_e	13.36	12.69

The presence of anomalously strong lines of C, N and O in the spectra of DQ Her at light maximum may be considered as a confirmation of this conclusion. The abnormal intensity of these lines cannot be due to an abnormal excitation of atoms of C, N and O, since the absorption lines of other elements which have approximately the same potentials of excitation, (Si II, Mg II) give quite normal abundance.

The appearance of very intense bands of CN just after light maximum in the spectrum of DQ Her confirms also the conclusion about the abnormal abundances of C and N in the envelope of the Nova. The appearance of bands of CN in the spectrum of DQ Her and in the spectra of some other novae immediately after light maximum is very interesting. These bands are practically not discernible in the spectra of usual supergiants of the same spectral types. An investigation by Antipova (1969), carried out on the base of spectrograms of the Mount Wilson Observatory, has shown that the appearance of intensive bands of CN in the spectra of novae is connected with: (1) the abnormally high chemical abundances of O, C and N in these envelopes, (2) a drop of the temperature of the star after light maximum and (3) a very rapid compression of the envelope immediately after light maximum. A transformation of

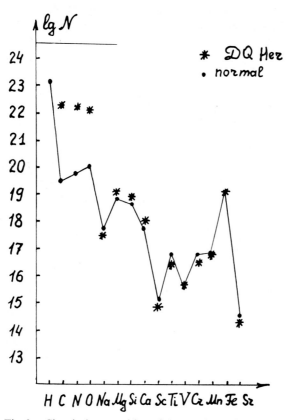

Fig. 3. Chemical composition of the envelope of DQ Her.

the pre-maximum spectrum into the principal one is also connected with this rapid compression of the envelope (Mustel, 1949, 1962).

A similar investigation of the chemical composition of the envelope of HR Del 1967 at light maximum was carried out by Antipova (1974). Spectrograms, obtained in the Haute Provence Observatory and kindly put at our disposal by Prof. Ch. Fehrenbach have been used in this study.

The nova HR Del is a peculiar nova because of its very slow increase and decrease of brightness. The spectrum of HR Del is the typical spectrum of an supergiant F-type with narrow absorption lines. The careful analysis of the profiles of these lines shows the absence of any asymmetry, which might speak for the presence of a velocity gradient in the envelope.

Half a year has passed between the explosion of the star and the light maximum. The computations show that at light maximum the envelope become so large in comparison with the radius of the 'photosphere', that we may neglect the influence of the expansion of the envelope on the profiles of the absorption lines (see above). Thus, there are reasons to conclude that turbulent motions and radiation damping (or collisional damping) are also the main factors for broadening the absorption lines in the spectrum of HR Del 1967.

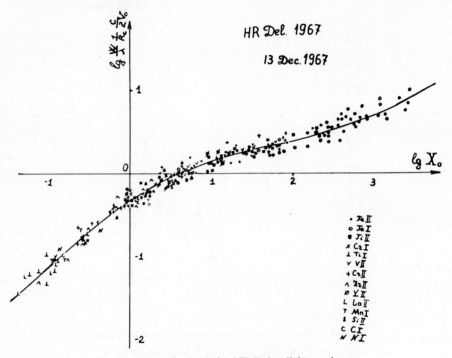

Fig. 4. Curve of growth for HR Del at light maximum.

Figure 4 shows the curve of growth for HR Del 1967 at light maximum (13 December, 1967). The scattering of the points in this figure is very small. This means, that the use of the Schwarzschild-Schuster model is a sufficiently good approximation for the interpretation of the spectrum of this star. It is found again that the excitation temperature in the envelope of HR Del 1967 depends on the excitation potential, similar to the case of DQ Her. This fact was taken into account in the analysis of chemical composition.

Figure 5 shows a comparison of the chemical composition of the envelope of HR Del 1967 with the composition of the solar atmosphere. We may conclude that, similar to the case of DQ Her 1934, the relative chemical composition of the envelope of HR Del

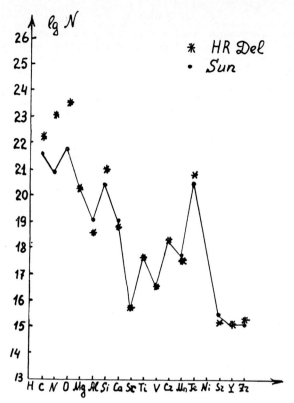

Fig. 5. Chemical composition of the envelope of HR Del.

coincides for the metals with the composition of the solar atmosphere, but the abundances of C, N and O in the envelope of DQ Her exceed the solar abundances by an appreciable factor.

2. The Emission Method

The chemical composition of the envelopes of novae can be determined also by using the emission lines in their spectra during the nebular stage. The methods which are usually used for the determination of the chemical composition of gaseous nebulae, can be also used in this case.

The number density of the atoms of any particular element relative to that of the hydrogen atoms can be determined from the ratio between the energies, emitted in the lines of this element and in the lines of Balmer series. In this case we have to know the ratio between the volumes, which emit in the lines of the element and in the Balmer lines. The evaluation of these volumes is generally the most serious difficulty in the problem of the determination of the chemical composition of the envelopes around novae. In addition, since the degree of ionisation during the nebular stage is rather high, it is necessary to know the temperature of the star, the electron density and the dilution factor.

TABLE II

Chemical composition of the envelopes of novae determined from the analysis of emission lines (according to Pottasch (1967))

Element	lg N		
	RS Oph.	Average five novae	Normal stars (C. W. Allen)
H	12	12	12
He	11.63	11.18	11.16
O	9.81	9.63	8.83
N	9.40	9.70	7.96

The determination of the chemical composition of the envelopes of five novae in the nebular stage has been carried out by Pottasch (1967). It was found, that the abundances of the O and N atoms relative to H is considerably higher, than that in the atmospheres of 'normal' stars (Table II).

The abundances of He, O and N atoms in the envelopes of several novae on the base of emission lines in the nebular stage were determined by Ruusalepp and Luud (1970). They concluded that the abundances of these elements in the envelopes of the novae exceed the normal ones. In addition, they found that the larger the rate of decrease of brightness, the larger is this excess; thus, this excess is less in 'slow' novae than in the 'fast' ones. The authors explained this result by assuming that the fast novae are more evolved objects than the slow ones.

References

Abhyankar, K. D.: 1964a, *Astron. J.* **140**, 1353.
Abhyankar, K. D.: 1964b, *Astron. J.* **140**, 1368.
Abhyankar, K. D.: 1965, *Astron. J.* **141**, 1056.
Allen, C. W.: 1955, *Astrophysical Quantities*, The Athlone Press, London.
Antipova, L. I.: 1969, *Astron. Zh.* **46**, 366.
Antipova, L. I.: 1971, *Astron. Zh.* **48**, 288.
Beer, A.: 1937, *Monthly Notices Roy. Astron. Soc.* **97**, 231.
Ciurla, T.: 1966, *Acta Astron.* **16**, 249.
Kubikowski, J. and Ciurla, T.: 1965, *Acta Astron.* **15**, 177.
McLaughlin, D. B.: 1937, *Michigan Obs. Publ.* **6**, 103.
Mustel, E. R.: 1945, *Astron. Zh.* **22**, 65, 185.
Mustel, E. R.: 1949, *Izv. Krymsk. Astrofiz. Obs.*, **4**, 23.
Mustel, E. R.: 1957, in G. H. Herbig (ed.), 'Non-Stable Stars', *IAU Symp.* **3**, 57.
Mustel, E. R.: 1962, *Astron. Zh.* **39**, 185.
Mustel, E. R. and Antipova, L. I.: 1971, *Nauch. Inform. Moscow* **19**, 32.
Mustel, E. R. and Baranova, L. I.: 1965, *Astron. Zh.* **42**, 42.
Mustel, E. R. and Boyarchuk, M. E.: 1959, *Astron. Zh.* **36**, 762.
Pottasch, S. R.: 1967, *Bull. Astron. Inst. Neth.* **19**, 227.
Ruusalepp, M. and Luud, L.: 1970, *Tartu Obs. Publ.* **39**, 89.

THE SHELL OF V603 Aql AND THE EARLY STAGES OF THE NOVA EVENT

H. WEAVER

Dept. of Astronomy and Radio Astronomy Laboratory,
University of California, Berkeley, Calif., U.S.A.

1. Introduction

Nova V603 Aql 1918 was uniquely suited for studies of the structure of its shell because of the very favorable orientation of the shell in space. Additionally, almost by chance, spectroscopic observations were made in such a way that they permitted derivation of a three dimensional model of the shell.

2. Symmetry, Anti-Symmetry in the Shell

V603 Aql, as we see from its light curve (Figure 1), was a typical fast nova which showed strong periodic fluctuations in light during the transition stage. The period of fluctuation was approximately 10 days, during which time the visible light changed by roughly 50%.

Approximately three and one half months after maximum light, Barnard (1919, 1920) observed that the nova was surrounded by a nebular shell which was later found to be expanding at the rate of $2''$ yr^{-1}. Thirteen and one-helf months after maximum, when the nebular shell was $2.2''$ in diameter, Wright (1919) used a slitless

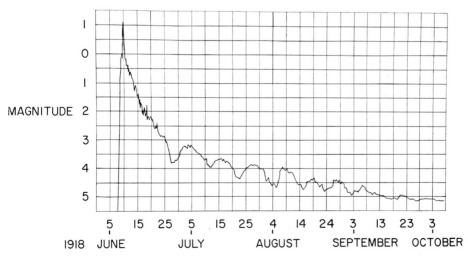

Fig. 1. The light curve of V603 Aql. Data are from Leon Campbell, *Ann. Harv. Coll. Obs.* **81**, 179, 1920.

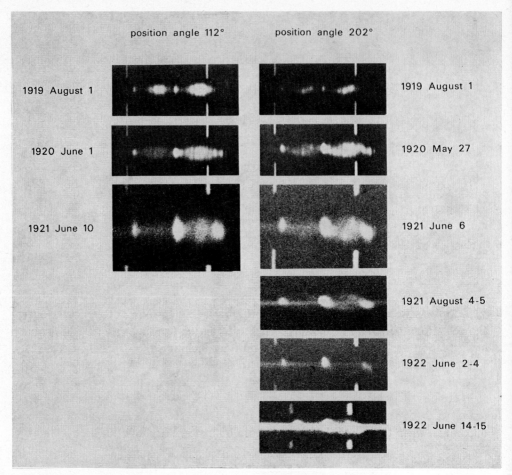

Fig. 2. The spectral development of V603 Aql during 1919–1922. The lines shown in each spectrogram are the N1 and N2 lines of [O III].

spectrograph to observe the N1 and N2 lines of [O III] originating in the shell. From a series of exposures made by rotating the spectrograph so that the refracting edge of the prism occupied different position angles across the shell, he demonstrated that the visible expanding shell was not spherically symmetrical. In position angle 112° the spectrographic images were symmetric; in position angle 202°, 90° different from the first, the spectrographic images were antisymmetric. Following Wright's discovery, Lick spectrograms of the nova were normally taken with the slit in one of these two position angles, and the nebular image was held stationary on the slit.

The spectral development of the nova shell during the period 1919–1922 is shown in Figure 2. In any one position angle the essential nature of the slit-spectrographic image of the shell, which was always complex in character, remained remarkably constant in form over the years. Change in any one feature within either the N1 or N2 line involved intensity and vertical extent of the feature – that is, expansion of the shell – but not velocity, which remained constant for any feature within a line. A high degree of symmetry, antisymmetry remained at all times the outstanding characteristic of the structure.

3. The Form of the Shell Produced by Different Types of Ejection

A shell of gas producing slit-spectrographic images having regularities as pronounced as those shown by V603 Aql must be a highly organized structure of complex geometrical form. To gain insight into the nature of slit-spectrographic images produced by shells of various geometrical forms we look at the results from a series of model calculations.

Figure 3a illustrates what we would see if we examined a completely filled, uniformly expanding spherical shell of optically thin gas with a slit spectrograph. The slit is assumed to be placed in each of three representative position angles, one after the other.

In Figures 3b through 3h we see slit-spectrographic images for an infinitely thin, uniformly expanding shell, a series of uniformly expanding gaseous rings, and a series of uniformly expanding cones. In the latter two cases the axis of the system is shown in three different positions; in every case the gas is assumed to be optically thin.

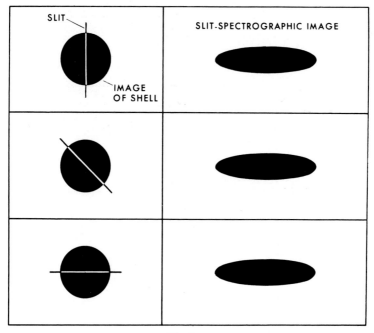

Fig. 3a.

Figs. 3a–h. Results for model calculation of slit-spectrographic images arising from various types of gas ejection. The velocity of ejection has always been taken to be isotropic. The gas is optically thin. (a) A completely filled sphere of gas. (b) An infinitely thin spherical shell of gas. (c, d, e) Rings of gas. The axis of the ring system is shown at three different tilt angles with respect to the line of sight. (f, g, h) Cones of gas. The axis of the cone system is shown at three different tilt angles with respect to the line of sight.

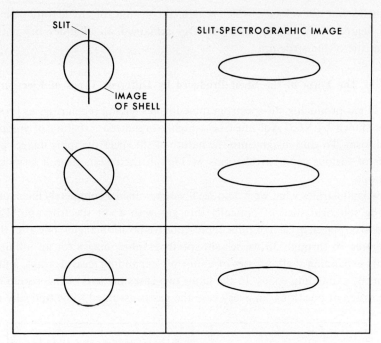

Fig. 3b.

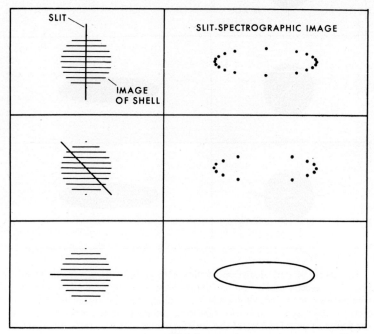

Fig. 3c.

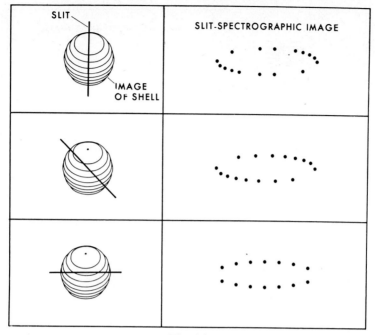

Fig. 3d.

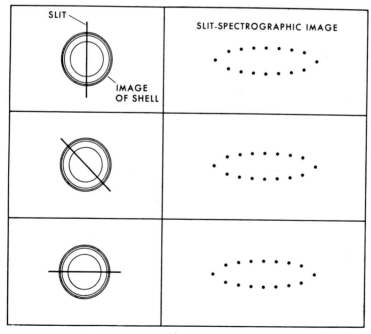

Fig. 3e.

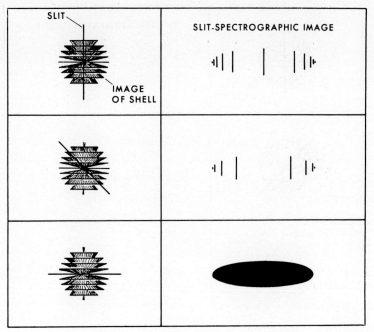

Fig. 3f.

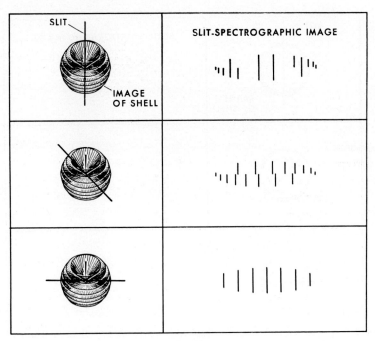

Fig. 3g.

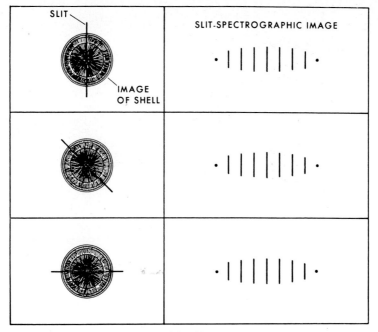

Fig. 3h.

All of the model calculations displayed in Figure 3 have been made for an isotropically expanding shell. If non-isotropic expansion were to take place so that an ellipsoidal shell would be produced rather than a spherical one, there would be no change in the character of the slit-spectrographic images pictured in Figure 3; they would simply be transformed by smooth stretching in the slit-spectrographic image plane.

4. Reconstruction of the Shell of V603 Aql

We can invert the problem illustrated in Figure 3. If for a real nova shell we have slit-spectrographic images taken with the slit in a number of different position angles, we can reconstruct the form of the expanding envelope provided:

(a) that all features originate at the same time, and

(b) that the motion of the gas giving rise to a specific feature in the line is rectilinear.

In the case of V603 Aql these requirements are met. We shall also assume that the expansion velocity is isotropic. This assumption is probably not correct for reasons to be mentioned later. Failure of the assumption means only that the reconstructed picture of the nova shell will be somewhat distorted in different directions, but not changed in character.

Figure 4 shows examples of slit-spectrographic images taken in 1919, approximately 15 months after maximum light. The lines shown are N1 and N2 of [O III]. Figure 5 shows slit-spectrographic images from the same period but at higher dispersion, approximately 10 Å mm^{-1}. From these and other slit-spectrographic images, the cross

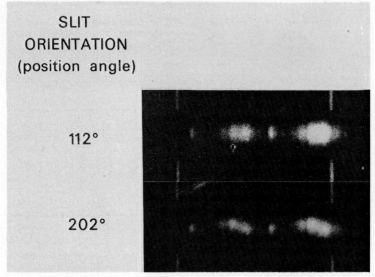

Fig. 4. Slit spectrographic images of V603 Aql taken in 1919. The lines shown are N1 and N2 of [O III].

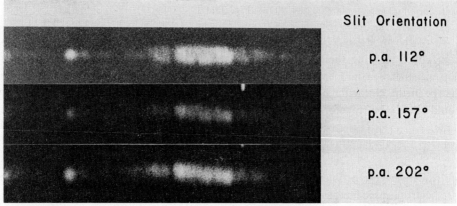

Fig. 5. Slit spectrographic images of V603 Aql taken in 1919; dispersion approximately 10 Å mm^{-1}. The line shown is N1 of [O III].

section of the nova shell was reconstructed for the symmetrical, 112°, position angle as shown in Figure 6a. The shell thickness on this and the drawings that follow must be regarded as schematic only, and is shown greater than it was in reality.

Figure 6b shows the reconstruction of the antisymmetric, 202° position angle cross section of the shell. It is particularly notable that this cross section has two diametrically opposite holes in it. These holes, combined with the particular skewed intensity pattern present in the shell in 1919, produced the antisymmetry observed by Wright in position angle 202°.

Figure 7 is a drawing of the reconstructed three-dimensional model of V603 Aql.

The shell is composed of truncated coaxial cones. That is, in 3-dimensional space the trajectory of each mass of gas in the shell is along a radius drawn outward from the nova. Motion along this radial trajectory remains constant in direction in space and in velocity, hence the observed radial velocity of any feature in the slit-spectrographic image remains constant in time.

Some cones are intense emitters because of their density and excitation, others are only very weak emitters. The very weakly emitting cones are the source of the fine dark 'lines' that cross the slit-spectrographic image of the nova. These dark 'lines' are not absorption features. The optical depth is $\ll 10^{-5}$ in the [O III] lines in which the observations were made.

The shell is distorted as is readily seen on the drawing. It bulges outward in the general regions of the holes that are present in the shell.

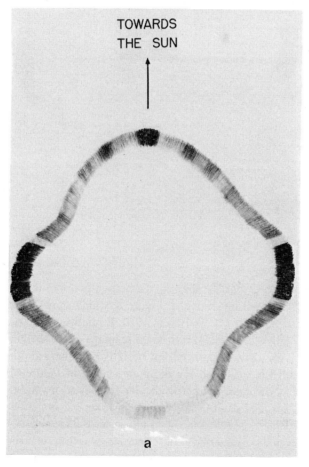

Fig. 6a–b. (a) Drawing of the cross section of the shell of V603 Aql observed when the slit of the spectrograph was in p.a. 112°. Intensity distribution in the shell is shown as it was in 1919. (b) Drawing of the cross section of the shell of V603 Aql observed when the slit of the spectrograph was in p.a. 202°. Intensity distribution in the shell is shown as it was in 1919.

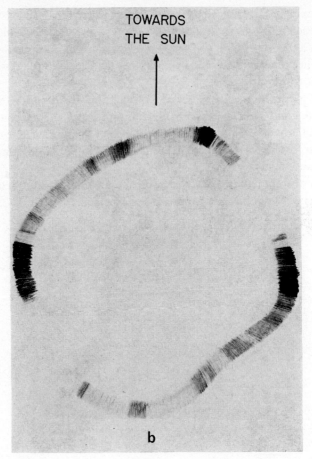

Fig. 6b.

There are, of course, two holes; they are diametrically opposite to each other in the shell. These holes are well illustrated in Figure 8, which shows a slit-spectrographic image of V603 Aql taken at Mt. Wilson in 1920. Radially symmetrical breaks in the quasi elliptical ring that is the slit-spectrographic image of the shell are clearly visible. These breaks are the counterparts of the radially symmetrical holes in the shell. Angular resolution was very high for the plate shown in Figure 8, hence the shell was well resolved in the direction perpendicular to dispersion.

The picture of the shell of V603 Aql shown in Figure 7, it should be emphasized, represents the shell as it appeared in the radiation of the N1 line of [O III].

Over the time period covered by the observations used in reconstruction of the shell no velocity changes were detected in any parts of the shell. Only changes in angular diameter of the shell and in the relative brightnesses of the truncated cones were observed to take place. The brightness changes observed over the period 1919–1922 follow the decay times of [O III] under conditions of different electron density.

Fig. 7. Drawing of a three dimensional model of V603 Aql derived from spectrograms taken at the Lick Observatory in 1919.

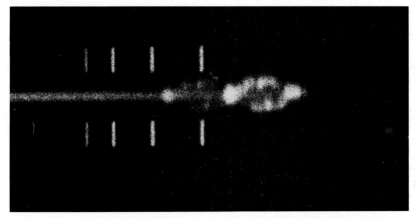

Fig. 8. Slit-spectrographic image of V603 Aql taken at Mt. Wilson on July 23, 1920. The slit was in p.a. 90°. (I am indebted to Dr R. Minkowski for this photograph.)

5. Tests of the Technique of Reconstructing the Shell

(a) Having reconstructed the shell, we are in a position to relate features in a slit-spectrographic image to specific regions of the shell. Of particular interest is the bright poleward spot marked out in Figure 9. From slit-spectrographic images with the slit in different position angles we predict that this condensation should appear on the sky in position angle $22°\pm3°$, at a distance 0.42 ± 0.02 from the center towards the edge of the shell. We here count the radius of the shell as unity. On a direct photograph of V603 Aql taken at Mt. Wilson by Baade in 1940 and shown in Figure 10, we see the shell and observe that there is a bright condensation in position angle $28°$ at distance 0.43. This condensation, visible on the direct photograph, produced the bright feature of highest negative velocity seen in the spectra taken 20–22 yr earlier. The plate taken in 1950 suffers from poor seeing. It is not certain that the condensation was visible at that time. Since 1950 both shell and condensation have faded and cannot now be photographed.

(b) We consider next the tilt angle of the axis of the cone system with respect to the line of sight. In Figure 11 we see a collection of Lick spectrograms of relatively high dispersion (10 Å mm^{-1}) taken in a number of different position angles. Such a set of spectrograms permits determination of the cone axis with high accuracy. The fact that the major narrow dark 'lines' and other features cross the images at the same velocity whatever the slit orientation means that the line of sight is very close to the axis of the cone system. Detailed measurements show that the axis of the cone system differs from the line of sight by less than a degree. It is this fortunate circumstance that makes the reconstruction of the shell of V603 Aql particularly simple and makes V603 a kind of Rosetta Stone in interpreting the early stages in the development of novae. Because of the near perfect alignment of the axis of the cone system and the line of sight, we can specify what part of the shell was involved in producing each portion of a spectral line even when the image was trailed along the slit as it was during the early observations of the nova.

6. Early Spectral Changes in the Nova Spectrum and Their Relationship to the Shell

I wish next to trace the early spectral stages of the nova and show how they relate to the light curve (Figure 1) and the three-dimensional model of the nova just described (Figure 7). The points to be emphasized in this discussion are as follows:

(a) The shell developed in a *very* short time at the very start of the nova process.

(b) The shell underwent acceleration, particularly near the time of maximum light.

(c) For only a few days near the time of maximum light and maximum shell acceleration, double absorption lines were present in hydrogen and other elements in the shell.

(d) The lines in the spectrum divide into two quite different velocity structures which we can identify with the shell and with the nova.

(e) A series of complex changes took place in the spectrum when the nova fluctuated

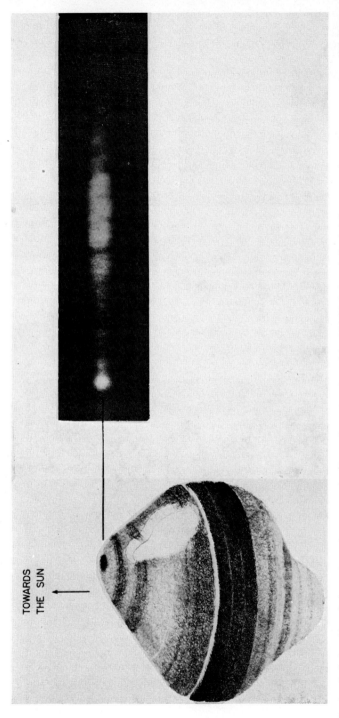

Fig. 9. Relationship of the poleward spot on the three dimensional model of V603 Aql and the bright spot of maximum negative radial velocity in the slit spectrographic images of the nova.

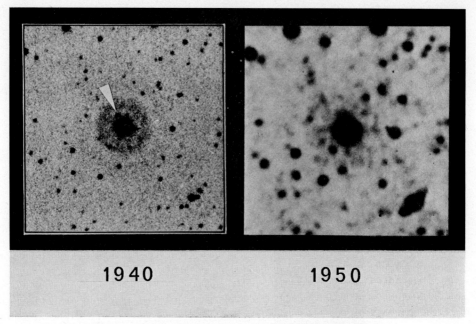

Fig. 10. Direct photographs of V603 Aql taken by W. Baade.

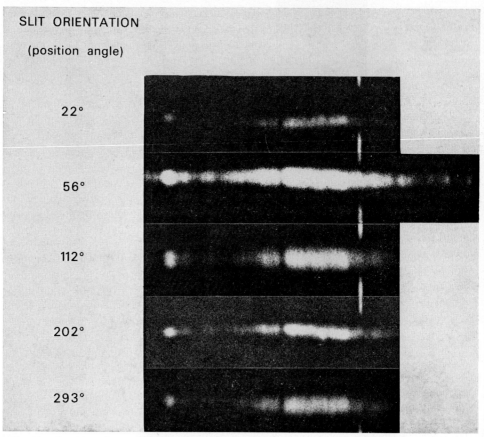

Fig. 11. A series of spectrograms of V603 Aql taken in 1919 with the Mills Spectrograph at a variety of positions angles as indicated. The feature shown is the N1 line of [O III].

in light during the transition stage. During these fluctuations the two velocity structures mentioned in (d) were clearly resolved.

Finally, I shall present a physical model of the nova that brings all these phenomena together and shows how the shell shown in Figure 7 was produced.

(i) The first spectrum of the group shown in Figure 12 was taken 0.9 days *before* maximum light. The nova was then approximately one day old. We see in the spectral range covered only a broad absorption line of $H\beta$. At the time the spectrum was taken the shell was of the order one astronomical unit in radius and similar in character to the atmosphere of an early supergiant. The opacity of the expanding gas was too high to permit visibility of the internal structural character of the shell.

(ii) By 0.1 to 0.9 days after maximum light, the opacity had decreased enough to permit the internal structure of the shell to be seen. By 0.9 day after maximum the characteristic bright equatorial part of the shell was visible in its counterpart, the bright central part of the spectral line. The equatorial structure of the shell must have been formed at the very onset of shell formation. It could not have been formed, for example, when the shell was 1–2 astronomical units distant from the nova.

Note that at 0.9 days after maximum light a second absorption feature at velocity greater than the shell expansion velocity had developed in the lines of hydrogen and Fe II.

Figure 13 shows a collection of early Mills spectrograms of V603 from the Lick collection. There are three features to be especially noted on these spectrograms.

(1) Double absorption lines in the shell. By June 10.8, 0.9 day after maximum, two systems of absorptions had developed in hydrogen as well as other elements making up the shell. This was noted earlier on the Mt. Wilson plates. One of the two absorptions was the 'normal' one arising from the most negative velocity of the expanding shell. Although this normal absorption evolved in velocity, (see Section 2 below), its evolution was relatively slow. It was the original feature in the spectrum; it lasted for as long as absorption was visible. The second absorption was transitory; it had a velocity higher than shell expansion velocity. Double lines originating in the shell existed only during the period of time covered by these spectrograms. The absorption feature at higher-than-shell expansion velocity remained visible for a period of 5–6 days; it was accompanied by emission extending out to its maximum velocity. After disappearance of the once-strong second absorption line, some very weak and broad features may have appeared briefly during the time period covered by these spectra, but no activity at velocities higher-than-shell expansion velocity was visible in lines arising from the shell at any later date.

(2) Acceleration of the shell. From the collection of spectra in Figure 12 we see that the shell of V603 Aql underwent acceleration. Acceleration was particularly large during the first few days after maximum light which occurred on June 9.9.

Note the change in radial velocity of the shell between June 10.8 and 11.9. At that time the force giving rise to the acceleration was approximately 3.5 orders of magnitude greater than the force of gravity experienced by the shell. Such force arising from radiation pressure and/or direct gas pressure in the form of gas streams or shells

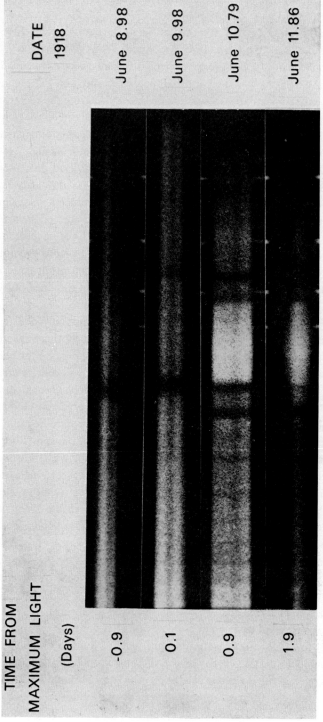

Fig. 12. Group of spectrograms of V603 Aql taken at the Mt. Wilson Observatory. (These spectra were provided by Dr A. H. Joy.)

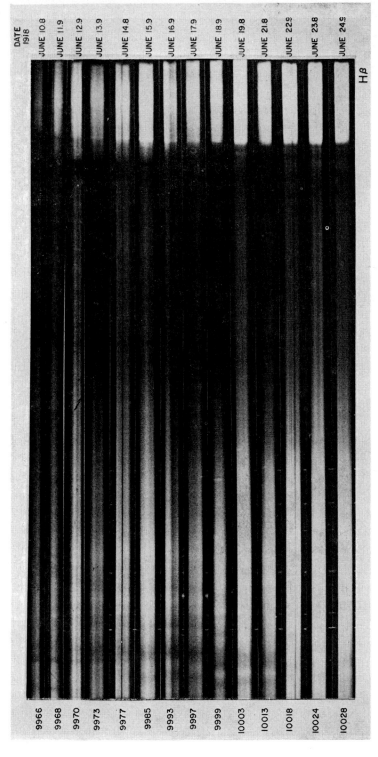

Fig. 13. Group of early spectrograms of V603 Aql taken with the Mills Spectrograph at the Lick Observatory.

ejected by the nova must have had a profound effect in accelerating and deforming the nova shell. The time at which this maximum pressure was exerted on the shell was also the time – the only time – at which a second absorption feature appeared in the spectral lines originating in the shell. The implication is strong that the shell was partly ruptured by the pressure that caused the acceleration. Rupture of the shell did not take place at the part of the shell pointed directly towards the Sun. There is no velocity, time discontinuity in the negative velocity edge of the hydrogen lines as would necessarily be the case if the shell were disrupted at the point giving rise to maximum negative velocity.

Ejection of gas from other parts of the shell would, by projection, give rise to a mass of gas appearing to move across the shell, away from its center, and producing absorption at a velocity higher than shell expansion velocity while it was between the shell and the observer. Given the velocity of expansion of the blown-out gas, one computes that the blown-out gas should remain visible as absorption for 2–5 days depending upon location and size of the hole blown out of the shell. The computed time is closely similar to the length of time the second hydrogen absorption feature was actually visible.

(3) The existence of two classes of lines in the spectrum.

(a) There are lines with well-defined edges and absorptions on the negative velocity side. Such lines define the shell expansion velocity. Hydrogen is the most prominent example.

(b) There are lines without defined edges and without absorption components on the negative-velocity side. These ill-defined lines show velocities much greater than shell expansion velocity. He II at $\lambda 4686$ and N III at $\lambda 4640$ are the most prominent lines in this category.

The 4686 and 4640 lines, the principal contributors to the extensive hazy feature on the left-hand side of the spectra in Figure 13, are extremely broad. They show no definite edges; they simply fade out. No absorption appears on the shortward side of the line. These very broad edgeless features are directly related to what has, in the literature, been termed 'Nitrogen Flaring'. The term is somewhat of a misnomer. The effect is not confined to nitrogen. He, O, Fe, and other elements are involved also. He and N are simply the elements with the most prominent lines showing the phenomenon. The effects might more appropriately be called simply 'flaring'. Flaring was conspicuous during the time period covered by these spectrograms; it was not present at all times. Flaring lines showed a totally different velocity structure from hydrogen, for example, which *never* flared.

As we shall see with increasing clarity as we look at the data in more detail, a nova spectrum consists of two sets of lines normally present at the same time. One set arises from the expanding shell, the other set arises from ejecta from the nova. These two objects, the shell and the nova, have quite different chemical compositions. The shell is essentially normal in chemical abundances. The nova is a hydrogen-exhausted white-dwarf-like object. The ejecta contain no hydrogen, only He, N, O, ... and other elements characteristic of a hydrogen exhausted object. This chemical

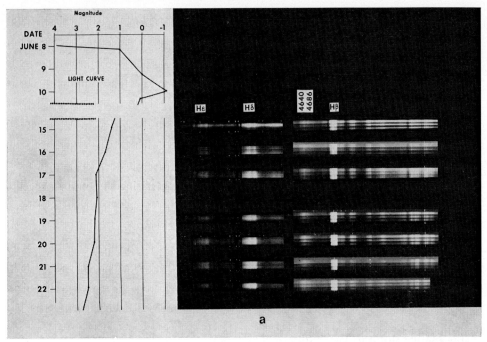

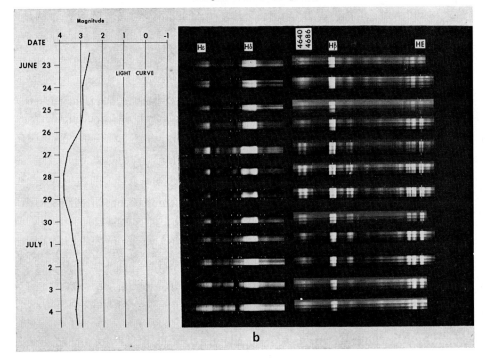

Fig. 14. (a, b) Spectra of V603 Aql taken at the Lick Observatory. The spectrograms showing the red end of the spectrum have been widened by joining three prints of slightly different exposures. The purpose in doing this was to produce prints of sufficient width and of appropriate density to show the spectral lines adequately.

difference shows in the totally different velocity structure of lines originating in the two objects.

Figure 14a is a further attempt to provide a comprehensive view of the spectral development of V603 Aql.

(a) The basic shell structure shows clearly in hydrogen and other lines with well-defined edges and shortward absorptions.

(b) The elements He, N, O, ... which characterize the nova are also present in the shell. Lines arising from these elements are mixtures of shell lines and nova lines. He II at $\lambda 4684$ and N III at $\lambda 4640$ are excellent examples. Note that the equatorial part of the shell has a clear counterpart in the lines, but there are no definite edges to the lines which extend to very large velocities characterizing the outflow of gas from the nova.

(c) With the appearance of the double absorption lines in the shell, there also appeared the N III doublet at rest wavelengths 4097 and 4103 Å. These lines arise from the Bowen He–O–N fluorescence process which was strong in the ejecta of the nova. The presence of this N III doublet always indicated gas flow from the nova.

Figure 14b continues the time sequence of spectra and covers the first light fluctuation. The sequence of events accompanying fading light in a fluctuation were as follows.

(a) Flaring stopped; gas outflow from the nova ceased. The N III doublet at $\lambda 4097$ and $\lambda 4103$ arising from Bowen fluorescence increased in velocity and faded away.

(b) At minimum light all lines exhibited well-defined edges. Hazy lines associated with nova gas flow disappeared. Only shell lines were visible. The lines at $\lambda\lambda 4684$ and 4640 showed well defined edges; only the shell components of these lines were visible during the minimum of a light fluctuation.

(c) At and near minimum light of a fluctuation the central sections of the shell lines – those sections arising from the equatorial region of the shell – faded rapidly relative to the portions of the lines arising from the mid-latitude sections of the shell. With decline of the nova activity and cessation of gas flow from the nova the source of excitation of the shell was cut off. Different electron densities existed in different parts of the shell; intensities in the different parts declined at different rates determined by the electron density. The electron density in the equatorial region of the shell was 10^7–10^8 cm^{-3}; the decay constant for the hydrogen and other elements there was therefore less than 24 h, in agreement with observation.

The situation for the forbidden lines was quite different because their excitation process is different. The decay constant for [N II] $\lambda 5755$ was much longer; it did not fade as did the hydrogen.

The changes in the N1 and N2 lines of [O III] were especially interesting, but the time allocated to this review does not allow discussion of their intensity changes, which were also controlled by the electron density and temperature present in different parts of the shell.

(d) As the nova brightened and started ejecting gas again, the N III pair at $\lambda\lambda 4097$ and 4103 first reappeared at high velocity; they grew in strength as their velocity decreased. The various changes discussed earlier reversed; $\lambda\lambda 4686$ and 4640 became hazy.

This entire process repeated each time a light fluctuation occurred.

7. A Physical Model of the Nova

Figure 15 presents a model of V603 Aql that brings all of these diverse phenomena together. The model is based largely on the important fundamental worker of Walker (1954, 1956) and of Kraft (1959, 1964) on Nova DQ Her.

V603 Aql is a binary consisting of a late dwarf that fills its inner Lagrangian surface and a hydrogen exhausted white dwarf that is the nova. The chemical composition of the late dwarf is normal. The observed binary period of V603 is $3^h 20^m$. The observed radial velocity variation is small; the tilt of the axis of rotation relative to the direction towards the Sun is therefore small as shown on the diagram.

The masses of the nova and the dwarf are equal, $\sim 0.25 \, M_\odot$.

The diameter of the nova is $\sim 10^8$ cm; the diameter of the inner Lagrangian surface of the dwarf is $\sim 3 \times 10^{10}$ cm.

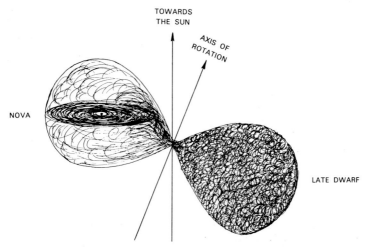

Fig. 15. Model of V603 Aql before the nova outburst.

Gas of normal chemical composition flows from the late dwarf into the inner Lagrangian surface of the nova. A disk of gas forms around the nova. The electron density in the disk is $\sim 3 \times 10^{14}$ cm^{-3}; its temperature is $\sim 4 \times 10^4$ K. Under normal conditions before the nova explosion, most of the light of the binary pair comes from this disk, which has a thickness of 10^7 or a few $\times 10^7$ cm.

The nova process starts, caused, possibly, by an excess of hydrogen from the disk mixing with the material of the hydrogen-exhausted white dwarf. The exploding gas encounters the disk, piling up against it and being forced to flow off in conically-shaped streams. Density irregularities in the radial structure of the disk will be transformed into latitude irregularities in the outflowing gas. These remarks are based on simple essentially intuitive calculations. A problem of the type encountered here appears never to have been solved by the aerodynamicists. It needs to be investigated.

The expanding gas from the nova thus sweeps up the disk, hence, as suggested by Sparks and Starrfield (1973), expansion is slowed in the equatorial region. The shell becomes prolate rather than spherical. The equatorial region of the shell has the highest density; it consists largely of the old swept-up disk.

The nova shell is formed largely, probably almost entirely, from gas which surrounded the nova and originally came from the dwarf. Its chemical composition is therefore essentially normal. The gas that surrounded the nova must have been swept up in the explosive outflow in a matter of minutes. At the velocity of 1000 km^{-1} s the gas passed beyond the edge of the disk in 100 s.

After the first explosive ejection, the gaseous outflow from the nova was pulsed in a period of ~ 10 days. These pulses caused the light fluctuations during the transition stage. The first pulse started at or shortly before maximum light. Gas was ejected from the nova at several times shell expansion velocity. The outflowing gas, which was not isotropic in distribution, impinged on the shell, causing the pressure that accelerated and distorted the shell. The resultant force acting on the shell (at maximum, 3 to 4 orders of magnitude greater than gravity) finally ruptured the shell at two places, causing the double absorption lines seen shortly after maximum light and creating the holes in the shell that were seen spectroscopically at a much later time after the shell had expanded to visible size.

The gas flow from the nova, very high velocity, low density material, provided the source of the broad, hazy, edgeless lines seen in the spectrum. The shell lines, on the other hand, showed sharp edges since they arose from a gas shell of compact, well-defined velocity characteristics.

Cessation of the first nova pulse of flowing gas caused the first decrease of light amounting to $\sim 50\%$ of the visible light and started the sequence of spectral changes described earlier.

Approximately 10 nova emission pulses were counted in the first 116 days of the nova lifetime. They decreased in intensity with time. It is porbable that the nova gas impinging on the shell was an important source of shell excitation, particularly when the shell was of relatively small radius.

Time does not permit specification of further details of the model. The model does, however, account for all the spectral phenomena observed. It also accounts for the shape of the shell which bulges outward in the regions of the holes. It provides a logical explanation of the observed holes in the shell. The model offers a possible explanation of the ejected truncated cones forming the shell, but here a detailed hydrodynamical calculation needs to be made.

8. Generality of the Model

The shell of V603 Aql can be reconstructed in very great detail from the early spectra.

It is a very complex structure, but it can be accounted for on the basis of a rather simple model. Is V603 unique? Not at all. It is likely that all novae are similar to what has been described here. Mustel and Boyarchuk (1970) have long pointed out the equatorial ring and polar cap structure of the shell of DQ Her. Hutchings (1972) has devised cone and ring models for N. Delphini 1967, N. Velpeculae 1968 (1), and N. Serpentis 1970. Molakpur (1973) has also proposed a model for N. Delphini consisting of polar caps and equatorial rings. This type of structure appears to be the general case for novae.

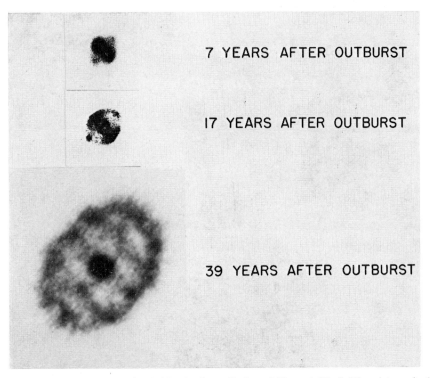

Fig. 16. Direct photographs of DQ Her in the radiation of Hα and [N II]. The pictures in 1942 and 1951 were taken by W. Baade. The picture from 1973 was taken by Ford and Jenner. I am grateful to them for permission to show this remarkable photograph. The three pictures are not all to the same scale; that is, the ratio of image size to shell diameter is not the same for all three pictures.

Finally, in Figure 16 I show two early photographs of Nova DQ Her taken by Baade in 1942 and 1951 along with a most remarkable photograph of DQ Her taken in May 1973 by Ford and Jenner with the 120″ at the Lick Observatory. I am grateful to them for permitting me to show this picture.

From these photographs, particularly the remarkable one by Ford and Jenner, there can be no doubt that a nova shell is a most remarkable geometrical structure having a very high order of spatial symmetry.

References

Barnard, E. E.: 1919, *Astrophys. J.* **49**, 199.
Barnard, E. E.: 1920, *Monthly Notices Roy. Astron. Soc.* **80**, 582.
Hutchings, J. B.: 1972, *Monthly Notices Roy. Astron. Soc.* **158**, 177.
Kraft, R. P.: 1959, *Astrophys. J.* **130**, 110.
Kraft, R. P.: 1964, *Astrophys. J.* **139**, 469.
Malakpur, I.: 1973, *Astron. Astrophys.* **24**, 125.
Mustel, E. R. and Boyarchuk, A. A.: 1970, *Astrophys. Space Sci.* **6**, 183.
Sparks, W. M. and Starrfield, S. G.: 1973, *Monthly Notices Roy. Astron. Soc.*, in press.
Walker, M.: 1954, *Publ. Astron. Soc. Pacific.* **66**, 230.
Walker, M.: 1956, *Astrophys. J.* **123**, 68.
Wright, W. H.: 1919, *Lick Obs. Bull.* **10**, 30.

SPECTROPHOTOMETRY OF SUPERNOVAE

R. P. KIRSHNER

Hale Observatories, California Institute of Technology, Carnegie Institution of Washington, Pasadena, Cal., U.S.A.

Abstract. Absolute spectral energy distributions for supernovae of both types I and II have been obtained. These observations demonstrate three facets of supernova spectra. First, both SN I's and SN II's have a continuum that varies slowly and uniformly with time, and which carries the bulk of the radiated flux at early epochs. Second, some lines in both SN I's and SN II's have P Cygni profiles: broad emissions flanked on their violet edges by broad absorptions. Third, some lines are common to SN I's and SN II's and persist throughout the evolution of the spectrum. The continuum temperatures for both SN I's and SN II's are about 10 000 K at the earliest times of observation and drop in one month's time to about 6000 K for SN II's and about 7000 K for SN I's. After several months, the continuum may cease to carry the bulk of the flux, which might be in emission lines, but continues to exist, as shown by the presence of absorption lines. The P Cygni line profiles indicate expansion velocities of 15 000 km s^{-1} in SN II's and 20 000 km s^{-1} in the SN I 1972e in NGC 5253. Line identifications for SN II's include Hα, Hβ, H and K of Ca II, the Ca II infrared triplet at λ8600, the Na I D-lines, the Mg I b-lines at λ5174, and perhaps Fe II. The [O I] lines $\lambda\lambda$6300, 6363 and [Ca II] lines $\lambda\lambda$7291, 7323 appear after eight months. For SN I's, the lines identified are H and K of Ca II, the infrared Ca II lines, the Na I D-lines, and the Mg I b-lines. There is some evidence that Balmer lines are present two weeks after maximum. The strong and puzzling λ4600 features drifts with time from λ4600 near maximum light to λ4750 after 400 days.

1. Introduction

Although supernova explosions are at the root of many modern astrophysical theories for nucleosynthesis, cosmic rays, pulsars, black holes, and the interstellar medium, the study of the actual event has not progressed rapidly since Minkowski's (1939) classic work. This valuable investigation of the SN I 1937c, recently rendered even more useful by Greenstein and Minkowski (1973), still provides the observational evidence used to analyze the supernova event empirically.

Supernovae appear at random, and are only found by time-consuming searches. Even though they may be of 13th mag. at first, within a month spectroscopic observations become quite difficult. Thus, it is no surprise that the available information has been limited.

Recently, the multichannel spectrometer (Oke, 1969) has been employed by J. B. Oke, L. Searle, M. V. Penston, J. E. Gunn, and J. L. Greenstein, to obtain series of spectral energy distributions for supernovae of both type I and type II. This instrument, attached to the 200-in. (508 cm) telescope, produces a quantitative spectrum covering the wavelength range $\lambda\lambda$3200–10000. The resolution of 20 to 160 Å is well suited to the broad indistinct features of supernovae spectra.

This work was given impetus by the appearance of several relatively bright supernovae, especially SN I 1972e. For that supernova, an extensive series of observations has been obtained, using both the 200-in. telescope with multichannel spectrometer and the Palomar 60-in. (152 cm) telescope with a single channel scanner.

Much of this work has been reported recently (Kirshner *et al.* 1973a, b). Some of the observations reported there are illustrated in Figure 1, which shows scans for the SN II 1970g in M 101, and in Figures 3, 4, 5 of the SN I 1972e in NGC 5253. The purpose of this contribution is to give a brief summary of the principal results of these preliminary investigations.

2. Supernovae of Type II

2.1. Continuum

Figure 1 exhibits spectral scans for a type II supernova. The first scan of 1970g in M 101 was made soon after its discovery, and near maximum light. Here the continuum is well defined and resembles that of a blackbody at about 9500 K, although the blackbody which fits best between $\lambda 5000$ and $\lambda 10000$ is brighter than the supernova in the ultraviolet. The scans of 1970g for JD 837, made about 1 month after discovery, resembles a blackbody in the vicinity of 6000 K. Both at the earliest epoch and at age one month, the continuum transports most of the energy.

After 250 days, observations show strong emission bands, with some evidence for a continuum. The continued presence at these late dates of absorption features which are identical with those of early times argues that the continuum, though faint, is still present.

We interpret these observations as indicating that SN II's have optically thick and cooling photospheres. Comparison of the observed flux with the expected blackbody flux at a distance of 7 Mpc for M 101 gives a photospheric radius of order 1×10^{15} cm at age one month. Study of the expansion of the photosphere may lead to a method for determining the distances to supernovae by purely astrophysical methods.

2.2. Line shapes

When the continuum dominates the energy distribution, all of the strongest line features in the spectra of SN II's have P Cygni shapes. The broad emission bands are flanked on the short wavelength side by a broad absorption trough. The line profiles at $\lambda\lambda 8600, 6500, 4800, 4300$, and 3950 can all be characterized in this way, even though they are not identical.

The most clearly defined profile is at $\lambda 8600$, where the continuum is smooth. Figure 6, taken from the scan of SN 1970g on JD 837 shows that this line has nearly zero net equivalent width. One simple interpretation of this profile is that the line is formed by scattering photospheric radiation in a spherically symmetric and differentially expanding reversing layer. Then the violet shifted absorption is due to gas along our line of sight to the photosphere absorbing the radiation, and re-emitting it in another direction. Similarly, the emission comes from the re-emitted radiation from the rest of the expanding envelope. In the simplest cases, the net zero width is the natural consequence of pure scattering as from a resonance line.

The violet edge of absorption, at $v = -15000$ km s^{-1} is shifted about as much as the red edge of emission. For SN 1969l in NGC 1058, the observed shifts are also about 15000 km s^{-1}.

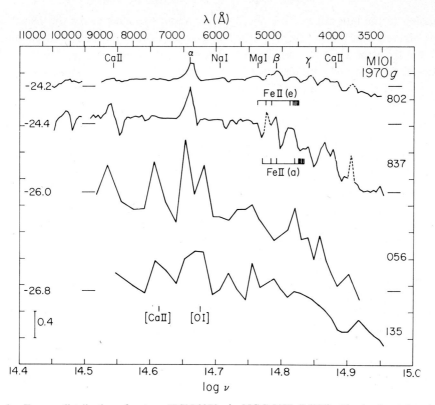

Fig. 1. Energy distributions for type II SN 1970g in NGC 5457 (M101). The horizontal scale is $\log \nu$ where ν is the observed frequency in Hz. A wavelength scale is shown at the top of the figure. The vertical scale is $\log f_\nu$ where f_ν is the observed flux in ergs s^{-1} cm^{-2} Hz^{-1}. The absolute flux level is indicated at the left of each energy distribution along with a tick mark which is repeated at the right. Each energy distribution is identified by the last three digits of the Julian Day when the observation was made. Representative standard deviation bars are shown where they are significant towards the ends of each scan. Errors in the center part of each energy distribution (i.e. $14.55 \leq \log \nu \leq 14.85$) are normally too small to be shown. The description above also applies to Figures 2–5. The dotted portions of the top two energy distributions are the emission lines $\lambda 3727$ of [O II] and $\lambda 5007$ of [O III] which come from the background H II region. For the bottom two energy distributions the background H II region [which had been observed previously (Searle, 1971)] has been subtracted as well as possible. Possible line identifications of the emission peaks are marked. The positions of the strongest expected Fe II lines are shown. The emission wavelengths are plotted without velocity corrections. The absorption wavelengths are displaced by 0.008 in $\log \nu$ to the violet corresponding to the velocity of other absorption minima. (By permission of the *Astrophysical Journal*.)

2.3. LINE IDENTIFICATIONS

2.3.1. *Balmer Lines*

As reported by many observers for many years, the Balmer series is certainly present in the spectra of SN II's, at all the epochs observed. The Hα profile, shown in Figure 6 for SN 1970g, indicates net emission, with the emission peak centered near the rest wavelength, in the rest frame of M 101. The width of this line, from the red edge of emission to the blue edge of the absorption is about the same as for the $\lambda 8600$ line. This suggests that the two lines are formed over the same volume, so that conditions inferred from the Hα line apply to the entire reversing layer.

On JD 802, the observed net flux from SN 1970g in M 101 was about 3.6×10^{12} erg cm^{-2} s^{-1}. Using a distance of 7 Mpc, this corresponds to 2×10^{39} erg s^{-1} emitted at the source. At about 10000K, this requires $n_e^2 V = 3.5 \times 10^{64}$ cm^{-3}. Since the reversing layer is somewhat bigger than the photosphere, we estimate a radius of 10^{15} cm, which implies an electron density $n_e \approx 3 \times 10^9$ cm^{-3}, and thus 3×10^{24} electrons cm^{-2} along the line of sight. Similarly, the net Hα emission for JD 837 combined with a velocity 1.5×10^9 cm s^{-1} and an age of about 40 days gives $n_e \simeq 3 \times 10^8$ and a column density of about 2×10^{24} cm^{-2}.

It may be significant that the column density is just adequate to make the reversing layer optically thick in electron scattering. A lower bound to the mass in the envelope comes from assuming the hydrogen is fully ionized. Then $M > 0.1\ M_\odot$.

2.3.2. *Lines of* Ca II, Na I, *and* Mg I

The continuum and Balmer lines provide information on the physical state of the expanding envelope. We deduced a differentially expanding cloud with $n_e \simeq 10^9$, irradiated by a photosphere of $T = 5000$–10000K, diluted by a factor of $W \approx 0.01$. The lines to be anticipated are those arising from the ground state or metastable states of ions that are abundant in such an environment. In general, lines present in A and F-type supergiants and in shells surrounding A and F stars might be expected to appear.

The strongest observed features, aside from the Balmer lines are at $\lambda 3950$ and $\lambda 8600$. Consideration of the physical state of the envelope leads to the suggestion that $\lambda 3950$ is due to the H and K lines of Ca II and that the $\lambda 8600$ feature is due to the Ca II infrared triplet $\lambda\lambda 8498, 8542, 8662$. The triplet arises from the metastable $3d\ ^2D$ level, and the upper level is the upper level of the H and K transitions.

The next strongest features are $\lambda 5890$ and $\lambda 5180$. The first could be either $\lambda 5876$ of He I or $\lambda 5890$ of Na I. The sodium seems most likely, as none of the other strong triplet lines of He I is apparent. Although most of the sodium will be ionized, the residual neutral sodium can account for the observed modest line strength.

Analogously, the resonance lines of Ca I $\lambda 4226$ should be present, even though most calcium is ionized. This line falls in a blended region of the spectrum, and could be as strong as the D-lines without being detected. We also expect the Mg I b-lines arising from metastable levels to be strong at $\lambda 5174$. This probably accounts for the feature at $\lambda 5180$.

2.3.3. *The Feature at* $\lambda 4600$

The prominent feature at $\lambda 4600$ is reminiscent of the similar feature in Nova Herculis (Stratton, 1934). There, the feature is attributed to Fe II. A similar situation is very likely to arise here. Accordingly, the emission and blueshifted absorptions for the relevant Fe II lines are plotted in Figure 1. The fit is not exact but does suggest that the feature at $\lambda 4600$ is at least partly due to Fe II. A similar conclusion, based only on the absorption features, has already been reached by Patchett and Branch (1972).

2.3.4. [O I]

In Figure 1, the scans obtained 8 months or more after maximum light show the appearance of two strong emission features at $\lambda 6300$ and $\lambda 7300$.

The feature at $\lambda 6300$ is likely to be the [O I] doublet at $\lambda 6300$ and $\lambda 6363$. Although blended with Hα in these scans, slit spectra at comparable times demonstrate the clear separation of the two lines.

The appearance of this line, and its strength relative to Hα, indicate that the envelope has both recombined and grown less dense since maximum light.

2.3.5. [Ca II]

There are two possible identifications for the strong emission line at $\lambda 7300$ which occurs in the later phases of SN II's. It could be $\lambda 7320 + \lambda 7330$ of [O II] or it could be $\lambda 7291 + \lambda 7323$ of [Ca II]. The excitation potential of the upper level of the $\lambda 7320$ transition in [O II] is 4 eV and it is unclear how this could be excited collisionally without producing at the same time other permitted and forbidden lines of the abundant once ionized metals with comparable intensities, as happens, for example, in the spectra of novae. The great strength of this line at $\lambda 7300$ in the later phases of SN II's and the fact that, in the case of SN II 1969l in NGC 1058, it is not accompanied by any other emissions that can be identified with other well-known forbidden lines, leads us to reject the identification with [O II].

The identification with $\lambda 7291$ and $\lambda 7323$, on the other hand, provides a natural explanation of the behavior of this line. The [Ca II] lines are an intersystem transition that connects the metastable lower level of the $\lambda 8600$ transitions with the ground state of the ion. In the first month after maximum light the $\lambda 8600$ line in SN II's shows a P Cygni profile with a zero net equivalent width. This implies that the depopulation of the $3d\,^2D$ level is almost exclusively by radiative reabsorptions in the $\lambda 8600$ line rather than by alternative processes, the most important of which are expected to be collisional and radiative transitions to the ground state and photoionizations resulting from absorption of Lα photons.

At some time between 1 and 8 months following maximum light, the energy density in the radiation field has dropped sufficiently that depopulations via the intersystem transition are more probable than radiative reabsorptions. When these reabsorptions eventually become negligible and the density has fallen to the point that radiative downward transitions from $3d\,^2D$ are more probable than collisionally induced ones, absorptions in H and K, plus collisions to the upper state, feed a cascade via $\lambda 8600$ and $\lambda 7300$. In these circumstances the lines at $\lambda 8600$ and $\lambda 7300$ would have very nearly the same intensity as observed. It will be of importance to follow in detail the development of the $\lambda 7300$ line in SN II's between 1 month and 8 months after maximum light.

3. Supernovae of Type I

3.1. Continuum

As shown in Kirshner *et al.* (1973a), the overall energy distributions from 2.2 μ to 0.33 μ of the type I SN 1972e in NGC 5253 can be well represented by blackbodies. The best fit at the earliest epochs has $T = 10\,000$ K, which falls smoothly and gradually to 7000 K in 2 weeks. Further decreases in temperature were quite slow for at least 3 weeks. The smooth and gradual evolution of the spectrum demonstrates that the type I spectrum is not a superposition of emission bands: just as in type II's the bulk of the energy is radiated in a continuum. As in type II, the photospheric radius in the first month is of order 10^{15} cm.

3.2. Line profiles

As Figure 2 demonstrates, the spectra of SN I's are more complicated than those of SN II's. There, scans of the two types are shown at comparable phases. It also demonstrates that the strongest and most persistent features of SN II's appear among the strongest and most persistent features in SN I's. The outstanding examples are the features near $\lambda 3950$ and $\lambda 8600$, which we identified in SN II's with Ca II: they are prominent in type I spectra during the entire interval covered by observations.

The lines in SN I's have a characteristic P Cygni profile, as shown in Figure 6.

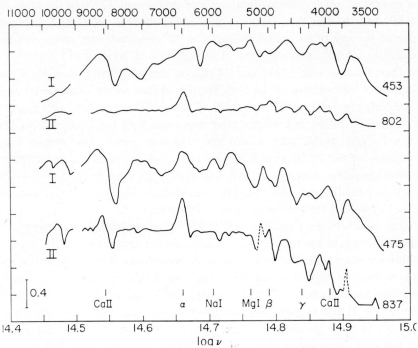

Fig. 2. Comparison of spectral energy distributions of type I and type II supernovae at two selected phases. The type I curves are from SN 1972e, while those for type II are from SN 1970g. (By permission of the *Astrophysical Journal*.)

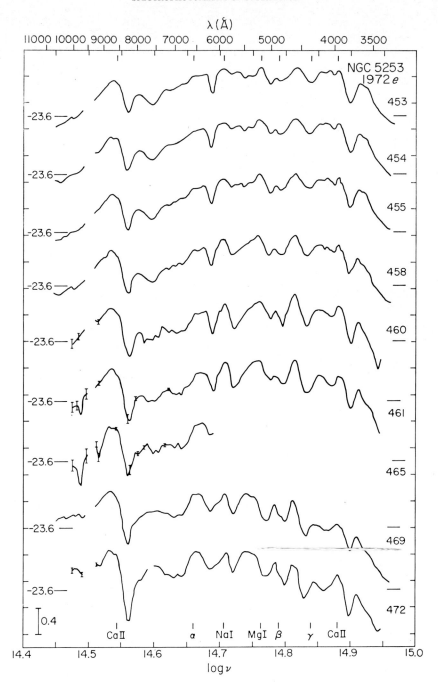

Fig. 3. Spectral energy distributions for SN 1972e in NGC 5253. Description is the same as in Figure 1. The smoothness of the distributions as drawn is real since the curves go through all measured points. Possible identifications of emission peaks are shown. The Balmer lines Hα, Hβ, and Hγ are included but do not necessarily imply the existence of these lines. (By permission of the *Astrophysical Journal*.)

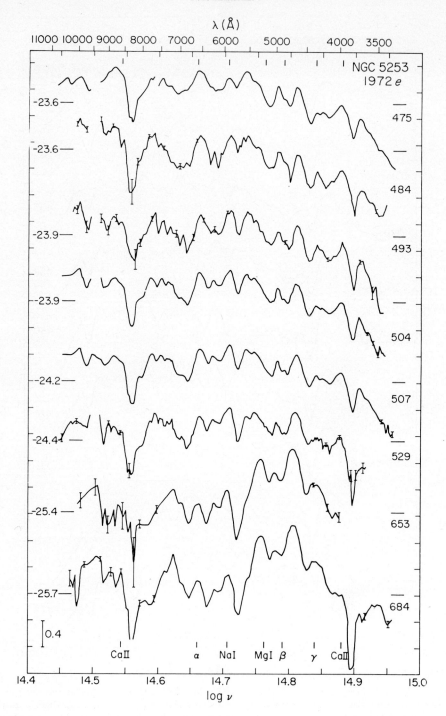

Fig. 4. Continuation of Figure 3. (By permission of the *Astrophysical Journal*.)

Here the $\lambda 8600$ line shape is somewhat irregular, but the essential feature of symmetrical emission near the rest wavelength and blueshifted absorption is present. The red edge of the emission peak indicates a recession velocity of about 20000 km s^{-1} while the blue edge of the absorption, though less well defined, shows a shift that is at least as large.

3.3. Line identifications in type I's

Four strong features in Figures 3 and 4 persist through the supernova's early evolution. Each consists of an emission peak and a blueshifted absorption. The emission peak wavelengths are at $\lambda 8700$, $\lambda 5890$, $\lambda 4600$, and $\lambda 3970$.

The striking fact that these four lines coincide with the strongest non-hydrogen features in the spectra of type II's leads to the conclusion that they are the *same* features and that there exists an underlying similarity to the two types of supernovae.

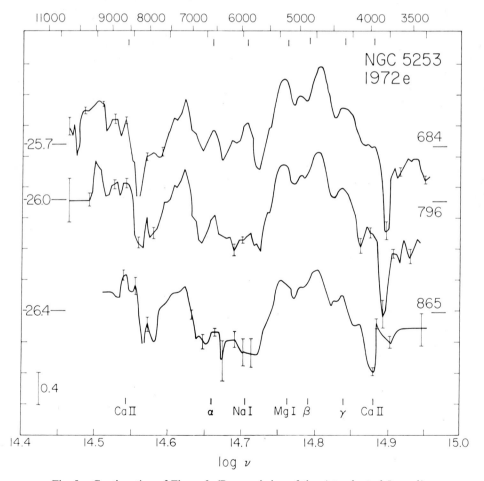

Fig. 5. Continuation of Figure 3. (By permission of the *Astrophysical Journal*.)

Real differences remain! Many other features, strong and weak, come and go in the confusing spectra of SN I's. The evidence for identifying these with well-known lines is far from compelling. The most puzzling feature is the emission at $\lambda 4600$ in the early phases of SN 1972e. Unlike other features, it drifts in wavelength to $\lambda 4700$ by JD 689, and has since been observed at $\lambda 4750$ on JD 865. Whether this can be explained in terms of Fe II is not clear.

In type II's, the Balmer series is clearly present, but in type I's, the question is more difficult. No traces of Balmer lines other than a possible weak Hα are seen

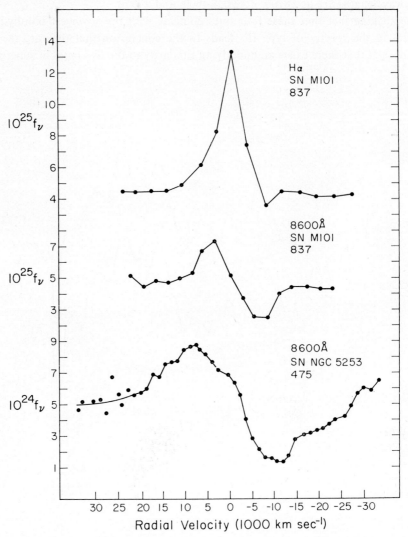

Fig. 6. Flux vs velocity shift (corrected to the rest wavelengths of each galaxy) for Hα and $\lambda 8600$ in the type II SN 1970g in M 101 on Julian Day 837, and for $\lambda 8600$ in the type I SN 1972e in NGC 5253 on Julian Day 475. (By permission of the *Astrophysical Journal*.)

in the first two weeks of the evolution of SN 1972e. However, on JD 460, a strong emission does appear in the expected position of Hα, and persists through JD 865. At about the same time, an emission feature appears in the position of Hγ. The spectra are too blended to determine whether Hβ is present. In Figure 2, the case for Balmer lines in type I is illustrated by comparison with type II. On the whole, the evidence in favor of Balmer lines seems good, but in a highly blended spectrum doubts must remain.

The conclusion seems inescapable that the main features which contribute to the spectra of SN II's are also prominent contributors to the spectra of SN I's. Generally line features are stronger and more numerous in SN I's. It is possible that a low hydrogen abundance in the envelopes of SN I's leads to more ions of common metals per free electron and hence stronger lines relative to the continuum.

3.4. Similarities among SN I's

Similarities among the light curves of SN I's are often cited as a striking property of these objects. (Zwicky, 1965). Our evidence from scans of four SN I's and comparison with SN 1937c is that the spectrum of any type I supernova can be identified with a particular epoch in SN 1972e. In this way, relative ages can be reliably-assigned.

3.5. Late phases

Figure 5 presents the observations obtained 7, 10, and 14 months after the discovery of SN 1972e. Perhaps the most striking feature of this late evolution is how little change takes place in the spectrum. The principal features from $\lambda 5500$ to $\lambda 4000$ remain nearly constant in shape, while fading in brightness. The P Cygni feature at $\lambda 5900$, attributed to Na I, does disappear by JD 865. Similarly, there is evidence that the Ca II absorptions at H and K and at $\lambda 8600$ are disappearing in that scan. Further observations are planned for Winter 1973, when the supernova will be about 20 months old.

4. Concluding Remarks

Absolute energy distributions help establish the physical setting of the supernova explosion. The temperature is immediately available from the continuum shape, the bulk motions from the line shapes, and for type II, electron density from emission strengths. In this physical setting, plausible identifications are possible which account for all the strong features in SN II's, and some of the same identifications seem appropriate for SN I's.

Acknowledgements

Much of the work described here was done jointly with J. B. Oke, L. Searle, and M. V. Penston. Many other colleages at Hale Observatories contributed generously of their time and enthusiasm, especially J. E. Gunn and J. L. Greenstein. I am grateful to the National Science Foundation both for a fellowship and for a travel grant. Part of this work was supported by Grant NGC-05-002-134 from the National Aeronautics and Space Administration.

References

Greenstein, J. L. and Minkowski, R.: 1973, *Astrophys. J.* **182**, 225.
Kirshner, R. P., Willner, S. P., Becklin, E. E., Neugebauer, G., and Oke, J. B.: 1973a, *Astrophys. J. Letters* **180**, L97.
Kirshner, R. P., Oke, J. B., Penston, M. V., and Searle, L.: 1973b, *Astrophys. J.* **185**, 303.
Oke, J. B.: 1969, *Publ. Astron. Soc. Pacific* **81**, 11.
Oke, J. B. and Schild, R.: 1970, *Astrophys. J.* **161**, 1015.
Patchett, B. and Branch, D.: 1972, *Monthly Notices Roy. Astron. Soc.* **158**, 375.
Searle, L.: 1971, *Astrophys. J.* **168**, 327.
Zwicky, F.: 1965, *Stars and Stellar Systems* **8**, 367.

ON THE PHYSICAL MODEL OF SUPERNOVAE CLOSE TO LIGHT MAXIMUM

E. R. MUSTEL

Astronomical Council, USSR Academy of Sciences, Moscow, U.S.S.R.

Abstract. One of the principal sources of information about supernovae are the spectra of these stars. Thus we are going to discuss mainly the spectra of type I and type II supernovae around light maximum ($t_{\max}$).

1. Type I Supernovae

The identification. During several decades it was accepted that the spectrum of a typical type I supernova around light maximum is composed mainly of a large number of very wide overlapping emission bands (Minkowski, 1939). However all the attempts to identify these bands were not successful. In this connection McLaughlin (1963) suggested that the principal element of the spectra of type I supernovae are the *absorptions* (intensity minima) but *not* the emissions. This hypothesis was confirmed by Pskowskij (1968) who identified several absorptions in the spectra of type I supernovae with certain sufficiently strong and heavily displaced absorption lines of the spectra of supergiants of the spectral classes B and A. A comparison of absorptions in the spectra of type I supernovae with the absorption lines in the spectra of closely related objects, namely Novae permitted to confirm the 'absorption hypothesis' of the origin of the spectra of type I supernovae and to explain the peculiar evolution of the spectra of these objects with time (Mustel, 1971a, 1972a; Mustel and Chugay, 1974). Then a comparison of the spectrum of a type I supernova 1966j in NGC 3198 (Chalonge and Burnichon, 1968) with the principal *absorption* spectrum of DQ Her permitted to explain practically all the features of the spectrum of this supernova, see Figure 1a, taken from the article of Mustel (1972b). The Doppler displacement $\kappa = \Delta\lambda/\lambda = v/c$ is equal for this supernova to: $\kappa = -0.0255$. Only the lines of the principal absorption spectrum of DQ Her were used which according to McLaughlin (1937) had intensities $I \geqslant 2$. Two exceptions from this rule were: Ba II 5854(1) and Fe II 5425(0).* The intensity of all the absorption lines in the spectrum of DQ Her is indicated in round brackets. Figure 1b gives a similar identification made on the base of photographic tracings of the spectrum of the supernova, reproduced in the paper of Chincarini and Perinotto (1968). This spectrum was also taken on the 9th January 1967. The displacement factor κ in Figure 1b is the same (-0.0255).

Finally it is necessary to mention an analysis of Branch and Patchett (1973) who also confirmed the 'absorption hypothesis' of type I supernovae.

All these studies permitted to determine the expansion velocities of the envelopes ejected by different type I supernovae. These velocities range from $\simeq 6000$ km s^{-1} to

---

* According to Mustel and Baranova (1965) the intensity of this line Fe II, 5425 Å in the spectrum of DQ Her was quite perceptible.

546 E. R. MUSTEL

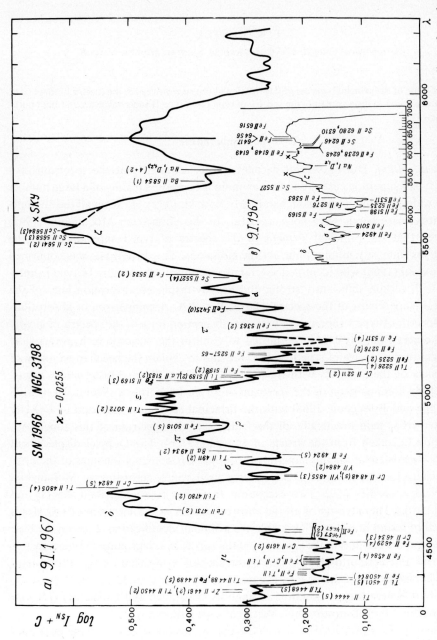

Fig. 1. Identification of absorptions in the spectra of type I supernova 1966j in NGC 3198 with relatively strong absorption lines of metals (Int ≥ 2) in the principal spectrum of DQ Her. It seems that the absorption d is spurious. The spectral energy distribution is taken from Chalonge et Burnichon (1968).

$\simeq 15000$ km s^{-1}. On the *average* the expansion velocity of the envelopes of type I supernovae is of the order of 10000 km s^{-1}. This magnitude of v coincides well with expansion velocities of the supernova *remnants*: shell sources MSH14-415, Tycho, Kepler (van den Bergh, 1973). As to the Kepler's remnant see in more detail Mustel (1974).

It is very important to mention that for the majority of spectra of type I supernovae the Doppler factor κ is practically constant for *all* the available ranges of wave-lengths.

The next very essential progress in the analysis of spectra of type I supernovae is connected with the bright supernova 1972e in NGC 5253 the epoch of light maximum of which was close to the 5th of May. I have in mind the large number of the absolute spectral energy distributions which were obtained for this supernova by Kirschner *et al.* (1973b) and by Kirschner (1974). These distributions were obtained in the range from 3200 to 11000 Å. The authors of this article write: "The smooth and gradual change with time of the overall shape of the *spectrum* is a very striking phenomenon and clearly shows that the SN I spectrum is not a superposition of emission bands but that just as in type II's the bulk of the energy radiated by the supernova is contained in a continuum; the net flux in line radiation is relatively insignificant." This conclusion agrees with the principal starting point of the absorption hypothesis. Moreover an identification of absorption lines in the spectrum of supernova in NGC 5253 carried out by Mustel (1973) also confirms this hypothesis.

However there is one point which deserves some discussion. Namely, Kirschner *et al.* (1973b) consider that the spectra of type I supernova contain not only absorption lines but also emissions and that these spectra are similar to the spectra of P Cygni. They write: "The hypothesis of blended absorption lines without associated emission lines cannot account for the fact that the minima in the spectra of SN I's are so frequently displaced from their neighbouring redward* maxima *by the same velocity shift*." However we have to take into account the fact that the *principal* property of spectra of type I supernovae are the extremely wide absorptions. Therefore in the majority of cases the intensity maxima are produced by *neighbouring* absorptions and it is extremely difficult to distinguish the true line emissions and these maxima! This statement follows from the analysis of the evolution of the absorption spectra of type I supernovae. It is shown in the paper of Mustel (1973) that many changes in the intensity maxima of the spectra of the supernova in NGC 5253 are mostly due to time-evolution of the *absorption* component of the spectra.

In this connection we give Figure 2 which is based on the absolute spectral energy distributions presented in the paper of Kirschner *et al.* (1973b). These distributions are here transformed to the usual system of coordinates (I_λ vs λ). The line-identifications are from the paper of Mustel (1973). The line of the continuous spectrum (dashed line) is drawn in accordance with the absorption hypothesis.

Before discussing Figure 2 we must mention the following fact. Observations show (see further) that the temperature of type I supernovae decreases rapidly after light

---

* The same is true for violetward 'maxima' and all this is connected with the mechanism of the origin of absorption lines in these stars, see further.

maximum* during 30–40 days and then begins again to increase slowly. Correspondingly the temperature of supernova for the moment $t = 2441453$ JD (first moment of observations) was relatively high whereas for the moment 2 441 475 it was considerably lower.

Now we shall consider the two intensity maxima ζ' and ζ''; the position of the second maximum ζ'' on Figure 2a is indicated by an arrow. This arrow is above region A with reduced intensity. This region is due to the fact that it contains many sufficiently strong absorption lines of S II (see Pskowskij, 1968; Mustel, 1973; and especially Figure 6 in the paper of Mustel, 1972a). These absorption lines are relatively strong in B-type stars and their intensity decreases rapidly with the decrease of the temperature of the star.** Then, a blend of lines of N II, λ_0 5680 Å was also observed in the same region A for the moment JD... 453 and this was also due to the high temperature of the supernova at this moment. Thus we may consider that the intensity maximum ζ' on Figure 2a was only a part of maximum ζ'', which was displayed a little later.

In connection with the further decrease of temperature the absorption lines of S II and N II completely disappear near the moment $t = 2441470$ and therefore it may be suggested that the maximum ζ'' on Figure 2b corresponds to the undisturbed continuous spectrum of the star.

Later on this maximum ζ'' suffered new disturbances. This was due to the same absorption blend of N II, $\lambda_0 = 5680$ Å which reappeared again and it was connected with the growth of the temperature of the star (see Figure 1b in the paper of Mustel, 1973).

All these transformations of the intensity maxima are more or less typical for the spectra of type I supernovae. Even the cases when an intensity maximum is keeping its position for a long period of time are easily explained. In those cases we deal with two neighbouring absorptions, the observed wave lengths and intensities of which are practically stable and as a result of this the space between them – the intensity maximum – occupies the same place.

Finally we should like to point out the following fact. From Figure 2 we see that absorption lines play a much more important role for the moment $t = ...$ 475 than for the moment $t = ...$ 453. This may be explained also from the point of view of the absorption hypothesis. In fact we know that the number of absorption lines and their depth grow rapidly when we go from hot B-stars to cooler stars, say stars of F-type. The same phenomenon is present in Figure 2.

Thus we may conclude that the principal feature of the spectra of type I supernovae are *very wide absorptions* and that in the majority of cases the line emission does not play an important role.

The next problem is the problem of a quantitative interpretation of the spectra of type I supernovae. This problem is discussed in the paper of Mustel and Chugay (1974). The first step is to choose a model of the supernova for the moments close to light maximum. Two of such models are shown in Figure 3. According to the first model

---

* Some observations indicate that the same takes place before light maximum.
** The blend of Si II $\lambda 5970$ also weakens fast after light maximum.

the supernova has a central star (central remnant) which is the principal source of the continuous spectrum. The line-absorption spectrum is produced by the expanding envelope of the supernova. It is not excluded that the central remnant (CR) has a very hot nucleus, indicated by a black circular spot. According to the second model the principal source of the continuous spectrum of the supernova are the inner layers of the expanding envelope. The high temperature of the central body, the black spot in Figure 3, may be the source of heating of these layers.

It seems that the first model is nearer to reality than the second one, though the 'possibilities' of the second model should be studied in more detail. Here we shall consider the first model.

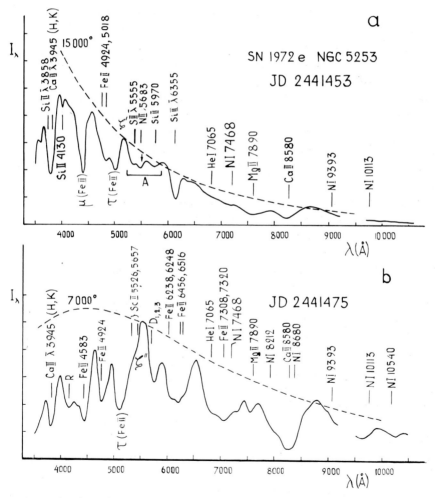

Fig. 2. Interpolated continuous spectrum and the suggested identifications for type I supernova 1972e in NGC 5253 for two moments. The region A in Figure 2a with reduced intensity is due to absorption lines of S II, Si II and N II. The absolute spectral energy distribution in this Figure is taken from the paper of Kirshner *et al.* (1973).

The first model was already discussed by Mustel (1971b). The available data permit to suggest that the emission from CR is approximately Planckian and that there is a correspondence between the temperature of CR and the line-absorption spectrum (the spectral class) of the supernova.

According to Pskowskij (1970) the change of $B-V$ and $U-B$ with time indicates a decrease of the temperature of a typical type I supernova after light maximum during 30–40 days. Then the temperature begins to increase again. The same results are obtained by Barbon et al. (1973). The type I supernova 1972e has shown a drop in temperature after light maximum also, see the article of Kirschner et al. (1973a) as well as Figure 2 of this paper. Finally we may mention the results of Holm et al. (1973) according to which the UV-luminosity (down to 1900 Å) of the supernova in NGC 5253 was decreasing during the first period of its evolution. Moreover several facts show that the changes in the line-absorption component of type I supernovae correspond more or less to the changes of their colour temperature. We may mention again the blend of N II, $\lambda_0 = 5680$ Å, which was sufficiently strong just after light maximum and approximately 140 days after it.

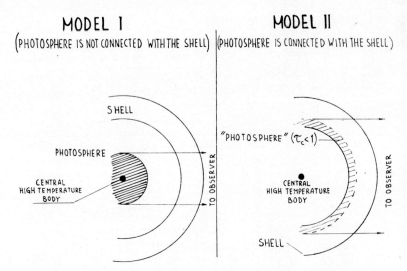

Fig. 3. Two possible models of type I supernovae close to light maximum.

It may be estimated that the average temperature of a typical type I supernova at light maximum is of the order of $10000°–15000°$. For supernova 1960f in NGC 4496 it gives a 'photospheric' radius R_p of the order of $20000 R_\odot$. The radius obtained from the expansion velocity of the envelope is 2 or 3 times larger (Mustel, 1971b).

Now we shall consider a very important question – the *profiles* of absorption lines in the spectra of type I supernovae. These profiles are very wide, 10–50 times wider than the absorption lines of the same elements in the spectra of novae. The very large width of the absorption lines in the spectra of type I supernovae was the *principal* reason of the difficulty in the interpretation of the spectra of these objects. In fact the overlapping of the neighbouring very wide absorption lines (mostly metallic lines) produces sometimes so strong blends that the recognition of these blends becomes very difficult. For example a very wide blend τ, produced by an accumulation of metallic lines (mostly Fe II) in the range of wave-lengths from λ_0 5169 (Fe II) to λ_0 5414 (Fe II) does not usually show in the spectra of type I supernovae individual absorptions belonging to metals. Only in the spectrum of supernova 1966j with the unusually sharp lines* we can see some traces of metallic lines in the blend τ (see Figure I).

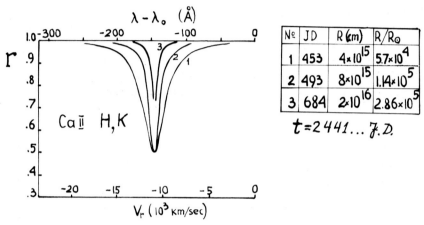

Fig. 4. An extrapolated decrease of the total width of the blend of absorption lines H and K, which is computed on the base of a simple model of an expanding envelope with velocity gradient. It is accepted that $N \sim R^{-2}$.

The effect of blending is especially strong in the *violet* parts of the spectra of the supernovae, where there are numerous strong absorption lines of once ionized metals; see for example the very noticeable depression in the spectrum of type I supernova 1972e (Figure 2). The same phenomenon of a strong crowding of absorption lines towards the ultraviolet part of the spectrum is observed in the spectra of normal stars of classes A, F, etc.

We cannot explain the very large width of absorption lines in the spectra of type I supernovae as a result of the very large mass of the envelopes of supernovae in comparison with the mass of the envelopes ejected by novae. It appears that due to the very large size of the envelopes around supernovae the mass of gas per cm^2 is approximately the same in both cases (Mustel and Chugay, 1974).

An analysis of different mechanisms which may be responsible for the strong widening of absorption lines in the spectra of type I supernovae is carried out in the article of Mustel and Chugay (1974). The results of this analysis are the following:

(a) The application of usual damping mechanisms leads to improbably high masses of the envelopes of type I supernovae (for example for Fe atoms) up to approximately 500 $M_\odot$.

* It seems that this is due to a relatively small velocity gradient inside this envelope; see further.

(b) Doppler profiles for the turbulent motions of the gases inside the envelope must be rejected too since these turbulent motions would lead to a very strong heating of the envelope due to the dissipation of the turbulence.

(c) The next simple mechanism in which the widening of absorption lines is due to the presence of a strong velocity gradient inside the envelopes meets also a difficulty. It cannot explain the fact that very often the outermost parts of *both* wings of the sufficiently strong absorption lines in the spectra of type I supernovae occupy the same position during a very long period of time.* On the contrary it is expected that the width of all the absorption lines must diminish very rapidly, see Figure 4. This figure shows an estimated change of the profile of Ca II (H+K)-absorption lines in the spectrum of supernova 1972e. The computations are carried out for the case when the number of the absorbing atoms (per 1 cm^2 of the envelope) decreases according to the law: $N \sim R^{-2}$. The *first starting* profile (for the moment $t = \ldots 453$) is taken from the observations. This difficulty takes place also in case of widening mechanism (a).

(d) In spite of the difficulty of the previous mechanism (c) and taking into account very serious difficulties of mechanisms (a) and (b) we are forced to accept a model with a large velocity gradient. In order to explain in this case a very slow evolution of the profiles of strong absorption lines (Figure 4) we should modify the previous mechanism (c) and admit that during a long period of time the optical depth τ_v of the envelope is very large,** even in the wings of the absorption lines. Here the total width of the absorption line will remain constant also for a long period of time. However for the explanation of the observed profiles of the strong absorption lines we introduce the following additional assumptions: (1) The mass of the envelope must be sufficiently large (in comparison with the mass in the model (c) in order to fulfill the requirement $\tau_v \gg 1$ during many months.† (2) Inside the envelope there must be a considerable inhomogeneity. This 'patchiness' of the envelope explains *noticeable* observed residual intensity r_v in the profiles of strong absorption lines. (3) Besides, in order to explain the variation r_v inside the profiles, we should suggest that the filling factor (see, for example, Peimbert, 1971) is a maximum in the *middle* part of the envelope and decreases in both (opposite) radial directions. All these suggestions are quite natural.

Now we may try to describe the time-evolution of the spectrum of a typical type I supernova; see also the article of Mustel (1972a). This evolution is due to the following factors:

(a) The *temperature changes of* CR. As we mentioned above, the temperature of CR at light maximum and before it is rather high and decreases after the light maximum, during a period Δt_0 of 30–40 days. Then it begins to rise again. Thus immediately after light maximum the ionization of elements is expected to drop, to grow again

---

* In other words the total width of a strong absorption lines may be practically constant during a very long period of time.
** In this case the total width of strong absorptions is determined by the velocity gradient. This explains the fact that the wave length distance between minima and neighbouring 'maxima' in the spectra of supernova 1972e corresponded for different absorptions to the same velocity shift, see page 547.
† Calculations show that this mass is considerably less than that in the damping mechanism (a).

after the period Δt_0. We already described such a case – the behaviour of a blend of absorption lines of N II, $\bar\lambda_0 = 5680$ Å. It is very interesting to note that certain sufficiently clear and intense lines of N I appeared *after* the disappearance of this blend of N II. Probably there are other similar examples but the effects of blending are so strong that any definite conclusions are inadequate. Moreover it seems that only the nitrogen atoms have more or less clear absorption lines for two stages of ionization: N I and N II.

The changes of T (CR) produce also certain changes in the state of atomic excitation. But it is necessary to take into account the fact that the central remnant is a star with quite unusual properties and it is expected from analogy with similar objects that this star should emit anomalously high radiation in the regions of high frequencies, in the far ultra-violet regions of the spectrum. It seems that the lines of He I 5875 and 7065 in the spectrum of supernova 1972e are connected with this high-frequency emission. The correctness of this suggestion is confirmed by the following considerations. It is described in a paper of Mustel (1972a) that there is usually a good agreement between observed wave-lengths of the absorptions in the spectra of type I supernovae and the calculated positions of the corresponding spectral lines of different elements. Only the absorption with $\lambda \simeq 5700$ shows in some cases relatively small velocity v. This effect may be naturally explained if we admit that the principal contributor of the absorption at $\lambda \simeq 5700$ Å is the He I line D_3 and not the $D_{1,2}$ lines of Na I. In fact the strongest influence of high-frequency radiation must be for the *internal* (relatively transparent) parts of the envelope which expand with the smallest velocity and therefore the deviations from the calculations and observations may be due to the fact that the absorption $\lambda \simeq 5700$ Å is formed in these internal parts of the envelope.

(b) *The expansion of the envelope of the supernova.* This produces a continuous decrease in the gas density inside the envelope and a decrease of the dilution factor W. Thus we expect that there must be a continuous accumulation of atoms on metastable levels. This prediction is completely confirmed by observations. For example, at light maximum we observe mostly the lines of Fe II, similar to the lines 5018, 4924, the low level of which has an E.P. ≈ 2.9 eV. At the same time in ten or fifteen days after the light maximum the lines of multiplets N73 and 74 of Fe II appear, for which E.P. $\simeq$ $\simeq 3.9$ eV. A similar situation is observed with the Sc II lines 5526 and 5657. They attain their maximum intensity only after the light maximum and their behaviour is the same as the behaviour of these lines in the spectra of DQ Her.

The last question is the question of the chemical composition of the envelopes of type I supernovae. This question is discussed in more detail in the papers of Mustel (1973, 1974) and we shall reproduce some principal results. First, it is necessary to say that from a quantitative point of view this question is a very difficult one: (1) Owing to a very strong blending of the absorption lines the line of the continuous spectrum is very uncertain and this uncertainty may introduce considerable errors into the equivalent widths W_λ of the absorption lines; (2) There are only a few sufficiently clear and relatively isolated absorption lines in the spectra. In particular it is very difficult to identify weak absorption lines and therefore the usual method of the curve of growth is practically inapplicable here. Therefore it is necessary to work out a special method for the chemical analysis of the envelopes of type I supernovae. It must take into account the presence of a strong velocity gradient inside the envelopes of these stars and other considerations, as described above.

The main semiquantitative results of the papers mentioned above are the following:

(a) There is abundance of metals in the envelopes of these supernovae; see for example Figure 1.

(b) There are reasons to state that some of the intense absorption lines in the spectra of type I supernovae are due to He I. Since the E.P. of the low levels of the corresponding transitions in the He I atoms is high ($\simeq 20$ eV) it is concluded that the abundance of He I in the envelopes of these supernovae is relatively high.

(c) It is concluded that nitrogen is noticeably more abundant than carbon and oxygen.

(d) Observations do not show any hydrogen lines in the spectra of type I supernovae. In particular the absence of the absorption lines of the Balmer series and the simultaneous presence of relatively strong lines of N I, show that the abundance of H atoms is noticeably lower than the abundance of N atoms. An analysis of the supernova remnants confirms the excessive abundance of nitrogen in the envelopes of supernovae.

2. Type II Supernovae

The spectra of type II supernovae are *similar* to the spectra of common novae. The spectral lines of these supernovae have usually absorption and emission components. Here the problem of the identification was not so difficult as the problem for type I supernovae. In particular hydrogen (absorption + emission) plays an important role in the spectra of type II supernovae. Nevertheless there are many specific properties in the spectra of these objects:

(a) For type II supernovae different observers find different velocities of expansion, ranging from $v \approx 5000$ km s^{-1} to 10000 km s^{-1}. It may be that this is partly due to the fact that for example in the spectra of the type II supernovae 1969l in NGC 1058 there was a progressive drift with time toward the red of all absorption and emission features (Ciatti *et al.*, 1971). After light maximum the mean expansion velocity was decreasing during two months (!) from $v \simeq 9500$ km s^{-1} to 5500 km s^{-1}. It would be very important to find out whether this phenomenon (even on a smaller scale) is inherent to all type II supernovae. In addition it may be mentioned that different groups of absorption lines (for example hydrogen and metals) show somewhat different velocities (Ciatti *et al.*, 1971; Patchett and Branch, 1972).

(b) The time-evolution of the spectra of type II supernovae is much faster than the time-evolution for type I supernovae, see for example Figure 5 taken from the article of Kirschner *et al.* (1973b). We see that the changes in the spectrum of the supernova 1970g in M101 from JD 2440802 to JD 2441056 are very strong; for this second moment the spectrum is already practically an emission spectrum.

(c) The problem of the temperature of type II supernovae and generally the problem of the continuous spectrum of these objects is very complex. According to Arp (1961) the temperature of the supernova of 1959 in NGC 7331 at light maximum was

$\approx 25000°$ and the radius R of the envelope for this moment was $\approx 6 \times 10^{14}$ cm. According to Ciatti et al. (1971) the pattern of the supernova 1969l in NGC 1058 in the first 30 days after light maximum followed rather closely that of a supergiant with decreasing temperature from 15 000° downward. According to Kirschner et al. (1973b) the continuum of type II spectra at light maximum resembles that of a blackbody at about 9500°. They found that when a supernova ages the continuum becomes fainter and the temperature of the best fitting blackbody drops rapidly to a value of about 5000°.

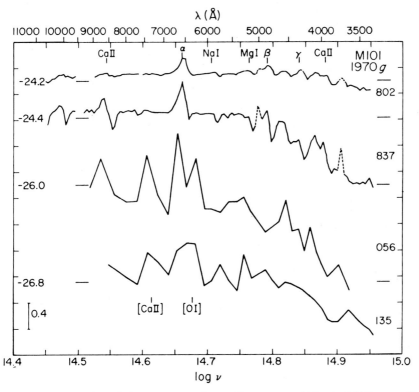

Fig. 5. The time-evolution of the spectrum of type II supernova 1970g in M101 according to Kirshner et al. (1973b).

In order to understand these observational results let us consider again Figure 5. The energy distribution for the moment $t = 2440802$ (JD) is more or less 'flat' and shows a noticeable extension into the ultraviolet. This fact was already mentioned by many observers. But at the moment $t = 2440837$ the situation is quite different. The energy distribution in the spectrum for $\lambda \geqslant 5000$ Å remains practically the same, but for the region with $\lambda < 5000$ we observe a very noticeable reduction of intensity I_v. This reduction is reflected in the difference between the U and V-light curves of type II supernovae. The drop in the U-brightness immediately after light maximum is much steeper than the drop in the V-brightness (see, for example, Ciatti et al., 1971). It seems that this phenomenon of such a noticeable intensity reduction in the region $\lambda < 5000$ Å is mostly responsible for the 'apparent' temperature decrease mentioned above. Another specific property of the 'short' wave-length part of the spectra of type II supernovae ($\lambda < 5000$ Å) is that at intermediary stages of evolution* this part of the spectrum becomes more or less similar to the spectra of type I supernovae, of course, for the same wave-lengths; see the energy distribution for the moment $t = 2440837$ on Figure 5.

We may suggest the following explanation for all these effects. It is expected that the *outer* layers of the massive and semi-transparent envelopes of type II supernovae are rapidly cooling, more rapidly than the internal parts of these envelopes which are exposed directly to radiation emitted by the relatively hot 'central remnant' of the supernova. As a result of this cooling many absorption metallic lines due to Fe II, Cr II, Ti II, Ca II etc. are expected to appear in the short wavelength part of the spectra of supernovae where these lines are especially numerous, see for example Figure 1. These lines will originate mostly in the outer directly observable layers of the envelope and this explains the temporary similarity between the spectra of type I and type II supernovae.

Now the strong blending of wide metallic lines in the short wavelength part of the spectrum ($\lambda < 5000$ Å) will produce practically a continuous opacity. On the contrary the envelope for $\lambda \geqslant 5000$ Å will remain more or less transparent** and we shall be able to see the 'central remnant'. Therefore, approximately the same energy distribution in the long-wave-length part of the continuous spectrum ($\lambda \geqslant 5000$ Å) of the supernova in M101 suggests for the first two moments on Figure 5 that the 'central remnant' of a supernova keeps practically the same temperature during this period. This may explain the apparent 'contradiction' between the appearance of relatively strong emission lines and the relatively low 'apparent' colour temperature of $\approx 5000°$ for type II supernovae during the later stages of their evolution (see the two last moments on Figure 5).

In terminating this discussion on the spectra of type II supernovae we may mention two facts: (a) the radioemission from supernova 1970g in M101 (Goss *et al.*, 1973); (b) An infrared excess observed for the first time in supernova 1969l in NGC 1058 (Ciatti *et al.*, 1971).

There is no detailed chemical analysis of the envelopes of type II supernovae. In contradistinction to type I supernovae the spectra of type II supernovae have sufficiently strong absorption and emission lines of hydrogen. The article of Kirschner *et al.* (1973b) contains the identification of lines of Ca II. The identification of lines of He I, He II, N II, N III is made in the article of Ciatti *et al.* (1971). The absorption lines of Fe II in the spectra of type II stars are identified in the article of Patchett and Branch (1972).

The supernova remnants give another possibility to study the chemical analysis of gases ejected by supernovae (Mustel, 1974).

* Later on these spectra become purely emission spectra.
** Except for a few relatively isolated absorptions.

References

Arp, H.: 1961, *Astrophys. J.* **133**, 883.
Barbon, R., Ciatti, F., and Rosino, L.: 1973, *Astron. Astrophys.* **25**, 241.
Branch, D. and Patchett, B.: 1973, *Monthly Notices Roy. Astron. Soc.* **161**, 71.
Chalonge, D. and Burnichon, M. L.: 1968, *J. Obs.* **51**, 5.
Ciatti, F., Rosino, L., and Bertola, F.: 1971, *Asiago Obs. Contr.*, No. 255.
Cincarini, G. and Perinotto, M.: 1968, *Asiago Obs. Contr.*, No. 205.
Goss, W. N., Allen, R. J., Ekers, R. D., and de Bruyn, G.: 1973, *Nature Phys. Sci.* **243**, 42.
Holm, A. V., Wu, C. C., and Caldwell, J. J.: 1973, *Bull. Am. Astron. Soc.* **5**, No. 1, Part 1, 28.
Kirschner, R. P.: 1974, this volume, p. 533.
Kirschner, R. P., Willner, S. P., Becklin, E. E., Neugebauer, G., and Oke, J. B.: 1973a, *Astrophys. J.* **180**, L97.
Kirschner, R. P., Oke, J. B., Penston, M. V., and Searle, L.: 1973b, *Astrophys. J.* **185**, 303.
McLaughlin, D. B.: 1937, *Michigan Obs. Publ.* **6**, 107.
McLaughlin, D. B.: 1963, *Publ. Astron. Soc. Pacific* **75**, 133.
Minkowski, R.: 1939, *Astrophys. J.* **89**, 156.
Mustel, E. R.: 1971a, *Astron. Zh.* **48**, 3; *Soviet Astron.* **15**, 1, 1971.
Mustel, E. R.: 1971b, *Astron. Zh.* **48**, 665; *Soviet Astron.* **15**, 527, 1972.
Mustel, E. R.: 1972a, *Astron. Zh.* **49**, 15; *Soviet Astron.* **16**, 10, 1972.
Mustel, E. R.: 1972b, *Astron. Tsirk.*, No. 674.
Mustel, E. R. and Baranova, L. I.: 1965, *Astron. Zh.* **42**, 42.
Mustel, E. R.: 1973, *Astron. Zh.* **50**, 1121.
Mustel, E. R. and Chugay, N. N.: 1974, *Astrophys. Space Sci.*, in press.
Mustel, E. R.: 1974, in R. J. Tayler (ed.), 'Late Stages of Stellar Evolution', *IAU Symp.* **66**, in press.
Patchett, B. and Branch, D.: 1972, *Monthly Notices Roy. Astron. Soc.* **158**, 375.
Peimbert, M.: 1971, *Astrophys. J.* **170**, 261.
Pskowskij, Yu, P.: 1968, *Astron. Zh.* **45**, 942; *Soviet Astron.* **12**, 750, 1969.
Pskowskij, Yu. P.: 1970, *Astron. Zh.* **47**, 994; *Soviet Astron.* **14**, 798, 1971.
Van den Bergh, S.: 1973, *Publ. Astron. Soc. Pacific* **85**, 335.

SUPERNOVA REMNANTS

S. VAN DEN BERGH
David Dunlap Observatory, Richmond Hill, Ontario, Canada

Abstract. The present structure and recent changes in Cas A are discussed. On the deepest available exposures the optical remnant of this supernova is seen to consist of an almost complete shell. The southern part of this shell is outlined by knots that have developed during the last decade. It is pointed out that moving nebulosity in which [O III] radiation is particularly strong is distributed differently from nebulosity in which the [O III]/[S II] ratio is more nearly normal. Observation of the motions of individual knots suggests a model in which dense blobs of matter, which have a highly anomalies composition, become luminous as they plough through a stationary interstellar (or circumstellar) medium.

1. Introduction

Observational data on all known optical supernova remnants have recently been summarized in 'An Atlas of Galactic Supernova Remnants' by van den Bergh *et al.* (1973). A more detailed discussion of the supernovae of the second millennium A.D. (Lupus 1006, Crab 1054, Tycho 1572, Kepler 1604 and Cas A 1667) is given by van den Bergh (1973).

In the present paper I shall confine myself to a brief discussion of some observations of Cassiopeia A that have recently been obtained with the 200-in. telescope. The first observations of Cas A are described by Baade and Minkowski (1954). Subsequent discussions of the optical observations of this supernova are given by van den Bergh and Dodd (1970) (expansion), van den Bergh (1971a, b) (photometry, spectroscopy), Peimbert and van den Bergh (1971), Peimbert (1971) (composition), and by Searle (1971) (reddening).

2. Structure of Cas A

Figure 1 shows a 200-min exposure of Cas A obtained through an [S II] interference filter that transmit light in the range 6650–6830 Å. This deep exposure shows the following:

(a) The optical remnant of Cas A forms an almost complete shell with inner and outer radii of $\sim 80''$ and $\sim 120''$ respectively.

(b) The structure of the bright and faint nebulosity differs. The brightest moving nebulosity consists of crisp knots, whereas the faintest detectable nebulosity has a much more diffuse character.

(c) Figure 2 shows the distribution of all moving nebulosity in which [S II] appeared brighter on a 200-min $\lambda\lambda 6650$–6830 (S II interference filter + 098-02) exposure than did [O III] on a 120-min $\lambda\lambda 4700$–5500 (GG7 + 103aJ) exposure. In Figure 2 hatching represents the faintest nebulosity which is only marginally visible on the 200-min exposure.

[O III] emission seems to be entirely absent from the knots labelled A, B and

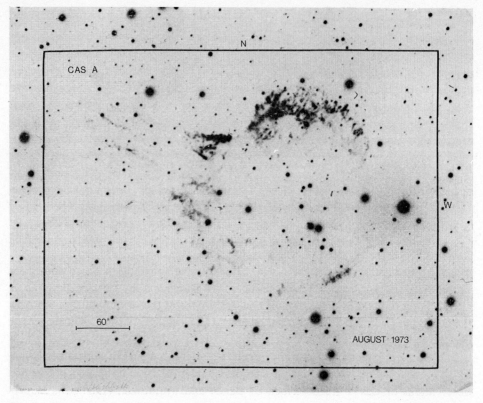

Fig. 1. 200-min exposure of Cas A with the Hale telescope through an [S II] interference filter.

C. The streak marked D appears to be a luminous but stationary conduit through which moving knots are seen to pass. Possibly this object resembles the streaks in the 'jet' that is located in the NE part of the remnant of Cas A.

(d) Nebulosity in which [O III] was brighter on a 120-min exposure than was [S II] on a 200-min exposure is marked in Figure 3. This figure shows that the distribution of the knots that are particularly bright in [O III] $\lambda\lambda 4959, 5007$ differs radically from that of the other moving nebulosity. In the northern part of Cas A this [O III] nebulosity seems to outline a well defined filament. In the eastern part of this filament individual knots are *invisible* on a deep 098-02+RG2 ($\lambda\lambda 6300$–6900) exposure. The western part of the same filament is easily visible on this same plate but is invisible on a deep interference filter plate covering the wavelength range $\lambda\lambda 6650$–6830. This suggests that the western part of this filament probably radiates strongly in the range $\lambda\lambda 6300$–6650. Since no other moving knots have yet been observed to show Hα or [N II] it follows that [O I] $\lambda\lambda 6300, 6364$ is a logical candidate for the emission in these knots. Nevertheless it seems remarkable that a knot should emit [O I] and [O III] but not [S II]. Clearly it would be very desirable to obtain spectra of these knots.

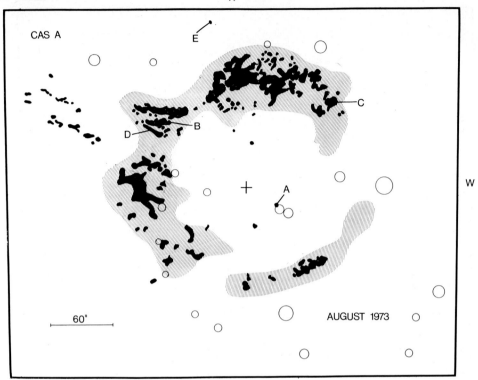

Fig. 2. Distribution of nebulosity in which [S II] is strong relative to [O III]. No [O III] is visible in the knots labelled A, B and C. The centre of expansion of Cas A (van den Bergh and Dodd, 1970) is marked by a cross.

Inspection of Figure 3 suggests that the filament of nebulosity in which [O III] is particularly strong might be physically connected with filament No. 1 of Baade and Minkowski.

(e) The knot marked E in Figure 2 was clearly visible on the first red plate of Cas A that Walter Baade took 22 years ago. During the intervening period this isolated knot has travelled $\sim 10''$, i.e. approximately ten times its own diameter. This observation suggests that such fast-moving knots are produced by dense blobs of matter that are ploughing through a more or less stationary interstellar (or circumstellar) medium. The anomalous composition of these blobs (Peimbert and van den Bergh, 1971; Peimbert, 1971) shows that they were ejected from the Cas A supernova.

3. Changes in Cas A

Figure 1 shows the appearance of Cas A in August of 1973. Comparison of this plate with plates taken 20 years ago (see Figure 2 of Baade and Minkowski, 1954) shows the following:

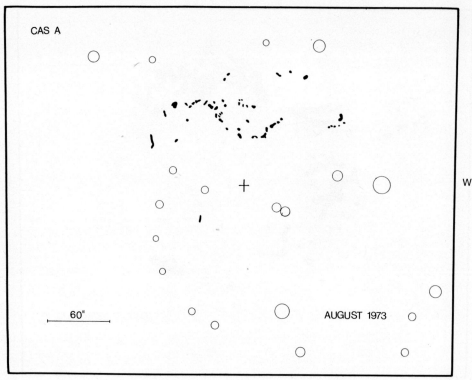

Fig. 3. Distribution of knots in which [O III] is particularly strong. Note that most of this nebulosity appears to be distributed in a single filament.

(a) A number of bright knots have recently developed along the southern rim of this supernova remnant. The intensity ratio of [S II] to [O III] in these new knots is similar to that observed in the majority of moving knots. A spectrum of knots 11a–11d (see van den Bergh, 1971b) shows that the internal velocity dispersion in these new knots is similar to that observed in moving knots in other parts of the remnant.

(b) Typical moving knots have lifetimes ~ 10 yr. Only a few of the brightest knots have remained visible during the entire 22 year time space covered by available 200-in. plates. Small faint knots typically have lifetimes of only a few years.

(c) The overall position of the bright arc of nebulosity situated to the NNW of the centre of expansion has *not* changed during the last 20 years despite the fact that individual knots in this arc have moved $\sim 8''$. This observation suggests that moving knots in this region light up as they pass through a stationary interstellar (or circumstellar) cloud bank.

(d) The brightest filament in the NE part of the remnant (Baade and Minkowski's No. 1) is gradually breaking up into individual knots. Radial velocity observations (van den Bergh, 1971b) show that individual knots in this filament have velocities that differ by up to ~ 3000 km s^{-1}. The origin of these large velocity differences remains entirely obscure.

It is a pleasure to thank the Hale Observatories for their generosity in making observing time available during the last six years for the continued study of the evolution of this enigmatic object.

References

Baade, W. and Minkowski, R.: 1954, *Astrophys. J.* **119**, 206.
Peimbert, M.: 1971, *Astrophys. J.* **170**, 261.
Peimbert, M. and van den Bergh, S.: 1971, *Astrophys. J.* **170**, 261.
Searle, L.: 1971, *Astrophys. J.* **168**, 41.
Van den Bergh, S.: 1971a, *Astrophys. J.* **165**, 259.
Van den Bergh, S.: 1971b, *Astrophys. J.* **165**, 457.
Van den Bergh, S.: 1973, *Publ. Astron. Soc. Pacific* **85**, 335.
Van den Bergh, S. and Dodd, W. W.: 1970, *Astrophys. J.* **162**, 485.
Van den Bergh, S., Marscher, A. P., and Terzian, Y.: 1973, *Astrophys. J. Suppl.* **26**, 19 (No. 227).

SOFT X-RAY OBSERVATIONS OF SUPERNOVA REMNANTS

J. C. ZARNECKI, J. L. CULHANE, A. C. FABIAN*,
C. G. RAPLEY, and R. L. F. BOYD

Mullard Space Science Laboratory, University College London, England

and

J. H. PARKINSON** and R. SILK

Physics Dept., Leicester University, England

Abstract. Observations of a number of supernova remnants have been carried out with the low energy X-ray telescope on the Copernicus satellite. Data are presented on the X-ray structure of the remnants Cassiopeia A and Puppis-A. Marginal detections or new upper limits are reported for the remnants IC443, DR4, MSH15-52A, Downes 83, Downes 84 and 3C392.

1. Introduction

In the past, supernova remnants have been recognised and studied as extended sources of non-thermal radio emission. A few of these objects exhibit pronounced filamentary structures at optical wavelengths. About 100 supernova remnants have been identified in the Galaxy. Recently a number of them have been identified as X-ray sources but most of the X-ray data have been obtained with either mechanically collimated detectors or with one dimensional reflecting systems having fields of view of about 0.2° by 10°. Use of the grazing incidence paraboloidal reflectors on Copernicus has allowed us to examine, for the first time, the structure of several supernova remnants with fields of view that range from 3' to 12' in size.

2. Studies of Individual Supernova Remnants

Following a supernova explosion, a number of phases may be distinguished in the evolution of a remnant as the shock wave moves out from the explosion site. The propagation of a shock wave in the interstellar medium has been discussed by Taylor (1950) and by Sedov (1959). Up to several hundred years after the explosion, the mass of ejected gas is greater than the mass swept up from the interstellar medium. At this time also the radio observations suggest that much of the energy is in the form of relativistic electrons, which radiate by the synchrotron process. After several thousand years however, the swept up interstellar gas, which has been heated by the passage of the shock front, dominates the appearance of the remnant. Preliminary analysis of the data from Copernicus has provided new information on the structures of both young and old supernova remnants.

* Now at the Institute of Astronomy, Cambridge.
** Now at Mullard Space Science Laboratory.

2.1. CASSIOPEIA-A

The structure of this remnant was observed with the 1.4 to 4.2 keV telescope using the 3′ field of view. The data obtained are shown in Figure 1 together with the UHURU error box for the source and the equivalent beamwidth of the Copernicus telescope. The data points have been input to a simple contour plotting programme but the contours produced represent a convolution of the telescope impulse response and the X-ray source structure. In order to unfold the impulse response, various trial source distributions were convolved with the impulse response and the resulting convolutions were χ^2 fitted to the 13 data points. The validity of this procedure was first tested on data from the point source GX2+5. The results of this work showed that the chosen impulse response adequately represented the properties of the telescope.

The symmetry of the 'contour map' as well as the shape of the radio maps of Cas A suggested trial source distributions in the form of a point source, annuli of

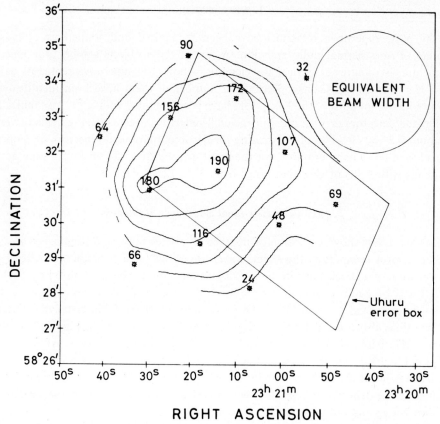

Fig. 1. The numbers of counts registered in the 1.2 to 4.6 keV band during the various samples of data from Cassiopeia A. The UHURU error box, the Copernicus equivalent beam width and a simple contour map are also shown.

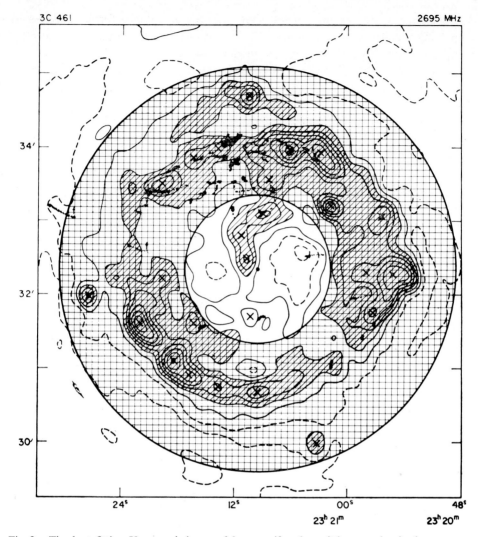

Fig. 2. The best fitting X-ray emission model – a uniformly emitting annulus is shown superimposed on the Cambridge radio map.

various inner and outer radii (approximating to a shell), and discs of varying diameter. The convolution was performed by a Fourier transform method and the χ^2 fitting was done by moving an array of values representing the source, where elements of the array were separated by 0.5′ intervals. Source position, intensity and size were altered and χ^2 values computed for each case. The number of degrees of freedom is equal to 13 minus the number of free parameters.

A point source gave the worst fit to the data. We suggest therefore that Cas A is not a compact X-ray emitting object. While none of the uniform extended source models gave a really good fit, it is clear that χ^2 is minimum for an object of about

5.5′ diameter. For extensions of less than 5.5′, the χ^2 value is further reduced by employing a uniform annular source. For extensions greater than 5.5′, the need for annular structure becomes less and disc models can give acceptable fits. Of the models tried, the data are best fitted by a uniform annular or disc source of outer diameter 5.5′. Such an annular source has been superimposed on the Cambridge 2695 MHz radio data (Rosenberg, 1970) in Figure 2.

Since all of the extended trial sources were of uniform surface brightness and since this is unlikely to be true in practice, we believe that the discrepancies in the χ^2 fits, even for the best source models, arise due to the existence of structure in the X-ray emission. This is supported by the fact that further studies of the remnant, carried out with smaller fields of view, suggest some non-uniformity in surface brightness in addition to confirming the overall extension discussed in the previous paragraph.

In summary, the results of our preliminary analysis (1) Identify the UHURU source 2U 2321+58 with the radio source Cas A, (2) Suggest that the X-ray emission arises in an annular source of outer diameter 5.5′±1.5′ and inner diameter 2.0′±2.0′, (3) Indicate that the surface brightness of the annulus is non-uniform and (4) Indicate that the bulk of the X-ray emission does not come from a compact object.

The data are suggestive of a shell source. The X-ray emission may originate in thermal Bremsstrahlung from a blast wave, or the synchrotron mechanism. Further studies of this object with better spatial and spectral resolution may allow us to discuss the radiation production mechanism in greater detail.

2.2. THE X-RAY STRUCTURE OF PUPPIS-A

The Puppis-A supernova remnant is much older than Cas A (3×10^4 as against 3×10^2 yr). Because of its age, its emission is probably from the shock heated plasma of the interstellar medium. This conclusion is supported by earlier observations which suggest that the plasma temperature is about 4×10^6 K (Burginyon *et al.*, 1973b). While a number of other workers have observed the remnant with collimated proportional counters and one dimensional X-ray optical systems, the data presented here are the first to be obtained with high spatial resolution in two dimensions.

The remnant has been mapped with the 0.5–1.5 keV telescope system by sampling the intensity at 27 discrete points both inside and outside the radio shell using the nominal 10′ field of view. The duration of each sample was approximately thirty of the instrument's 62.5 S integration periods. The mean background rate during these observations was approximately 90 counts per sample. In Figure 3 are shown the corrected number of X-ray counts in each sample. These data have been processed using a standard computer contour routine to produce the map of Figure 4. For comparison we show selected contours of the radio map of Milne (1971) chosen because of its similar spatial resolution. The X-ray map represents a convolution of the telescope impulse response with the real X-ray source distribution. It is clear that the bulk of the X-ray emission originates from within the radio shell and it is likely that any apparent emission from outside the shell is entirely due to the effect of the

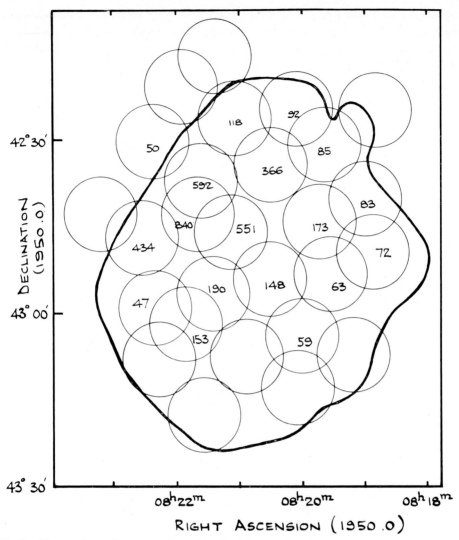

Fig. 3. The numbers of counts registered in the 0.5 to 1.5 keV band during the various samples of data from Puppis A. The outermost radio contour is shown as a solid line. Empty circle denote samples where the significance of the data was less than 3σ.

convolution. The X-ray emission is strongly peaked and does not coincide with any of the obvious radio or optical features. Furthermore the strong radio regions do not appear to emit significantly in soft X-rays although we note that the X-ray emission is concentrated in the more intense half of the radio shell. We find that over 50% of the observed counts originate from the central 20% of our contoured region which has a total extent of ~ 0.5 sq deg. We note that the peak of the distribution lies within the UHURU error box. However, the source is extended with some evidence of elongation towards the north-west and south-east. An attempt to unfold our data using a minimum χ^2 technique has shown that no simple combination of point and disc sources will adequately describe the X-ray distribution.

If it is assumed that the site of the original Puppis-A explosion lies within the present X-ray emitting region, current supernova theory can be invoked to at least qualitatively explain the features of the X-ray and radio emission. The asymmetry of the radio distribution with respect to the X-ray region can be accounted for by assuming the existence of density gradients in the surrounding interstellar medium, the density being greater on the north-east side of the remnant which is nearest the galactic plane. The application of the models of Stevens (1973) which attribute the X-radiation to thermal emission can then lead to an explanation of the separation between the edges of the X-ray and radio emission in the southern and western sections of the remnant. In discussing recent observations of the Cygnus Loop, Stevens has shown that the passage of a low velocity (~ 150 km s^{-1}) shock wave through a uniform interstellar medium can lead to a separation of X-ray and radio emission similar to that now observed in Puppis-A. Recent calculations by Lada and

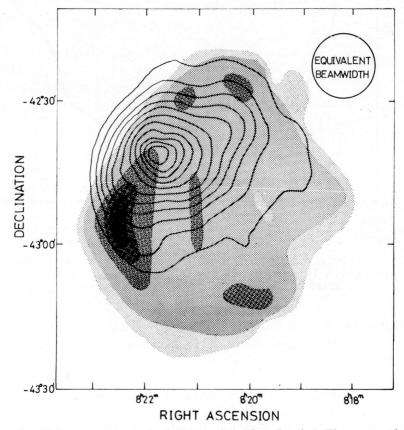

Fig. 4. The solid lines are contours of soft X-ray emission from Puppis A. They are superimposed on the radio map of Milne.

Straka (1973) lead to a similar conclusion. The low radial velocities of the optical filaments led support to such theories.

The existence of a bright compact X-ray emitting region leads to speculation on the possible presence of a compact object. If such an object does exist it is noteworthy that it still lies within the radio shell and the extended X-ray region and hence must have a low runaway velocity. It is worth emphasizing that our data are not consistent with a simple point source and disc model.

We may summarise the significant features of the observations as follows:

(1) All the 0.5–1.5 keV X-ray emission is contained within the radio shell.

(2) The distribution of X-ray emission is very different from that of the radio emission and does not correlate well with radio or optical features.

(3) The X-ray emission is extended but strongly peaked.

(4) The possibility of existence of a compact object is of great interest but the present evidence is not conclusive.

(5) In several parts of the remnant, particularly in the south and west, there exists a marked separation between the edges of the X-ray and radio emission.

(6) The observed radio and X-ray features can be accounted for by current supernova models in which the interstellar gas is heated to a temperature of several million degrees by the passage of a shock wave.

2.3. Observations of other Supernova Remnants

X-ray observations of a number of remnants have been carried out in order to confirm results obtained by other workers. These sources are weak ones and so either marginal detections of X-ray flux or upper limits have been achieved.

2.3.1. *IC443*

The radio structure and size of this remnant are very similar to those of Puppis-A. Unlike Puppis-A however, this object has pronounced and well developed optical filaments but only weak X-ray emission. The diameter of the remnant is 45'. Seven observations were carried out with the 10' telescope field of view. By combining data from all of these samples, the source was detected a significance level of 5.3σ thereby confirming the identification of this remnant with the source 3U 0620+23 as suggested in the UHURU catalogue (Giacconi *et al.*, 1972). There is some suggestion in the data that the X-ray emission is associated with the brightest filament but further observations of greater significance are required to confirm this. A 3σ upper limit of *0.02 photons cm^{-2} s^{-1} (2–6 keV)* can be assigned to the emission from the nearby pulsar.

2.3.2. *DR4*

This remnant has been suggested by Burginyon *et al.* (1973a) as a candidate for a source of low energy X-ray emission which they detected in Cygnus. The flux registered in their observation was 0.06 photons cm^{-2} s^{-1} (0.5–1.5 keV). From our observations of DR4 with the low energy telescope, we find a 3σ upper limit to the flux in the same band of *0.01 photons cm^{-2} s^{-1}*. A power law spectrum of photon number index -3 was assumed in deriving this limit. There are a number of other supernova remnants in the region observed by Burginyon *et al.* that could be sources of the emission which these workers detect.

2.3.3. *MSH15–52A*

This source was observed for 100 minutes by the Copernicus telescopes. For an assumed power law spectrum with a photon number index of -3, a 3σ upper limit of *4.5 × 10^{-11} erg cm^{-2} s^{-1}* was obtained for the 2–6 keV band. The UHURU source 3U 1510–59 has been suggested as a possible candidate for association with this remnant. However, the UHURU catalogue flux is *1.1 × 10^{-10} erg cm^{-2} s^{-1} in the same band*.

2.3.4. *Downes No. 83 and 84*

A possible detection of these sources has been claimed by Schwartz *et al.* (1972). Copernicus observations have established a 3σ upper limit of *0.01 photons cm^{-2} s^{-1}* for the 2–6 keV band.

2.3.5. *3C392*

We have obtained a signal of 5σ significance in the 3–9 keV band. This result confirms a marginal identification by Schwartz *et al.* of this source as an X-ray emitter. However, our result was obtained with the 3° field of view proportional counter. So while the nearest UHURU source is more than 3° array, a further observation is probably required for final confirmation. No low energy signals were detected from this source.

References

Burginyon, G. A., Hill, R., Palmieri, T., Scudder, J., Seward, F. D., Stoering, J., and Toor, A.: 1973a, *Astrophys. J.* **179**, 615.
Burginyon, G. A., Hill, R. W., Seward, F. D., Tarter, C. B., and Toor, A.: 1973b, *Astrophys. J.* **180**, L175.
Giacconi, R., Murray, S., Gursky, H., Kellogg, E., Schreier, E., and Tananbaum, H.: 1972, *Astrophys. J.* **178**, 281.
Lada, C. J. and Straka, W. C.: 1973, *Bull. Am. Astron. Soc.* **5**, 12.
Milne, D. K.: 1971, *Australian J. Phys.* **24**, 429.
Rosenberg, I.: 1970, *Monthly Notices Roy. Astron. Soc.* **147**, 215.
Schwartz, D. A., Bleach, R. P., Boldt, E. A., and Holt, S. S.: 1972, *Astrophys. J.* **173**, L51.
Sedov, L. I.: 1959, *Similarity and Dimensional Methods in Mathematics*, Academic Press, New York.
Stevens, J. C.: 1973, Thesis, California Institute of Technology.
Taylor, G. I.: 1950, *Proc. Roy. Soc.* **A201**, 159.

REPORT ON THE LECCE CONFERENCE ON SUPERNOVAE

F. BERTOLA
Dept. of Physics, The University of Lecce, Italy

From May 7 through May 11, 1973 an International Conference on Supernovae was held in Lecce, organized by the local University. About one hundred participants, from eighteen countries, attended it. The basic topics covered were the following:

results and techniques of supernova surveys,
photometric studies of supernovae,
spectra of supernovae and their interpretation,
statistics of supernovae,
supernova remnants,
theories on supernovae and supernova remnants.

The number of contributed papers was forty-six and the proceedings have been published in Cosmovici (1974).

Since the previous conference dealing with the same subject was held in September 1963 at the Haute Provence Observatory, the Lecce Conference marks the enormous progress made in the past decade.

The reviews of the supernova surveys, which are carried out at different observatories, have shown that a more careful planning of the fields to be surveyed could lead to more useful data. In order to use the supernovae as distance indicators, premaximum observations are of extreme importance. This can be obtained by a closer spacings of the patrol plates. The need of discovery of fainter supernovae and of studying spectra and light curves to fainter magnitudes has been stressed.

Until a few years ago only two types of supernovae were recognized. Recently Zwicky introduced the additional types III, IV and V. At the Lecce Conference we learned that several subclasses are present within these types. As an example the differences in the light curves of type I supernovae, found at Asiago, has to be quoted. The fact becomes very important when it is taken into account that these differences are probably related to the stellar population associated with the supernova.

The successful attempt to identify the lines in the spectra of type I supernovae is one of the main results achieved in the recent years, together with the estimation of the matter ejected and of the physical conditions in the supernova shell.

Completely new types of observations of supernovae were presented in Lecce. The report of the first detection of strong radio emission from a supernova which appeared two years ago in another galaxy has opened a new field of research, which could lead to a better understanding of the pulsar phenomenon.

The study of the supernova remnants has received great stimulation by the X-ray data, which, together with the optical and radio data, give a new look into the problem. The analysis of the structure of some remnants with the satellite Copernicus currently in orbit, seems very promising. At the same time the elucidation of the detailed structure of magnetic fields and cosmic ray particles in supernova remnants from the measurements made with new powerful radiotelescopes is a recent contribution.

On the theoretical side the following has to be mentioned: (i) the first beginnings of an understanding of the physics of the supernova explosion and the kind of stars in which it occurs; (ii) the construction of very detailed models of the structure of supernova remnants and their radio emission; (iii) the construction of more detailed models for the formation of pulsars and the way in which pulsars accelerate cosmic rays and generate magnetic fields which energize the supernova remnants.

Reference

Cosmovici, C. B. (ed.): 1974, *Supernovae and Supernova Remnants*, D. Reidel Publ. Co., Dordrecht.